Electrical Power Systems: Engineering Essentials

Electrical Power Systems: Engineering Essentials

Edited by
Isla Cooper

MURPHY & MOORE
www.murphy-moorepublishing.com

Published by Murphy & Moore Publishing,
1 Rockefeller Plaza,
New York City, NY 10020, USA

ISBN: 978-1-63987-182-7

Cataloging-in-Publication Data

Electrical power systems : engineering essentials / edited by Isla Cooper.
p. cm.
Includes bibliographical references and index.
ISBN 978-1-63987-182-7
1. Electric power systems. 2. Electric power production. 3. Electrical engineering. I. Cooper, Isla.
TK1001 .E34 2022
621.31--dc23

For information on all Murphy & Moore Publications
visit our website at www.murphy-moorepublishing.com

MURPHY & MOORE

Contents

Preface

This book aims to highlight the current researches and provides a platform to further the scope of innovations in this area. This book is a product of the combined efforts of many researchers and scientists, after going through thorough studies and analysis from different parts of the world. The objective of this book is to provide the readers with the latest information of the field.

A network of electrical components that are assembled to supply, use and transfer electric power is called an electrical power system. It is an integration of several components, such as power sources, conductors, reactors, capacitors, loads, etc. A power system needs to be designed in a manner that it maintains the value of the voltage, power and frequency supplied to loads, otherwise it can lead to damage to synchronous machines and other appliances. This remains the most challenging aspect of such systems. Power systems often use three-phase AC power, which is the standard for large scale power distribution and transmission. Some power systems are not dependent on three-phase AC power, such as those used in ocean liners, automobiles, electric rail systems and aircrafts. This book is a valuable compilation of topics, ranging from the basic to the most complex advancements in electrical power systems. It presents this complex subject in the most comprehensible and easy to understand language. The extensive content of this book provides the readers with a thorough understanding of this domain.

I would like to express my sincere thanks to the authors for their dedicated efforts in the completion of this book. I acknowledge the efforts of the publisher for providing constant support. Lastly, I would like to thank my family for their support in all academic endeavors.

Editor

1

Restraining the Demand Side Power Fluctuation of Active Distribution Network Using 0°/180° Phase Controlled Electric Spring

Yixi Chen, Gang Ma, Guchao Xu, Huaiyi Chen, and Hang Zhang

School of Electrical & Automation Engineering, Nanjing Normal University, Nanjing 210042, China

Correspondence should be addressed to Gang Ma; nnumg@njnu.edu.cn

Academic Editor: Daniela Proto

In active distribution network, the random power output by wind/solar distributed generation may cause the stochastic fluctuation of demand side power, which will bring difficulties to power dispatching. In this paper, a method of restraining the demand side power fluctuation of active distribution network is proposed, in which a new power electronic device—electric spring—is applied by 0°/180° phase control strategy. Firstly, the basic principles of electric spring are introduced. Secondly, the reason for demand side power fluctuation of active distribution network is analyzed. After that, the 0°/180° phase control strategy of electric spring is proposed to restrain the demand side power fluctuation of active distribution network and the selection basis of noncritical loads is also obtained. The simulation results show that the method proposed in this paper is effective and the obtained selection basis of noncritical loads is reasonable.

1. Introduction

Energy saving and emission reduction have jointly promoted the development of renewable energy generation [1, 2]. As one of the main forms of renewable energy generation, distributed generation (DG) outputs fluctuant power due to the intermittent characteristic of renewable energy [3, 4].

Active distribution network (ADN) represents the distribution network containing DG, the demand side power of which mainly includes the output power of DG, and the input power of loads. The fluctuant power output by DG may cause the fluctuation of demand side power, and therefore the power dispatching will be difficult [5].

In order to control the demand side power effectively, some demand side management methods such as direct load control, real-time pricing, charging, and discharging by energy storage devices have been applied in distribution networks [6]. However, these methods have some weaknesses. The method direct load control switches on/off loads directly according to the amount of power supply and power demand, which is intrusive to consumers [7, 8]. The method real-time pricing can shape the load curve as desired at the macrolevel, but it could not dispose of the demand side power fluctuation caused by DG [9, 10]. The method using energy storage devices can eliminate peak and fill valley for demand side power, but this approach needs to add much storage capacity when the fluctuation of DG is big, which is uneconomical [11, 12].

As a novel power electronic device, electric spring (ES) was at first used to restrain the bus voltage fluctuation caused by the access of DG, and the frequency adjustment function can be obtained simultaneously. Chen X. et al. [13] analyzed the function of ES to mitigate voltage and frequency fluctuation in microgrids. Lee CK. et al. [14] drew a conclusion that using ES to regulate mains voltage can reduce the energy storage requirements in future smart grid. Wang Q. et al. [15] proposed a novel δ control strategy to improve the steady-state performance of ES. Yan S. et al. [16] proposed a new topology of ES by using back-to-back converter in order to extend its operating range. Moreover, the criteria and conditions for minimizing the average and oscillating power of the 3-ph ES are introduced in [17]. ES is evaluated for improving the fault-ride through capability of the wind turbine generation system with fixed speed induction generator in [18]. In general, although there have been some research results of ES, however the potential of ES has not been fully tapped. As (1) current analysis of ES is chiefly focused on the voltage stabilizing function, more functions of ES should be derived. (2) Current use of ES is restricted to ensure the reactive power compensation function firstly; however, 0°/180° phase control which is without the consideration of reactive power has not been studied and analyzed deeply.

In this paper, a method of restraining the demand side power fluctuation of ADN using 0°/180° phase controlled ES is proposed. Firstly, the basic principles of ES are introduced. Then, the reason for demand side power fluctuation of ADN is analyzed. After that, the 0°/180° phase control strategy of ES is proposed to restrain the demand side power fluctuation of active distribution network and the selection basis of noncritical loads is also obtained. At last, simulations are conducted to verify the effectiveness of the method proposed in this paper.

2. Analysis of ES Principles

Inspired by Hooke's law of mechanical spring, Hui SYR. et al. [19] proposed the concept of ES. Similar to the supporting function of mechanical spring, ES can stabilize the controlled variable effectively by controlling the power flow [20, 21].

As shown in Figure 1(a), ES consists of measurement module, control loop, PWM power inverter, and low-pass filter (i.e., filter inductance L_f and filter capacitance C_f) [22, 23]. The measurement module measures voltage, current, and other real-time data of the distribution network and then sends them to the control loop. The control loop decides the required ES voltage based on the real-time data measured by the measurement module and sends it to the inverter as a modulation signal. Then the inverter generates the corresponding ES voltage according to the modulation signal.

What needs to be explained here is that the first generation of ES (ES-1) does not involve battery storage. It uses reactive power to stabilize the bus voltage. Version two (ES-2) utilizes batteries as energy storage, which allows the ES to output/absorb both active and reactive power. ES-3 applies bidirectional AC/DC converter with battery storage and without association with noncritical load [24]. For ES-1, as it does not have the ability of active power control, it is not applicable for this paper. For ES-3, as noncritical load is not applied, it can only increase the capacity of ES by simply increasing the capacity of batteries. For ES-2, as noncritical load is applied, it can increase the capacity of ES not only by increasing the capacity of batteries, but also by increasing the number of noncritical loads, which is obviously more suitable for this paper. Therefore, all of the 'ESs' mentioned in this paper are ES-2.

Loads can be divided into noncritical loads and critical loads according to different voltage stability requirements. For example, electric heaters, lighting systems, and small motors with no stalling problems (e.g., fans, ovens, dish washes, and dryers) can tolerate a relatively large voltage variation with little or no noticeable impact on the consumers. These loads can be set as noncritical loads [25]. Loads which require well-regulated mains voltage such as life-supporting medical equipment and precision instrument can be set as critical loads [26]. As shown in Figure 1(b), ES is in series with a noncritical load to form a smart load, where $\mathbf{V}_{\mathbf{ES}}$ is ES voltage, $\mathbf{V}_{\mathbf{NC}}$ is noncritical load voltage, $\mathbf{V}_{\mathbf{S}}$ is bus voltage (i.e., critical load voltage), and $\mathbf{I}_{\mathbf{SL}}$ is smart load current.

As shown in Figure 1(b), the relationship of $\mathbf{V}_{\mathbf{ES}}$, $\mathbf{V}_{\mathbf{NC}}$, and $\mathbf{V}_{\mathbf{S}}$ can be obtained as

$$\mathbf{V}_{\mathbf{ES}} + \mathbf{V}_{\mathbf{NC}} = \mathbf{V}_{\mathbf{S}} \quad (1)$$

With the fluctuation of power output by DG, ES will change its voltage $\mathbf{V}_{\mathbf{ES}}$ dynamically in order to maintain the stability of the controlled variable; therefore $\mathbf{V}_{\mathbf{NC}}$ will change dynamically according to (1). That is to say, ES stabilizes the controlled variable by transferring its fluctuation to $\mathbf{V}_{\mathbf{NC}}$.

Now, ES has been deeply researched on in order to restrain the bus voltage fluctuation caused by the access of DG [27, 28]. In this application, (1) can be expressed by vector diagrams as shown in Figure 2, taking the noncritical load whose impedance angle is φ as an example. The phase angle difference between $\mathbf{V}_{\mathbf{ES}}$ and $\mathbf{I}_{\mathbf{SL}}$ is generally controlled to be ±90°, in order that ES inputs/outputs reactive power only. When $\mathbf{V}_{\mathbf{ES}}$ leads $\mathbf{I}_{\mathbf{SL}}$ 90°, ES works in inductive mode, as shown in Figure 2(a); in this mode, ES can input reactive power to depress the overhigh bus voltage. When $\mathbf{V}_{\mathbf{ES}}$ lags $\mathbf{I}_{\mathbf{SL}}$ 90°, ES works in capacitive mode, as shown in Figure 2(b); in this mode, ES can output reactive power to raise the overlow bus voltage.

Current use of ES is restricted to ensure the reactive power compensation function firstly. When ES is used to control active and reactive power simultaneously, the power compensated by ES can be illustrated by Figure 3 [29]. S_{max} is the maximum apparent power ES can compensate, which is restricted by the maximum value of $\mathbf{V}_{\mathbf{ES}}$ amplitude. According to Figure 3, on the condition that the apparent power compensate value is constant, when the reactive power compensation amount of ES is large, it can compensate less active power (as S_2 in Figure 3); when the reactive power compensation amount of ES is small, it can compensate more active power (as S_1 in Figure 3). At present, as STATCOM can effectively adjust reactive power in distribution network, ES is not used to regulate reactive power in this paper; therefore the active power compensation range of ES can be maximized.

3. Reason for the Demand Side Power Fluctuation in ADN

3.1. Analysis. In order to study the method of restraining the demand side power fluctuation of ADN, the reason for demand side power fluctuation of ADN must be analyzed first.

Compared with the traditional distribution network, the demand side power of ADN not only includes the input power of loads but also includes the output power of DG. Due to the intermittent characteristic of renewable energy, the output power of DG will be fluctuant; therefore the demand side power of ADN can be expressed as

$$P_{\mathrm{DEM}} = P_{\mathrm{Load}} - \left[P_{\mathrm{DG\text{-}REF}} + \Delta P_{\mathrm{DG}}(t)\right] \quad (2)$$

where P_{Load} represents the input power of loads, $P_{\mathrm{DG\text{-}REF}}$ represents the reference value of the output power of DG,

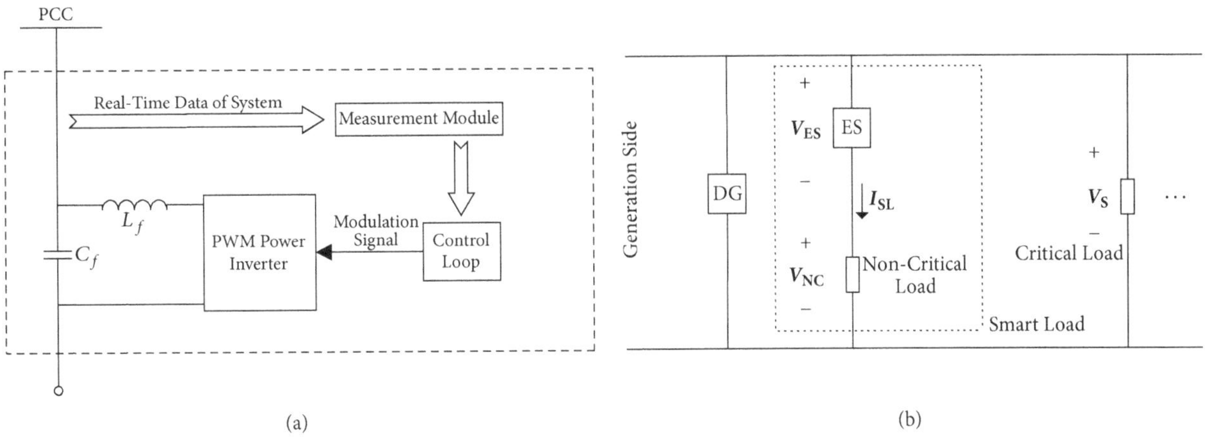

FIGURE 1: The basic structure and connection of ES. (a) Structure of ES. (b) Connection of ES.

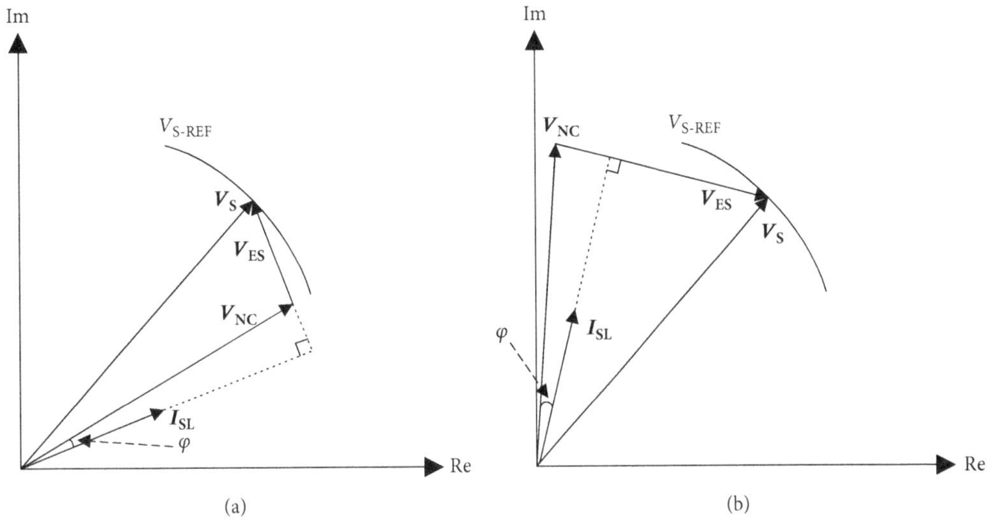

FIGURE 2: The vector diagram when ES is applied to restrain the bus voltage fluctuation caused by DG. (a) Inductive mode. (b) Capacitive mode.

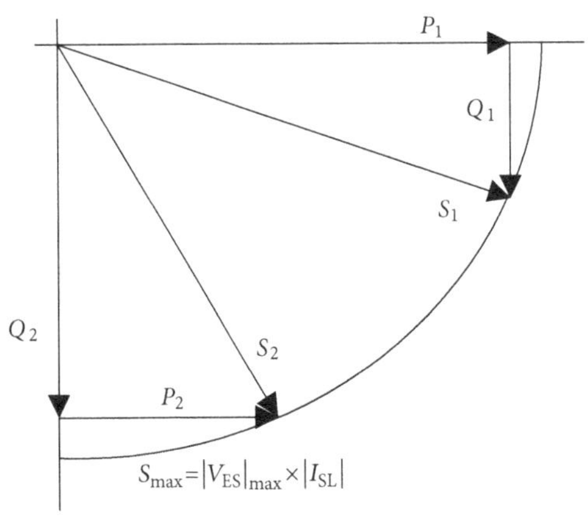

FIGURE 3: Illustration of power compensation by the ES.

and $\Delta P_{DG}(t)$ represents the fluctuation quantity of DG output power comparing with $P_{DG\text{-REF}}$. In order to simplify the analysis, the line loss which accounts for a relatively small proportion in the demand side power is not considered in this paper.

The selection range of $P_{DG\text{-REF}}$ in (2) is

$$P_{DG\text{-REF}} \in [0, P_{DG\text{-MAX}}] \tag{3}$$

where $P_{DG\text{-MAX}}$ represents the maximum power DG can output.

Let

$$P_{DEM\text{-REF}} = P_{Load} - P_{DG\text{-REF}} \tag{4}$$

where $P_{DEM\text{-REF}}$ represents the reference value of the demand side power of ADN.

Therefore, substituting (4) into (2) gives

$$P_{DEM} = P_{DEM\text{-REF}} - \Delta P_{DG}(t) \tag{5}$$

As shown in (5), due to the fluctuation quantity $\Delta P_{DG}(t)$ of DG output power, the demand side power of ADN P_{DEM} will fluctuate randomly around $P_{DEM\text{-REF}}$, which will bring difficulties to power dispatching.

3.2. Idea for Solution. According to (2), the fluctuation of DG output power is $\Delta P_{\text{DG}}(t)$. In order to eliminate the influence of this fluctuation on demand side power, the same fluctuation $\Delta P_{\text{DG}}(t)$ can be exerted to loads power P_{Load}, which can balance out the fluctuation quantity $\Delta P_{\text{DG}}(t)$ caused by DG. According to this, (2) can be modified as

$$P_{\text{DEM}} = [P_{\text{Load}} + \Delta P_{\text{DG}}(t)] - [P_{\text{DG-REF}} + \Delta P_{\text{DG}}(t)] \quad (6)$$

Substituting (4) into (6) gives

$$P_{\text{DEM}} = P_{\text{DEM-REF}} \quad (7)$$

Comparing (7) with (5), it can be found that the fluctuation of demand side power caused by DG can be eliminated if the above idea is achieved and the demand side power of ADN can be stabilized at the reference value $P_{\text{DEM-REF}}$. Moreover, according to this idea, the power consumption will increase when the quantity of DG output power is large, while the power consumption will decrease when the quantity of DG output power is small; therefore the new control paradigm where demand follows generation can be realized.

4. 0°/180° Phase Control Strategy of ES

In order to achieve the above idea, ES can be applied. As shown in Figure 1(b), ES is in series with a noncritical load to form a smart load. With the dynamic voltage $\mathbf{V}_{\text{ES}}$ generated by ES, the input power of smart load can be changed dynamically; therefore the loads power P_{Load} can be altered according to the idea above.

Different from the application of restraining bus voltage fluctuation, when restraining the fluctuation of demand side power caused by DG in ADN, ES can be set to control active power only. Therefore, the phase angle difference between $\mathbf{V}_{\text{ES}}$ and $\mathbf{I}_{\text{SL}}$ can be controlled to be 0°/180°. When the phase angle difference is 0°, ES works in resistive mode, while when the phase angle difference is 180°, ES works in negative resistive mode.

4.1. ES in Series with Resistive Noncritical Load. If ES is used to restrain the fluctuation of demand side power in ADN when it is in series with a resistive noncritical load, (1) can be expressed by vector diagrams as shown in Figure 4, where the dashed vectors $\mathbf{V}_{\text{NC}}$ and $\mathbf{I}_{\text{SL}}$ represent the noncritical load voltage and smart load current respectively on the condition that ES is not applied and the vectors $\mathbf{V}_{\text{NC}}'$ and $\mathbf{I}_{\text{SL}}'$ represent the noncritical load voltage and smart load current, respectively, on the condition that ES is applied.

As shown in Figure 4(a), when ES works in resistive mode, $\mathbf{V}_{\text{NC}}'$ is less than $\mathbf{V}_{\text{NC}}$; accordingly $\mathbf{I}_{\text{SL}}'$ is less than $\mathbf{I}_{\text{SL}}$ and the input power of smart load is decreased. As shown in Figure 4(b), when ES works in negative resistive mode, $\mathbf{V}_{\text{NC}}'$ is greater than $\mathbf{V}_{\text{NC}}$; accordingly $\mathbf{I}_{\text{SL}}'$ is greater than $\mathbf{I}_{\text{SL}}$ and the input power of smart load is increased.

According to Figure 4, if ES is not applied, the input power of smart load can be expressed as

$$P = \mathbf{V}_{\text{NC}} \cdot \mathbf{I}_{\text{SL}} = |\mathbf{V}_{\text{NC}}| \times \left|\frac{\mathbf{V}_{\text{NC}}}{R_{\text{NC}}}\right| \quad (8)$$

If ES is applied, the input power of smart load can be expressed as

$$P' = \mathbf{V}_{\text{NC}}' \cdot \mathbf{I}_{\text{SL}}' + \mathbf{V}_{\text{ES}} \cdot \mathbf{I}_{\text{SL}}' = |\mathbf{V}_{\text{NC}}| \times \left|\frac{(\mathbf{V}_{\text{S}} - \mathbf{V}_{\text{ES}})}{R_{\text{NC}}}\right| \quad (9)$$

Therefore, the adjustment of smart load input power produced by ES can be obtained as

$$\begin{aligned} \Delta P = P - P' &= |\mathbf{V}_{\text{NC}}| \times \left(\left|\frac{\mathbf{V}_{\text{NC}}}{R_{\text{NC}}}\right| - \left|\frac{(\mathbf{V}_{\text{S}} - \mathbf{V}_{\text{ES}})}{R_{\text{NC}}}\right|\right) \\ &= |\mathbf{V}_{\text{NC}}| \times \left(\frac{(|\mathbf{V}_{\text{NC}}| - |\mathbf{V}_{\text{S}} - \mathbf{V}_{\text{ES}}|)}{|R_{\text{NC}}|}\right) \end{aligned} \quad (10)$$

where $|\mathbf{V}_{\text{NC}}|$ is the noncritical load voltage if ES is not activated, which is equal to the bus voltage (the bus voltage is assumed to be constant in this paper). $\Delta P > 0$ means that the phase angle difference between $\mathbf{V}_{\text{ES}}$ and $\mathbf{I}_{\text{SL}}$ is 0° and the input power of smart load is decreased by ES, while $\Delta P < 0$ means that the phase angle difference between $\mathbf{V}_{\text{ES}}$ and $\mathbf{I}_{\text{SL}}$ is 180° and the input power of smart load is increased by ES.

4.2. ES in Series with Resistive-Inductive Noncritical Load. If ES is used to restrain the fluctuation of demand side power in ADN when it is in series with a resistive-inductive noncritical load, (1) can be expressed by vector diagrams as shown in Figure 5, where the impedance angle of noncritical load is φ and θ is the difference angle between $\mathbf{V}_{\text{NC}}$ and $\mathbf{V}_{\text{S}}$.

Similar to the above analysis, according to Figure 5, the input power of smart load, if ES is not applied, can be expressed as

$$P = \mathbf{V}_{\text{NC}} \cdot \mathbf{I}_{\text{SL}} = |\mathbf{V}_{\text{NC}}| \times \left|\frac{\mathbf{V}_{\text{NC}}}{Z_{\text{NC}}}\right| \times \cos\varphi \quad (11)$$

If ES is applied, the input power of smart load can be expressed as

$$\begin{aligned} P' &= \mathbf{V}_{\text{NC}}' \cdot \mathbf{I}_{\text{SL}}' + \mathbf{V}_{\text{ES}} \cdot \mathbf{I}_{\text{SL}}' \\ &= |\mathbf{V}_{\text{NC}}'| \times \left|\frac{(\mathbf{V}_{\text{S}} - \mathbf{V}_{\text{ES}})}{Z_{\text{NC}}}\right| \times \cos\varphi + |\mathbf{V}_{\text{ES}}| \\ &\quad \times \left|\frac{(\mathbf{V}_{\text{S}} - \mathbf{V}_{\text{ES}})}{Z_{\text{NC}}}\right| \\ &= |\mathbf{V}_{\text{S}}| \times \left|\frac{(\mathbf{V}_{\text{S}} - \mathbf{V}_{\text{ES}})}{Z_{\text{NC}}}\right| \times \cos(\varphi - \theta) \end{aligned} \quad (12)$$

Therefore, the adjustment of smart load input power produced by ES can be obtained as

$$\begin{aligned} \Delta P &= P - P' \\ &= |\mathbf{V}_{\text{NC}}| \times \left|\frac{\mathbf{V}_{\text{NC}}}{Z_{\text{NC}}}\right| \times \cos\varphi - |\mathbf{V}_{\text{S}}| \times \left|\frac{(\mathbf{V}_{\text{S}} - \mathbf{V}_{\text{ES}})}{Z_{\text{NC}}}\right| \\ &\quad \times \cos(\varphi - \theta) \end{aligned}$$

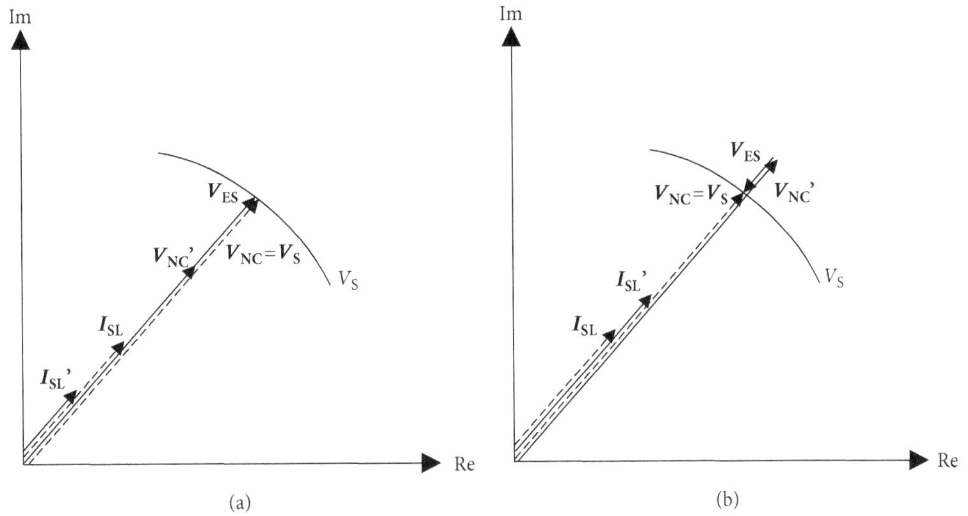

FIGURE 4: The vector diagram when ES is applied to restrain the demand side power fluctuation while noncritical load is resistive. (a) Resistive mode. (b) Negative resistive mode.

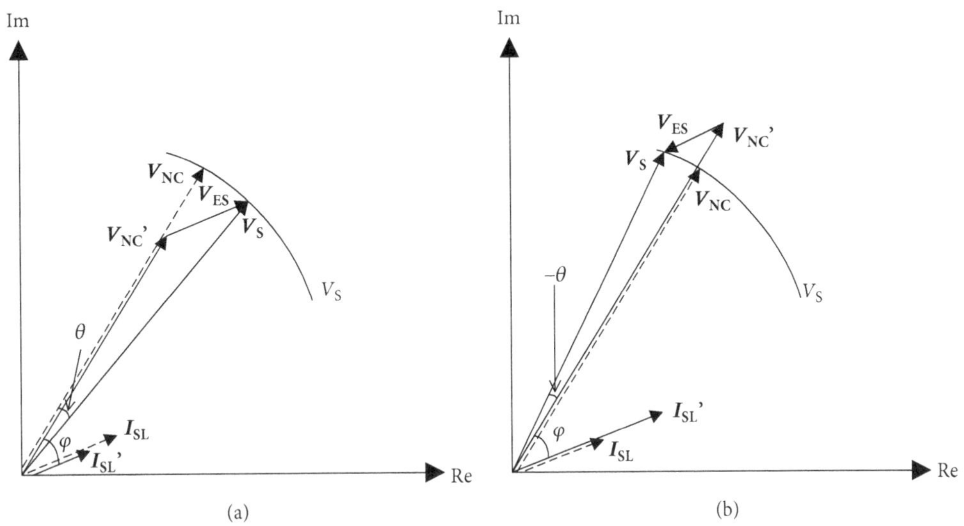

FIGURE 5: The vector diagram when ES is applied to restrain the demand side power fluctuation while noncritical load is resistive-inductive. (a) Resistive mode. (b) Negative resistive mode.

$$= |\mathbf{V}_{\mathrm{NC}}| \times \left(\left| \frac{\mathbf{V}_{\mathrm{NC}}}{Z_{\mathrm{NC}}} \right| \times \cos\varphi - \left| \frac{(\mathbf{V}_{\mathrm{S}} - \mathbf{V}_{\mathrm{ES}})}{Z_{\mathrm{NC}}} \right| \times \cos(\varphi - \theta) \right)$$
$$= |\mathbf{V}_{\mathrm{NC}}| \times \left(\frac{(|\mathbf{V}_{\mathrm{NC}}| \cos\varphi - |\mathbf{V}_{\mathrm{S}} - \mathbf{V}_{\mathrm{ES}}| \cos(\varphi - \theta))}{|Z_{\mathrm{NC}}|} \right) \tag{13}$$

Same as (10), $\Delta P > 0$ means that the input power of smart load is decreased by ES, while $\Delta P < 0$ means that the input power of smart load is increased by ES.

4.3. ES in Series with Resistive-Capacitive Noncritical Load. If ES is used to restrain the fluctuation of demand side power in ADN when it is in series with a resistive-capacitive noncritical load, (1) can be expressed by vector diagrams as shown in Figure 6, where the impedance angle of noncritical load is -φ and θ is the difference angle between $\mathbf{V}_{\mathrm{S}}$ and $\mathbf{V}_{\mathrm{NC}}$.

Same as the analysis of "*ES in series with resistive-inductive noncritical load,*" according to Figure 6, it can be obtained that the input power of smart load, if ES is not applied, can also be expressed as (11), the input power of smart load, if ES is applied, can also be expressed as (12), and the adjustment of smart load input power produced by ES can also be expressed as (13).

Based on the analysis above, (10) can be summarized as a special case of (13) (φ=θ=0°); that is, no matter what kinds of noncritical load is in series with ES to form a smart load, the adjustment of smart load input power produced by ES can all be expressed as (13). According to (13), ES can produce an

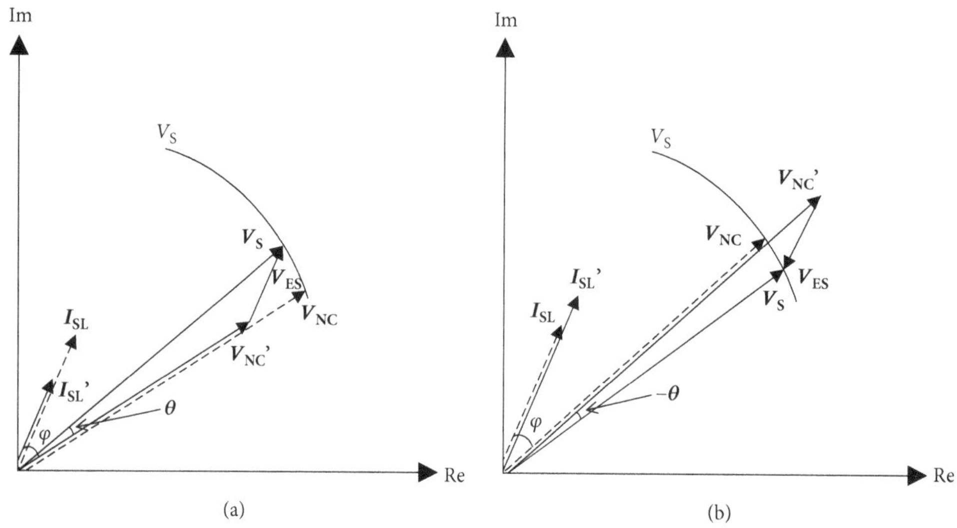

FIGURE 6: The vector diagram when ES is applied to restrain the demand side power fluctuation while noncritical load is resistive-capacitive. (a) Resistive mode. (b) Negative resistive mode.

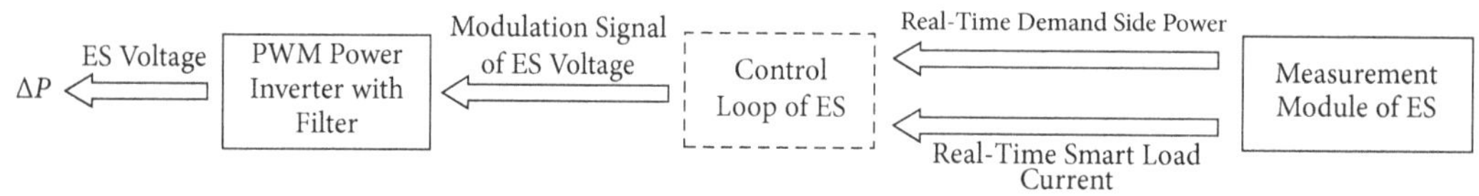

FIGURE 7: The flowchart of using ES to restrain the fluctuation of demand side power in ADN.

adjustment ΔP for smart load input power by generating the voltage $\mathbf{V}_{\mathbf{ES}}$. Assume that the input power of loads, if ES is not applied, is P_{Load}; therefore the input power of loads, if ES is applied, should be P_{Load}-ΔP and (2) will change to be

$$P_{\text{DEM}} = [P_{\text{Load}} - \Delta P] - [P_{\text{DG-REF}} + \Delta P_{\text{DG}}(t)] \tag{14}$$

If ΔP equals to -$\Delta P_{\text{DG}}(t)$, (14) can be equivalent to (6) and (7), the fluctuation of demand side power caused by DG can be eliminated, the demand side power of ADN P_{DEM} can be stabilized at the reference value $P_{\text{DEM-REF}}$, and the difficulties of power dispatching caused by the output power fluctuation of DG can be relieved.

4.4. Selection Basis of Noncritical Loads. According to the 0°/180° phase control strategy above, when ES is used to restrain the fluctuation of demand side power caused by DG, no matter what kinds of noncritical load is in series with ES, the adjustment of smart load input power produced by ES can all be expressed as (13), which can be transformed to be

$$|\mathbf{V}_\mathbf{S} - \mathbf{V}_{\mathbf{ES}}| = \frac{[|\mathbf{V}_{\mathbf{NC}}| \times \cos\varphi - (\Delta P/|\mathbf{V}_{\mathbf{NC}}|) \times |Z_{\text{NC}}|]}{\cos(\varphi - \theta)} \tag{15}$$

According to (15), to achieve same ΔP, bigger impedance modulus of noncritical load means that ES needs to generate bigger voltage, on the condition that the impedance angle of noncritical load is constant. However, according to the basic principles of PWM inverter, the value range of ES voltage is constrained by the magnitude of DC voltage. Therefore, as the DC voltage is constant, in order to achieve stronger ability to restrain the fluctuation of demand side power caused by DG, noncritical loads which have smaller impedance modulus should be considered to be in series with ES firstly.

5. Control Loop Design

Based on the analysis above, the flowchart of using ES to restrain the fluctuation of demand side power in ADN can be expressed as Figure 7. As shown in Figure 7, the control loop of ES is the core part of the whole framework. In order to obtain a reasonable ES voltage modulation signal based on the real-time data, specific control loop needs to be designed.

5.1. Control Loop of the Amplitude of ES Voltage. According to the "*idea for solution*" and the 0°/180° phase control strategy analyzed above, when the adjustment of smart load input power ΔP equals to -$\Delta P_{\text{DG}}(t)$, the fluctuation of demand side power caused by DG can be eliminated and the demand side power of ADN can be stabilized at reference value $P_{\text{DEM-REF}}$. Therefore, the control target of control loop can be set as

$$\Delta P = -\Delta P_{\text{DG}}(t) \tag{16}$$

Substituting (5) into (16) gives

$$\Delta P = P_{\text{DEM}} - P_{\text{DEM-REF}} \tag{17}$$

Getting the absolute value of (17) gives

$$|\Delta P| = |P_{\text{DEM}} - P_{\text{DEM-REF}}| \tag{18}$$

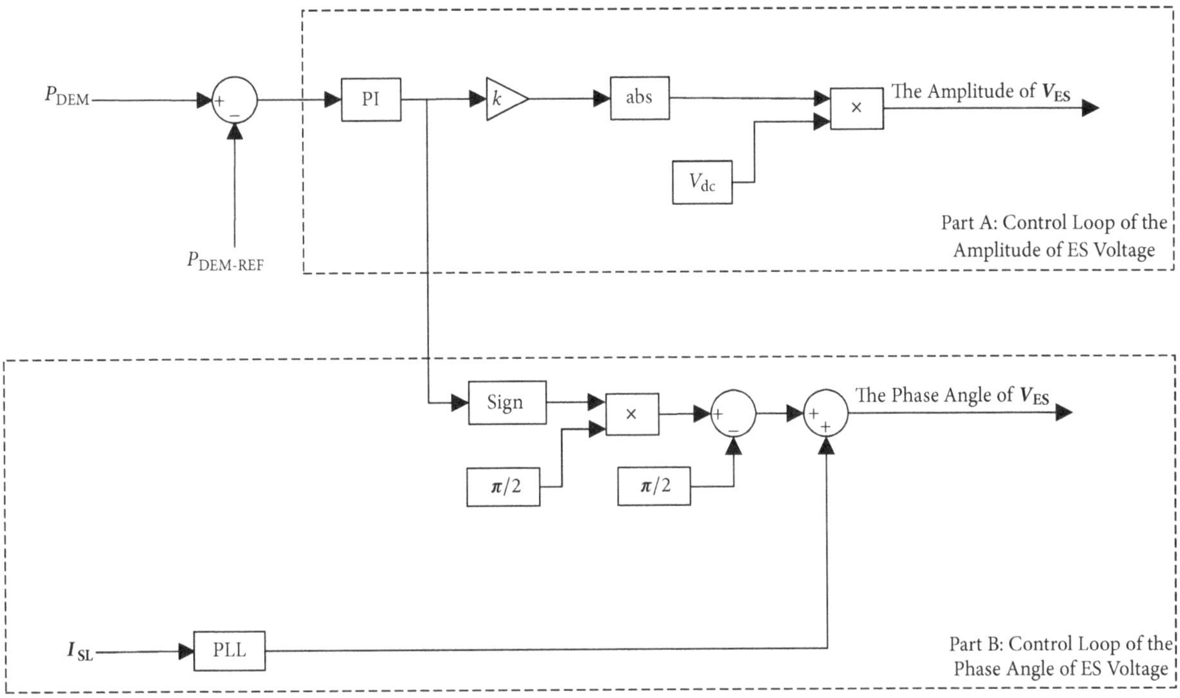

FIGURE 8: The control loop of ES voltage.

As a result, the absolute value of ΔP should have a positive correlation with the absolute value of P_{DEM}-$P_{DEM\text{-}REF}$.

According to (13), it can be obtained that the amplitude of $\mathbf{V_{ES}}$ has a positive correlation with the absolute value of ΔP. Therefore, combined with the conclusion above, the amplitude of $\mathbf{V_{ES}}$ should have a positive correlation with the absolute value of P_{DEM}-$P_{DEM\text{-}REF}$.

Accordingly, the amplitude of $\mathbf{V_{ES}}$ can be decided according to the absolute value of difference between the measured P_{DEM} and the given $P_{DEM\text{-}REF}$.

As a kind of linear controller, PI controller can ensure dynamic performance and steady-state performance simultaneously. Therefore, PI controller is used in this paper to control the amplitude of $\mathbf{V_{ES}}$. Combined with the analysis above, the control loop of the amplitude of $\mathbf{V_{ES}}$ can be obtained as Part A shown in Figure 8, where k is the PI postgain, abs is the absolute value function, and V_{dc} is the DC voltage of PWM inverter.

5.2. Control Loop of the Phase Angle of ES Voltage. According to the control target (17), ΔP should satisfy

$$\begin{aligned} &\Delta P > 0 \quad P_{DEM} > P_{DEM\text{-}REF} \\ &\Delta P < 0 \quad P_{DEM} < P_{DEM\text{-}REF} \end{aligned} \tag{19}$$

According to the 0°/180° phase control strategy analyzed above, when the phase angle difference between $\mathbf{V_{ES}}$ and $\mathbf{I_{SL}}$ is 0°, ES works in resistive mode, $\Delta P > 0$, and when the phase angle difference between $\mathbf{V_{ES}}$ and $\mathbf{I_{SL}}$ is 180°, ES works in negative resistive mode, $\Delta P < 0$. Therefore, based on (19), the phase angle difference between $\mathbf{V_{ES}}$ and $\mathbf{I_{SL}}$ should satisfy

$$\begin{aligned} &\arg(\mathbf{V_{ES}}) = \arg(\mathbf{I_{SL}}) + 0^\circ \quad P_{DEM} > P_{DEM\text{-}REF} \\ &\arg(\mathbf{V_{ES}}) = \arg(\mathbf{I_{SL}}) + 180^\circ \quad P_{DEM} < P_{DEM\text{-}REF} \end{aligned} \tag{20}$$

Based on (20), the 0°/180° phase control loop of the phase angle of $\mathbf{V_{ES}}$ can be obtained as Part B shown in Figure 8, where PLL is a phase-locked loop which can output the phase of $\mathbf{I_{SL}}$.

6. Simulation Studies

In order to verify the effectiveness of the method proposed in this paper, simulations are conducted by MATLAB/Simulink. The simulation model is shown as Figure 9, where the generation side of ADN is simulated by an AC voltage source and the random output power of DG is simulated by a PWM power inverter with prerecorded random data. There are 18 load nodes in the simulation model and the DG is set to be connected with node 15. The total loads of the 18 nodes is 15kW, in which a 10Ω load connected to node 8 is chosen as the noncritical load while other loads are critical loads. Specific parameters of the simulation model are shown in Table 1.

6.1. ES Is Not Activated. Switch on S in Figure 9 and conduct a 400s simulation; the demand side power waveform can be obtained as Figure 10, the output power of DG can be obtained as Figure 11, the input power waveform of noncritical load can be obtained as Figure 12(a), and the input power waveform of critical loads can be obtained as Figure 12(b). In this paper, positive value means input power, while negative value means output power.

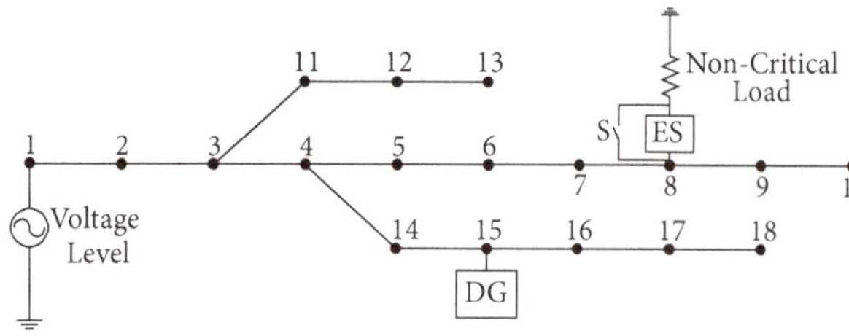

FIGURE 9: The simulation model of ADN.

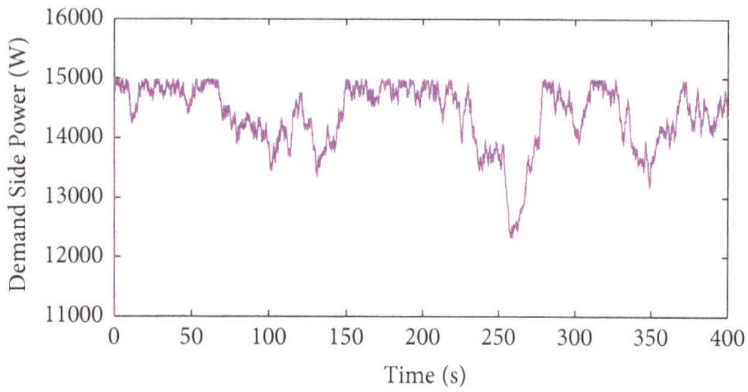

FIGURE 10: Demand side power waveform when ES is not activated.

TABLE 1: Parameters of the simulation model.

Voltage level	220V
DC voltage (V_{dc})	200V
Total loads (P_{Load})	15kW
Non-critical load (R_{NC})	10Ω
Inductance of the low-pass filter (L_f)	0.33mH
Capacitance of the low-pass filter (C_f)	22μF
Switching frequency (f_S)	20kHz
kp of PI	55
ki of PI	20
PI post-gain k	1/350000

As shown in Figures 10, 11, and 12, when ES is not activated, although the input power of loads is constant, the random output power of DG will cause the fluctuation of demand side power, which will bring difficulties to power dispatching.

6.2. ES Is Activated. As shown in Figure 11, the DG output power approximately fluctuates around 1kW. Therefore, $P_{DG\text{-}REF}$ can be set as 1kW according to (3), and the reference value of the demand side power $P_{DEM\text{-}REF}$ can be set as 14kW according to (4). Switch off S in Figure 9 and conduct a 400s simulation in the same environment as before; the demand side power waveform can be obtained as Figure 13, the output power of DG can be obtained as Figure 14, the smart load current waveform can be obtained as Figure 15(a), the input power waveform of smart load can be obtained as Figure 15(b), the input power waveform of noncritical load can be obtained as Figure 16(a), the input power waveform of critical loads can be obtained as Figure 16(b), the amplitude of ES voltage can be obtained as Figure 17(a), and the phase difference between ES voltage and smart load current can be obtained as Figure 17(b).

As shown in Figures 13–17, ES can change the smart load current and the input power of loads dynamically by generating dynamical ES voltage. In this way, the fluctuation of demand side power caused by DG which is shown in Figure 10 can be balanced out and the demand side power can be stabilized approximately at the reference value 14kW as shown in Figure 13. Therefore, the difficulties of power dispatching caused by the output power fluctuation of DG can be relieved.

Moreover, as shown in Figures 14 and 15(b), the power consumption increases when the quantity of DG output power is large, while the power consumption decreases when the quantity of DG output power is small. That is, the new control paradigm where demand follows generation has been achieved by applying ES to restrain the fluctuation of demand side power in ADN.

6.3. ES in Series with Noncritical Loads of Different Impedance Modulus. The simulations above have verified the effectiveness of the method applying 0°/180° phase controlled ES to restrain the fluctuation of demand side power in ADN, taking ES in series with a 10Ω noncritical load as an example. However, the selection basis of noncritical loads has been obtained in this paper, which concludes that "non-critical loads which have smaller impedance modulus should be considered to be in series with ES." In order to verify this selection basis and provide a reference for the selection of noncritical loads in practical application, further simulations based on Figure 9 should be conducted.

Firstly, same as the above simulations, the noncritical load is set as 10Ω. Conduct an 800s simulation; during 0-400s, ES is not activated, while during 400-800s, ES is activated on the condition that the fluctuation produced by the output power of DG is same as 0-400s. The demand side power waveform can be obtained as Figure 18(a) and the amplitude of ES voltage can be obtained as Figure 18(b). The reference value $P_{DEM\text{-}REF}$ in this simulation is set as 14kW, which is the same as previous simulations.

Then, rechoose a noncritical load which is 9Ω; the demand side power waveform can be obtained as Figure 19(a) and the amplitude of ES voltage can be obtained as Figure 19(b).

Similarly, rechoose a noncritical load which is 11Ω; the demand side power waveform can be obtained as Figure 20(a) and the amplitude of ES voltage can be obtained as Figure 20(b).

As shown in Figures 18–20, the fluctuation of demand side power caused by DG can be restrained effectively no matter which noncritical load is in series of ES. However, the impedance modulus of noncritical load is bigger and the voltage required to be generated by ES is bigger. According to the basic principles of PWM inverter, the value range of ES voltage is constrained by the magnitude of DC voltage. Therefore, as the DC voltage is constant, in order to achieve stronger ability to restrain the fluctuation of demand side power caused by DG, noncritical loads which have smaller impedance modulus should be considered to be in series with ES firstly.

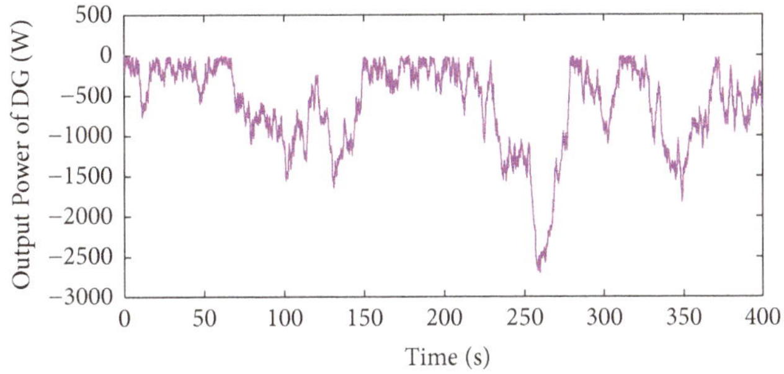

FIGURE 11: The output power waveform of DG.

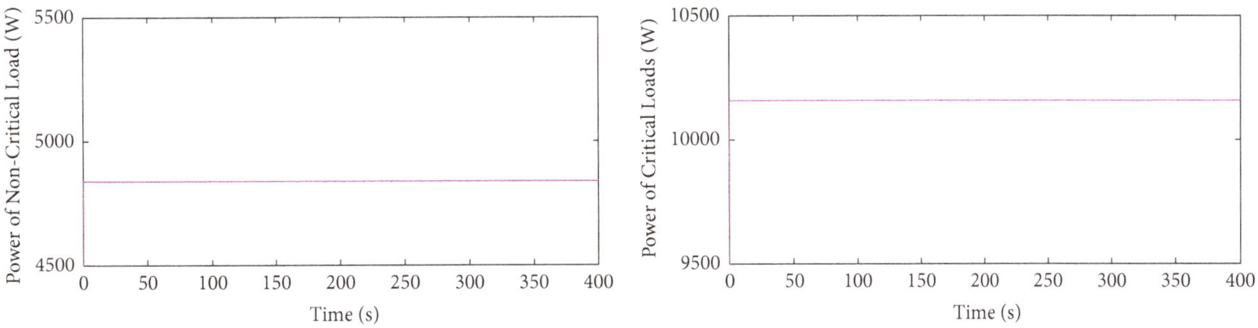

(a)

(b)

FIGURE 12: The input power waveform of loads when ES is not activated. (a) Noncritical load. (b) Critical loads.

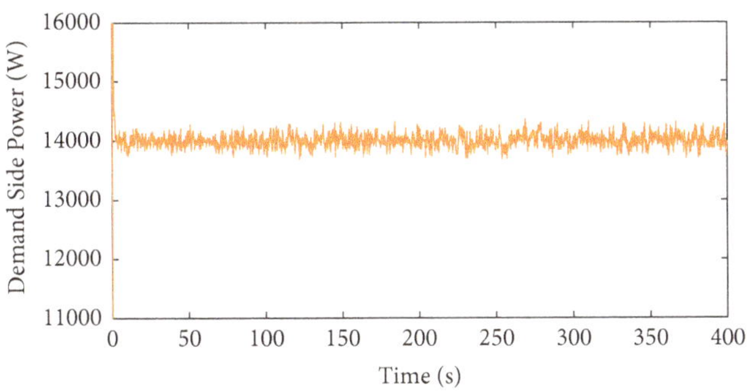

FIGURE 13: Demand side power waveform when ES is activated.

6.4. Simulation with the Consideration of Loads Change. As shown in the simulations above, the demand side power fluctuation caused by DG can be restrained effectively when ES is applied. However, the loads power in the above simulations is constant (15kW), which does not correspond with the actual condition in active distribution network. Therefore, the fourth simulation is conducted with the consideration of loads change, which will show the practicability of the proposed method when it is combined with the load forecasting technology in the future. The simulation model is also same as Figure 9, and the parameters are same as Table 1 except the total loads power, which is set as the solid line as shown in Figure 21. $P_{\text{DG-REF}}$ is set as 1kW which is same as former simulations and, therefore according to (4), the curve of $P_{\text{DEM-REF}}$ can be obtained as the dashed line shown in Figure 21.

Firstly, ES is not activated and the demand side power waveform can be obtained as shown in Figure 22(a). Secondly, ES is activated and the demand side power waveform can be obtained as shown in Figure 22(b).

As shown in Figure 22, if ES is not activated, there is a great discrepancy between the shape of demand side power waveform and the shape of load curve as the existence of the DG output power fluctuation will bring difficulties to power dispatching. However, when ES is activated, the demand side power waveform will have quite a similar shape with the load curve and the influence of the fluctuation of DG output power can be relieved. Therefore, with the application of ES, the power dispatch can be conducted easily according to the load curve only.

7. Conclusion

In this paper, a method of restraining the fluctuation of demand side power in ADN is proposed, in which a new power electronic device-ES is applied by the proposed 0°/180° phase control strategy. The control strategy has been proposed concretely, the control loop has been designed in detail, and the selection basis of noncritical loads is also obtained. Simulation results have indicated that adopting the method proposed in this paper can restrain the demand side power fluctuation caused by DG effectively, and therefore the difficulties of power dispatching caused by the output power fluctuation of DG can be relieved.

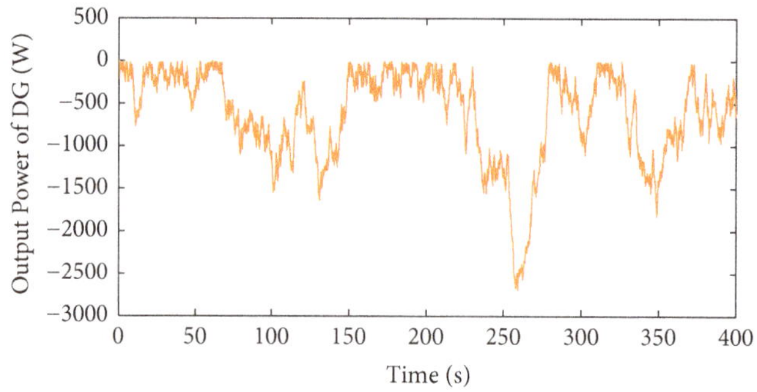

FIGURE 14: The output power waveform of DG.

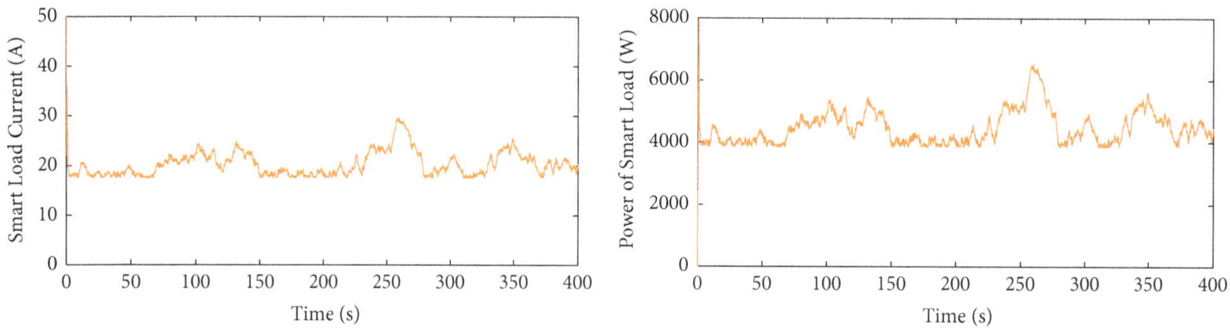

(a) (b)

FIGURE 15: The current and input power waveform of smart load. (a) Smart load current. (b) The input power of smart load.

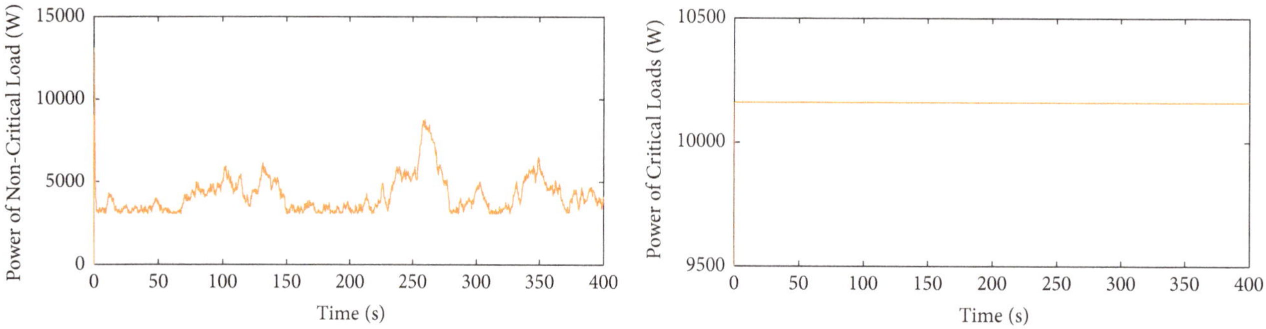

(a) (b)

FIGURE 16: The input power waveform of loads when ES is activated. (a) Noncritical load. (b) Critical loads.

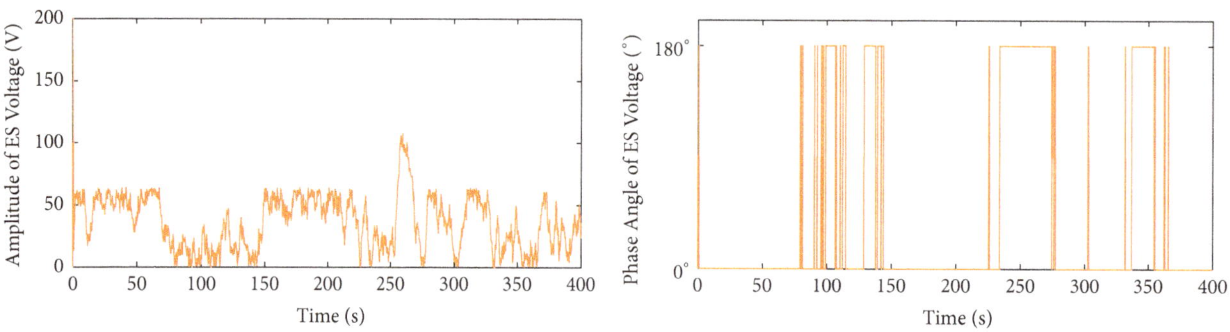

(a) (b)

FIGURE 17: The amplitude and phase angle condition of ES voltage. (a) The amplitude of ES voltage. (b) The phase angle difference (0° or 180°).

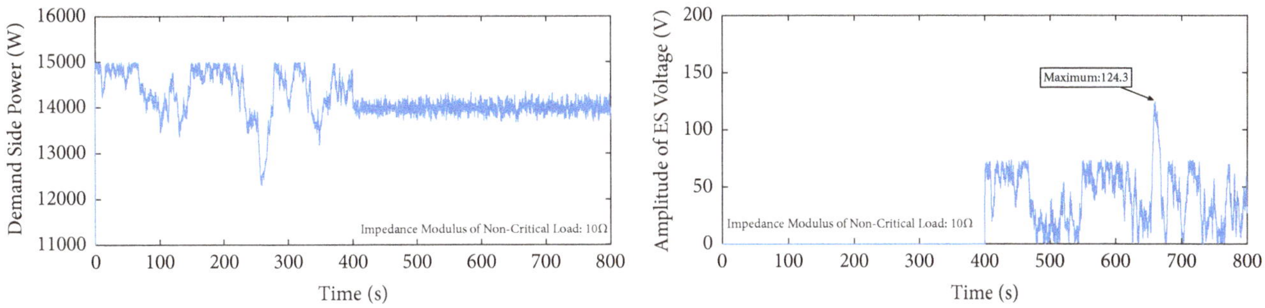

(a) (b)

FIGURE 18: ES in series with a 10Ω noncritical load. (a) Demand side power. (b) The amplitude of ES voltage.

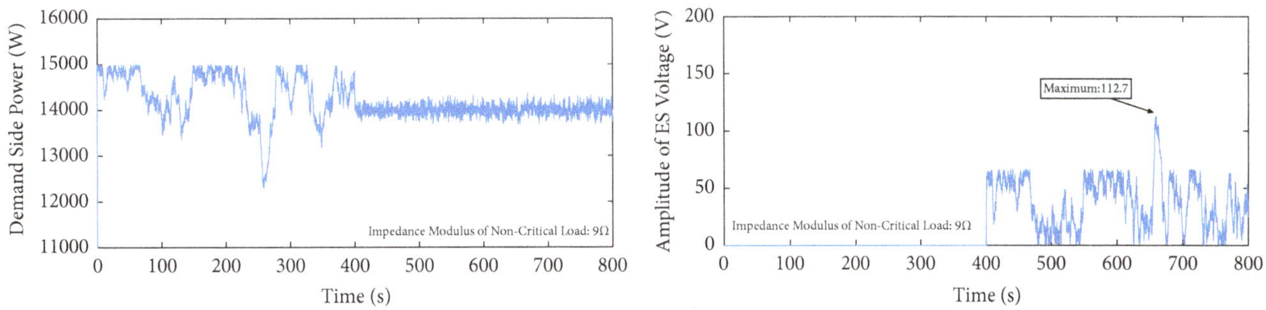

(a) (b)

FIGURE 19: ES in series with a 9Ω noncritical load. (a) Demand side power. (b) The amplitude of ES voltage.

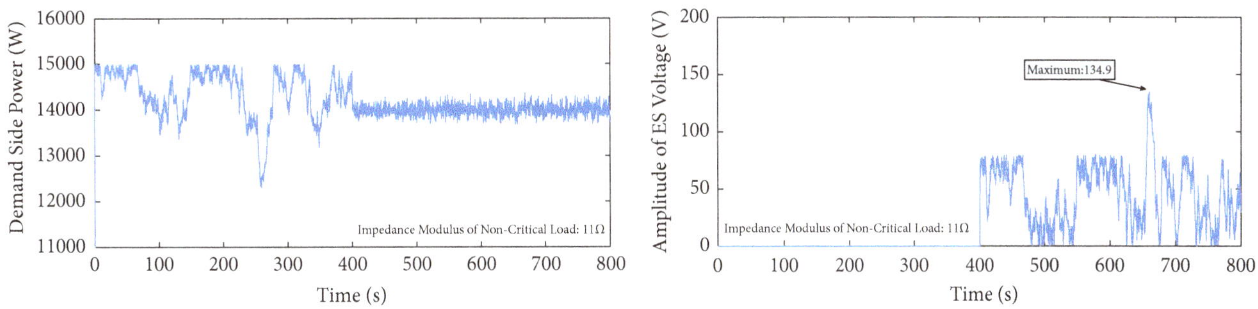

(a) (b)

FIGURE 20: ES in series with a 11Ω noncritical load. (a) Demand side power. (b) The amplitude of ES voltage.

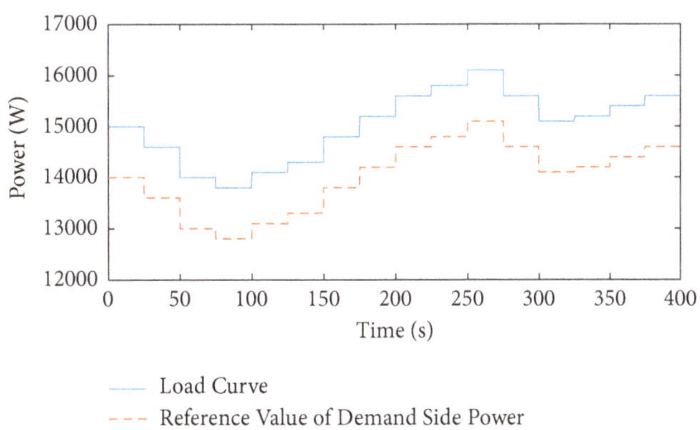

FIGURE 21: The total load curve and the reference value curve of demand side power.

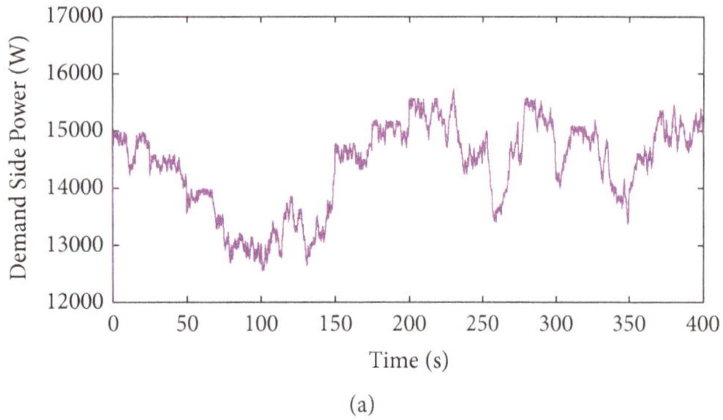

FIGURE 22: Demand side power waveform with loads change. (a) ES is not activated. (b) ES is activated.

The method proposed in this paper does not need complex communication technologies and requires easier operation. When the fluctuation of DG is big, the noncritical loads connected with ES can be added, which is more economical than the traditional way of increasing the energy storage capacity only. Besides, the new control paradigm where demand follows generation can be achieved by adopting this method. However, only one smart load is set in this paper; therefore the fluctuation of noncritical load power is not small as shown in the simulation result. To consummate the method proposed in this paper, multi smart loads should be applied to share the power fluctuation in the future, and the coordinated control strategy between them should be studied deeply.

In addition, this paper mainly uses ES to stabilize the demand side active power fluctuation in ADN. Therefore, in order to maximize the active power adjustment range, ES is not involved in the reactive power regulation. In the future, based on power decoupling technology, ES active power regulation and STATCOM reactive power regulation can be conducted simultaneously, in order to make the demand side power more flexible to be controlled.

Conflicts of Interest

There are no conflicts of interest regarding the publication of this paper.

Acknowledgments

This research was supported by the National Natural Science Foundation of China (51607093).

References

[1] G. Ma, G. Xu, Y. Chen, and R. Ju, "Multi-objective optimal configuration method for a standalone wind-solar-battery hybrid power system," *IET Renewable Power Generation*, vol. 11, no. 1, pp. 194–202, 2017.

[2] T. Vigneysh and N. Kumarappan, "Grid interconnection of renewable energy sources using multifunctional grid-interactive converters: A fuzzy logic based approach," *Electric Power Systems Research*, vol. 151, pp. 359–368, 2017.

[3] A. Soroudi, M. Ehsan, R. Caire, and N. Hadjsaid, "Possibilistic evaluation of distributed generations impacts on distribution networks," *IEEE Transactions on Power Systems*, vol. 26, no. 4, pp. 2293–2301, 2011.

[4] S. Lumbreras, A. Ramos, and F. Banez-Chicharro, "Optimal transmission network expansion planning in real-sized power systems with high renewable penetration," *Electric Power Systems Research*, vol. 149, pp. 76–88, 2017.

[5] S. S. Reddy and P. R. Bijwe, "Real time economic dispatch considering renewable energy resources," *Journal of Renewable Energy*, vol. 83, pp. 1215–1226, 2015.

[6] C. K. Lee, B. Chaudhuri, and S. Y. Hui, "Hardware and control implementation of electric springs for stabilizing future smart grid with intermittent renewable energy sources," *IEEE Journal of Emerging and Selected Topics in Power Electronics*, vol. 1, no. 1, pp. 18–27, 2013.

[7] G. C. Heffner, C. A. Goldman, and M. M. Moezzi, "Innovative approaches to verifying demand response of water heater load control," *IEEE Transactions on Power Delivery*, vol. 21, no. 1, pp. 388–397, 2006.

[8] A. Brooks, E. Lu, D. Reicher, C. Spirakis, and B. Weihl, "Demand dispatch," *IEEE Power & Energy Magazine*, vol. 8, no. 3, pp. 20–29, 2010.

[9] E. Salies, "Real-time pricing when some consumers resist in saving electricity," *Energy Policy*, vol. 59, pp. 843–849, 2013.

[10] H. Allcott, "Rethinking real-time electricity pricing," *Resource and Energy Economics*, vol. 33, no. 4, pp. 820–842, 2011.

[11] R. Zamora and A. K. Srivastava, "Controls for microgrids with storage: Review, challenges, and research needs," *Renewable & Sustainable Energy Reviews*, vol. 14, no. 7, pp. 2009–2018, 2010.

[12] M. Aneke and M. Wang, "Energy storage technologies and real life applications – A state of the art review," *Applied Energy*, vol. 179, pp. 350–377, 2016.

[13] X. Chen, Y. Hou, S.-C. Tan, C.-K. Lee, and S. Y. R. Hui, "Mitigating voltage and frequency fluctuation in microgrids using electric springs," *IEEE Transactions on Smart Grid*, vol. 6, no. 2, pp. 508–515, 2015.

[14] C. K. Lee and S. Y. R. Hui, "Reduction of energy storage requirements in future smart grid using electric springs," *IEEE Transactions on Smart Grid*, vol. 4, no. 3, pp. 1282–1288, 2013.

[15] Q. Wang, M. Cheng, Z. Chen, and Z. Wang, "Steady-State Analysis of Electric Springs with a Novel δ Control," *IEEE Transactions on Power Electronics*, vol. 30, no. 12, pp. 7159–7169, 2015.

[16] S. Yan, C.-K. Lee, T. Yang et al., "Extending the operating range of electric spring using back-to-back converter: Hardware implementation and control," *IEEE Transactions on Power Electronics*, vol. 32, no. 7, pp. 5171–5179, 2017.

[17] S. Yan, M. Wang, T. Yang, S. Tan, B. Chaudhuri, and S. Y. Hui, "Achieving Multiple Functions of Three-Phase Electric Springs in Unbalanced Three-Phase Power Systems Using the Instantaneous Power Theory," *IEEE Transactions on Power Electronics*, vol. 33, no. 7, pp. 5784–5795, 2018.

[18] S. Yan, M. Wang, J. Chen, and S. Y. Hui, "Smart Loads for Improving the Fault-Ride-Through Capability of Fixed-Speed Wind Generators in Microgrids," *IEEE Transactions on Smart Grid*.

[19] S. Y. Hui, C. K. Lee, and F. F. Wu, "Electric springs - A new smart grid technology," *IEEE Transactions on Smart Grid*, vol. 3, no. 3, pp. 1552–1561, 2012.

[20] X. Luo, Z. Akhtar, C. K. Lee, B. Chaudhuri, S.-C. Tan, and S. Y. R. Hui, "Distributed voltage control with electric springs: Comparison with STATCOM," *IEEE Transactions on Smart Grid*, vol. 6, no. 1, pp. 209–219, 2015.

[21] S. Yan, S.-C. Tan, C.-K. Lee, B. Chaudhuri, and S. Y. R. Hui, "Use of Smart Loads for Power Quality Improvement," *IEEE Journal of Emerging and Selected Topics in Power Electronics*, vol. 5, no. 1, pp. 504–512, 2017.

[22] N. R. Chaudhuri, C. K. Lee, B. Chaudhuri, and S. Y. R. Hui, "Dynamic modeling of electric springs," *IEEE Transactions on Smart Grid*, vol. 5, no. 5, pp. 2450–2458, 2014.

[23] G. Ma, G. Xu, Y. Chen, and R. Ju, "Voltage stability control method of electric springs based on adaptive PI controller," *International Journal of Electrical Power & Energy Systems*, vol. 95, pp. 202–212, 2018.

[24] T. Yang, K. T. Mok, S. Tan, C. Lee, and S. R. Hui, "Electric springs with coordinated battery management for reducing voltage and frequency fluctuations in microgrids," *IEEE Transactions on Smart Grid*, vol. 9, no. 3, pp. 1943–1952, 2018.

[25] Z. Akhtar, B. Chaudhuri, and S. Y. R. Hui, "Smart Loads for Voltage Control in Distribution Networks," *IEEE Transactions on Smart Grid*, vol. 8, no. 2, pp. 937–946, 2017.

[26] C. K. Lee, K. L. Cheng, and W. M. Ng, "Load characterisation of electric spring," in *Proceedings of the 5th Annual IEEE Energy Conversion Congress and Exhibition, ECCE 2013*, pp. 4665–4670, usa, September 2013.

[27] Q. Wang, M. Cheng, Y. Jiang, W. Zuo, and G. Buja, "A Simple Active and Reactive Power Control for Applications of Single-Phase Electric Springs," *IEEE Transactions on Industrial Electronics*, vol. 65, no. 8, pp. 6291–6300, 2018.

[28] Y. Zheng, D. J. Hill, K. Meng, and S. Y. Hui, "Critical Bus Voltage Support in Distribution Systems with Electric Springs and Responsibility Sharing," *IEEE Transactions on Power Systems*, vol. 32, no. 5, pp. 3584–3593, 2017.

[29] S.-C. Tan, C. K. Lee, and S. Y. Hui, "General steady-state analysis and control principle of electric springs with active and reactive power compensations," *IEEE Transactions on Power Electronics*, vol. 28, no. 8, pp. 3958–3969, 2013.

2

Cost-Benefit Analysis for the Concentrated Solar Power in China

Shuxia Yang, Xianguo Zhu, and Weishang Guo

School of Economics and Management, North China Electric Power University, Beijing 102206, China

Correspondence should be addressed to Weishang Guo; 547180981@qq.com

Academic Editor: Gorazd Stumberger

In 2016, the first batch of concentrated solar power (CSP) demonstration projects of China was formally approved. Due to the important impact of the cost-benefit on the investment decisions and policy-making, this paper adopted the static payback period (SP), net present value (NPV), net present value rate (NPVR), and internal rate of return (IRR) to analyze and discuss the cost-benefit of CSP demonstration plants. The results showed the following. (1) The SP of CSP systems is relatively longer, due to high initial investment; but the cost-benefit of CSP demonstration plants as a whole is better, because of good expected incomes. (2) Vast majority of CSP projects could gain excess returns, on the basis of meeting the profitability required by the benchmark yield of 10%. (3) The cost-benefit of solar tower CSP technology (IRR of 12.33%) is better than that of parabolic trough CSP technology (IRR of 11.72%) and linear Fresnel CSP technology (IRR of 11.43%). (4) The annual electricity production and initial costs have significant impacts on the cost-benefit of CSP systems; the effects of operation and maintenance costs and loan interest rate on the cost-benefit of CSP systems are relatively smaller but cannot be ignored.

1. Introduction

1.1. Background. Concentrating solar power (CSP) is an important solar energy generation power technology that has attracted increasing attention in many countries. Spain, America, South Africa, India, Morocco, Italy, Greece, Portugal, and others have issued a series of policies to promote the development of the CSP industry. At present, only the CSP plants in Spain and the United States realized commercial operation. The cumulative installations of both plants were about 4,079 MW in 2016, accounting for more than 80% of the global total installed capacity. However, the growth of new installed capacity of CSP in the United States and Spain started to slow down since 2015, due to the cancellation of preferential policies. The CSP industry is still in the early commercialization demonstration stage in some countries and even in the initial stage in other countries. But those countries have ambitious plans for the CSP development. For example, South Africa and Morocco were to install, respectively, 3.3 GW and 2 GW CSP plants by 2022 and Saudi Arabia planned to install 25 GW CSP projects by 2040.

As the world's largest carbon emission country, China has proposed ambitious emission reduction targets. In order to reduce the carbon emission effect of CSP, Chinese government began to pay great attention to the development of CSP industry. According to the 2050 solar roadmap of China [1], in order to achieve the large-scale commercialization operation, the development of CSP industry was divided into four stages, namely, the initial experimental stage (before 2015), demonstration phase (2015–2020), large-scale development phase (2020–2030), and large-scale commercial development stage (2030–2050). With the development of CSP, its role in the power system would change from the middle load of peak regulation to the power source of basic power load in the third stage. In recent years, China has made a series of policy reforms in order to create the CSP Roadmap plan. For example, in 2016, the CSP installations target of 2020 was set to 5 GW in China's 13th Five-Year electric power plan. Notice on the benchmark feed-in tariff (FIT) policy of CSP was issued by National Development and Reform Commission (NDRC), and the on-grid price was determined to be RMB1.15/kWh (including tax) [2]. The first twenty CSP pilot plants have been approved officially by National Energy Administration (NEA) and the total installed capacity was 1,349 MW. The pilot projects of CSP will play an important demonstration effect for the development of CSP industry in

China. Thus it can be expected that the CSP development will enter a rapid development stage in the future.

1.2. Literature Review. There are four available technologies for CSP, namely, parabolic trough (PT), linear Fresnel (LF), solar tower (ST), and parabolic dish (PD). The heat transfer fluids mainly include water, thermal oil, molten salt, all solid state concrete, and air. Over the past few decades, the thermal oil PT-CSP has been the most commercially attractive option among the existing CSP technologies and has widely been installed in the United States and Spain. ST-CSP technology has not been adopted widely in the existing CSP plants, but it has a better development potential because of the higher thermodynamic efficiency [3, 4]. The high initial costs, which were estimated to vary between RMB27.73/W and RMB57.44/W (Exchange Rates on Dec. 19, 2017: $1 = RMB6.602), have been obstructing the development of CSP [5]. However, with the advanced technology and the expansion of market scale, further reductions in the costs of CSP are expected in the future to finally reach a better power grid parity than the conventional fossil fuel power technologies [6, 7].

Different from the photovoltaic (PV) power, CSP is able to effectively solve the issues on the instability and intermittency of solar energy by installing thermal energy storage (TES) systems. The application of TES is of significance for CSP development and it has been an important research direction in recent years. For example, Shinnar and Citro (2006) analyzed the prominent impact of TES on CSP systems [8]; Tian and Zhao (2013) provided a review for the latest improvement of the ETS applications [9]; Sioshansi and Denholm (2010) researched the profits obtained by CSP with TES in the electricity markets and concluded that TES has the potential to add to the values of CSP [10]; Madaeni et al. (2012) evaluated the cost-benefit of CSP plants with TES, and the installation of TES proved to be a reliable option [11]; International Renewable Energy Agency (2014) pointed out that CSP plants with TES usually required higher costs but improved the capacity factor of CSP system, resulting in lower levelized cost of energy (LCOE, the discounted uniform unit cost of energy in the whole life-cycle) [12]; REN 21 (2015) believed that TES has a large cost-reduction potential and provides a perfect opportunity to overcome the impediment of high initial investment for CSP plants [13]. The PV-CSP hybrid systems could also reduce the costs of CSP and improve both the reliability and efficiency of CSP systems [14–16]. Ju et al. (2017) held that PV-CSP hybrid system has several advantages compared to the PV-alone or CSP-alone systems [17]. Parrado et al. (2016) calculated the LCOE of PV-CSP hybrid plant based on the Blue Map scenario and Roadmap scenario. Between 2014 and 2050, the LCOE values obtained varied from RMB0.97/kWh to RMB0.57/kWh and from RMB0.92/kWh to RMB0.51/kWh [15]. Besides, there are many factors affecting the CSP costs. Caldés et al. (2009) concluded that the total costs of CSP mostly depended on TES capacity, labor costs, power plant size, and so on [5]. Izquierdo et al. (2010) studied the effects of solar multiple capacity factor and TES capacity on the costs of CSP [18]; Hernández-Moro and Martínez-Duart (2012) analyzed the main factors impacting the CSP costs, including cumulative installed capacity, electricity production, land cost, discount rate, operation and maintenance (O&M) costs, insurance cost, and life-time [19].

In view of the importance of cost-benefit to the investment policy-making of power generation industry, the cost-benefit analysis has been the research focus of energy electricity industry. About the solar power, it is divided into three types of power generation technologies, namely, PV power, solar chimney power plant (SCPP, one technology for the large-scale utilization of solar energy) [20], and CSP (the mechanism of a conventional SCPP is as follows [20]: sunlight transmits through the solar collector cover and heats the ground below; then ambient air enters the solar collector and is heated as it flows along the collector toward its centre, where a chimney is located; then the airflow enters the chimney, due to the pressure created by the density difference between the warm airflow and ambient cold air; finally, with the power conversion unit, the kinetic energy of the airflow is transferred into electrical power). PV power has achieved large-scale commercial development and its cost-benefit has been discussed widely in large numbers of literatures. For example, Zhao et al. (2015) calculated the static payback period (SP), net present value (NPV), and internal rate of return (IRR) of distributed PV plants and found that the cost-benefit of PV power varied widely in different regions, with IRR of about 7%–21% [21]; Yuan et al. (2014) adopted the LCOE to analyze the economy of the different scale distributed PV systems and concluded that their LCOE varied between RMB1.16/kWh and RMB1.29/kWh [22]. Solar chimney power plant (SCPP) is still in the early experiment stage, but some scholars have begun to focus on the cost-benefit of SCPP. For example, Li et al. (2014) analyzed the cost-benefit of reinforced concrete SCPP by calculating the total NPV of four different phases, with the total NPV rate of 0.390, and held that it has very good application prospect in the future [23]; Guo et al. (2017) estimated the LCOE of the SCPP of RMB0.4178/kWh by taking a 10-MW SCPP of Yinchuan China as an example and concluded that the cost-benefit of SCPP could compete with the wind power and PV electricity [24]. CSP achieved the commercial development in Spain and in the United States but is still in commercialization demonstration stage in other countries. Studies on the cost-benefit of CSP mainly focus on the LCOE. According to Parrado et al. (2016), from 2014 to 2050, the LCOE of CSP decreased from RMB1.01/kWh to RMB0.60/kWh for the Blue Map scenario and was between RMB0.98/kWh and RMB0.50/kWh for the Roadmap scenario [15]; Hernández-Moro and Martínez-Duart (2012) used the life-cycle cost method to discuss the LCOE of CSP in the period of 2010–2050 [19]. Zhu et al. (2015) found that the LCOE values of CSP in China varied between RMB1.2/kWh and RMB2.7/kWh [25], while Zhao et al. (2017) pointed out that the LCOE of SCP in China was approximately RMB1.44/kWh [26]. It can be seen that there were large numbers of investigations discussing the cost-benefit of SCP from the aspect of LCOE but they have not involved financial indexes such as SP, NPV, and IRR. Therefore, this study focused on calculating the static payback period (SP), the net present value (NPV), the net present value rate (NPVR), and the internal rate of return (IRR) of CSP and discussed the sensitivity factors of IRR, based on the first batch of CSP demonstration projects in order to analyze the cost-benefit of CSP in China comprehensively.

2. Methodology

This research used SP, NPV, NPVR, and IRR to study the cost-benefit of CSP in China. For CSP projects, the cash outflows (CO_i) include initial investment (II), operation and maintenance (O&M) costs, insurance expenses, loan interest, and taxes costs. The cash inflows (CI_i) involve electricity generation incomes and government tax subsidies (namely, value-added tax (VAT) refund); that is,

$$\mathrm{CO}_i = \begin{cases} \mathrm{II}_i, & i = 1, \dots, T_0 \\ (\mathrm{OM} + I)\, C + R_i + \mathrm{TAX}_i, & i = T_0 + 1, \dots, T \end{cases}$$
$$\mathrm{CI}_i = \frac{(P \times E_i)(1 - \alpha)}{(1 + t_{\mathrm{VAT}})} + \pi_i, \quad i = T_0 + 1, \dots, T, \tag{1}$$

where T_0 and T are, respectively, construction period and life-time; C represents the construction costs; OM is the ratio between annual O&M costs and construction costs; I represents the proportion of annual insurance expenses and construction costs; R_i and TAX_i are, respectively, annual interest payments and annual taxes costs; P and E_i represent the on-grid price and annual electricity production, respectively; α is the proportion of the self-consumption electricity to total electricity production; t_{VAT} is added-value tax (VAT) rate; π_i represents government tax subsidies.

The taxes costs for CSP projects mainly involve income tax and additional taxes; that is,

$$\mathrm{TAX}_i = \text{income tax}_i + \text{additional tax}_i, \tag{2}$$

where the income tax can be expressed as

$$\text{income tax}_i = \left[\frac{(P \times E_i)(1 - \alpha)}{(1 + t_{\mathrm{VAT}})} + \pi_i - (\mathrm{O\&M} + I)\, C - R_i - \mathrm{DE}_i\right] \times t_{\mathrm{income}}, \tag{3}$$

where DE_i is the annual depreciation expense and t_{income} represents the income tax rate.

VAT is not a cash outflow but directly determines the government tax subsidies and additional taxes. According to the notice of NEA on reducing the tax and fees burdens for the renewable energy companies, CSP in China could enjoy the taxes preference policy to refund 50% of VAT [27]. Thus we can obtain

$$\pi_i = 50\% \times \mathrm{VAT}_i$$
$$\text{additional tax}_i = \mathrm{VAT}_i \times t_{\mathrm{additional}}. \tag{4}$$

(1) Payback Period. The equation on SP is

$$\sum_{i=1}^{\mathrm{SP}} (\mathrm{CI} - \mathrm{CO})_i = 0. \tag{5}$$

The SP can be obtained by calculating the cumulative net cash flows, based on the cash flow statement of CSP project investment. SP is the time point where the cumulative net cash flows turn from negative value to positive value. Equation (5) can be expressed as

$$\mathrm{SP} = T_1 - 1 + \frac{\left|\sum_{i=1}^{T_1 - 1} (\mathrm{CI} - \mathrm{CO})_i\right|}{(\mathrm{CI} - \mathrm{CO})_{T_1}}, \tag{6}$$

where T_1 is the number of years when the cumulative net cash flows are positive or zero for the first time.

(2) Net Present Value. NPV is the dynamic index to evaluate the profit ability of CSP project during the calculation period. The formula of NPV is

$$\mathrm{NPV} = \sum_{i=1}^{T} (\mathrm{CI} - \mathrm{CO})_i (1 + r_e)^{-i+1}, \tag{7}$$

where r_e is benchmark yield or discount rate.

NPV is the absolute index to evaluate the profit ability of CSP projects. When NPV $>$ 0, it indicates that CSP projects could get excess earnings on the basis of meeting the profitability that is required by the benchmark yield; that is, the investment of the CSP projects has strong economic feasibility. If NPV = 0, it illustrates that the CSP projects are only able to meet the profitability required by the discount rate, which means that the investment for the CSP projects is basically feasible or to be improved, while NPV $<$ 0 indicates that the CSP projects cannot achieve the profitability of benchmark yield, and their investment is not feasible.

Besides, the magnitude of NPV depends not only on the profitability but also on the total investment or projects scale. If NPV $>$ 0, the NPV is positively correlated with the total investment; that is, the bigger the total investment, the bigger the NPV. If NPV $<$ 0, the NPV is negatively correlated with the total investment. Therefore, in general, NPV cannot be used to compare the profitability between different CSP projects.

(3) Net Present Value Rate. NPVR is the ratio of the NPV of a project to the present value of its total investment. It reflects the unit of investment profitability. NPVR can be used as the auxiliary evaluation index of NPV and evaluate the profitability between different CSP projects. The formula of NPVR is as follows:

$$\mathrm{NPVR} = \frac{\mathrm{NPV}}{\mathrm{II}_p}$$
$$\mathrm{II}_p = \sum_{i=1}^{T_0} \mathrm{II}_i (1 + r_e)^{-i}, \tag{8}$$

where II_p represents the present value of total investment.

(4) Internal Rate of Return. In the engineering economic analysis, IRR is the main dynamic evaluation index to investigate the profitability of projects. IRR reflects the profitability of the unrecovered funds occupied by the CSP projects, namely,

Table 1: Design parameters of CSP demonstration plants.

Number	Projects	DNI ($kWh/m^2/yr$)	Total investment (million RMB)	Annual power generation (GWh/yr)	TES capacity	TES medium
ST_1 [32]	Qinghai Delingha 50 MW MS ST-CSP plant	1,976	1,050	136	6 h	MS
ST_2 [33]	Dunhuang 100 MW MS ST-CSP plant	2,000	3,040	483	11 h	MS
ST_3 [34]	Qinghai Gonghe 50 MW MS ST-CSP plant	1,900	1,222	156.9	6 h	MS
ST_4 [35]	Xinjiang Hami 50 MW MS ST-CSP plant	1,920	1,580	198.4	8 h	MS
ST_5 [32]	Delingha 135 MW DSG ST-CSP plant	1,900	–	–	3.7 h	MS
ST_6 [36]	Gansu Jinta 100 MW MS ST-CSP plant	1,900	2,500	323.2	8 h	MS
ST_7 [37]	Yumen 50 MW MS ST-CSP plant	1,800	1,790	216	9 h	MS
ST_8 [38]	Yumen 100 MW MS ST-CSP plant	1,800	3,000	388	10 h	MS
ST_9 [32]	Shangyi 50 MW DSG ST-CSP plant	1,600	–	–	12 h	MS
PT_1 [39]	Yumen East Town 50 MW TO PT-CSP plant (Royal tech CSP Co., Ltd.)	1,800	1,344.77	169.3	9 h	MS
PT_2 [32]	Gansu Akesai 50 MW MS PT-CSP plant	2,056.5	1,986	256	15 h	MS
PT_3 [32]	Yumen East Town 50 MW TO PT CSP plant (Rayspower Energy Group Co., Ltd.)	1,800	–	–	9 h	MS
PT_4 [40]	Urat Middle Banner 100 MW TO PT-CSP plant	2,025	2,800	350	10 h	MS
PT_5 [32]	Delingha 50 MW TO PT-CSP plant	1,976	1,938.38	225	9 h	MS
PT_6 [32]	Gansu Gulang 100 MW TO PT-CSP plant	1,913	–	–	7 h	MS
PT_7 [41, 42]	Zhangjiakou 64 MW MS PT-CSP plant	1,700	1,800	300	16 h	MS
LF_1 [43]	Dacheng Dunhuang 50 MW MS LF-CSP plant	2,000	1,680	201.5	13 h	MS
LF_2 [32]	Zhangbei 50 MW DSG LF-CSP plant	1,750	–	–	14 h	ASSC
LF_3 [44, 45]	Huaqiang Zhaoyang Zhangjiakou 50 MW DSG LF-CSP plant	1,750	1,800	250	14 h	ASSC
LF_4 [46]	Urat Middle Banner 50 MW TO LF-CSP plant	2,025	1,476	195.88	8 h	MS

Note. MS: molten salt; TO: thermal oil; ASSC: all solid state concrete; DSG: direct steam generation.

the profitability of the projects themselves. The higher the IRR, the better the economy of projects. For CSP projects, IRR is the discount rate that makes the NPV zero and the mathematical expression is

$$\text{NPV}(\text{IRR}) = \sum_{i=1}^{T} (\text{CI} - \text{CO})_i (1 + \text{IRR})^{-i+1}. \tag{9}$$

It indicates that the calculation of IRR determined by (9) has to be solved for the higher order equation. This process is very tedious and difficult. Therefore, in actual projects, the linear interpolation is usually used to calculate the approximate value of IRR, and the linear interpolation formula of IRR is as follows:

$$\text{IRR} = r_1 + \frac{\text{NPV}(r_1)}{\text{NPV}(r_1) + |\text{NPV}(r_2)|}(r_2 - r_1), \tag{10}$$

where $\text{NPV}(r_1) > 0$, $\text{NPV}(r_2) < 0$, $r_2 > r_1$, and the value of $(r_2 - r_1)$ should be not more than 2%.

3. Results and Discussions

3.1. Data. Table 1 displays the basic information on the first twenty CSP demonstration projects of China, including nine ST-CSP projects, seven PT-CSP plants, and four LF-SCP projects.

According to Table 1, these demonstration projects are distributed in Gansu, Qinghai, Hebei, Inner Mongolia, and Xinjiang which are abundant in solar resources. The heat transfer fluids include molten salt, thermal oil, and water/steam, with the molten salt and thermal oil primarily. The TES medium includes molten salt and all solid state concrete. Table 1 shows that the initial costs of CSP demonstration plants vary between RMB21.0/W and RMB39.7/W and the annual effective operation hours change from 2720 h to 5120 h, due to the impacts of regions conditions, direct normal irradiation (DNI), CSP technologies, TES capacity, and so on (given the lack of design parameters for ST_5, ST_9, PT_3, PT_6, and LF_2 projects, the analysis and discussion for the costs and benefits of CSP will not involve these five projects in the study). Simply from the CSP technologies perspectives, the average TES capacity of ST, PT and LR CSP systems are respectively 8 h, 10 h and 12 h. Correspondingly, the average initial costs and annual effective operation hours are RMB28.6/W and 3750 h for ST-CSP systems, RMB31.4/W and 4140 h for PT-CSP systems, and RMB33.0/W and 4320 h for LF-CSP systems. Solely from the regions perspectives, the average TES capacity of CSP systems in Hebei, Gansu, Inner Mongolia, and Qinghai is 15 h, 10 h, 9 h, and 7 h, respectively. Accordingly, the average initial costs and annual effective operation hours of CSP systems are RMB31.6/W and 4825 h in Hebei, RMB30.7/W and 4074 h in Gansu, RMB28.5/W and 3693 h in Inner Mongolia, RMB28.5/W and 3693 h in Qinghai. Thus it can be seen that the TES capacity has crucial influence on the initial costs and annual electricity production of CSP. Besides, from projects perspectives, the TES capacity has also an important impact on the CSP systems. In general, the larger the TES capacity, the higher the initial costs and annual electricity generation. Taking the ST_1 project, for example, the TES capacity is the smallest but both the initial costs and annual effective operation hours are also lower than those of other CSP projects. For the PT_2 project, its TES is higher than that of other CSP projects apart from the PT_7 project and both its initial costs and annual effective operation hours are the highest.

According to the notice on organizing the construction of CSP demonstration projects, the National Energy Administration (NEA) required that the equity capital ratio cannot be less than 20%; the loan rate could be a reference to the actual commercial loan interest rate; the construction period and operation period were set as 2 years and 25 years, respectively [28]. In 2016, the long-term loan interest rate of China was 4.9%. The initial costs are divided into land cost and construction costs and the proportions of both costs are 4% and 96%, respectively [26]. Due to no fuel consumption, the O&M costs are very low and usually account for about 2% of construction costs [29]. Besides, it is necessary for CSP projects to adopt an insurance measure to decrease the risks caused by high initial investment and immature technology; the ratio between insurance expense and construction costs is considered to be 0.5% [14, 30].

The proportion of the self-consumption electricity in total electricity production was usually set at 10%; both repayment period and depreciation period were 15 years [31]. In China, CSP demonstration plants could enjoy the income tax concession with full exemption in the first three years; then it was half exemption for the income tax in the second three years, and the income tax rate is reduced from 25% to 15%. The rates of value-added tax (VAT) and additional taxes are 17% and 8%, respectively [22]. Therefore, this paper could obtain the values of parameters and data related to the CSP systems of China as described in Table 2.

Table 2: Parameters values and data related to the CSP systems of China.

Basic parameter	Values
Benchmark FIT (P) [2]	RMB1.15/kWh
Initial investment (II)	See Table 1
O&M costs ratio (OM) [29]	2%
Operation period ($T - T_0$) [29]	25 years
Repayment period [31]	15 years
Long-term loan rate (r)	4.9%
VAT rate (t_{VAT}) [22]	17%
Additional taxes ratio ($t_{additional}$) [22]	8%
Proportion of self-consumption electricity (α) [31]	10%
Annual generation electricity (E_i)	See Table 1
Construction costs ratio (C) [26]	96%
Insurance expenses ratio (I) [14, 30]	0.5%
Construction period (T_0) [28]	2 years
Depreciation period [22]	15 years
Discount rate (r_e) [31]	10%
Income tax rate (t_{income}) [22]	15%
Loan ratio [28]	80%

3.2. Cost-Benefit Evaluation. The cost-benefit is to examine the economic efficiency of CSP projects and it has an important effect on the investment decisions and policy-making of CSP industry. At present, the CSP in China has entered the stage of commercial demonstration development. The first CSP demonstration plants of China were approved formally by the National Energy Administration (NEA). With the start-up of CSP demonstration projects, more and more renewable energy investors began to pay attention to the CSP development of China. The costs and benefits of CSP demonstration plants have a direct impact on the investors' willingness to invest in the CSP industry and the results can provide scientific basis for the policy makers to improve the industrial policies.

The first CSP demonstration projects of China involved the ST, PT, and LF-CSP technologies, which are encouraged by NEA [28]. Therefore, in order to fully analyze the cost-benefit of CSP in China, this study selected SP, NPV, NPVR, and IRR as measured indicators to calculate and discuss the cost-benefit of ST, PT, and LF-CSP technologies, respectively.

3.2.1. Solar Tower CSP. Table 3 shows the cost-benefit for the ST-CSP demonstration plants of China. The SP of ST-CSP projects vary from 7.3 years to 16.4 years. Apart from ST_2 project, the SP of other ST-CSP plants are all more than 13 years. Besides, the average SP of ST-CSP technology is about 12.0 years. It means that the investors of ST-CSP plants have to take long time to recover the initial capital due to the huge upfront investment. The NPV of ST-CSP systems are between RMB−52.0 million and RMB676.8 million. Further analysis shows that the NPV of ST_7 project is less than zero and the NPV of the remaining ST plants are all more than zero. It means that only the investment for ST_7 project cannot meet the profitability required by the benchmark yield of 10% and other ST-CSP plants are all able to get excess earnings on the basis of realizing the benchmark profitability.

The NPVR of ST-CSP plants are between −0.15 and −1.11. The average NPVR of ST-CSP technology is approximately 0.30; the IRR of ST-CSP systems are between 8.89% and

TABLE 3: Cost-benefit for the ST-CSP technology of China.

	ST_1	ST_2	ST_3	ST_4	ST_6	ST_7	ST_8	Mean
DNI (kWh/m^2/yr)	1,976	2,000	1,900	1,900	1,900	1,800	1,800	–
Initial costs (RMB/W)	21	30.4	24.44	31.6	25	35.8	30	–
TES capacity (h)	6	11	6	8	8	9	10	–
SP (years)	13.3	7.3	13.7	14.6	13.4	16.4	13.4	12.0
NPV (Million RMB)	30.7	676.8	26.7	5.0	69.1	−52.0	84.0	–
NPVR	0.15	1.11	0.11	0.02	0.14	−0.15	0.14	0.30
IRR	11.14%	19.22%	10.85%	10.12%	11.08%	8.89%	11.09%	12.33%

TABLE 4: Cost-benefit for the PT-CSP technology of China.

	PT_1	PT_2	PT_4	PT_5	PT_7	Mean
DNI (kWh/m^2/yr)	1,800	2,056.5	2,025	1,976	1,700	–
Initial costs (RMB/W)	26.90	39.72	28.00	38.77	28.13	–
TES capacity (h)	9	15	10	9	16	–
SP (years)	14.5	13.5	14.8	17.6	6.5	12.6
NPV (Million RMB)	7.2	49.9	−1.6	−115.0	493.0	–
NPVR	0.03	0.13	0.00	−0.30	1.37	0.22
IRR	10.21%	10.98%	9.98%	7.75%	21.50%	11.72%

19.22%, and the average IRR of investment for ST-CSP technology is about 12.33%. Thus it can be seen that the cost-benefit of ST-CSP technology is better since its average NPVR is more than zero, and its average IRR is higher than the benchmark yield of 10%. Further analysis shows that the cost-benefit of ST_2 plant, with NPVR of 1.11 and IRR of 19.22%, is better than that of other ST-CSP plants due to higher DNI and larger TES capacity. The ST_1 plant is second, with NPVR of 0.15 and IRR of 11.14%, because of its higher DNI and lower initial costs. The ST_8 project is next with NPVR of 0.14 and IRR of 11.09% due to its larger TES capacity, followed by the ST_6 project with NPVR of 0.14 and IRR of 11.08% due to lower initial costs. The cost-benefit of ST_7 project, with NPVR of −0.15 and IRR of 8.89%, is the worst of all the ST-CSP projects because of its lower DNI and higher initial costs; followed by ST_4 project with NPVR of 0.02 and IRR of 10.12% because of its higher initial costs; and finally ST_3 project with NPVR of 0.11 and IRR of 10.85%, because of smaller TES capacity.

3.2.2. Parabolic Trough CSP. Table 4 shows the cost-benefit for the PT-CSP demonstration plants of China. The SP of PT-CSP projects are between 6.5 years and 17.6 years and the SP of other PT-CSP plants are all more than 13.5 years except for PT_7 project. In addition, the average SP of investment for PT-CSP technology is approximately 12.6 years and longer than that of PT-CSP technology. Thus it can be seen that the investors of PT-CSP plants need to take a longer time to recover the initial investment. The NPV of PT-CSP plants vary from RMB-115.0 million to RMB493.0 million; in which, the NPV of both PT_4 and PT_5 projects are less than zero, and the remaining PT-CSP plants are more than zero. It illustrates that the investment for both PT_4 and PT_5 projects cannot meet the profitability required by the benchmark yield of 10%, but PT_1, PT_2, and PT_7 projects could obtain excessive profits based on achieving the benchmark profitability.

The NPVR of PT-CSP projects vary between −0.30 and 1.37, and the average NPVR for PT-CSP technology is approximately 0.22. The IRR of investment for PT-CSP projects are in the range of 7.75%–21.50% and the average IRR of investment for PT-CSP technology is approximately 11.72%. It means that the cost-benefit for some PT-CSP plants is worse, but overall it is good for LF-CSP technology because its average NPVR is more than zero and its average IRR is higher than the benchmark yield of 10%. Besides, the cost-benefit of ST-CSP technology is better than that of PT-CSP technology since both the average NPVR and IRR of the ST-CSP are more than that of PT-CSP. The cost-benefit of PT_7 project, with NPVR of 1.37 and IRR of 21.50%, is better than that of other PT-CSP plants, because of its larger TES capacity. The PT_2 project is the second project, with NPVR of 0.13 and IRR of 10.98%, due to the higher DNI and larger TES capacity; followed by PT_1 project with NPVR of 0.03 and IRR of 10.21% because of the lower initial costs. In addition, the cost-benefit of PT_5 project, with NPVR of −0.30 and IRR of 7.75%, is the last of the PT-CSP projects because of its higher initial costs and smaller TES capacity.

3.2.3. Linear Fresnel CSP. Table 5 shows the cost-benefit for the LF-CSP demonstration plants of China. The SP of LF-CSP projects vary between 10.7 years and 16.7 years and the average SP for LF technology is about 13.0 years. The payback period of LF-CSP technology is longer than that of ST and PT-CSP technologies. The NPV of LF-CSP systems are between RMB-56.9 million and RMB163.7 million; the NPV for LF_1 project is less than zero and for both LF_3 and LF_4 plants it is more than zero. It indicates that the investment for LF_1 plant cannot achieve the profitability required by the benchmark

Table 5: Cost-benefit for the LF CSP of China.

	LF_1	LF_3	LF_4	Mean
DNI (kWh/m^2/yr)	2,000	1,750	2,025	–
Initial costs (RMB/W)	33.60	36.00	29.52	–
TES capacity (h)	13	14	8	–
TES medium	M-S	A-S-S-C	M-S	–
SP (years)	16.7	10.7	12.3	13.0
NPV (Million RMB)	−56.9	163.7	74.1	–
NPVR	−0.17	0.45	0.25	0.18
IRR	8.70%	13.61%	11.97%	11.43%

yield of 10%. But the investment of LF_3 and LF_4 plants can obtain excess earnings on the basis of meeting the benchmark profitability.

The NPVR of LF-CSP projects vary from −0.17 to −0.45 and the average NPVR for LF-CSP technology is approximately 0.18. The IRR of investment for LF-CSP systems vary from 8.70% to 13.61%, and the average IRR of investment for LF-CSP technology is about 11.43%. Thus it can be seen that the cost-benefit of LF-CSP technology overall is good but worse than that of ST and PT-CSP technologies. Further analysis shows that the cost-benefit of LF_3 plant, with NPVR of 0.45 and IRR of 13.61%, is better than that of other two LF-CSP projects, due to the larger TES capacity and the application of advanced TES medium (i.e., all solid state concrete), followed by LF_4 plant with NPVR of 0.25 and IRR of 11.97%, because of the higher DNI and lower initial costs. The cost-benefit of LF_1 plant, with NPVR of −0.17 and IRR of 8.70%, is the lowest among all the LF-CSP plants, due to the higher initial costs.

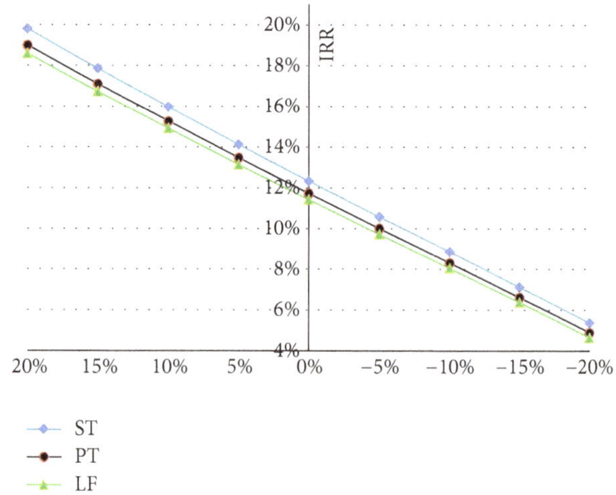

Figure 1: The sensitivity analysis results of the IRR of CSP to the annual electricity production.

3.3. Sensitivity Analysis. The cost-benefit analysis for CSP requires comprehensively considering the impacts of sensitivity factors, mainly involving annual electricity production, initial investment, O&M expenses, and interest rate. The study selected IRR as the evaluation index to discuss the impacts of these sensitivity factors on the cost-benefit of ST, PT, and LF-CSP technologies respectively; and the variation range is set as ±20%.

3.3.1. Annual Electricity Production. Figure 1 shows the effects of annual electricity production on the IRR of three different CSP technologies. An increase by 20% of the annual electricity production will cause the IRR to increase from 12.33% to 19.80% for ST-CSP technology, from 11.72% to 18.99% for PT-CSP technology, and from 11.43% to 18.60% for LF-CSP technology. However, if the annual electricity production is decreased by 20%, the IRR of ST, PT, and LF-CSP technologies will decrease to 5.38%, 4.89%, and 4.65%, respectively. Thus, it can be seen that increasing the annual electricity production could effectively improve the cost-benefit of CSP systems, but reducing it will have a very negative effect on the cost-benefit of CSP systems. Therefore, increasing the annual electricity production is crucial for the improvement of the cost-benefit of CSP systems.

The annual electricity production determines the incomes directly and has an important effect on the cost-benefit. Annual electric generation capacity of CSP systems is determined by the thermal conversion efficiency, DNI, TES capacity, and so on. For the three CSP technologies involved in the first CSP demonstration projects of China, the thermodynamic efficiency of ST-CSP technology is higher than that of other two CSP technologies, in view of the higher temperatures that the heat transfer fluids in the ST systems could achieve; so the ST systems could significantly increase the thermal conversion efficiency of ST-CSP plants in relatively lower costs. PT-CSP technology is second. Although it has been widely adopted among the existing CSP plants, its potential for development may be limited by the low thermodynamic efficiency. LF-CSP technology is the simplified technique of PT-CSP systems, and its thermal conversion efficiency is less than that of other two CSP technologies; meaning that the LF systems need a relatively higher cost to increase the thermodynamic efficiency of LF-CSP systems. Accordingly, the average IRR of investment for ST-CSP technology of 12.33% is higher than that of PT-CSP technology of 11.72% and LF-CSP technology of 11.43%. DNI and TES technology have also important effects on the annual generation capacity of CSP. Excluding the impacts of other factors, a higher DNI implies a better electricity production

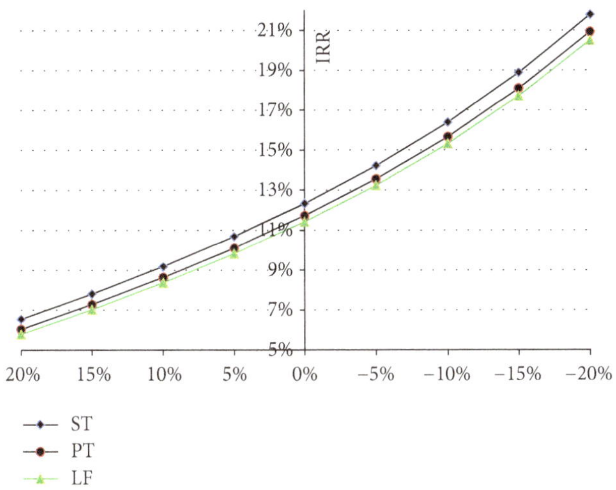

FIGURE 2: The sensitivity analysis results of the IRR of CSP to the initial costs.

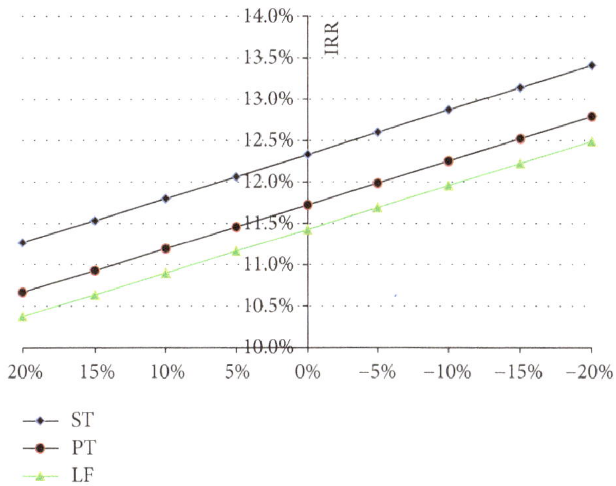

FIGURE 3: The sensitivity analysis results of the IRR of CSP to the operation and maintenance costs.

of CSP systems in a year. So it is necessary for CSP plants to select the sites with higher DNI, in order to increase the annual electricity production and improve the cost-benefit of CSP. Besides, the installation of TES systems adds to the initial investment but could improve the capacity factor of CSP systems and increase the annual electricity production.

3.3.2. Initial Costs. Figure 2 displays the effects of initial costs on the IRR of three different CSP technologies. When the initial costs are reduced by 20%, the IRR will increase from 12.33% to 21.78% for ST-CSP technology, from 11.72% to 20.92% for PT-CSP technology, and from 11.43% to 20.50% for LF-CSP technology. If increased by 20%, the IRR of ST, PT, and LF-CSP technologies will decline to 6.54%, 6.04%, and 5.79%, respectively. Thus, it can be seen that the effects of initial costs reduction on the IRR of CSP are greater than that of initial costs increase. Reducing the initial costs can effectively improve the cost-benefit of CSP, while its increase has negative effect on the cost-benefit of CSP. Therefore, reducing the initial costs is crucial for improving the cost-benefit of CSP systems.

High initial investment has been the main obstacle for the development of CSP. As seen in Table 1, the initial costs of CSP vary between RMB21/W and RMB39.72/W and they are higher than that of traditional energy generation technologies. With growth of cumulative installed capacity and technology progress, the initial costs of CSP will be reduced greatly in the future [47]. On the one hand, with the expansion of projects scale and CSP market, larger CSP plants will be designed and constructed. The mass production of the equipment and materials and the optimization of CSP systems could dramatically reduce the initial costs. On the other hand, the experience gained by professional technical personnel, engineers, and contractors through the accumulation of the construction of CSP projects will improve the design and construction scheme of CSP plants to reduce the initial costs. This experience is usually characterized by learning curves. Based on the learning curves theory, the decrease in the initial costs is proportional to the learning rate with the growth of total installed capacity. More specifically, if the learning rate of CSP industry is 10%, the initial costs could be reduced by about 10% when the total installed capacity double.

3.3.3. Operation and Maintenance Costs. Figure 3 shows the impacts of O&M costs on the IRR of three different CSP technologies. If the O&M costs are increased by 20%, the IRR of ST, PT, and LF-CSP technologies will decrease from 12.33%, 11.72%, and 11.43% to 11.27%, 10.67%, and 10.38%, respectively. When decreased by 20%, the IRR will increase to 13.41% for ST-CSP technology, 12.79% for PT-CSP technology, and 12.49% for LF-CSP technology. Different from initial investment, the O&M costs of CSP systems are relatively lower because of no fuel consumption. Accordingly, the impacts of O&M costs on the IRR of CSP systems are relatively small. The IRR of CSP systems only increase by about 1.07% when the O&M costs are reduced by 20%.

With the construction and operation of CSP demonstration plants, the O&M costs of CSP systems are expected to be reduced remarkably in the future. Therefore, the improvement of cost-benefit, which is caused by the reduction of O&M costs, should not be ignored.

3.3.4. Interest Rate. Figure 4 shows the impacts of interest rate on the IRR of three different CSP technologies. When the interest rate is increased by 20%, the IRR of ST, PT, and LF-CSP technologies will decrease from 12.33%, 11.72%, and 11.43% to 11.03%, 10.46%, and 10.18%, respectively. If reduced by 20%, the IRR will increase to 13.76% for ST-CSP technology, 13.11% for PT-CSP technology, and 12.80% for LF-CSP technology, respectively. Although the effects of interest rate on the IRR of CSP systems are relatively low, it cannot be ignored.

CSP is a capital intensive industry. The investment and financing are crucial to the start-up of CSP projects due to high initial costs. The long-term commercial loan has been the main sources for the funding of CSP projects in China

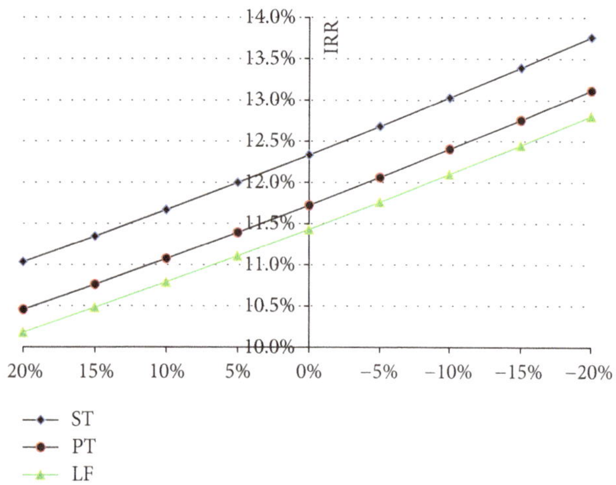

FIGURE 4: The sensitivity analysis results of the IRR of CSP to the interest rate.

and the loans ratio in the total investment could achieve 80%. Therefore, the loan interest is also an important expenditure for the construction and operation of CSP plants. The interest rate determines the capitals cost directly and affects the equity capital benefit. The lower interest rate could decrease the loans interest and increase the IRR of CSP plants. Therefore, preferential loans play an important role in improving the cost-benefit of CSP technologies. In China, the low-interest loans can be obtained from policy banks such as China investment bank and Asian investment bank.

4. Conclusions

In 2016, with the implementation of the first batch CSP demonstration projects, China made ambitious development goals for CSP industry. The cost-benefit has an important impact on the investment discussions and policy-making; the cost-benefit analysis for CSP demonstration plants is therefore of great significance for the development of CSP industry in China. Based on the related data and design parameters of the first CSP demonstration projects, this study adopted financial indicators, including SP, NPV, NPVR, and IRR, to calculate and discuss the cost-benefit of different CSP technologies in China. The results showed the following.

Firstly, Analysis from Projects Perspective. The investment payback periods of most of CSP demonstration projects are more than 13 years, meaning that the CSP systems usually take a longer time to recover the initial investment, because of the huge upfront input. However, with the formulation and implementation of the benchmark FIT of CSP industry, the CSP demonstration projects could obtain good expected incomes in the operation period; so most of CSP projects are able to gain excess returns on the basis of realizing the benchmark profitability, which is required by the discounted rate of 10%. That is, under the current technical level, policy system, and resource endowment of China, vast majority of CSP projects have a strong economic feasibility. Besides, the cost-benefit among CSP demonstration projects has large differences, due to the combined influence of region conditions, CSP technologies, DNI, initial costs, TES capacity, and so on. However, generally speaking, the larger TES capacity usually could bring a good cost-benefit just from the CSP projects perspectives.

Secondly, Calculation from the Perspective of Technologies. The average SP of ST, PT, and LF-CSP technologies are about 12 years, 12.6 years, and 13.0 years, respectively. Their average NVPR are, respectively, 0.30, 0.22, and 0.18. Their average IRR of investment are 12.33%, 11.72%, and 11.43%, respectively. This indicates that the cost-benefit of ST-CSP technology is better than that of other two CSP technologies, because of the higher thermal conversion efficiency, closely followed by the PT-CSP technologies, which has been widely adopted among the existing CSP plants. The cost-benefit of LF-CSP technology is relatively worse, due to the lower thermodynamic efficiency. Thus it can be seen that the impacts of the types of technology on the cost-benefits of CSP are more important than those of TES capacity.

Thirdly, Discussions from the Perspective of Influencing Factors. The annual electricity production and initial costs have significant impacts on the cost-benefit of CSP systems. For each increase of 5% of annual electricity production, the IRR can be increased by about 1.87% for ST, 1.82% for PT, and 1.79% for LF-CSP technologies. Conversely, their IRR will be decreased by approximately 1.74%, 1.72%, and 1.70%, respectively. For each decrease of 5% of initial costs, the IRR can be increased by approximately 2.36% for ST, 2.30% for PT, and 2.27% for LF-CSP technologies. Conversely, their IRR will be reduced by about 1.45%, 1.42%, and 1.41%, respectively. Besides, the effects of O&M costs and loan interest rate on the cost-benefit of CSP systems are relatively less than the effects of the annual electricity production and initial costs, but cannot be ignored, especially the loan interest rate.

Conflicts of Interest

The authors declare that they have no conflicts of interest.

Acknowledgments

This paper is supported by National Social Science Foundation (Grant no. 16BJY055).

References

[1] Sino-Danish renewable energy development project administration office. The China Renewable Energy Development Roadmap of 2050 [Internet], C2014, National renewable energy center, 2014, http://www.docin.com/p-1151220292.html.

[2] National Development and Reform Commission (NDRC), NDRC issued "Notice on the benchmark feed-in tariff price policy of CSP" [Internet], C2016, NDRC, http://www.gov.cn/xinwen/2016-09/01/content_5104496.htm.

[3] I. Purohit, P. Purohit, and S. Shekhar, "Evaluating the potential of concentrating solar power generation in Northwestern India," *Energy Policy*, vol. 62, pp. 157–175, 2013.

[4] H. L. Zhang, J. Baeyens, J. Degrève, and G. Cacères, "Concentrated solar power plants: review and design methodology," *Renewable & Sustainable Energy Reviews*, vol. 22, pp. 466–481, 2013.

[5] N. Caldés, M. Varela, M. Santamaría, and R. Sáez, "Economic impact of solar thermal electricity deployment in Spain," *Energy Policy*, vol. 37, no. 5, pp. 1628–1636, 2009.

[6] International Energy Agency (IEA), *World energy outlook 2013: renewable energy outlook [dissertation]*, Paris, France, 2013, OGCDIIGA.

[7] F. Dinter and D. M. Gonzalez, "Operability, reliability and economic benefits of CSP with thermal energy storage: First year of operation of ANDASOL 3," in *Proceedings of the International Conference on Solar Power and Chemical Energy Systems, SolarPACES 2013*, pp. 2472–2481, USA, September 2013.

[8] R. Shinnar and F. Citro, "A road map to U.S. decarbonization," *Science*, vol. 313, no. 5791, pp. 1243-1244, 2006.

[9] Y. Tian and C. Y. Zhao, "A review of solar collectors and thermal energy storage in solar thermal applications," *Applied Energy*, vol. 104, pp. 538–553, 2013.

[10] R. Sioshansi and P. Denholm, "The value of concentrating solar power and thermal energy storage," *IEEE Transactions on Sustainable Energy*, vol. 1, no. 3, pp. 173–183, 2010.

[11] S. H. Madaeni, R. Sioshansi, and P. Denholm, "How thermal energy storage enhances the economic viability of concentrating solar power," *Proceedings of the IEEE*, vol. 100, no. 2, pp. 335–347, 2012.

[12] International Renewable Energy Agency (IRENA), *Renewable power generation costs in 2014 [dissertation]*, IRENA, 2015.

[13] Renewable Energy Policy Network for the 21st Century (REN 21), The first decade: ten years of renewable energy progress: 2004-2014 [dissertation], REN 21, 2015.

[14] C. Parrado, A. Marzo, E. Fuentealba, and A. G. Fernández, "2050 LCOE improvement using new molten salts for thermal energy storage in CSP plants," *Renewable & Sustainable Energy Reviews*, vol. 57, pp. 505–514, 2016.

[15] C. Parrado, A. Girard, F. Simon, and E. Fuentealba, "2050 LCOE (Levelized Cost of Energy) projection for a hybrid PV (photovoltaic)-CSP (concentrated solar power) plant in the Atacama Desert, Chile," *Energy*, vol. 94, pp. 422–430, 2016.

[16] A. R. Starke, J. M. Cardemil, R. A. Escobar, and S. Colle, "Assessing the performance of hybrid CSP + PV plants in northern Chile," *Solar Energy*, vol. 138, pp. 88–97, 2016.

[17] X. Ju, C. Xu, Y. Hu, X. Han, G. Wei, and X. Du, "A review on the development of photovoltaic/concentrated solar power (PV-CSP) hybrid systems," *Solar Energy Materials & Solar Cells*, vol. 161, pp. 305–327, 2017.

[18] S. Izquierdo, C. Montañs, C. Dopazo, and N. Fueyo, "Analysis of CSP plants for the definition of energy policies: The influence on electricity cost of solar multiples, capacity factors and energy storage," *Energy Policy*, vol. 38, no. 10, pp. 6215–6221, 2010.

[19] J. Hernández-Moro and J. M. Martínez-Duart, "CSP electricity cost evolution and grid parities based on the IEA roadmaps," *Energy Policy*, vol. 41, pp. 184–192, 2012.

[20] F. Cao, Q. Liu, T. Yang, T. Zhu, J. Bai, and L. Zhao, "Full-year simulation of solar chimney power plants in Northwest China," *Journal of Renewable Energy*, vol. 119, pp. 421–428, 2018.

[21] X. Zhao, Y. Zeng, and D. Zhao, "Distributed solar photovoltaics in China: Policies and economic performance," *Energy*, vol. 88, pp. 572–583, 2015.

[22] J. Yuan, S. Sun, W. Zhang, and M. Xiong, "The economy of distributed PV in China," *Energy*, vol. 78, pp. 939–949, 2014.

[23] W. Li, P. Wei, and X. Zhou, "A cost-benefit analysis of power generation from commercial reinforced concrete solar chimney power plant," *Energy Conversion and Management*, vol. 79, pp. 104–113, 2014.

[24] P. Guo, Y. Zhai, X. Xu, and J. Li, "Assessment of levelized cost of electricity for a 10-MW solar chimney power plant in Yinchuan China," *Energy Conversion and Management*, vol. 152, pp. 176–185, 2017.

[25] Z. Zhu, D. Zhang, P. Mischke, and X. Zhang, "Electricity generation costs of concentrated solar power technologies in China based on operational plants," *Energy*, vol. 89, pp. 65–74, 2015.

[26] Z.-Y. Zhao, Y.-L. Chen, and J. D. Thomson, "Levelized cost of energy modeling for concentrated solar power projects: A China study," *Energy*, vol. 120, pp. 117–127, 2017.

[27] CSPPLAZA, Notice of NEA on reducing the tax and fees burdens of the companies of renewable energy field [Internet], c2017, CSPPLAZA, http://www.cspplaza.com/article-10455-1.html.

[28] NEA, Notice of NEA on organizing the construction of CSP demonstration projects [Internet], NEA, c2017, http://zfxxgk.nea.gov.cn/auto87/201509/t20150930_1968.htm.

[29] J. Hernández-Moro and J. M. Martínez-Duart, "Analytical model for solar PV and CSP electricity costs: Present LCOE values and their future evolution," *Renewable & Sustainable Energy Reviews*, vol. 20, pp. 119–132, 2013.

[30] P. Viebahn, Y. Lechon, and F. Trieb, "The potential role of concentrated solar power (CSP) in Africa and Europe-A dynamic assessment of technology development, cost development and life cycle inventories until 2050," *Energy Policy*, vol. 39, no. 8, pp. 4420–4430, 2011.

[31] Inner Mongolia Electric Power Survey and Design Institute, *Pre-feasibility study report on the Inner Mongolia 50MW trough solar thermal power demonstration project [dissertation]*, 2008.

[32] CSPPLAZA, The brief introduction for the first batch twenty CSP demonstration projects [Internet], CSPPLAZA, c2016, http://guangfu.bjx.com.cn/news/20160926/775762.shtml.

[33] Shouhang energy saving co. LTD, The reply of Shouhang and INDUSTRIAL SECURITIES on "Feedback on the non-public offering of shares application documents of Beijing Shouhang Ihw Energy Saving Technology Co., Ltd" [Internet], Shouhang energy saving co. LTD, c2016, http://q.stock.sohu.com/cn,gg,002665,2279028837.shtml.

[34] CSPPLAZA, Foundation laying for Gonghe 50MW molten salt ST-CSP project of Northwest Institute of survey and design [Internet], CSPPLAZA, c2017, http://www.cspplaza.com/article-9937-1.html.

[35] CSPPLAZA, Northwest Electric Power Design Institute won the EPC of Hami 50MW ST-CSP project of CHINA POWER ENGINEERING [Internet], CSPPLAZA, http://www.cspplaza.com/article-9367-1.html.

[36] Jinta100MW molten salt ST-CSP project of China Three Gorges New Energy Co., Ltd [Internet], c2017, http://www.bidnews.cn/xiangmu/zhaobiao-67992.html.

[37] CSPPLAZA, Yumen Xinneng 50MW molten salt ST-CSP project obtained the construction permit [Internet], CSPPLAZA, c2017, http://www.cspplaza.com/article-9536-1.html.

[38] China state Department of Environmental Protection, Environmental impact assessment report publicity: Environmental impact assessment report on Yumen 100MW molten salt

ST-CSP project of Guohua [Internet], CSPPLAZA, c2017, http://www.doc88.com/p-5496378980290.html.

[39] China state Department of Environmental Protection, Environmental impact assessment report publicity: Environmental impact assessment report on Yumen East Town 50MW thermal oil PT-CSP project [Internet], China state Department of Environmental Protection, c2017, http://www.docin.com/p-1871037172.html.

[40] CSPPLAZA, WuLate Middle Banner 100MW thermal oil PT-CSP project of China National Nuclear Longtium was started officially [Internet], CSPPLAZA, c2017, http://www.cspplaza.com/article-9458-1.html.

[41] CSPPLAZA, Foundation laying for Zhangjiakou 64MW molten salt PT-CSP plant of Zhongyang [Internet], CSPPLAZA, c2014, http://www.cspplaza.com/article-3735-1.html.

[42] CSPPLAZA, Zhangjiakou 64MW molten salt PT-CSP plant of Zhongyang would be centralized bidding next month, financing was almost complete [Internet], CSPPLAZA, c2017, http://www.cspplaza.com/article-9397-1.html.

[43] China state Department of Environmental Protection, Environmental impact assessment report publicity: Environmental impact assessment report on 50MW molten salt LF-CSP project [Internet], China state Department of Environmental Protection, c2017, http://www.docin.com/p-1910872195.html.

[44] Hebei people's government, Thirteen projects with a total investment of 45 billion Yuan in Zhangbei were started at the same time [Internet], Hebei people's government., c2017, http://news.xinhuanet.com/local/2017-04/20/c_129555705.htm.

[45] CSPPLAZA. The first power generation test on the Conventional Island of Zhangjiakou 15MW CSP plant of Huaqiang Zhaoyang Energy Co. Ltd [Internet], CSPPLAZA, c2017, http://www.cspplaza.com/article-9955-1.html.

[46] China state Department of Environmental Protection, Environmental impact assessment report on Inner Mongolia WuLate Middle Banner 50MW thermal oil LF-CSP project of Northern United Electric Power [Internet], China state Department of Environmental Protection, c2017, http://www.docin.com/p-1883226646.html.

[47] China National CSP industry technology innovation strategic alliance, Research Report on the CSP industry policy of China, Special report 3: Analysis for the incentive mechanism of solar thermal power industry in China [Internet], China National CSP industry technology innovation strategic alliance, c2015, https://wenku.baidu.com/view/84d14e9d4a7302768e9939b2.html.

3

A Comparative Study of Symmetrical Cockcroft-Walton Voltage Multipliers

Mohsen Ruzbehani

Laser & Optics Research School, NSTRI, Tehran 14399-51113, Iran

Correspondence should be addressed to Mohsen Ruzbehani; mruzbehani@yahoo.com

Academic Editor: George S. Tombras

Decades after invention of the Cockcroft-Walton voltage multiplier, it is still being used in broad range of high voltage and ac to dc applications. High voltage ratio, low voltage stress on components, compactness, and high efficiency are its main features. Due to the problems of original circuit, reduction of output ripple and increase of accessible voltage level were the motivations for scientist to propose new topologies. In this article a comparative study on these voltage multipliers was presented. By simulations and experimental prototypes, characteristics of the topologies have been compared. In addition to the performances, components count, voltage stress on the components, and the difficulty and cost of construction are other factors which have been considered in this comparison. An easy to use table which summarized the characteristics of VMs was developed, which can be used as a decision mean for selecting of a topology based on the requirements. It is shown that, due to the application, sometimes a simple and not very famous topology is more effective than a famous one.

1. Introduction

High voltage dc power supplies are widely used for many applications, such as particle accelerators, X-ray systems, electron microscopes, photon multipliers, electrostatic systems, lasers systems, and electrostatic coating [1–5]. There are several approaches in producing high voltage dc sources depending on the desired voltage and current levels. A more usual choice when a power supply with high voltage and low current is needed is the Cockcroft-Walton voltage multiplier (CWVM). High voltage ratio, low voltage stress on the diodes and capacitors, compactness, and high efficiency are the main reasons for this choice [3, 5, 6]. Furthermore it can be used in medium voltage applications when the current is low and simple circuit is the matter, like portable pulsed power applications and handheld devices [7–9]. The CWVM has the unique characteristic of imposing equal voltage stresses on every stage. Its construction is also simple and easy to implement [10]. Usually, in modern types of voltage multipliers there are three stages; a high frequency inverter which produces fast dynamic voltage source with controllable duty cycle to control the current and voltage level, a transformer, and a CW voltage multiplier [11]. In this article only the topologies of VMs, regardless of inverter construction, are studied.

Historically the original idea was proposed by Greinacher in 1921. However, it did not get attention for a long time until Cockcroft and Walton performed their experiment using this circuit in 1932 [12]. Large output voltage ripple and output voltage drop were the main problems of the original half-wave voltage multiplier. Consequently, to overcome these problems, a symmetrical voltage multiplier (SVM) was developed by Heilpern in 1954 by adding an extra oscillating column of capacitors and a stack of rectifiers [1, 13] Because voltage ripple in a half-wave rectifier due to the charge and discharge of capacitors has almost a triangular-like shape, by using two voltage sources with 180° phase difference, the overall voltage ripple magnitude can be reduced greatly. If the ripple was an isosceles triangle shaped waveform, the opposite phase could exactly cancel out the voltage peaks so that the voltage ripple would be disappeared in the output voltage. However, this is not the case, so the ripple will not be removed completely, and the result is only reduction of ripple amplitude. This is the main feature of a symmetrical voltage multiplier. Based on

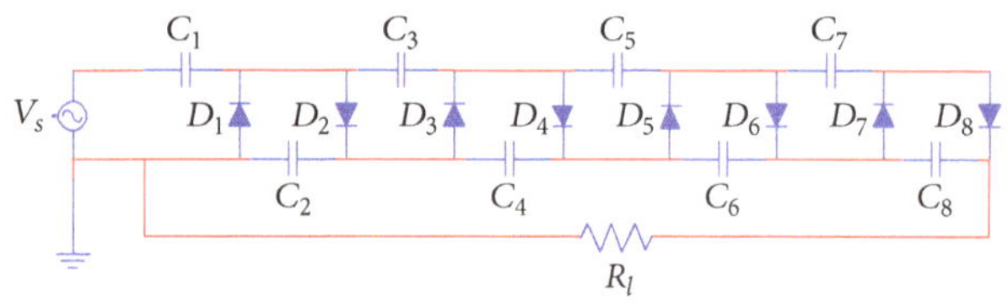

FIGURE 1: Schematic circuit of a basic 4-stage Cockcroft-Walton voltage multiplier.

this idea, recently a three-phase symmetrical voltage multiplier with six oscillating columns and one smoothening column has been proposed [2, 14]. Furthermore, there are other topologies which based on this idea have lower voltage ripple compared to the conventional one, which will be studied in this article. Other types of voltage multipliers based on CWVM have been developed recently which a CWVM is fed by a matrix converter [6]. The matrix converter generates an adjustable frequency and adjustable-amplitude current, which is injected into the CWVM to regulate the dc output voltage and smooth its ripple. However, in this article, the CWVM is the subject, regardless of how it may be fed.

The comparison of these VMs is important because selecting a topology for an application due to many factors involved is not a straightforward task. The range of output voltage level, necessary safety factors, ground-returned or bipolar output, and speed of response and so on are the main factors which should be considered in such selection. To the best of author's knowledge, until now such comparison has not been made.

This paper is organized as follows. The mathematical model of the CWVM is reviewed in Section 2. Symmetrical voltage multipliers are introduced in Section 3. Comparisons between the symmetrical VMs by simulation are presented in Section 4. In Section 5 experimental low voltage prototypes are constructed to justify the results of simulations. Finally conclusion is given in Section 6.

2. Basic Voltage Multiplier

Figure 1 shows a basic 4-stage Cockcroft-Walton voltage multiplier circuit. It consists of two capacitor columns, namely, oscillating and smoothening columns. Oscillating column capacitors (C_1, C_3, C_5, and C_7) are charged in half cycle by upward odd numbered diodes (D_1, D_3, D_5, and D_7, respectively) and in next half cycle the smoothening column capacitors (C_2, C_4, C_6, and C_8) are charged by downward even numbered didoes (D_2, D_4, D_6, and D_8). In steady state, in no-load condition, every capacitor in smoothening column is charged to $2V_{\max}$, that is, two times of maximum input voltage magnitude. Therefore, the maximum value of output voltage is $2nV_{\max}$ in which n is the number of multiplier stages (here $n = 4$).

However, this is not the case in presence of a load, which output voltage ripple exists and there is a drop in voltage due to the load current. To be clear, in Figure 2, simulation result of a basic half-wave Cockcroft-Walton voltage multiplier circuit, by Saber Synopsys software, is given. In the figure, the key parameters have been shown: ΔV_o is the voltage drop compared to the ideal case and $2\delta V$ is the peak-to-peak value of output voltage ripple due to the load current. To calculate the voltage ripple, we should consider the voltage ripple on every capacitor in smoothening column due to the load current. Let q be the charge transferred from C_{2n} to the load per cycle with a ripple voltage at capacitor C_{2n} [15–17]:

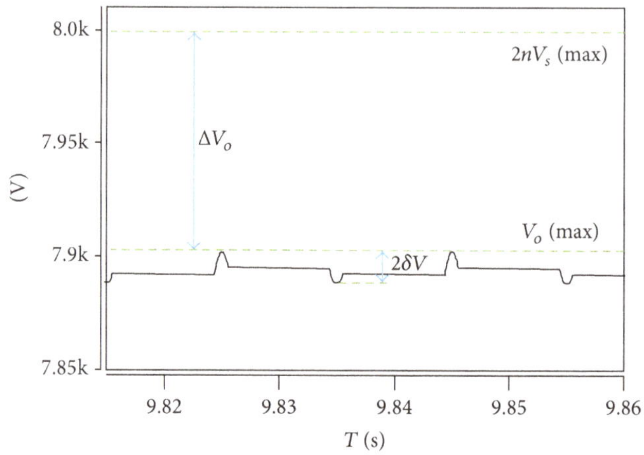

FIGURE 2: Simulation of a conventional 4-stage ($2n = 8$) Cockcroft-Walton voltage multiplier output voltage. In this example $V_s(\max) =$ 1000 V and capacitors are 1000 μF.

$$2\delta V_{2n} = \frac{q}{C_{2n}}. \tag{1}$$

Simultaneously C_{2n-2} transfers charge q to the load and q to C_{2n-1}. So the voltage ripple at capacitor C_{2n-2} is

$$2\delta V_{2n-2} = \frac{2q}{C_{2n-2}}. \tag{2}$$

By intuition,

$$2\delta V_{2n-2i} = \frac{(i+1)\,q}{C_{2n-2i}}. \tag{3}$$

Because ripples on all smoothening capacitors are in-phase, the total voltage ripple on load can be calculated by summation of all ripples values; that is,

$$2\delta V = q\left(\frac{1}{C_{2n}} + \frac{1}{C_{2n-2}} + \cdots + \frac{1}{C_2}\right). \tag{4}$$

Equal value capacitors (C) results in

$$2\delta V = \frac{q}{C}\left(\frac{n(n+1)}{2}\right) = \frac{I}{fC}\left(\frac{n(n+1)}{2}\right), \tag{5}$$

in which I is the load current and f is the frequency of input voltage. In addition to the voltage ripple, there is a voltage drop compared to no-load condition which is the consequences of uncompleted charging process of capacitors. From (3) it can be understood that C_2 loses nq charge during each cycle, so capacitor C_1 has to replenish it. Therefore, C_1 and C_2 charge up to ($V_s(\max) - nq/CJ$) and ($2V_s(\max) - nq/q$)

instead of $V_s(\max)$ and $2V_s(\max)$, respectively. In the same manner the voltage drop in all smoothening capacitors could be calculated. In general we will have [15]

$$\Delta V_2 = \frac{q}{C} n,$$
$$\Delta V_4 = \frac{q}{C} [2n + (n-1)],$$
$$\vdots$$
$$\Delta V_{2n} = \frac{q}{C} [2n + 2(n-1) + \cdots + 2 \times 2 + 1]. \quad (6)$$

Adding all n voltage drops gives the total voltage drop on load:

$$\Delta V_{\text{total}} = \frac{q}{C}\left(\frac{2}{3}n^3 + \frac{1}{2}n^2 - \frac{n}{6}\right) = \frac{I}{fC}\left(\frac{2}{3}n^3 + \frac{1}{2}n^2 - \frac{n}{6}\right). \quad (7)$$

Recently with more detailed analysis it has been shown that the above formula needs a small correction [18]. However the difference is of secondary importance for practical use, because it does not concern the leading order in n. From (7) and (5) it is clear for a given value of n, by increasing load current, voltage drop and voltage ripple increase proportionally. To compensate this effect the value of $f \times C$ should be increased. Usually increasing frequency has lower cost compared to increasing the capacitors values. Moreover from these relations it is also clear that increasing number of stages will increase these drop voltage and ripple amplitude. Therefore, there is an optimum value for n which gives the maximum accessible output voltage when other parameters are fixed. This optimum value is as follows [13]:

$$n_{\text{opt}} = \sqrt{\frac{V_s(\max) fC}{I}}. \quad (8)$$

Hence, the maximum accessible voltage is as follows.

$$V_o(\max) = 2n_{\text{opt}} V_s(\max) - \frac{I}{fC}\left(\frac{2}{3}n_{\text{opt}}^3 + \frac{1}{2}n_{\text{opt}}^2 - \frac{n_{\text{opt}}}{6}\right). \quad (9)$$

To exceed the restriction of (9) and also to decrease the output voltage ripple value, several configurations have been proposed which are the subject of next section. It is worth mentioning that a proper selection of nonequal value capacitors ($C_{2i} = C_{2i-1} = (n+1-i)C$) will result in better response compared to the classical case ($C_{2i} = C_{2i-1} = C$) [19]. The result can be extended to the symmetrical cases with minor modifications.

3. Symmetrical Versions of Voltage Multiplier

3.1. Conventional Symmetrical Voltage Multiplier. As mentioned in previous section, by increasing number of stages, the ripple amplitude and drop voltage increase significantly, which make the voltage multiplier inefficient. To solve this problem a symmetrical voltage multiplier (SVM) was developed by Heilpern in 1954 by appending an additional oscillating column of capacitors and a stack of rectifiers [1, 13]. As can be seen in Figure 2, ripple in half-wave CWVM is periodic with almost symmetrical shape relative to horizontal line; that is, peak and valley are almost equal with 180° phase difference. In SVM output voltage is the summation of two waveforms with 180° phase differences. If the shape of ripple was an isosceles triangle or any symmetrical waveform, the summation of two opposite phase could exactly cancel out the peak and valley of ripple. However, in practice, the ripple waveform is slightly distorted due to the discharge pattern or nonlinearity, and its periodical transient shape is formed depending on the time constant of the circuit [20]. Therefore the ripple still exists; however, its magnitude decreases greatly. This is the main reason of lower level of ripple in all symmetrical and multiphase versions of voltage multipliers. In Figure 3, the schematic diagram of a symmetrical voltage multiplier has been given. In fact in SVM, V_{s1} and V_{s2} have same magnitude with no phase difference, so V_{yo} has 180° phase shift related to V_{xo}. In Figure 4 the output voltage ripple for two different conditions has been shown. As can be seen, when V_{yo} and V_{xo} have 180° phase difference, that is, V_{s1} and V_{s2} are in-phase, amplitude of ripple is lower than the case in which V_{yo} and V_{xo} are in-phase.

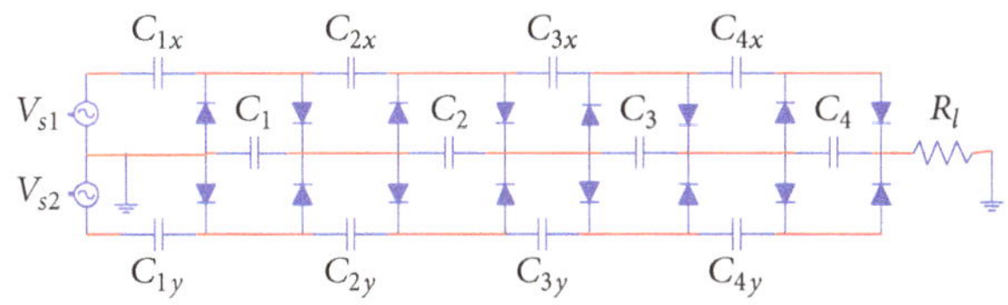

FIGURE 3: Schematic diagram of a 4-stage SVM.

It is shown that the ripple in symmetrical case compared to the original VM can be reduced approximately by the following formula [20]:

$$2\delta V_s \cong \frac{I}{Cf}\frac{n(n+1)}{2}\left(1 - \frac{\pi}{4\sin^{-1}\left(\left(V_{\max} - \delta V\right)/V_{\max}\right)}\right), \quad (10)$$

in which δV_s is the ripple of symmetrical VM and δV is the ripple of conventional VM.

In practice, any kind of asymmetry may result in higher value of ripple. It is known that the asymmetry of both input voltages and circuit elements can cause the ripple [21]. Also the circuit asymmetry causes odd harmonics, while the asymmetry in charge-discharge process of capacitors, which may arise from asymmetry in transformer outputs, causes the even harmonics [5]. For a symmetrical CW circuit, the fundamental harmonic due to the asymmetry of driving voltage is inevitable because of the difficulty of completely removing the symmetry deviation [22].

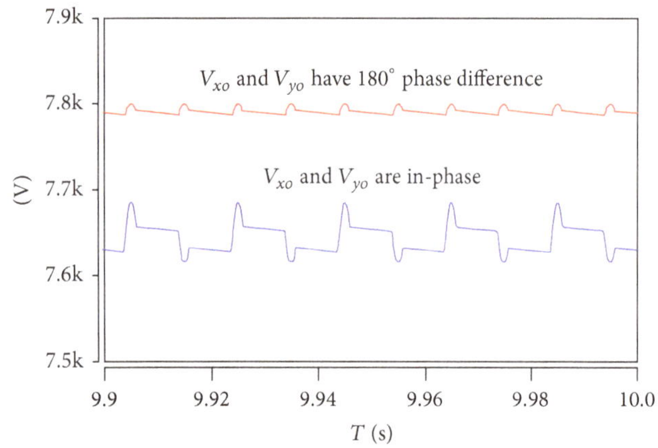

FIGURE 4: Output voltage ripple of a symmetrical voltage multiplier for two cases of in-phase and 180° out of phase power sources.

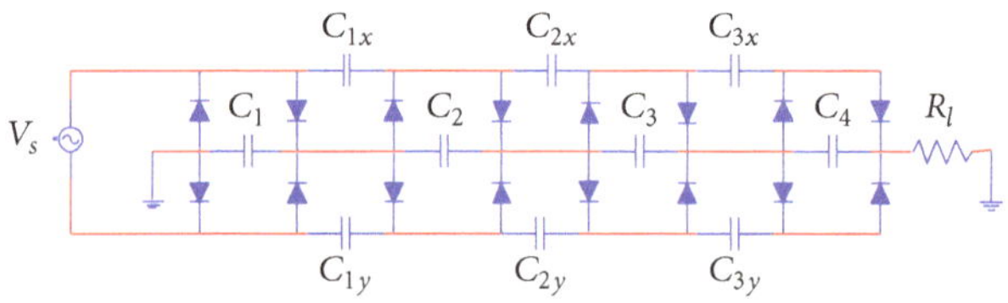

FIGURE 5: Schematic diagram of a 4-stage HSVM.

3.2. Hybrid Symmetrical Voltage Multiplier. Recently a new symmetrical voltage multiplier has been proposed [1]. Figure 5 shows the circuit diagram of 4 stages named Hybrid Symmetrical Voltage Multiplier, HSVM. It consists of a bridge rectifier with capacitor C_1 and a 3-stage SVM. As the proposed topology is a combination of diode-bridge rectifier and SVM, it is named HSVM. The first stage of the proposed topology does not have coupling capacitors; therefore, it saves two high voltage capacitors. This is the main advantage of this topology compared to the conventional SVM. Furthermore, due to its inventor claim, its faster transient response at startup compared with conventional SVM is its other superiority [1]. However, its main problem which makes it less effective is the lack of transformer ground. As can be seen in Figure 5, both x and y terminals are hot and so cannot be grounded. In SVM the center tap of the transformer is grounded; however, for HSVM, this is not the case. The author has not discussed this matter [1]. Due to this problem, in very high voltage applications, HSVM cannot be used for safety issue. In no-load condition we have $V_o = nV_s$. So to have same level of voltage scaling, the amplitude of input voltage which usually is the secondary of a transformer should be twice of that in conventional voltage multiplier. It means, for a given transformer, to have same level of output voltage, the number of elements is about two times of ordinary CWVM. This matter will be discussed in more detail in next sections.

3.3. Series-Connected Voltage Multiplier. As mentioned earlier the main features of a symmetrical voltage multiplier are lower voltage ripple and higher voltage level due to the complementary action of each half section of the multiplier. It means each half section compensates ripple of the other half due to 180° phase differences between the resulting voltage ripples. It is possible to have such condition with two series-connected voltage multipliers. In Figure 6 schematic diagram of such a series-connected voltage multiplier (SCVM) is shown. V_{s1} and V_{s2} usually are secondary windings of a high voltage transformer. With opposite sign of voltage sources, 180° phase difference between sources has been shown. Compared to a conventional one (Figure 1) it has two voltage sources (two secondary windings) and the same number of capacitors and diodes. However, based on previously mentioned symmetrical behavior, it has lower voltage ripple and drop values. Although it should be noticed that the voltage differences between two secondary windings (V_{s1} and V_{s2} in Figure 6) are equal to half of the output voltage and so when the output is a high voltage, this matter should be considered for insulators in construction of the transformer. This is the main problem of this topology that makes it ineffective especially in very high voltage.

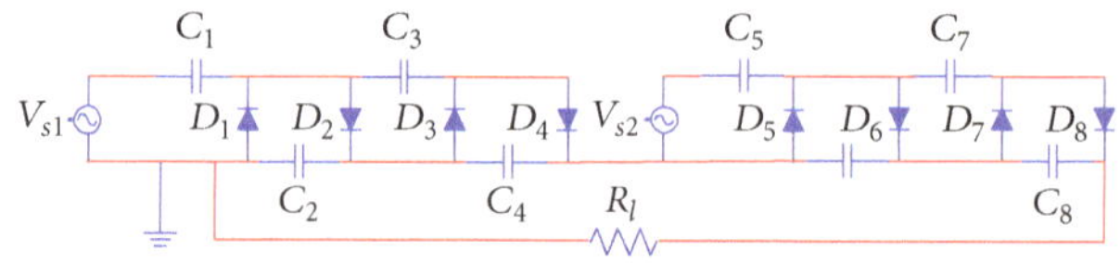

FIGURE 6: Schematic diagram of a series-connected voltage multiplier (SCVM).

Based on this idea, it is possible to build a three-phase (or even higher) VM with cascading three (or more) VMs whose power sources have 120° (360°/number of phases) phase differences with each other. The ripple of such 3-phase VM will be lower than that of two-phase symmetrical VMs. However three separate transformers and several times of components are needed, which make the circuit more costly and bulky [2, 14].

3.4. Series-Connected Positive-Negative Voltage Multiplier (SPNVM). When we cascade two conventional voltage multipliers with 180° out of phase voltage sources as Figure 6, the necessity of high voltage insulation between windings is a problem. Furthermore, compared to the conventional VM, two secondary windings with same voltage amplitude are needed. It is possible to solve this problem by another topology as shown in Figure 7 [23]. As shown, two positive and negative voltage multipliers with opposite direction have been connected in series to build a VM with higher output voltage and reduced ripple value. In this scheme, only one source (one secondary winding) is needed and due to the opposite direction of charge and discharge in positive and negative sections, output voltage ripple of each section compensates those of the other. So the resulting ripple is less than that of conventional VM. Compare to the SVM, this topology has lower number of components and problem of asymmetry of transformer does not exist; consequently the resulting harmonics, as discussed earlier [21, 22], have lower amplitude.

In Figure 7, C_1–C_4 consist of the positive section and C_5–C_8 consist of the negative section of SPNVM. If in this configuration we ground the negative side of the load, it will

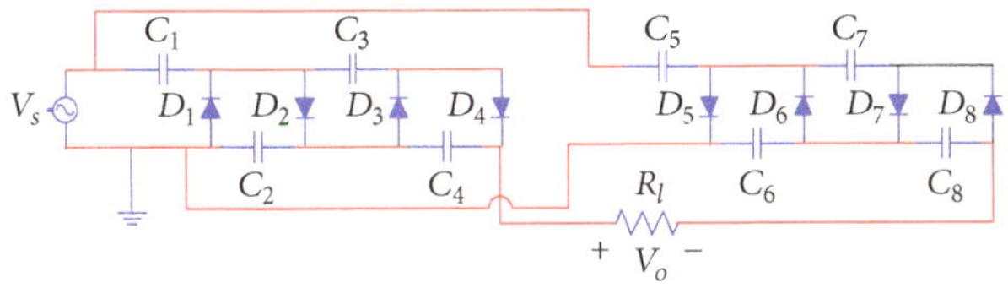

Figure 7: Schematic diagram of a series-connected positive-negative voltage multiplier.

impose a high value dc voltage on the source which usually is a transformer. To avoid this matter, it is better to ground the transformer. Therefore, in this VM, the output is not grounded. This matter restricts this topology to the cases where bipolar output voltage, and not ground-returned one, is needed. Therefore this topology can be used in portable and handheld apparatuses and applications with moderate output voltage, where grounding of load is not nessesary [7, 8].

4. Simulation Results

In this section, by Saber Synopsys software, simulation results of previously discussed VMs are compared. Input voltage and number of stages are chosen to get almost equal output voltage from all VMs. Therefore by consideration of devices count and output specifications of voltage multipliers, comparison is done. Nonideality of the components that are considered in simulations is ESR of capacitors (0.1 Ω), diodes forward voltage drop (0.7 V), input resistance of voltage sources (50 Ω), and inductance of transformer. To shorten the article, simulation results are given only for a case in which the parameters are the same as those for experimental work. It is clear that the operation of the VMs depends on the capacitors value, frequency of operation, number of stages, and load resistor. So, despite the difficulties and cost of high voltage prototypes, in performance comparison, there is no difference between several kilovolts or few volts VMs.

The structure of transformers of VMs is a matter which should be discussed in more detail. Some of the VMs use transformers with two secondary windings, such as SCVM and SVM. If we want to use the same transformer for all VMs, to have the same condition for comparison, a transformer with two secondary windings should be used. Then, in topologies that use one voltage source, series connection of secondary windings could be used. Therefore, for those topologies which need one secondary winding, the voltage would be twice of those with two secondary windings. However, in this case, the stress on components in SVM and SCVM is equal to half of that for other topologies and the number of stages should be doubled to have similar value of output voltage. HSVM also has the same condition and therefore, to decrease the number of stages, a transformer whose secondary voltage is twice of those for BVM and SPNVM could be used. It means there are two choices: first to use the same transformer for all VMs and second to consider the same stress on components. In simulations it is found that SCVM response is almost similar to that of SVM. Furthermore its transformer has the mentioned problem of extra voltage imposed between windings. So in simulations and experiments the SCVM was omitted.

In simulations and experiments, based on abovementioned matter about transformer, two versions of HSVM and SVM are considered. HSVM1 is an eight-stage VM with 10 V input. HSVM2 is a four-stage VM with 20 V input. SVM1 is an eight-stage VM with two 5 V inputs. And SVM2 is a four-stage VM with two 10 V inputs. SPNVM and BVM are four stages with 10 V input. It must be emphasized that the voltage stress on components in topologies depends on the transformer voltage. That is, the magnitude of voltage stress on HSVM1 and SVM1 components is half of that for others. So in components count we should consider this matter. We know that a capacitor with C value and V_b breakdown voltage can be built by four capacitors with C value and $V_b/2$ breakdown voltage, that is, parallel connection of two series-connected ones. Furthermore two diodes with $V_b/2$ breakdown voltage can build a diode with V_b breakdown voltage. Therefore, if the stress on components of VM_1 is twice of that for VM_2, in components count we can consider capacitors and diodes of VM_2 as basic components and multiply the number of capacitors of VM_1 by four and its diodes by two, to find out equivalent number of similar components. With this method we can correctly compare the number of components, that is, the cost of construction. This matter is considered in comparison between VMs. The discussed matter of this paragraph should be understood carefully to know why the proposed comparison in this article is fair.

The frequency of voltage sources (transformers output) is 2 KHz and capacitors in all VMs are 1.6 μF and load is a 51 kΩ resistor. In Figure 8 the transient responses and maximum value of VMs are given. Rise time of VMs was measured according to 10–90% of output voltage by software. We know that speed of VMs especially is an important factor when a VM wanted to be use in pulse mode. The maximum achievable frequency of operation clearly depends on rise time of the VM. As can be seen HSVM2 with highest value of output voltage and SPNVM with fastest response are the best. Furthermore, HSVM1 and SVM1 do not have satisfactory responses compared to others.

Figure 9 shows the ripples of voltage multipliers. It is true that SVM1 has the lowest absolute value of ripple. However if we calculate ratio of ripple to maximum value, that is, ripple factor for all VMs, it can be seen that SPNVM, SVM2, and HSVM2 with 0.007 have the best ripple factor and BVM, as expected, with 0.053 being the worst one.

5. Experimental Comparison

To verify the simulation results, prototypes of VMs have been constructed. BVM, SPNVM, HSVM1, HSVM2, SVM1, and SVM2 have been constructed for comparison. Input is a 2 kHz sinusoidal source with variable amplitude which combined with a ferrite core transformer provides the requested voltages. As in simulation a purely resistive 51 kΩ load is used. Capacitors are in range of 1.62 μF–1.72 μF and have been chosen among many, to be nearly equivalent. Reduction of error produced by nonequal value of capacitors when we compare the topologies was the reason of such selection. In

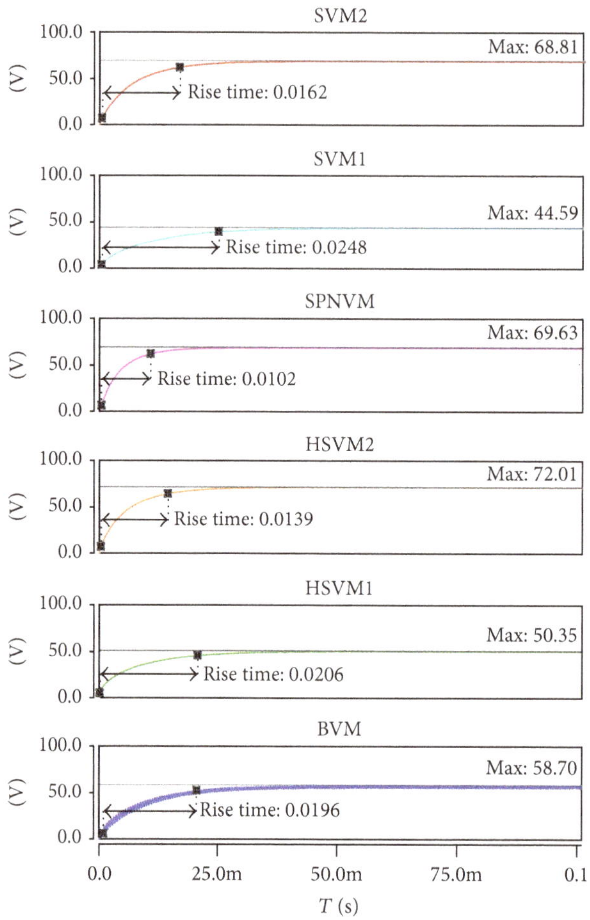

FIGURE 8: Simulation results: transient and maximum values of output voltages of VMs.

fact, because nonidealities of components have been considered in simulation, experimental prototypes have almost the same specifications as simulation.

In Figure 10, transient responses of VMs are shown. As can be seen SPNVM with a rise time of 14.39 ms has the fastest response and HSVM2 with 18.45 ms is in the second rank. The difference between rise time of VMs in simulation and experiments arises from this fact that the signal generator has a delay in its startup which could not be measured. Maximum output values in steady state are also shown. As simulation, HSVM2 with 70.4 V has the maximum output value.

In Figure 11 variations of VMs are shown. As expected, BVM has maximum peak-to-peak value of ripple with 3.04 V and SPNVM with 456 mV has the lowest value of ripple. By calculation, the ripple factor of SPNVM with 0.006 is the lowest one.

In Table 1 summary of simulation and experimental results are given. Some terms used in the table should be explained. The number of capacitors and diodes in first and second row shows the actual number of components, while base-cap number and base-diode number show the number of components with regard to voltage stress as discussed in simulation section. It means we have considered capacitors and diodes with lowest voltage stress as basic components. Then in VMs with twice voltage stress the number of capacitors is multiplied by four and that of diodes by two. For example, despite the higher actual value of components of HSVM1 relative to HSVM2, its base components number is less. Another term represents wire consumption in transformer construction and is defined as secondary wire factor.

6. Conclusion

Decades after invention of Cockcroft-Walton voltage multiplier, in broad range of high voltage and ac to dc conversion applications, it is still being used and has no competitor. In this paper a review over and a comparison between symmetrical versions of VMs were carried out. The features of different VMs as rise time, ripple, and output voltage are compared to each other with regard to the complexity of the topologies and number of components. In this comparison, some factors as voltage stress on components, number of components, and transformer construction cost are considered. In fact, the voltage stress on components can be used as a factor to evaluate the effective number of components and consequently the price of VMs.

Ground problem of transformer in some VMs makes them ineffective especially in very high voltage usage. However, in mobile apparatus and some low voltage applications, they still can be used. For example, SPNVM with

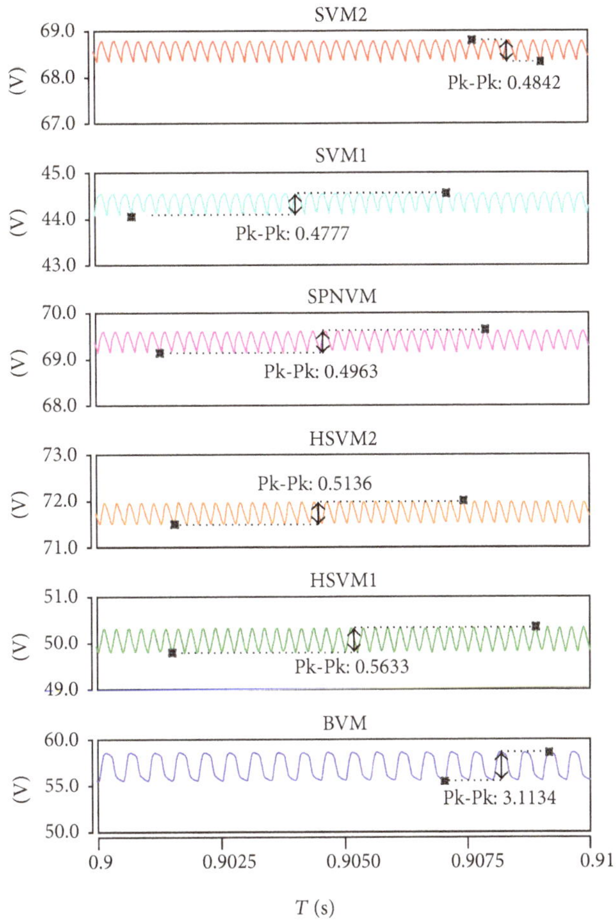

FIGURE 9: Simulation results: output ripples of voltage multipliers at steady state.

TABLE 1: Summary of VMs specifications.

Voltage Multiplier		BVM	HSVM1	HSVM2	SPNVM	SVM1	SVM2
Number of capacitors		8	22	10	8	24	12
Number of diodes		8	32	16	8	32	16
Number of secondary windings		one	one	one	one	Two	Two
Transformer windings voltage (V): V_{peak} (V)		10	10	20	10	5	10
Secondary wire factor		1	1	2	1	1	2
Voltage stress on components (V)		20	10	20	20	10	20
Base-cap number		32	22	40	32	24	48
Base-diode number		16	32	32	16	32	32
Simulation results							
$C = 1.6\,\mu$F $R_l = 51\,\text{k}\Omega$ $f = 2\,\text{kHz}$	V_{max} (V)	58.7	50.35	**72.01**	69.63	44.59	68.81
	Ripple (V)	3.11	0.563	0.513	0.496	0.477	0.484
	Ripple factor*	0.052	0.011	**0.007**	**0.007**	0.010	**0.007**
	Rise time (ms)	19.6	20.6	13.9	**10.2**	24.8	16.2
Experimental results							
$C = 1.62\text{–}1.72\,\mu$F $R_l = 51\,\text{k}\Omega$ $f = 2\,\text{kHz}$	V_{max} (V)	57.6	50.4	**70.4**	68.0	44.0	68.0
	Ripple (V)	3.04	0.648	0.500	0.456	0.560	0.480
	Ripple factor	0.052	0.012	0.007	**0.006**	0.012	0.007
	Rise time (ms)	22.7	21.1	18.45	**14.39**	27.32	20.23

*Ratio of ripple magnitude to the maximum output value.

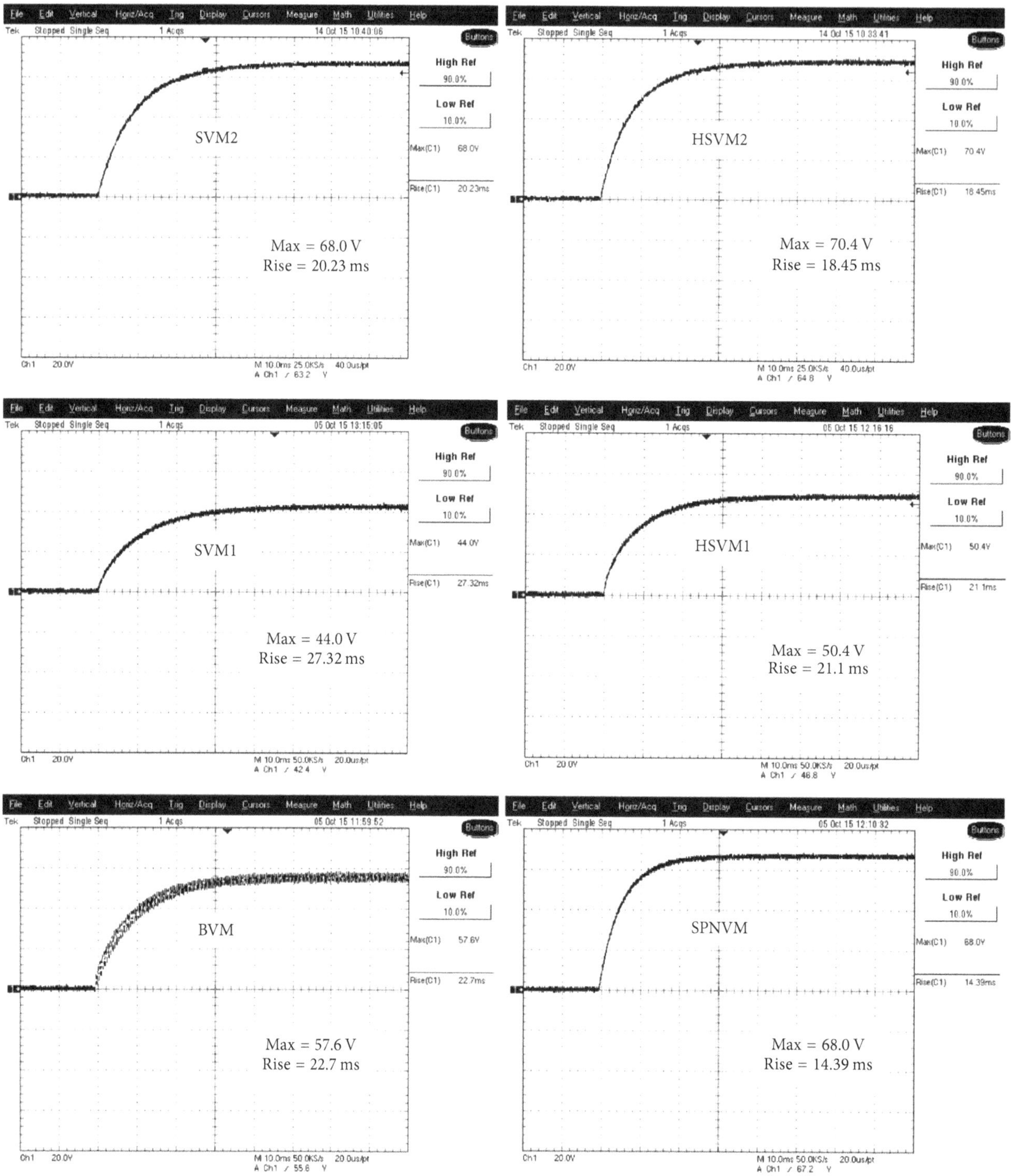

FIGURE 10: Experimental results: transient responses of VMs. SPNVM with 14.39 ms rise time has the fastest response and HSVM2 with 70.4 V has the maximum output value compared to others.

low components count and acceptable characteristics is an attractive choice which should not be ignored in such applications. With simulations and experimental prototypes, the responses of VMs were compared. The measured values of the prototype's waveforms agreed well with their counterparts simulations. In summary it is not possible to choose one topology as a perfect one. In fact, in each application based on the level of output voltage and therefore components price, with similar method used in Table 1, it is possible to compare the VMs and, therefore, to choose which of them is suitable for the proposed application. Furthermore, with this method, it is possible to understand if a VM has better performance compared to others and how much it costs for designer.

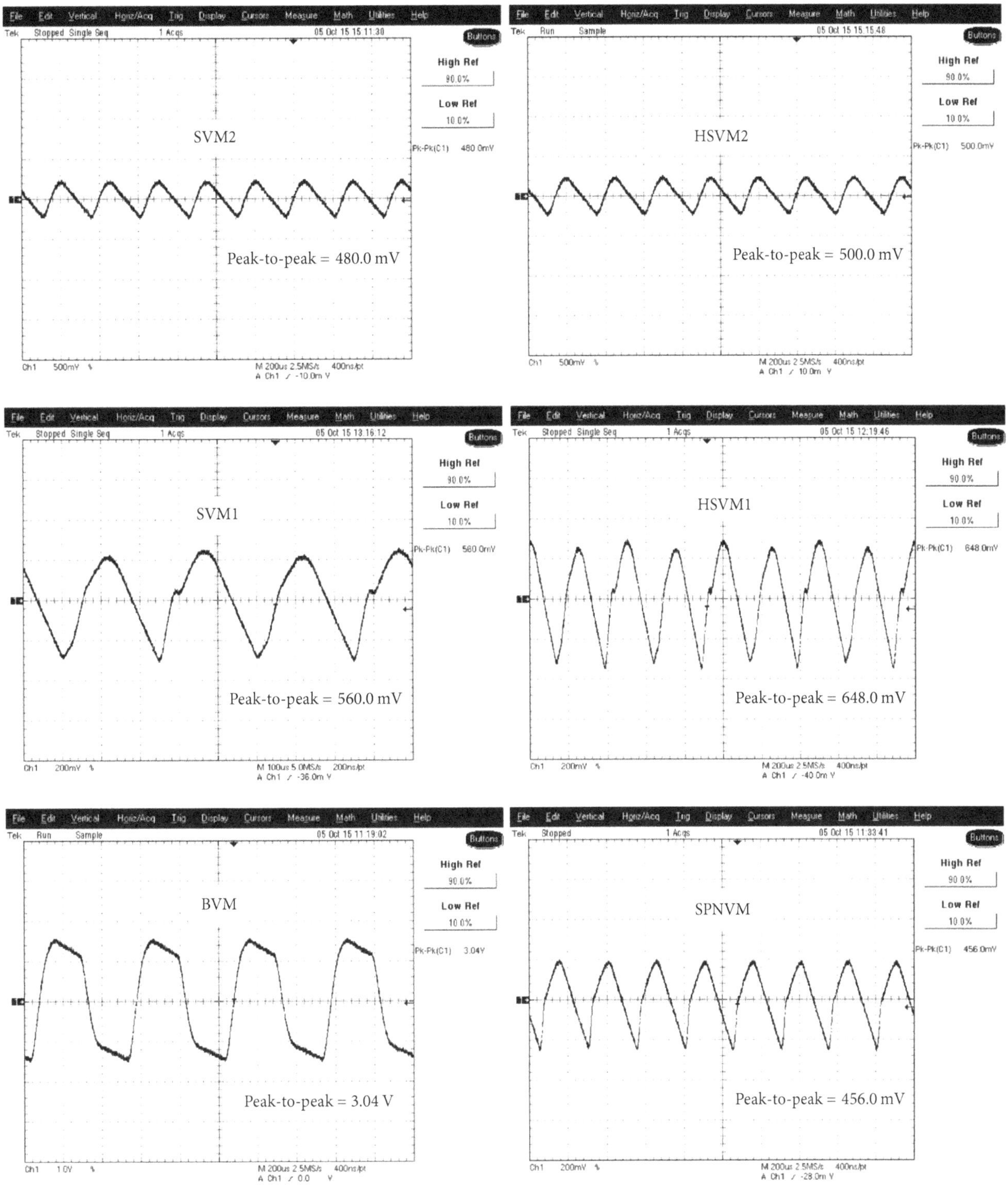

FIGURE 11: Experimental results: peak-to-peak value of ripple of VMs. SPNVM with 456 mV ripple has the smoothest response compared to others.

Competing Interests

The author declares that there is no conflict of interests regarding the publication of this paper.

Acknowledgments

The author would like to thank his colleagues Mr. M. Daneshyan and Mr. B. N. Mortazawi for their help in preparing experimental tests. This work was supported by project from "Laser & Optics Research School" (Code no. PRD-L2-93-007).

References

[1] S. Iqbal, "A hybrid symmetrical voltage multiplier," *IEEE Transactions on Power Electronics*, vol. 29, no. 1, pp. 6–12, 2014.

[2] S. Iqbal, R. Besar, and C. Venkataseshaiah, "Single/three-phase symmetrical bipolar voltage multipliers for X-ray power supply," in *Proceedings of the 2nd International Conference on Electrical Engineering (ICEE '08)*, pp. 1–6, IEEE, Lahore, Pakistan, March 2008.

[3] C.-M. Young, M.-H. Chen, S.-H. Yeh, and K.-H. Yuo, "A Single-phase single-stage high step-Up AC-DC matrix converter based on cockcroft-walton voltage multiplier with PFC," *IEEE Transactions on Power Electronics*, vol. 27, no. 12, pp. 4894–4905, 2012.

[4] T.-L. Su, Y.-M. Zhang, S.-W. Chen, Y.-T. Liu, H.-Y. Lv, and J.-T. Liu, "A 600kV 15mA Cockcroft-Walton high-voltage power supply with high stability and low-ripple voltage," *Nuclear Instruments and Methods in Physics Research Section A: Accelerators, Spectrometers, Detectors and Associated Equipment*, vol. 560, no. 2, pp. 613–616, 2006.

[5] Z.-F. He, J.-L. Zhang, Y.-H. Liu, Y.-T. Zhang, and Y. Zhang, "Characteristics of a symmetrical Cockcroft-Walton power supply of 50 Hz 1.2 MV/50 mA," *Review of Scientific Instruments*, vol. 82, no. 5, Article ID 055116, 2011.

[6] C.-M. Young, H.-L. Chen, and M.-H. Chen, "A cockcroft-walton voltage multiplier fed by a three-phase-to-single-phase matrix converter with PFC," *IEEE Transactions on Industry Applications*, vol. 50, no. 3, pp. 1994–2004, 2014.

[7] D. F. Spencer, R. Aryaeinejad, and E. L. Reber, "Using the Cockroft-Walton voltage multiplier with small photomultipliers," *IEEE Transactions on Nuclear Science*, vol. 49, no. 3, pp. 1152–1155, 2002.

[8] D. F. Spencer, R. Aryaeinejad, and E. L. Reber, "Using the cockcroft-walton voltage multiplier design in handheld devices," in *Proceedings of the IEEE Nuclear Science Symposium Conference Record*, pp. 746–749, San Diego, Calif, USA, November 2001.

[9] M. Rezanejad, A. Sheikholeslami, and J. Adabi, "High-voltage pulsed power supply to generate wide pulses combined with narrow pulses," *IEEE Transactions on Plasma Science*, vol. 42, no. 7, pp. 1894–1901, 2014.

[10] L. Müller and J. W. Kimball, "Dual-input high gain DC-DC converter based on the cockcroft-walton multiplier," in *Proceedings of the IEEE Energy Conversion Congress and Exposition (ECCE '14)*, pp. 5360–5367, IEEE, Pittsburgh, Pa, USA, September 2014.

[11] Z. Cao, M. Hu, N. Fröhleke, and J. Böcker, "Modeling and control design for a very low-frequency high-voltage test system," *IEEE Transactions on Power Electronics*, vol. 25, no. 4, pp. 1068–1077, 2010.

[12] E. Everhart and P. Lorrain, "The Cockcroft–Walton voltage multiplying circuit," *Review of Scientific Instruments*, vol. 24, no. 3, pp. 221–226, 1953.

[13] G. Reinhold, K. Truempy, and J. Bill, "The symmetrical cascade rectifier an accelerator power supply in the megavolt and milliampere range," *IEEE Transactions on Nuclear Science*, vol. 12, no. 3, pp. 288–292, 1965.

[14] S. Iqbal, "A three-phase symmetrical multistage voltage multiplier," *IEEE Power Electronics Letters*, vol. 3, no. 1, pp. 30–33, 2005.

[15] M. Abdel-Salam, *High-Voltage Engineering: Theory and Practice, Revised and Expanded*, CRC Press, Boca Raton, Fla, USA, 2000.

[16] J. Kuffel, E. Kuffel, and W. S. Zaengl, *High Voltage Engineering Fundamentals*, Newnes, 2000.

[17] C. L. Wadhwa, *High Voltage Engineering*, New Age International, New Delhi, India, 2007.

[18] C. G. H. Maennel, "Improvement in the modelling of a half-wave Cockroft-Walton voltage multiplier," *Review of Scientific Instruments*, vol. 84, no. 6, Article ID 064701, 2013.

[19] I. C. Kobougias and E. C. Tatakis, "Optimal design of a half-wave Cockcroft–Walton voltage multiplier with minimum total capacitance," *IEEE Transactions on Power Electronics*, vol. 25, no. 9, pp. 2460–2468, 2010.

[20] F. Hwang, Y. Shen, and S. H. Jayaram, "Low-ripple compact high-voltage DC power supply," *IEEE Transactions on Industry Applications*, vol. 42, no. 5, pp. 1139–1145, 2006.

[21] H. Zhang and A. Takaoka, "Ripple due to asymmetry in symmetrical Cockcroft–Walton cascade rectifier circuit," *Review of Scientific Instruments*, vol. 65, no. 12, pp. 3864–3865, 1994.

[22] H. Zhang and A. Takaoka, "Fundamental harmonic of ripples in symmetrical Cockcroft-Walton cascade rectifying circuit," *Review of Scientific Instruments*, vol. 67, no. 9, pp. 3336–3337, 1996.

[23] S. Iqbal and R. Besar, "A bipolar Cockcroft-Walton Voltage Multiplier for gas lasers," *American Journal of Applied Sciences*, vol. 4, no. 10, pp. 795–801, 2007.

4

A Decoupling Control Method for Shunt Hybrid Active Power Filter based on Generalized Inverse System

Xin Li[1,2] **and Bo Li**[2]

[1]*Key Laboratory of Optoelectronic Technology and Intelligent Control, Ministry of Education, Lanzhou Jiaotong University, Lanzhou 730070, China*
[2]*School of Automation & Electrical Engineering, Lanzhou Jiaotong University, Lanzhou 730070, China*

Correspondence should be addressed to Xin Li; lxfp167@163.com

Academic Editor: Ephraim Suhir

In this paper, a novel decoupling control method based on generalized inverse system is presented to solve the problem of SHAPF (Shunt Hybrid Active Power Filter) possessing the characteristics of 2-input-2-output nonlinearity and strong coupling. Based on the analysis of operation principle, the mathematical model of SHAPF is firstly built, which is verified to be invertible using interactor algorithm; then the generalized inverse system of SHAPF is obtained to connect in series with the original system so that the composite system is decoupled under the generalized inverse system theory. The PI additional controller is finally designed to control the decoupled 1-order pseudolinear system to make it possible to adjust the performance of the subsystem. The simulation results demonstrated by MATLAB show that the presented generalized inverse system strategy can realise the dynamic decoupling of SHAPF. And the control system has fine dynamic and static performance.

1. Introduction

In recent years, many nonlinear electrical components are widely used in our life; harmonics produced by Power System electronic devices have a serious impact on power quality. So it is important to find equipment to suppress harmonics and compensate reactive power for the grid.

The traditional compensation equipment is passive filter (PPF); it is inexpensive, but it cannot solve the impedance deviation caused by the mutation of the system, and it can only compensate fixed harmonics. Active power filter (APF) can be used to compensate each of the order harmonics of the grid, but it is not so feasible in large-capacity situation [1].

As a new compensation device, HAPF combines the advantages of PPF and APF; it can solve almost all the matters that PPF and APF have. Harmonics and reactive power are compensated by PPF. The main function of APF is to improve the filtering characteristics of passive filters and overcome the shortcomings that the PPF has which can easily resonate with the grid impedance. The capacity of APF occupies only about 2%–5% of the capacity of the harmonic source load, and most harmonics are compensated by PPT. So SHAPF can be used in large-capacity situations with relatively small-capacity, and it can improve the cost effectiveness of the system. So the use of SHAPF is becoming a significant practical direction for grid harmonic governance.

The design of the controllers seriously affects the harmonics compensation. Many controllers have been designed to make the APF decoupling. Reference [1] designed a double closed-loop PI controller to achieve the decoupling control. Reference [2] designed a fuzzy controller to achieve the decoupling control, but the settings of controllers mainly depend on experience, and the controllers cannot react to the mutation of the system. In [3], inverse system method is presented to achieve the decoupling of SHAPF. The compensation effect is better, but the traditional analytic inverse system method largely depends on the accurate mathematical model of the system, and it is unable to configure the poles of the system to achieve the expected performance of the system. In this paper, the traditional inverse system method is improved and a new generalized inverse decoupling control system is built. Finally we can ensure the stability of the

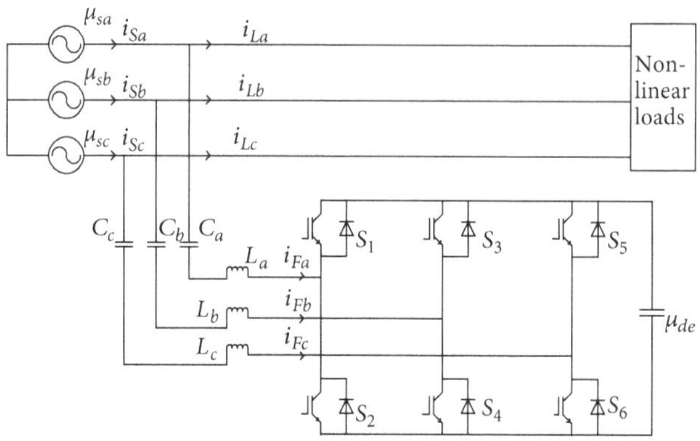

FIGURE 1: Mathematical model of SHAPF [1].

pseudolinear system and reduce the difficulty of additional controller design, by adjusting the generalized inverse parameters to configure the pole of the system.

2. Working Principle of SHAPF

SHAPF consists of command current arithmetic circuit, current tracking control circuit, driving circuit, and the main circuit (PWM converter). The structure of SHAPF is shown in Figure 1. By detection of the harmonic current on grid, compensated current signal is firstly calculated by compensated amplifier circuit. Then PWM converter generates compensated current by driving circuit; compensated current can offset the harmonic current on grid. Finally, expected current is obtained by PWM generator. The core device of SHAPF main circuit is IGBT (Insulated Gate Bipolar Transistor), it is widely used because of its high frequency inverter characteristic and automatic switching current capability, and it can transform the direct current (DC) into alternating current (AC). SHAPF is equivalent to the controlled current source in the main circuit; it is used to offset the harmonics and reactive power on the grid [4].

3. Mathematical and Generalized Inverse of SHAPF

The mathematical model of SHAPF is shown as Figure 1. v_{sj} $(j = a, b, c)$ is grid voltage. v_{cj} $(j = a, b, c)$ is the voltage of C_j $(j = a, b, c)$. v_{lj} $(j = a, b, c)$ is the voltage of point of common coupling. L_s is the equivalent inductance of the grid. C_{de} is the DC capacitor and the voltage is v_{de}.

From [5], we get the formula of SHAPF in d-q coordinate system:

$$L_F i'_{Fd} = -R_F i_{Fd} + \omega L_F i_{Fq} - v_{Cd} - u_d v_{dc} + v_{Ld} \tag{1}$$

$$L_F i'_{Fq} = -R_F i_{Fq} - \omega L_F i_{Fd} - v_{Cq} - u_q v_{dc} + v_{Lq} \tag{2}$$

$$C_{de} v'_{dc} = u_d i_{Fd} + u_q i_{Fq}. \tag{3}$$

For the facilitation of the controller design, nonlinear affine expression of SHAPF is built as follows:

$$\begin{aligned} x' &= f(x) + g(x) u \\ y &= h(x). \end{aligned} \tag{4}$$

Providing that $x = [i_{Fd}\ i_{Fq}\ v_{dc}]^T$, $u = [u_d\ u_q]^T$, and $y = [y_1\ y_2]^T = [i_{Fd}\ i_{Fq}]^T$, we can get the derived formula from formula (1)-(2):

$$x'_1 = \frac{(-R_F x_1 + \omega L_F x_2 - v_{Cd} - u_1 x_3 + v_{Ld})}{L_F} \tag{5}$$

$$x'_2 = \frac{(-R_F x_2 - \omega L_F x_1 - v_{Cq} - u_2 x_3 + v_{Lq})}{L_F} \tag{6}$$

$$x'_3 = u_1 x_1 + u_2 x_2. \tag{7}$$

In formula (1)–(6), ω is the fundamental angular frequency. In other words, it is the rotation angular velocity of the d-q axis coordinate; u_1 and u_2 are the switching function control quantities in d-q axis coordinate; v_{Ld} and v_{Lq} are the voltage at the connection point between AC and SHAPF in d-q axis coordinate. i_d and i_q are the compensation current of SHAPF in d-q axis coordinate. We can get the following formula from (9)–(11). From formula (5)-(6), it is easy to see that the system is a 2-input-2-output coupling and nonlinear system. In this paper, the generalized inverse system method is used to design the controller.

4. Generalized Inverse Controller Design for SHAPF

4.1. Invertible Analysis. Here, we use inverse system to analyze the invertibility of SHAPF. We first calculate the derivative of the output (y) until the input (u) appears explicitly [6]. In this paper, calculating $y = h(x)'$,

$$\begin{aligned} h_1(x)' &= \frac{(-R_F x_1 + \omega L_F x_2 - v_{cd} - u_1 x_3 + v_{Ld})}{L_F} \\ h_2(x)' &= \frac{(-R_F x_2 - \omega L_F x_1 - v_{cq} - u_2 x_3 + v_{Lq})}{L_F}. \end{aligned} \tag{8}$$

The Jacobi matrix is $\begin{bmatrix} -v_{de}/L_F & 0 \\ 0 & -v_{de}/L_F \end{bmatrix}$.

The Jacobi matrix is nonsingularity; relative degree vector of the system is $\alpha = [\alpha_1, \alpha_2]^T = [1, 1]^T$. For $\alpha_1 = \alpha_2 = 1$, $\alpha_1 + \alpha_2 = 2$, the order is less than the order of the SHAPF system, so the SHAPF system is invertible. It is confirmed in [7].

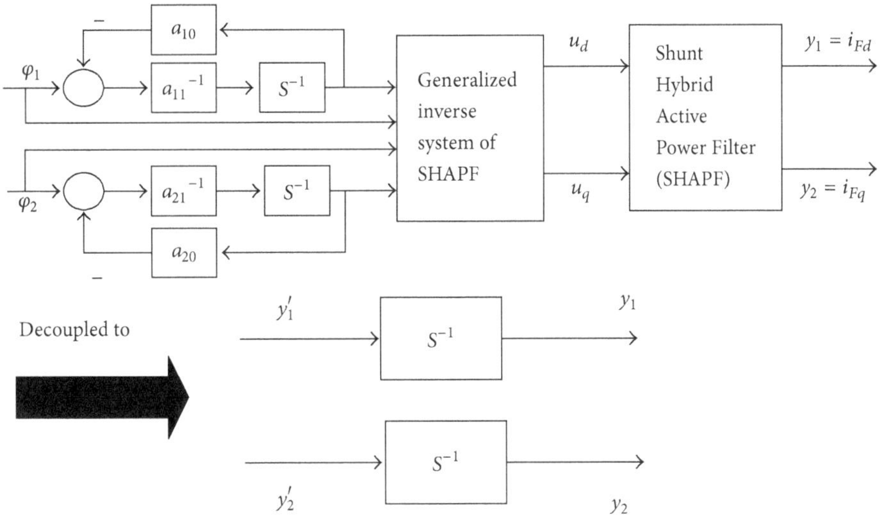

FIGURE 2: Feedback linearization and decoupling of SHAPF with generalized inverse system method.

4.2. Formula Derivation with Generalized Inverse Theory. As is proved in the previous Section 4.1, the system is invertible. In inverse system, the input vector of the inverse system is $\varphi = y'$; the output vector of the inverse system is $[y_1 \ \ y_2]^T$ and the state feedback vector of the inverse system is $[x_1 \ \ x_2]^T$. By generalized inverse system, the input vector changes to $\varphi_j = a_{j0}y_0 + a_{j1}y_1' + a_{j2}y_2'' + \ldots + a_{jn}y_{jn}^{(n)}$.

It is obvious that, compared with the inverse system, generalized inverse system increases some parameters, so the pseudolinear system can configure the poles to anywhere by adjusting parameters. It cannot be resolved by inverse system. So the decoupling effect by generalized inverse system is better than the traditional inverse system.

In this paper, ordering the input vector φ is

$$\varphi_1 = a_{11}y_1' + a_{10}y_1 \tag{9}$$

$$\varphi_2 = a_{21}y_2' + a_{20}y_2. \tag{10}$$

We transform (9)-(10) into

$$y_1' = \frac{(\varphi_1 - a_{10}y_1)}{a_{11}} \tag{11}$$

$$y_2' = \frac{(\varphi_2 - a_{20}y_2)}{a_{21}}. \tag{12}$$

From (5)-(6), we can get the expression of u_d and u_q:

$$u_d = \frac{\left(v_{Ld} - R_F x_1 + \omega L_F x_2 - v_{Cd} - L_F y_1'\right)}{x_3}$$
$$u_q = \frac{\left(v_{Lq} - R_F x_2 - \omega L_F x_1 - v_{Cq} - L_F y_2'\right)}{x_3}. \tag{13}$$

Substituting (11)-(12) into equation (13):

$$u_d = \frac{(v_{Ld}a_{11} - R_F x_1 a_{11} + \omega L_F x_2 a_{11} - v_{cd}a_{11} + L_F a_{10} x_1 - L_F \varphi_1)}{(a_{11}x_3)}$$
$$u_q = \frac{\left(v_{Lq}a_{21} - R_F x_2 a_{21} - \omega L_F x_1 a_{21} + L_F a_{21} x_2 - L_F \varphi_2\right)}{(a_{21}x_3)}. \tag{14}$$

Equation (14) is the duty cycle function of PWM generator for the new generalized inverse system. Feedback linearization and decoupling of SHAPF with generalized inverse system method diagram are shown in Figure 2. We can control the compensated current by adjusting the duty cycle function of PWM inverter. By this way, the control of 2-input-2-output system is transformed into the control of two linear systems, so the 1-order linear system is easy to control.

4.3. Design of Controller of Pseudolinear System. With generalized inverse system method, the controller is designed to achieve the desired effect. In this paper, generalized inverse PI composite controller is designed to achieve the decoupling control; two PI additional controllers are used to control the two decoupled 1-order pseudolinear systems which can reduce steady-state error of the system and configure the poles reasonably. The capacitor in DC side is controlled by PI control. Simulation flow chart of generalized inverse system is shown in Figure 3.

4.4. Adjustment of Generalized Inverse System Parameters. The composite system is a second-order system, structure diagram of SHAPF control system is shown in Figure 4. The transfer function of the system is easy to obtain. It is calculated that the poles of the composite system are all on the left side of the s-plane, so the composite system is stable. The generalized inverse system parameters $(a_{10}, a_{11}, a_{20}, a_{21})$

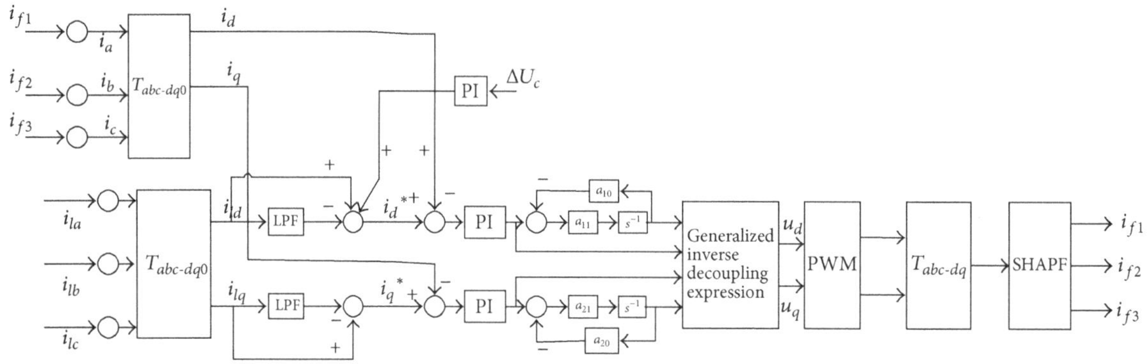

FIGURE 3: Simulation flow chart of generalized inverse system.

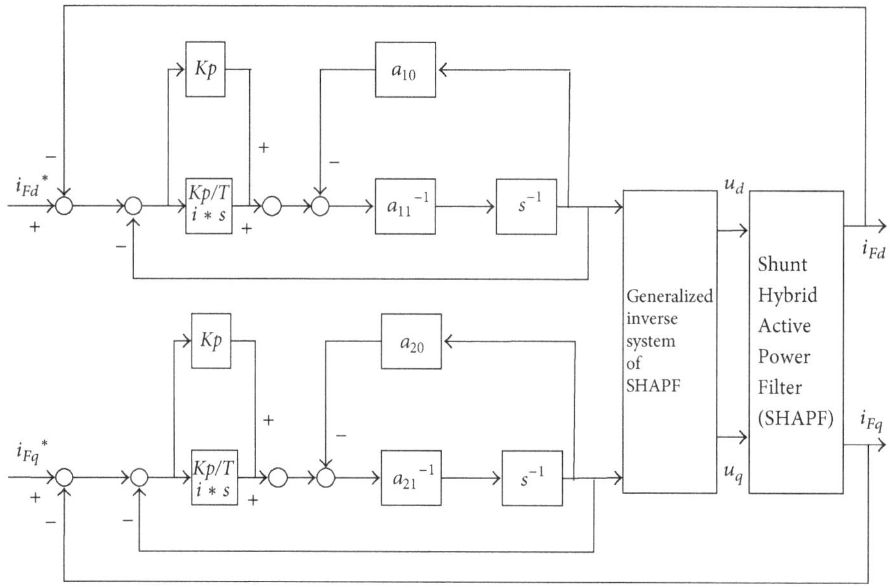

FIGURE 4: Structure diagram of SHAPF control system.

are set to make the damping ratio (ξ) of this composite subsystem close to 0.7. (The damping ratio is calculated by the transfer function.) The operating frequency range can reach to the maximum value, and the subsystem has the best overall dynamic performance. This best damping ratio value (ξ) is confirmed by Chen in [8]. In Figure 4, i_{Fd} and i_{Fq} are actual compensation current compensated by SHAPF; i_{Fd}^* and i_{Fq}^* are reference currents calculated from Figure 3.

5. Simulation Research

In order to verify the effectiveness of the generalized inverse control policy, we use the MATLAB simulation platform to build a simulation study. And we do the comparison with feedforward PI control and inverse control. System parameters for simulation are shown in Table 1.

The SHAPF model is built as Figure 1 and the controller is applied as shown in Figure 3. The load current waveform is shown in Figure 5.

TABLE 1: SHAPF and controller parameter.

System parameters	value
Virtual value of grid voltage (V_m) and frequency	220 V/50 Hz
Equivalent inductance on grid (L_s)	0.01 mH
Passive filter parameters C_f, L_f, R_f	10^{-4} F/1 mH/0.1 Ω
DC side capacitor (C_{dc}) and its voltage (V_{dc})	0.01 F/480 V
PWM frequency	10000 Hz
PI parameter of generalized inverse system ($k_{p1}, k_{i1}, k_{p2}, k_{i2}$)	10000/2000, 10000/2000
Generalized inverse parameter ($a_{11}, a_{10}, a_{21}, a_{20}$)	0.01, 1.41, 0.01, 1.41
PI controller parameter in DC side (k_p, k_i)	0.01/0.002
Three-phase bridge load	10 Ω/5 mH

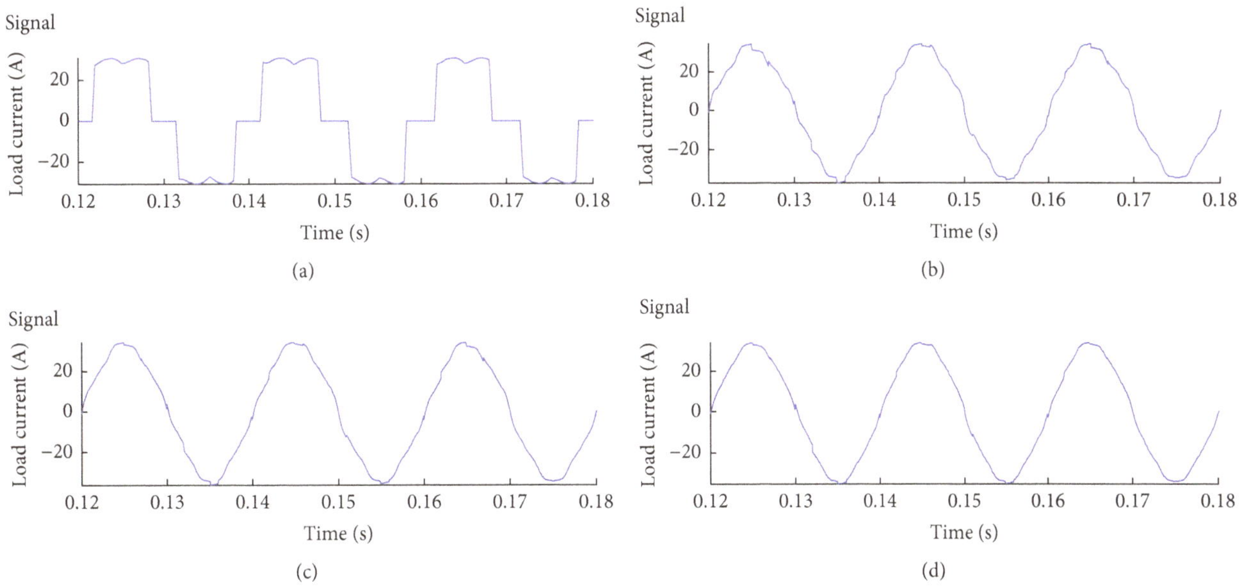

Figure 5: Load current waveform comparison. (a) Load current waveform without decoupling method. (b) Load current waveform with feedforward decoupling method. (c) Load current waveform with inverse system decoupling method. (d) Load current waveform with generalized inverse system decoupling method.

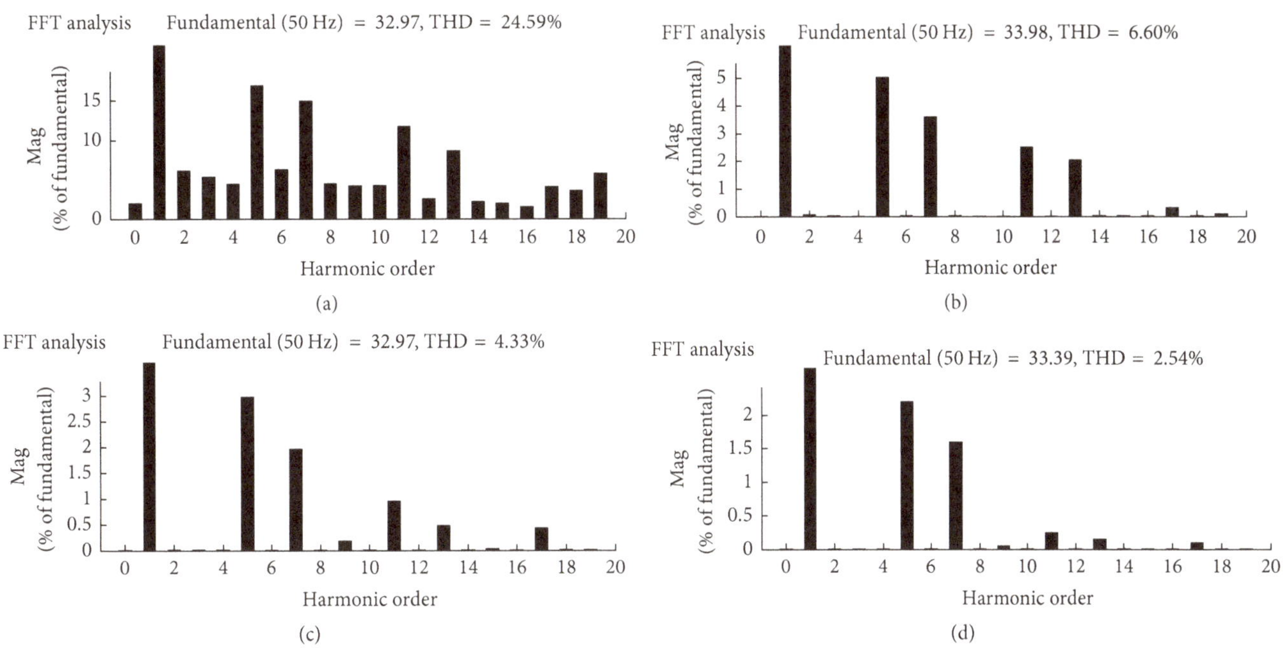

Figure 6: Spectrum analysis comparison. (a) Spectrum analysis of load current without decoupling method. (b) Spectrum analysis of load current with feedforward decoupling method. (c) Spectrum analysis of load current with inverse system decoupling method. (d) Spectrum analysis of load current with generalized inverse system decoupling method.

It is obvious that the load compensated waveform with generalized inverse system method is most close to ideal sine wave. The spectrum analysis is shown in Figure 6.

In Figure 6, the THD (total harmonic distortion) of the system without compensation device is 24.59%. The THD of the system with traditional feedforward PI controller is 6.60%. The THD of the system with inverse PI controller is 4.33%. The THD of the system with generalized inverse PI controller is 2.54%; each controller can compensate the harmonics on grid, but the generalized inverse controller has

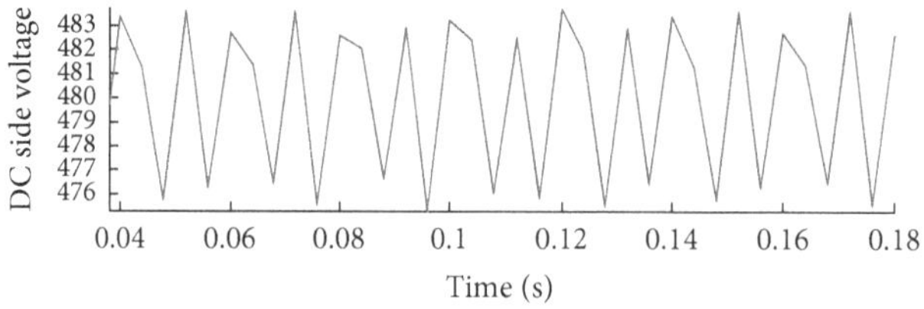

FIGURE 7: Voltage of DC side capacitor (V_{dc}).

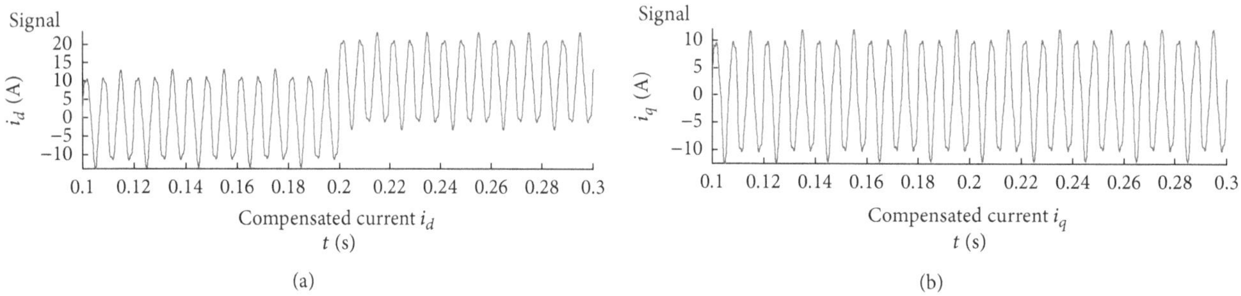

FIGURE 8: Compensated current in d-q coordinate. It is proved that i_d and i_q are completely decoupled in d-q coordinates.

the best effect: it can suppress the 5-order harmonic less than 2% of fundamental. Voltage of DC side capacitor is shown in Figure 7.

In Figure 7, the fluctuation of voltage of DC side capacitor is less than 4 V with PI controller. In order to prove that the system is completely decoupled, disturbance step signal is applied to i_d at time of 0.2 seconds of the simulation. Waveform of i_d and i_q is shown in Figure 8.

It is easy to see that i_q is not affected by the disturbance step signal on i_d. So the complex system is completely decoupled. The generalized inverse system decoupling method for SHAPF is effective.

6. Conclusion

In this paper, a novel method is presented to achieve the decoupling control of SHAPF; generalized inverse theory is applied to the system so that we can configure the pole parameters to adjust the dynamic performance of the system. The generalized inverse system is connected in series with SHAPF and the pseudolinear system can be decoupled into two linear systems to achieve the decoupling control. Decoupled subsystems can be easily bridled by PI controller. The simulation research proves that the system is completely decoupled and the compensation effect is better than the traditional PI control system and inverse system. This method can be used in reality for further research.

Conflicts of Interest

The authors declare that they have no conflicts of interest.

Acknowledgments

This work was supported by the National Natural Science Foundation of China under Project 61165006, the Science Foundation of Gansu Province of China under Project 145RJZA182, the Large Electric Drive System and State Key Laboratory of Equipment Technology Openness Foundation of China under Project SKLLDJ022016015, and a Project Funded by the Colleges Fundamental Research under Project 214144.

References

[1] Z. A. Wang and J. J. Liu, *Harmonic Suppression and Reactive Power Compensation*, China Machine Press, 2015.

[2] A. Abouloifa, F. Giri, I. Lachkar, F. Z. Chaoui, M. Kissaoui, and Y. Abouelmahjoub, "Cascade nonlinear control of shunt active power filters with average performance analysis," *Control Engineering Practice*, vol. 26, pp. 211–221, 2014.

[3] Z. B. Zou and L. J. Cai, "Active power filter nonlinear decoupling control," *Automation of Electric Power Systems*, vol. 14, no. 1, pp. 37–41, 2004.

[4] W. Luo and Y. Zhang, "Power Filter Based on current decouple control," *Light Technology*, vol. 8, no. 1, pp. 42–46, 2014.

[5] M. Benchouia, I. Ghadbane, A. Golea, K. Srairi, and M. Benbouzid, "Implementation of adaptive fuzzy logic and PI controllers to regulate the DC bus voltage of shunt active power filter," *Applied Soft Computing*, vol. 28, no. 1, pp. 125–131, 2015.

[6] X. Z. Deng, *Neural Network Inverse Control Method for Multi-variable Nonlinear Systems*, Science Press, 2004.

[7] W. Lu, C. W. Li, and C. B. Xu, "Inverse system method based decoupling control of shunt hybrid active power filter," *Control Theory & Applications*, vol. 9, no. 1, pp. 1145–1152, 2013.

[8] B. S. Chen, *Electric Power Drag Automation Control System-Motion Control System*, China Machine Press, 2003.

5

Investigation of Improved Methods in Power Transfer Efficiency for Radiating Near-Field Wireless Power Transfer

Hesheng Cheng[1] **and Huakun Zhang**[2]

[1]*School of Computer, Hefei Normal University, Hefei 230601, China*
[2]*Hefei University of Technology School of Instrument Science and Opto-electronics Engineering, Hefei, China*

Correspondence should be addressed to Hesheng Cheng; chenghesheng238@126.com

Academic Editor: Jit S. Mandeep

A metamaterial-inspired efficient electrically small antenna is proposed, firstly. And then several improving power transfer efficiency (PTE) methods for wireless power transfer (WPT) systems composed of the proposed antenna in the radiating near-field region are investigated. Method one is using a proposed antenna as a power retriever. This WPT system consisted of three proposed antennas: a transmitter, a receiver, and a retriever. The system is fed by only one power source. At a fixed distance from receiver to transmitter, the distance between the transmitter and the retriever is turned to maximize power transfer from the transmitter to the receiver. Method two is using two proposed antennas as transmitters and one antenna as receiver. The receiver is placed between the two transmitters. In this system, two power sources are used to feed the two transmitters, respectively. By adjusting the phase difference between the two feeding sources, the maximum PTE can be obtained at the optimal phase difference. Using the same configuration as method two, method three, where the maximum PTE can be increased by regulating the voltage (or power) ratio of the two feeding sources, is proposed. In addition, we combine the proposed methods to construct another two schemes, which improve the PTE at different extent than classical WPT system.

1. Introduction

Research and engineering application for WPT technology has been achieved much attention since the first WPT experiments were carried out by Tesla in the early 20th century [1, 2]. Research in wireless power transmission (WPT) technologies can be carried out by electromagnetic radiation in the radiating far-field region [3–8] and can by resonant coupling [9–18] or inductive coupling [19–26] techniques in the coupled mode near-field region. However, it is not very much that researchers' attention centers on the radiating near-field region [27–30]. Compared to the radio waves transceiver in the far-field, WPT in the radiating near-field region has less dissipated energy and obtains higher PTE. And compared to inductive coupling techniques, WPT in the radiating near-field region can transfer farther by efficient electromagnetic radiation. Using resonant coupling for WPT in the coupled mode near-field region, very high PTE can be achieved. However, the distance for the coupled mode near-field region is very short in terms of wavelength. And not only the optimal load impedance [15–17], but also the resonant frequency and the input impedance of the transmitter are influenced by antenna spacing [11, 12, 18]. It makes difficulty for design of an optimal scheme in the coupled mode near-field region [17].

Following the coupled mode region is the radiating near-field region, where the two antennas are weakly coupled and the PTE decreases rapidly with distance. The authors of [30] used spherical mode theory to derive a theoretical upper bound for wireless power transfer between two antennas in free space. It is pointed out that high antenna radiation efficiency is needed to maximize the power transfer in this region. In [27], two highly efficient folded cylindrical helix (FCH) dipole were used to demonstrate the PTE bound in [30]. Two transmitting and one receiving antennas were investigated in [28] in this region. The authors of the paper found that a stable region in PTE can be created for sufficiently close spacing between two transmitters, in contrast to the monotonic decay of the single-transmitter case. In addition, the authors of [29] investigated the PTE regarding

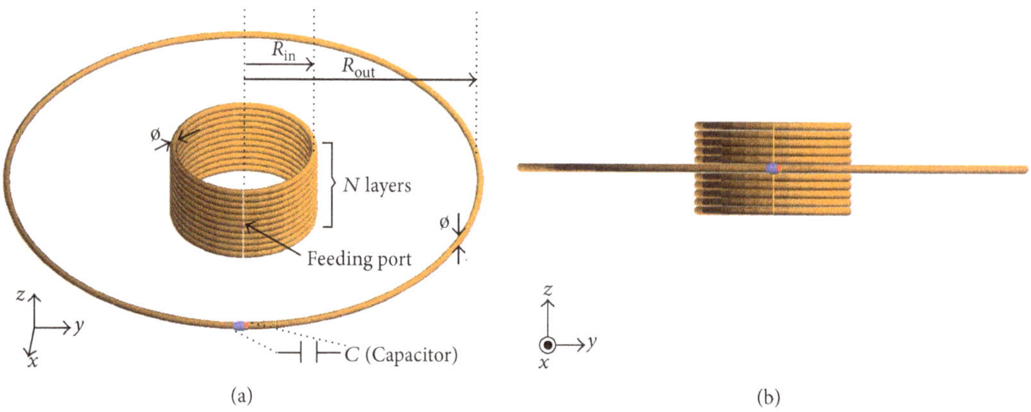

FIGURE 1: Antenna geometry and detailed specification of each design variable. (a) 3D view and (b) 2D lateral view.

lossy dielectric materials effects on WPT based on numerical simulations and experimental measurements.

In the present paper, firstly we design a metamaterial-inspired highly efficient electrically small antenna. The proposed antenna is shown to be well matched to a 50 Ω source. The purpose of this design is to obtain high PTE in radiating near-field region with the 50 Ω optimal load impedance. In the following sections, introduction of an antenna as a retriever to a classical WPT system, phase difference, and voltage ratio adjusting methods are proposed successively. Finally, two WPT schemes are constructed by combining the proposed methods to improve the PTE in varying degrees. The performance of the proposed antenna and WPT systems are characterized with FEldberechnung für Körper mit beliebiger Oberfläche (FEKO).

2. Efficient Electrically Small Antenna Design

We first define the power transfer efficiency for the usual WPT network in two cases: simultaneous conjugate matching and mismatching at feeding port of transmitting antenna:

$$\mathrm{PTE}_{\mathrm{mat}} = \frac{P_{R_L}}{P_{\mathrm{in}}}, \qquad \mathrm{PTE}_{\mathrm{mis}} = \frac{P_{R_L}}{P_T}, \tag{1}$$

where $\mathrm{PTE}_{\mathrm{mat}}$ is the power transfer efficiency in the case of simultaneous conjugate matching, and $\mathrm{PTE}_{\mathrm{mis}}$ is in mismatching case. P_{R_L} is the power dissipated in the load R_L of the receiving antenna, P_{in} is the input power accepted by the transmitting antenna, and P_T is the total input power, including accepted and reflected power by the transmitting antenna.

The feasibility of near-field WPT over a certain number of distances using two electrically small antennas was demonstrated [11, 12, 27–30]. When the WPT system performs in the radiating near-field region, it requires high antenna radiation efficiency to maximize the PTE between transmitting antenna and receiving antenna. In addition, without a matching network a traditional electrically small antenna is known for it is poorly matched to a given 50 Ω source. Thus, $\mathrm{PTE}_{\mathrm{mis}}$ decreases badly because there is a high reflective lossy at the feeding port of transmitting antenna of WPT system.

The so-called metamaterials has been utilized to realize the design of highly radiant efficiency electrically small antenna [31–33]. The metamaterial-based efficient electrically small antennas have been conceptualized with structures constructed from ideal double negative (DNG) or single negative (SNG) media. For example, an electrically small electric dipole and a loop antenna radiating in the presence of, respectively, an isotropic, homogenous lossless, and dispersive electrically small epsilon negative (ENG), and a mu-negative (MNG) spherical shell have been shown theoretically to produce a radiating element that is impedance matched to a specified source to obtain an efficient electrically small antenna system [31, 32]. The electrically small ENG (MNG) metamaterial spherical shell provides the necessary capacitance (inductance) to produce the matching mechanism. In [33], an efficient and electrically small antenna based on a three-dimensional (3D) structure was presented. This antenna is designed to be resonantly driven by a semicircular loop fed through a finite-sized ground plane with a coax feedline. For its large ground plane, the 3D antenna system is not suitable for wireless power transfer.

We present an efficient and electrically small antenna based on a circular loop that acts as an electrically small magnetic dipole. The loop is fed with the power source and surrounded by the other parasitic loop terminated in capacitive impedance, which served as negative magnetic permittivity (MNG) material. Figure 1 shows configuration of the proposed antenna made of copper wires. The copper conductivity value for this design is 5.8E7 Siemens/m. The outer loop efficiently captures and resonantly magnifies the magnetic flux generated by the inner loop that is driving it. A properly designed radius of outer loop and the lumped element capacitance can provide a capacitive element resonantly that matched to the highly inductive electrically small inner loop antenna. By varying the parameters of the proposed antenna, we can obtain the optimum dimensions of the copper wire and capacitance for 50 Ω input resistance and high radiation efficiency.

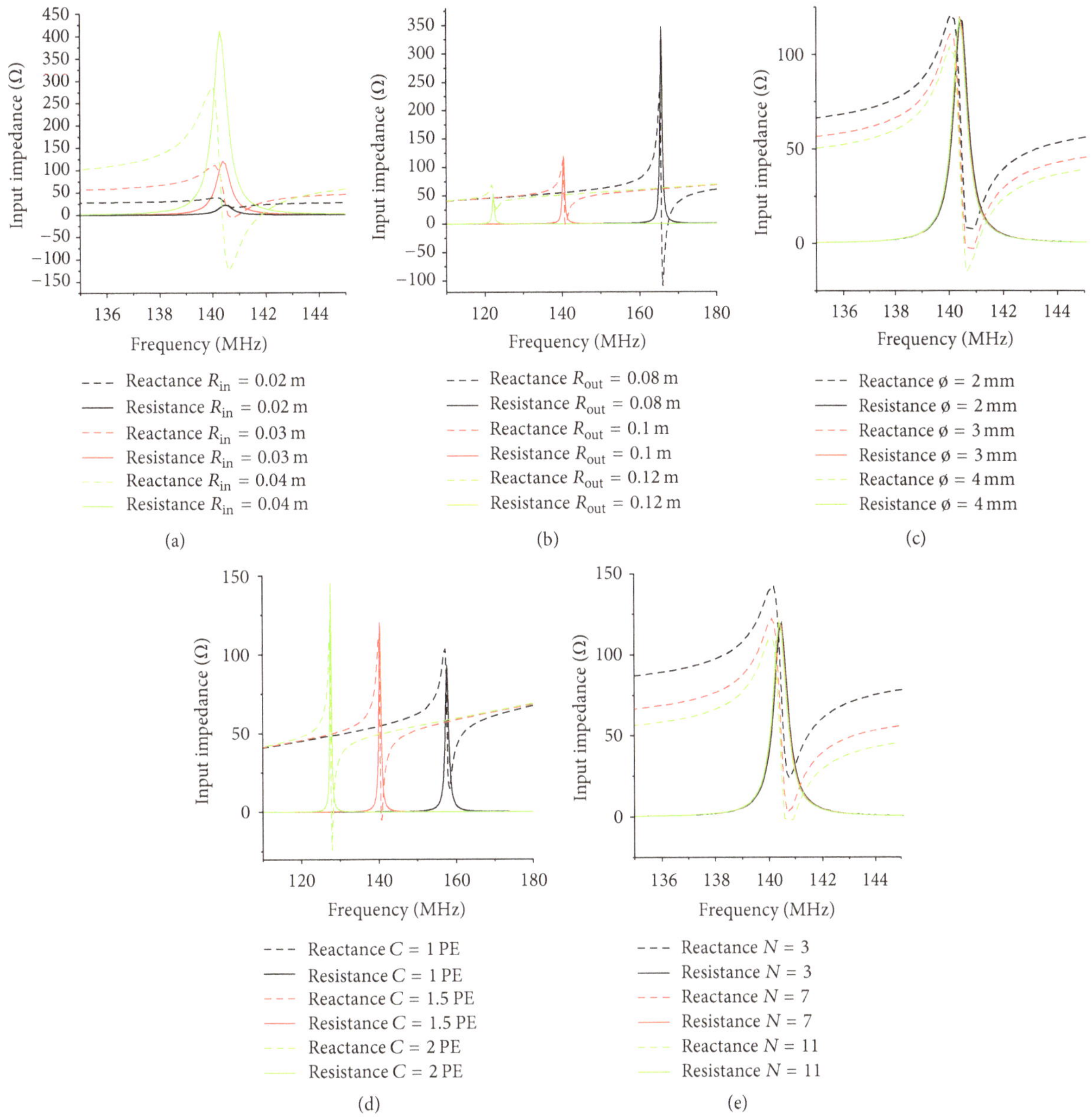

FIGURE 2: The complex input impedance response of proposed antenna against frequency with varied parameters. All the parameters, with exception of the variable to each corresponding subgraph, are fixed in Table 1. (a) ø = 3 mm, R_{out} = 100 mm, C = 1.5 pF, and N = 11. (b) ø = 3 mm, R_{in} = 30 mm, C = 1.5 pF, and N = 11. (c) R_{in} = 30 mm, R_{out} = 100 mm, C = 1.5 pF, and N = 11. (d) ø = 3 mm, R_{in} = 30 mm, R_{out} = 100 mm, and N = 11. (e) ø = 3 mm, R_{in} = 30 mm, R_{out} = 100 mm, and C = 1.5 pF.

Figure 2 plots the input impedance of the proposed antenna with respect to one variable under the other four fixed parameters (five parameters are showed in Figure 1). It can be seen that the radius of the inner loop has a major impact on the input impedance of radiating element but has impact faintly to the resonant frequency (Figure 2(a)). Actually, increasing the radius of the inner circular loop enhances the resonant coupling of the inner driving loop to the outer parasitic loop and thus enhances the resulting radiation resistance of the antenna. The outer loop termed with capacitor provides the major capacitance of this antenna scheme. The resonant frequency moves to lower frequency region with the increasing radius of the outer loop; meanwhile, the input resistance decreases (Figure 2(b)), that is, because the inductive capacitance turns down and the resonant coupling between the two loops goes into a decline. It is reasonable that resonant frequency moves down as increasing capacitance (Figure 2(d)). It is also found that when the diameters of conductive wire or numbers of inner loop are changed, the affected parameters are only the input resistance (Figures 2(c) and 2(e)). It is very useful for us to tailor the input impedance of antenna system to match to the feedline. Table 1 shows the final design parameters of the proposed antenna for WPT. Figure 3 shows the complex input impedance and reflection coefficient (S_{11}) values for this antenna. It is clear from Figure 3 that the antenna for

TABLE 1: The design parameters of the proposed antenna.

ø	R_{in}	R_{out}	C	N
3 mm	30 mm	100 mm	1.5 pF	11

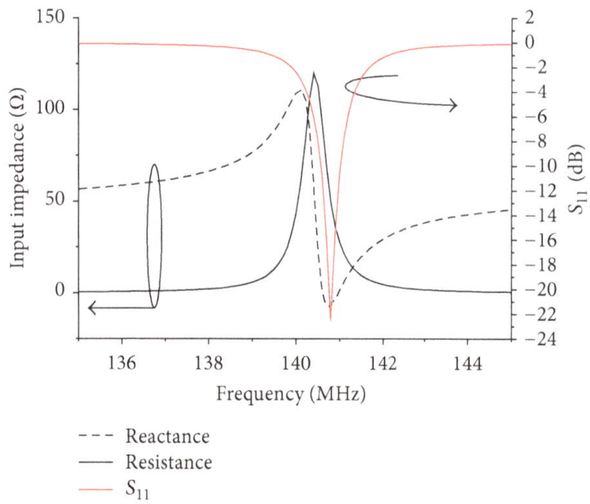

FIGURE 3: The complex input impedance response and S_{11} reflection coefficient values of proposed antenna for WPT against frequency. The reflection coefficient values are obtained using a $Z_{ch} = 50\,\Omega$ transmission line feed.

WPT is ant-resonant and matched to the feedline at about 140.8 MHz. Figure 4 presents that the radiation efficiency is 90.34% at 140.8 MHz.

3. Design Specifications for Efficient Power Transfer

3.1. Retriever-Based Scheme. In this section, a retriever-based system composed of proposed antennas shows in Figure 5(a). It consists of three antennas: a transmitter, a receiver, and a retriever. Only one power source is used to feed the transmitter of this system. The receiver is first fixed at a certain distance D from the transmitter, and then the distance D_t is optimized to achieve the maximum PTE. In this manner we can obtain the optimal D_t for maximum efficiency PTE_{mat} at each distance D of the receiver from the transmitter. The PTE_{mat} is calculated with a $50\,\Omega$ resistive load placed on port 2 of receiver. The frequency of maximum efficiency as a function of distance D_t for different values of separation D is plotted in Figure 6. It can be seen that the frequency for maximum efficiency is only affected by D when the distance D_t is long enough. This is reasonable since the effect of retriever grows fainter in the far distance. In the state D_t beyond 1.7 m, it deviates upward from the 140.8 MHz resonant frequency of an isolated antenna when D comes closely below 0.6 m. However, the frequency for maximum efficiency offsets obviously under the influence of shortened distance D_t. The transfer efficiency is calculated at the frequency shown in Figure 6 for various distance D_t and the corresponding spacing D. These results are illustrated in Figure 7. As demonstrated in this figure, the efficiency decays

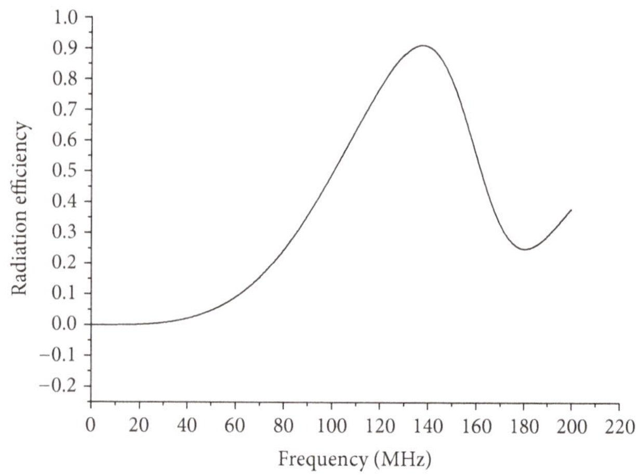

FIGURE 4: The radiation efficiency values of proposed antenna for WPT against frequency.

monotonously with the distance D between receiver and transmitter. For a certain D, the efficiency fluctuates feebly until it reaches a steady value with the incremental distance D_t and can achieve a maximum value at an optimal spacing D_t. This phenomenon is caused by phase difference between source field of transmitter and inductive field of retriever. In other words, when the fields from the transmitter arrive at the location of the receiver, differing from the phase of fields from the retriever by integral multiple of 360°, it is an in-phase condition and gives a relative maximum. Conversely, a minimum is obtained at an out-phase condition. Figure 8(a) shows the frequency and distance D_t for maximum efficiency PTE_{mat} extracted from Figures 6 and 7 against separation D. For the sake of comparison, we also calculate the efficiency of a classical WPT system made up of two proposed antennas (shown in Figure 5(b)). These results are illustrated in Figure 9. The frequency for maximum efficiency PTE_{mat} against separation D is shown in Figure 8(b). A relatively obvious improvement in the transfer efficiency can be seen in short distance D. For instance, the growth of PTE_{mat} from classical system to retriever-based scheme is 7.3% (from 75.8% to 83.1%), 5.2% (from 41.6% to 46.8%), and 1.86% (from 12.3% to 14.16%) for D at 0.2 m, 0.4 m, and 0.6 m, respectively.

3.2. Phase Difference Adjusting for WPT System under Two Transmitters. In this section, another power source is fed to the retriever in Figure 5(a). This setup, which is shown in Figure 10, consists of two transmitters and one receiver. The separation D_t between two antennas served as transmitters is fixed at 0.8 m. The following simulation results based on $V_2/V_1 = 1$ and load resistance $R_L = 50\,\Omega$ as the receiver are moved between the transmitters from 0.2 m to 0.6 m.

The operating frequency is drawn in Figure 11. From this picture, it can be seen that the frequency decreases with increasing D and varies in a small scale with dynamic phase difference for a certain D. The obtained power transmission efficiency as a function of phase difference for various distance D is shown in Figure 12. This figure shows that a relative maximum efficiency can be gained at optimum phase difference for a certain D. The amount of efficiency

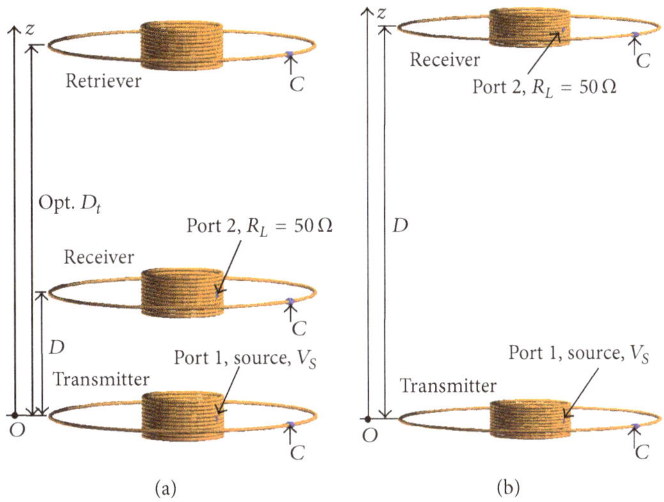

FIGURE 5: System schematic of retriever-based WPT. (a) WPT system with retriever, that is, retriever-based system. (b) WPT system without a retriever, that is, the classical WPT system.

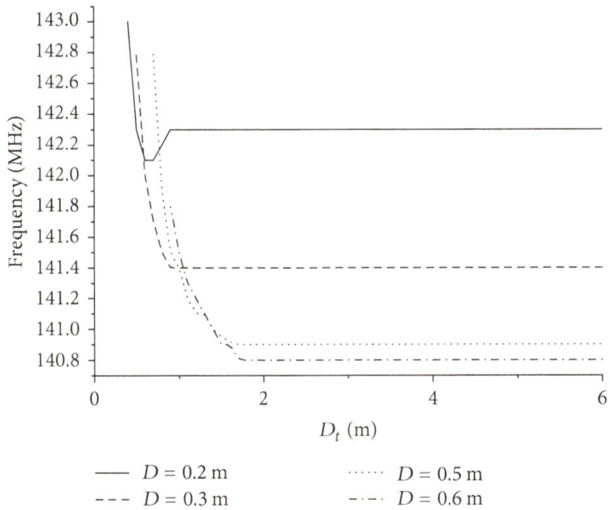

FIGURE 6: Operating frequency of PTE_{mat} against D_t for different D.

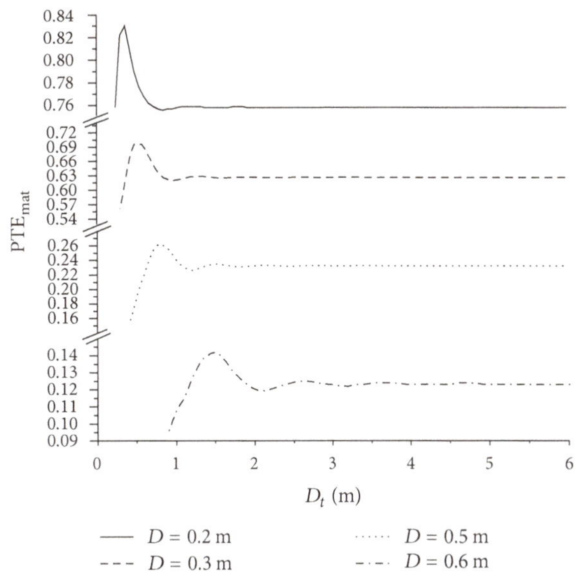

FIGURE 7: Efficiency of retriever-based WPT system at frequency shown in Figure 6 against D_t for different D.

variation changes with various phase difference by small degrees at close distance D. However, it displays a wide range of changes with various phase difference when the receiver converges towards the middle position ($D = 0.4$ m) of the two transmitters. Numerical examples show that the efficiency changes are 8.2% (from 75.7% at 162° to 67.5% at 320°) and 59.9% (from 59.9% at 0° to 0 at 180°) for D at 0.2 m and 0.4 m, respectively.

By this method, the calculated PTE_{mat} across distance D is shown in Figure 14. For the distance D from 0.2 m to 0.4 m, the PTE_{mat} result is achieved in the conditions of phase difference illustrated in Figure 12. When the receiver moves closer to transmitter 2 ($0.4\text{ m} \leq D \leq 0.6\text{ m}$), it is obtained by role reversal between the two power sources. The optimum frequency and phase difference for maximum PTE_{mat} plotted in Figure 14 is shown in Figure 13. For comparison, the efficiency of a classical WPT system is also plotted in Figure 14. From 0.2 m to 0.35 m, owing to high lossy of power transmitted from transmitter 2, the PTE_{mat} of phase difference adjusting scheme is lower than classical WPT system. Starting from approximately 0.35 m, the proposed method has an obvious advantage over the way of classical power transfer.

3.3. Voltage Ratio Adjusting for WPT System under Two Transmitters. We also use the WPT system shown in Figure 10. Two antennas served as transmitters with a separation D_t fixed at 0.8 m. The following simulation results are based on the condition of optimum phase difference investigated in the foregoing section and load resistance $R_L = 50\,\Omega$ as the receiver is moved between the transmitters from 0.2 m to 0.6 m.

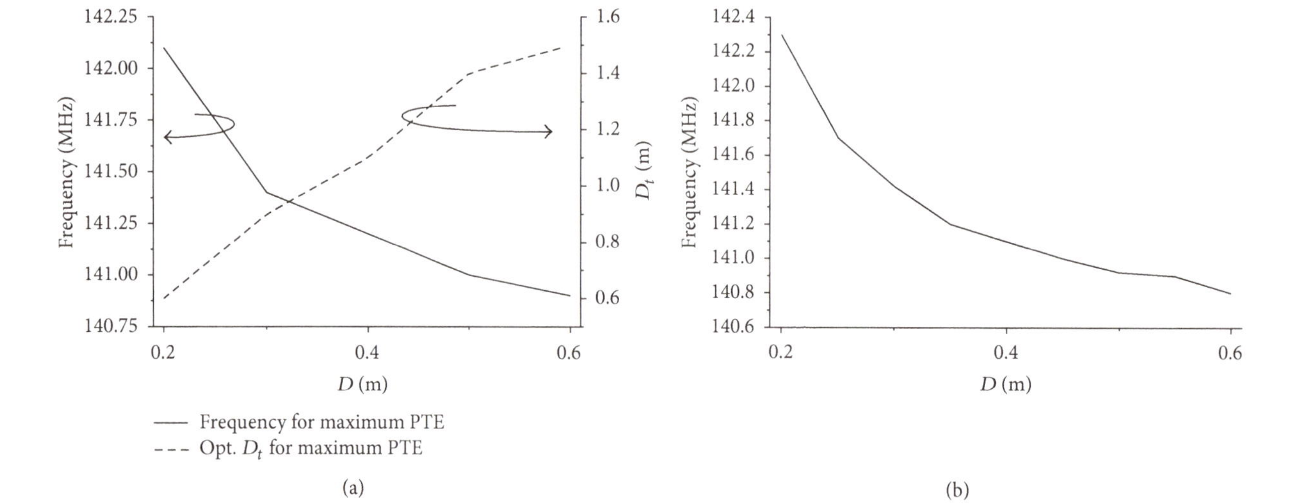

FIGURE 8: Optimum conditions for maximum efficiency PTE_{mat} against separation D (a) retriever-based system and (b) the classical WPT system.

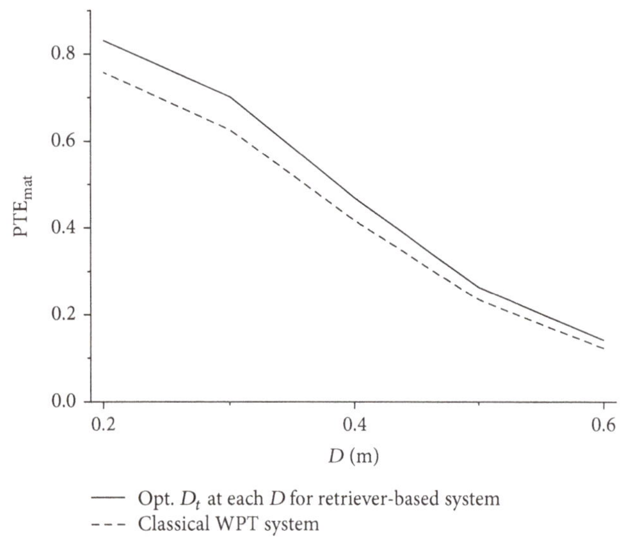

FIGURE 9: Efficiency of the considered WPT systems. Two different systems are considered: a WPT system with retriever and distance D_t optimized for each separation D (solid line); the classical WPT system shown in Figure 5(b) (dashed line).

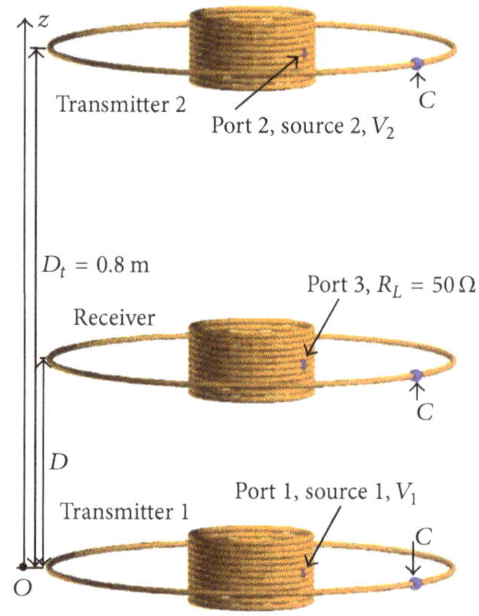

FIGURE 10: System schematic of two transmitters and one receiver.

The operating frequency of efficiency is drawn as in Figure 15. The transfer efficiency is calculated at the frequency shown in Figure 15 for various voltage ratio and the corresponding spacing D. These results are illustrated in Figure 16. This figure shows that a relative maximum efficiency can be gained at optimum voltage ratio for a certain D. The amount of efficiency variation changes with various voltage ratio by small degrees at close distance D. However, it displays a wide range of changes with various voltage ratio when the receiver converges towards the middle position ($D = 0.4$ m) of the two transmitters. Numerical examples show that the efficiency changes are 5.3% (from 80.99% at 0 to 75.7% at 1) and 20.8% (from 59.9% at 1 to 39.1% at 0) for D at 0.2 m and 0.4 m, respectively.

The calculated PTE_{mat} across distance D is shown in Figure 18. For the distance D from 0.2 m to 0.4 m, the PTE_{mat} result is achieved under the conditions of V_2/V_1, phase 2 − phase 1. When the receiver moves closely to transmitter 2 ($0.4\,\text{m} \leq D \leq 0.6\,\text{m}$), it is obtained by role reversal between the two power sources. The optimum frequency and voltage ratio, extracted from Figures 15 and 16, and phase difference for maximum PTE_{mat} are shown in Figure 17. It is observed that the frequency for maximum PTE_{mat} decreases gradually as the receiver approaches to the middle position of distance D_t between the two transmitters, whereas the voltage ratio for maximum PTE_{mat} increases progressively. PTE_{mat} of the two WPT systems mentioned above are also plotted in Figure 18. It is clear that the use of retriever, phase difference, and voltage ratio adjusting not only allows one to overcome the monotonic decay in efficiency observed in classical WPT

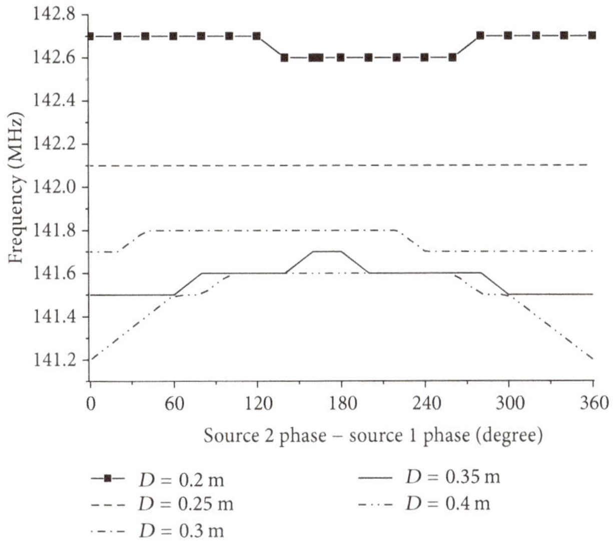

Figure 11: Operating frequency of PTE_{mat} against phase difference for different D.

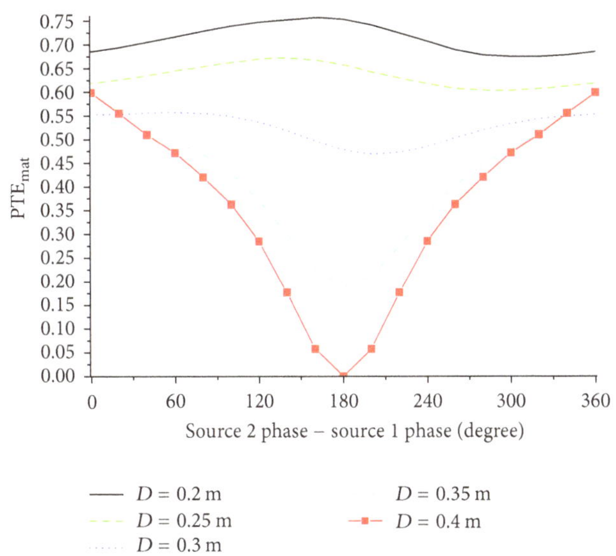

Figure 12: Efficiency of phase difference adjusting system against phase difference for different D.

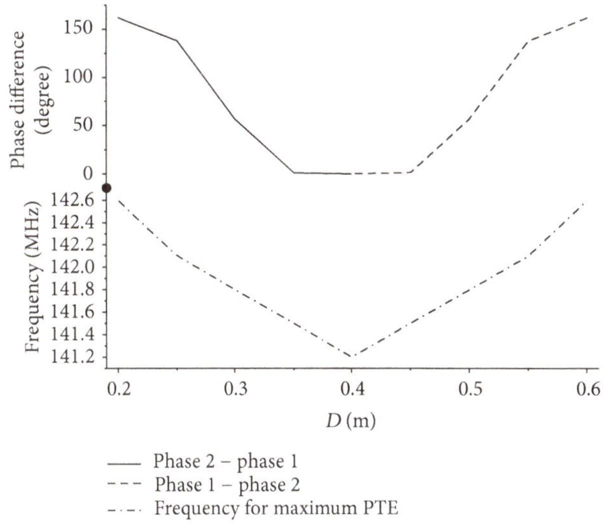

Figure 13: Optimum phase difference and frequency for maximum efficiency PTE_{mat} against separation D.

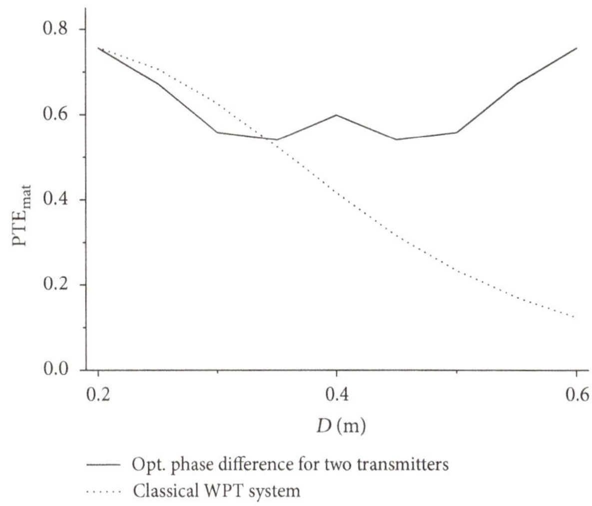

Figure 14: Efficiency of the considered WPT systems. Two different systems are considered: a WPT system with two transmitters and phase difference optimized for each separation D (solid line); the classical WPT system shown in Figure 5(b) (dotted line).

system but also increases efficiency across the whole range of separation D.

4. Integrated Utilization of the Proposed Methods for Efficient Power Transfer

In this section, two systematic frameworks for WPT based on the proposed methods are constructed. One is that phase difference and voltage ratio between the two sources are optimized to obtain maximum PTE_{mat} on the premise of optimum D_t of retriever-based system, that is, the retriever, shown in Figure 5(a), served as another transmitter, which another source fed with in this scheme. The configuration is shown in Figure 19(a). The processes operate in order as follows: (1) the optimum distances D_t for maximum PTE_{mat} are obtained in retriever-based system. (It has been done in Section 3.1). (2) Source 2 feeds to the retriever that served as transmitter 2. (3) Phase difference and voltage ratio between source 1 and source 2 are optimized to maximize the PTE_{mat}. Figure 20 shows the PTE_{mat} of this WPT scheme in solid line marked as case 3. Figure 21 shows the corresponding conditions of this scheme for maximum PTE_{mat}.

From Section 3.3, it is worthwhile to note that the maximum PTE_{mat} can be obtained under $V_2/V_1 = 1$, phase 2 − phase 1 = 0, and D in the middle of D_t. The other WPT system is therefore constructed with $D_t = 2D$, phase 2 − phase 1 = 0, and $V_2/V_1 = 1$ to obtain the maximum PTE_{mat}. The system schematic is shown in Figure 19(b). Figure 20 shows

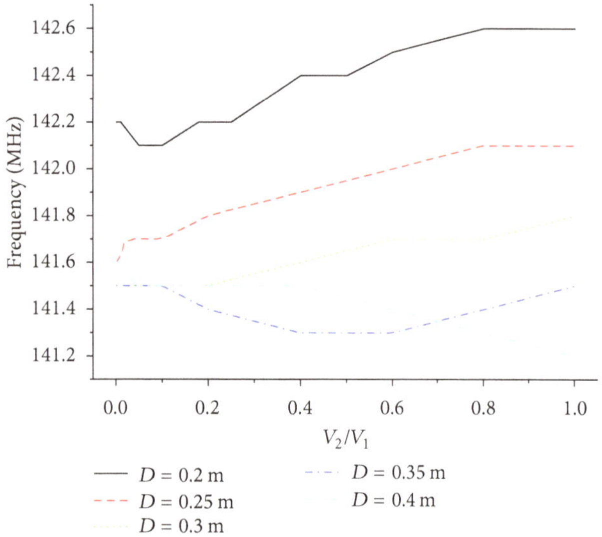

Figure 15: Operating frequency of PTE_{mat} against voltage ratio for different D.

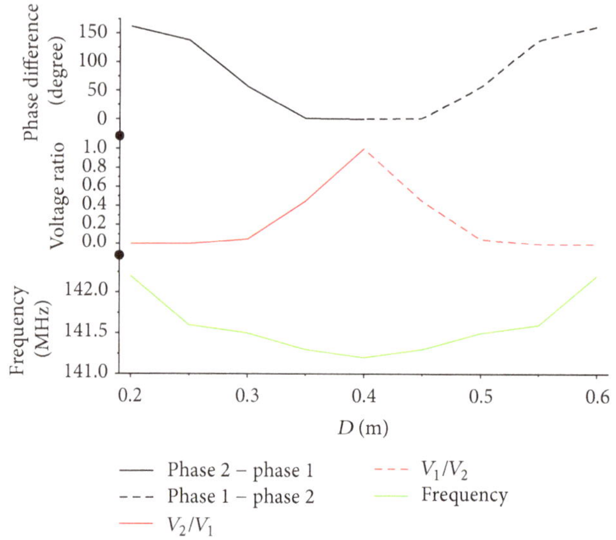

Figure 17: Optimum phase difference, voltage ratio, and frequency for maximum efficiency PTE_{mat} against separation D.

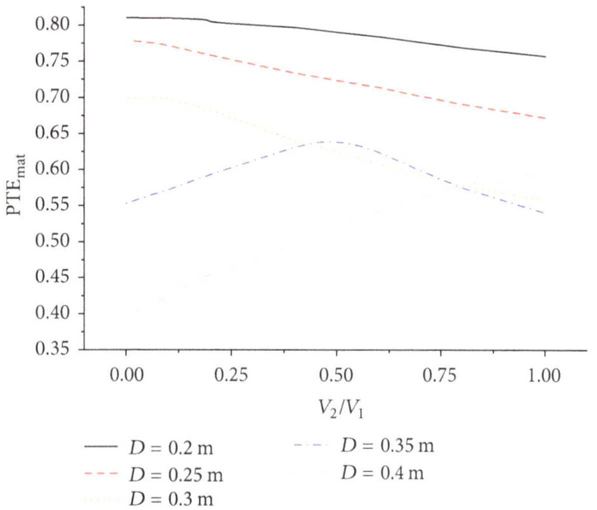

Figure 16: Efficiency of voltage ratio adjusting system at frequency shown in Figure 15 against voltage ratio for different D.

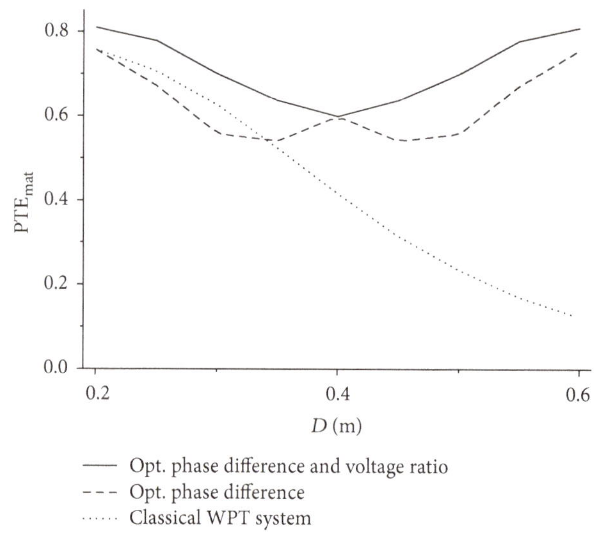

Figure 18: Efficiency of the considered WPT systems. Three different systems are considered: a WPT system with two transmitters, phase difference, and voltage ratio optimized for each separation D (solid line); a WPT system with two transmitters and phase difference optimized for each separation D (dashed line); the classical WPT system shown in Figure 5(b) (dotted line).

the PTE_{mat} of this WPT scheme in dot-dashed line marked as case 4. Figure 22 shows the corresponding operating frequency of this scheme for maximum PTE_{mat}.

For comparison purposes, the classical WPT system, retriever-based system and system with two transmitters fixed at $D_t = 0.8$ m, phase difference, and voltage ratio optimized for each separation D that have been discussed separately in the previous paragraph are also considered here. In Figure 20, the transfer efficiency of them is plotted with dotted line (marked as case 1), double-dot-dashed line (marked as case 2), and dashed line (marked as case 5), respectively.

Case 3 performs a little better than case 2 in the far distance region, which starts from approximately 0.4 m. This is because the optimum D_t is only conducive to retrieve power from transmitter 1 and unfavourable for use of power from transmitter 2. As shown in Figure 21, The power from transmitter 2 holds zero from 0.2 m to 0.3 m (actually, case 3 becomes case 2 in this situation). The proportion of power from transmitter 2 progressively increased for maximum PTE_{mat} from 0.3 m to 0.6 m. The efficiency in case 4 is more efficient than that in the first three cases and even higher than case 5 during the first half of the distance D. For the advantage of role reversal between the two power sources, case 5 performs well in the whole region.

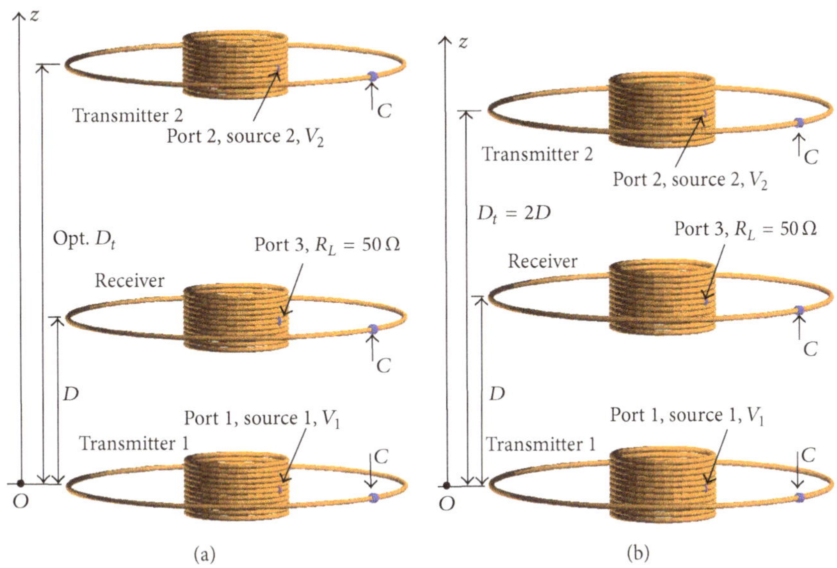

FIGURE 19: Integrated utilization of the proposed methods for WPT. (a) WPT system with two transmitters based on the optimum D_t of retriever-based system. (b) WPT system with $D_t = 2D$, phase 2 − phase 1 = 0, and $V_2/V_1 = 1$.

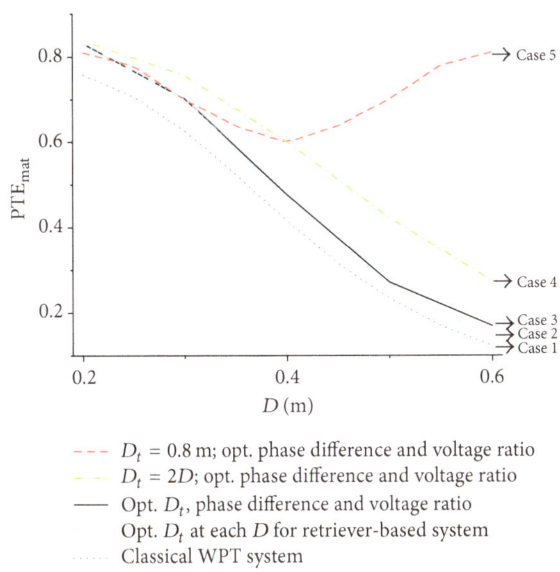

FIGURE 20: Efficiency of the considered WPT systems. Five different systems are considered: case 1: classical WPT system (single source; optimum frequency for maximum efficiency); case 2: retriever-based system (single source; optimum frequency and D_t for maximum efficiency); case 3: WPT system with two transmitters based on the optimum D_t of retriever-based system (two sources; optimum frequency, distance D_t, phase difference and voltage ratio for maximum efficiency); case 4: $D_t = 2D$ (two sources; optimum frequency, phase 2 − phase 1 = 0 and $V_2/V_1 = 1$ for maximum efficiency); case 5: $D_t = 0.8$ m (two sources; optimum frequency, phase difference, and voltage ratio for maximum efficiency).

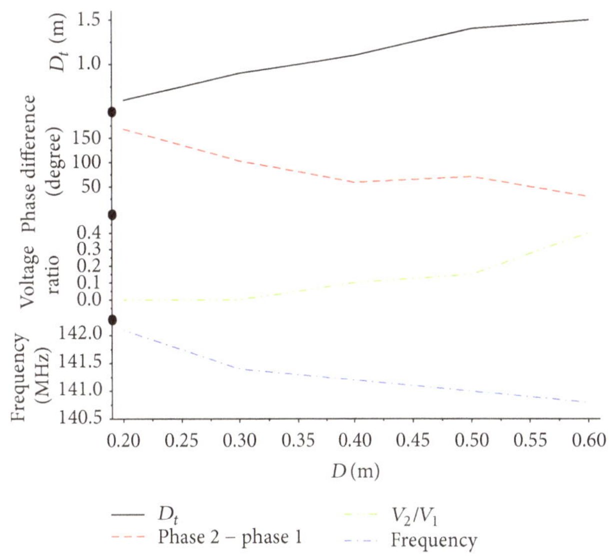

FIGURE 21: Optimum D_t, phase difference, voltage ratio, and frequency of case 3 for maximum efficiency PTE_{mat} against separation D.

5. Conclusion

This paper investigates several methods used to achieve efficient WPT over a radiating near-field region. Power transfer efficiency of the retriever-based system demonstrates some improvement compared with the classical system in the whole separation D. The performance of the WPT system with two transmitters based on the optimum D_t of retriever-based system is nearly the same as that of the retriever-based system in most region except for far separation region (from 0.5 m to 0.6 m). The proposed WPT system with two transmitters under condition of $D_t = 2D$, phase 2 − phase 1 = 0, and $V_2/V_1 = 1$ performs better than the classical and the first two proposed ones. Using phase difference, voltage ratio adjusting,and role reversal between the two transmitters, we propose a WPT system under fixed separation D_t ($D_t = 0.8$ m) to achieve a relatively high efficiency

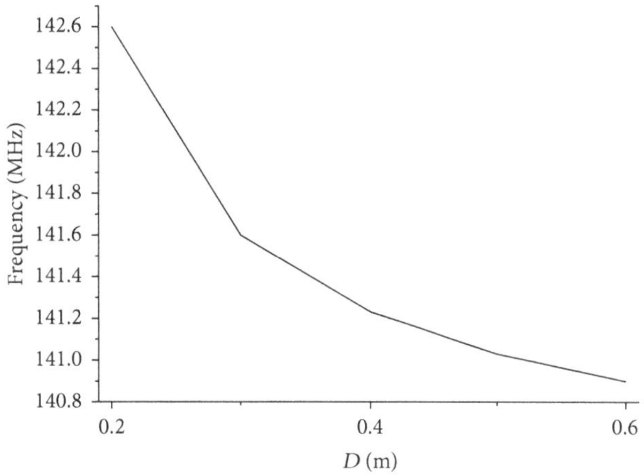

FIGURE 22: Optimum frequency of case 4 for maximum efficiency PTE_{mat} against separation D.

over the whole region and to overcome the monotonic decay of efficiency seen in classical WPT systems.

Competing Interests

The authors declare that they have no competing interests.

Acknowledgments

This research work was supported by Hefei Normal University Foundation of Anhui Province of China (2015QN12).

References

[1] N. Tesla, "The transmission of electric energy without wires," *Electrical World and Engineer*, vol. 5, p. 162, 1904.

[2] N. Tesla, *Experiments with Alternate Currents of High Potential and High Frequency: A Lecture Delivered before the Institution of Electrical Engineers*, McGraw Publishing Company, London, UK, 1904.

[3] W. C. Brown, "Adapting microwave techniques to help solve future energy problems," *IEEE Transactions on Microwave Theory and Techniques*, vol. 21, no. 12, pp. 753–763, 1973.

[4] W. C. Brown, "The history of power transmission by radio waves," *IEEE Transactions on Microwave Theory and Techniques*, vol. 32, no. 9, pp. 1230–1242, 1984.

[5] P. E. Glaser, "Power from the sun: its future," *Science*, vol. 162, no. 3856, pp. 857–861, 1968.

[6] A. Sample and J. R. Smith, "Experimental results with two wireless power transfer systems," in *Proceedings of the IEEE Radio and Wireless Symposium (RWS '09)*, pp. 16–18, IEEE, San Diego, Calif, USA, January 2009.

[7] A. Georgiadis, G. Andia, and A. Collado, "Rectenna design and optimization using reciprocity theory and harmonic balance analysis for electromagnetic (EM) energy harvesting," *IEEE Antennas and Wireless Propagation Letters*, vol. 9, pp. 444–446, 2010.

[8] F.-J. Huang, T.-C. Yo, C.-M. Lee, and C.-H. Luo, "Design of circular polarization antenna with harmonic suppression for rectenna application," *IEEE Antennas and Wireless Propagation Letters*, vol. 11, pp. 592–595, 2012.

[9] A. Kurs, A. Karalis, R. Moffatt, J. D. Joannopoulos, P. Fisher, and M. Soljačić, "Wireless power transfer via strongly coupled magnetic resonances," *Science*, vol. 317, no. 5834, pp. 83–86, 2007.

[10] A. Karalis, J. D. Joannopoulos, and M. Soljačić, "Efficient wireless non-radiative mid-range energy transfer," *Annals of Physics*, vol. 323, no. 1, pp. 34–48, 2008.

[11] Y. Kim and H. Ling, "Investigation of coupled mode behaviour of electrically small meander antennas," *Electronics Letters*, vol. 43, no. 23, pp. 1250–1252, 2007.

[12] Y. Tak, J. Park, and S. Nam, "Mode-based analysis of resonant characteristics for near-field coupled small antennas," *IEEE Antennas and Wireless Propagation Letters*, vol. 8, pp. 1238–1241, 2009.

[13] H. Hirayama, T. Ozawa, Y. Hiraiwa, N. Kikuma, and K. Sakakibara, "A consideration of electro-magnetic-resonant coupling mode in wireless power transmission," *IEICE Electronics Express*, vol. 6, no. 19, pp. 1421–1425, 2009.

[14] N. Inagaki and S. Hori, "Classification and characterization of wireless power transfer systems of resonance method based on equivalent circuit derived from even- and odd mode reactance functions," in *Proceedings of the 2011 1st IEEE MTT-S International Microwave Workshop Series on Innovative Wireless Power Transmission: Technologies, Systems, and Applications (IMWS-IWPT '11)*, pp. 115–118, IEEE, Uji, Japan, May 2011.

[15] W. Zhong, C. K. Lee, and S. Y. Ron Hui, "General analysis on the use of tesla's resonators in domino forms for wireless power transfer," *IEEE Transactions on Industrial Electronics*, vol. 60, no. 1, pp. 261–270, 2013.

[16] J. Moshfegh, M. Shahabadi, and J. Rashed-Mohassel, "Conditions of maximum efficiency for wireless power transfer between two helical wires," *IET Microwaves, Antennas and Propagation*, vol. 5, no. 5, pp. 545–550, 2011.

[17] J. Park, Y. Tak, Y. Kim, Y. Kim, and S. Nam, "Investigation of adaptive matching methods for near-field wireless power transfer," *IEEE Transactions on Antennas and Propagation*, vol. 59, no. 5, pp. 1769–1773, 2011.

[18] W. Fu, B. Zhang, and D. Qiu, "Study on frequency-tracking wireless power transfer system by resonant coupling," in *Proceedings of the IEEE 6th International Power Electronics and Motion Control Conference (IPEMC '09)*, pp. 2658–2663, Wuhan, China, May 2009.

[19] C.-S. Wang, G. A. Covic, and O. H. Stielau, "Power Transfer capability and bifurcation phenomena of loosely coupled inductive power transfer systems," *IEEE Transactions on Industrial Electronics*, vol. 51, no. 1, pp. 148–157, 2004.

[20] C.-S. Wang, G. A. Covic, and O. H. Stielau, "Investigating an LCL load resonant inverter for inductive power transfer applications," *IEEE Transactions on Power Electronics*, vol. 19, no. 4, pp. 995–1002, 2004.

[21] J. Boys, G. Covic, and A. W. Green, "Stability and control of inductively coupled power transfer systems," *IEE Proceedings-Electric Power Applications*, vol. 147, no. 1, pp. 37–43, 2000.

[22] O. Stielau and G. Covic, "Design of loosely coupled inductive power transfer systems," in *Proceedings of the International Conference on Power System Technology (POWERCON '00)*, pp. 85–90, Perth, Wash, Australia, 2000.

[23] C.-S. Wang, O. H. Stielau, and G. A. Covic, "Design considerations for a contactless electric vehicle battery charger," *IEEE*

Transactions on Industrial Electronics, vol. 52, no. 5, pp. 1308–1314, 2005.

[24] J. L. Villa, J. Sallán, A. Llombart, and J. F. Sanz, "Design of a high frequency inductively coupled power transfer system for electric vehicle battery charge," *Applied Energy*, vol. 86, no. 3, pp. 355–363, 2009.

[25] Y. P. Su, X. Liu, and S. Y. R. Hui, "Mutual inductance calculation of movable planar coils on parallel surfaces," *IEEE Transactions on Power Electronics*, vol. 24, no. 4, pp. 1115–1123, 2009.

[26] W. X. Zhong, X. Liu, and S. Y. R. Hui, "A novel single-layer winding array and receiver coil structure for contactless battery charging systems with free-positioning and localized charging features," *IEEE Transactions on Industrial Electronics*, vol. 58, no. 9, pp. 4136–4144, 2011.

[27] I.-J. Yoon and H. Ling, "Realizing efficient wireless power transfer using small folded cylindrical helix dipoles," *IEEE Antennas and Wireless Propagation Letters*, vol. 9, pp. 846–849, 2010.

[28] I.-J. Yoon and H. Ling, "Investigation of near-field wireless power transfer under multiple transmitters," *IEEE Antennas and Wireless Propagation Letters*, vol. 10, pp. 662–665, 2011.

[29] I.-J. Yoon and H. Ling, "Investigation of near-field wireless power transfer in the presence of lossy dielectric materials," *IEEE Transactions on Antennas and Propagation*, vol. 61, no. 1, pp. 482–488, 2013.

[30] J. Lee and S. Nam, "Fundamental aspects of near-field coupling small antennas for wireless power transfer," *IEEE Transactions on Antennas and Propagation*, vol. 58, no. 11, pp. 3442–3449, 2010.

[31] R. W. Ziolkowski and A. Erentok, "Metamaterial-based efficient electrically small antennas," *IEEE Transactions on Antennas and Propagation*, vol. 54, no. 7, pp. 2113–2130, 2006.

[32] A. Erentok and R. W. Ziolkowski, "Metamaterial-inspired efficient electrically small antennas," *IEEE Transactions on Antennas and Propagation*, vol. 56, no. 3, pp. 691–707, 2008.

[33] A. Erentok and R. W. Ziolkowski, "An efficient metamaterial-inspired electrically-small antenna," *Microwave and Optical Technology Letters*, vol. 49, no. 6, pp. 1287–1290, 2007.

6

Random Harmonic Detection and Compensation based on Synchronous Reference Frame

Yanbo Che, Zhaojing Yin, Shuyan Yu, and Qiang Sun

Key Laboratory of Smart Grid of Ministry of Education, Tianjin University, Tianjin 300072, China

Correspondence should be addressed to Yanbo Che; ybche@tju.edu.cn

Academic Editor: Ephraim Suhir

Algorithms for harmonic detection and compensation are important guarantees for an active power filter (APF) to achieve the harmonic control function and directly determine the overall performance. Existing algorithms usually need a large amount of computation, and the compensation effect of specified order harmonic is also limited. DC side capacitor voltage at sudden change of load is affected by the algorithm as well. This paper proposes a new algorithm for harmonic detection and compensation based on synchronous reference frame (SRF), in which a band-pass filter with center frequency of $6k$th harmonic is designed in fundamental frequency SRF to extract random harmonic current with two different frequencies of $(6k \pm 1)$th harmonic in stationary reference frame. This new algorithm can rapidly detect any specified harmonic, and it can adjust the power factor to compensate reactive power. Meanwhile, it has few impacts on DC side capacitor voltage under complicated operating conditions such as sudden change of load. The correctness and effectiveness of this new algorithm are verified by simulation and experiment.

1. Introduction

With the large-scale application of power electronics devices in recent years, the amount of various nonlinear loads has been increased rapidly. Current waveform distortion and harmonic problem is becoming more and more serious; hence the research and application of harmonic control devices have become a hot research topic at home and abroad [1–3]. At present, there are two main types of filter, passive power filter (PPF) and active power filter (APF). PPF removes harmonic or prevents harmonic spreading by adding extra impedance branches to a system. PPF is easy to be resonant with the power grid, and its compensation characteristic is easy to be affected by the impedance of the power network. The compensation accuracy is low and the flexibility of its application is poor. These drawbacks limit its application in some occasions with high requirements [4]. Adopting high-performance main circuit of power electronic converter and control system, APF can quickly detect changes in harmonic currents, then compensate harmonic, and adjust power quality according to different compensation targets. Being better than PPF in both compensation precision and flexibility of harmonic current detection, APF has gradually got more attention and research [5–8].

The detection and compensation performance of an APF is mainly related to the extraction algorithm of the system compensation current and the control strategy of the main power inverter. The performance of a compensation current extraction algorithm largely depends on the accuracy, fastness, and real-time of the current extraction. The key to controlling a main power inverter is to ensure a fast tracking of instruction current by APF output current and keep their consistency. Therefore, it is an important issue improving the accuracy of APF harmonic compensation to catch the instantaneous harmonic and reactive components from the power grid currents quickly and accurately. Meanwhile, with the rapid development of large-scale integrated circuit, especially digital signal processor technology, the technology of extracting harmonic and reactive current in APF system has gradually developed from analog to digital form, which greatly improves the speed of current extraction and the accuracy of compensation [9]. However, the traditional

TABLE 1: Comparison of different methods.

	Filter	Wavelet Transform	Neural Network	FFT	PQ	SRF
Computation	A	D	C	E	B	B
Response speed	E	D	B	C	B	A
Reliability	E	D	C	D	B	A
Compensation precision	E	D	C	A	B	A
Ability of random harmonic detection and compensation	No	No	No	Yes	No	No

harmonic detection and compensation methods which detect and compensate the content of harmonic cannot meet the functional requirements on variety to APF [10–12]. It is urgent to search a method that could detect the random harmonic current or voltage in load and then compensate them according to actual demands.

At present, the main methods for harmonic detection and compensation are classified as Filter method [13], Wavelet Transform method [14], Neural Network method [15, 16], FFT method [17, 18], PQ method [19], and SRF method [20]. The authors have deeply studied these methods by reading a large number of literatures [21–27] and evaluated them from computation, response speed, reliability, and compensation precision. As shown in Table 1, A to E stand for the grades from highest to lowest, and their abilities of random harmonic detection and compensation are also compared. This table indicates that the Filter method, Wavelet Transform method, Neural Network method, and FFT method only are suitable for some special occasions due to their shortcomings. Although the PQ method and the traditional SRF method, which are based on Instantaneous Reactive Power Theory [28], show good performance in all aspects and are widely used, they lack the function of detecting and compensating for random harmonics.

Different from traditional harmonic detection algorithm, random harmonic detection algorithm can detect and compensate specified order harmonic, improve processing speed, reduce work load, and thus get attention by scholars. Taking a three-phase three-wire parallel APF as research representative, this paper proposes a new SRF-based algorithm for random harmonic current detection and compensation. It can detect and compensate the specified harmonic, reduce the computation amounts, and have few impacts on DC side capacitor voltage when the loads mutate. Finally, simulation and experiment show the effectiveness and feasibility of this algorithm.

2. Basic Principle

2.1. Basic Principle of APF. The concept of APF was first proposed by Japanese scientists in the early 70s. The basic principle of APF is to generate a current that has the same amplitude and opposite phase with harmonic current by controllable power semiconductor devices and inject it into the grid, intentionally to cancel the total harmonic current and compensate in real-time. According to different applications, the active power filter can be divided into two categories: DC active power filter and AC active power filter.

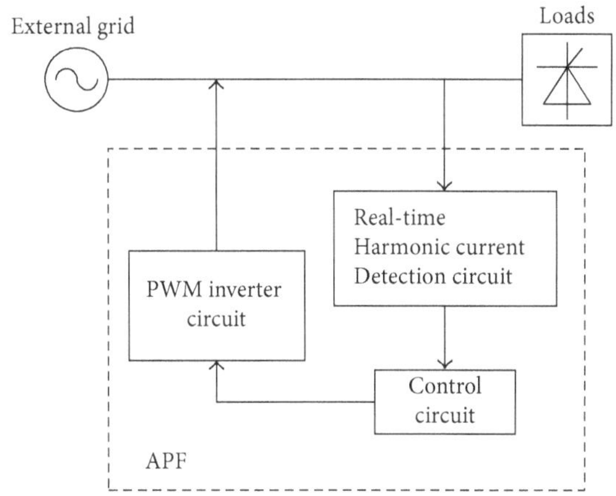

FIGURE 1: Schematic diagram of APF.

The former is mainly applied to HVDC system, and the latter (often cited as APF will be discussed in this paper) to AC power system of different voltage levels.

An APF usually consists of two functional parts, detecting harmonic signal and generating compensating component. As shown in Figure 1, distorted proportion of the current and voltage in grid are detected by the detection circuit of APF. Corresponding compensation current component is then generated from the power circuit under certain control strategy and injected into the grid to achieve the purpose of harmonic elimination.

2.2. Basic Idea of SRF-Based Harmonic Detection Algorithm. The SRF-based algorithm works by changing a fundamental positive sequence current into direct current by Park Transform, then filtering out the harmonic and reactive current through a low-pass filter (LPF), and finally getting the compensation current reference value [16], as shown in Figure 2.

Current in nonlinear load such as common rectifier bridge consists of the fundamental and $(6n \pm 1)$th harmonic components, respectively, corresponding to direct current (DC) and alternating current (AC) components in stationary reference frame when converted into fundamental frequency SRF. A LPF precisely designed in fundamental frequency SRF can greatly improve the percentage of fundamental component, and then the harmonic component can be obtained by subtracting the fundamental harmonic from the load waveform.

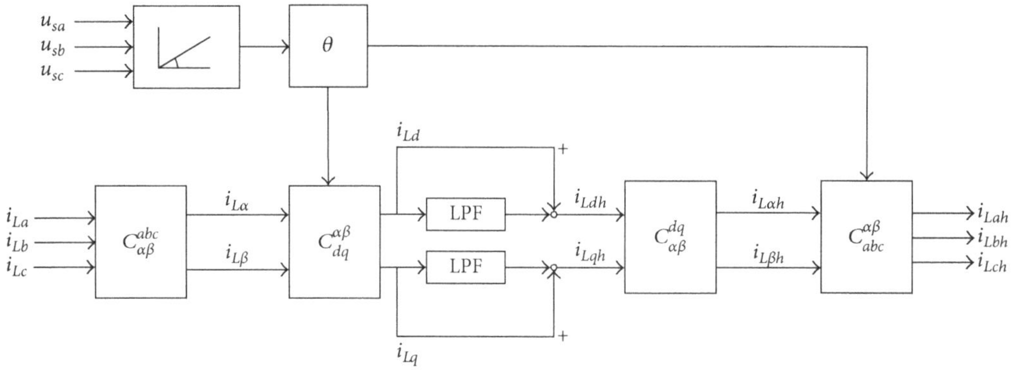

FIGURE 2: The basic principle of SRF.

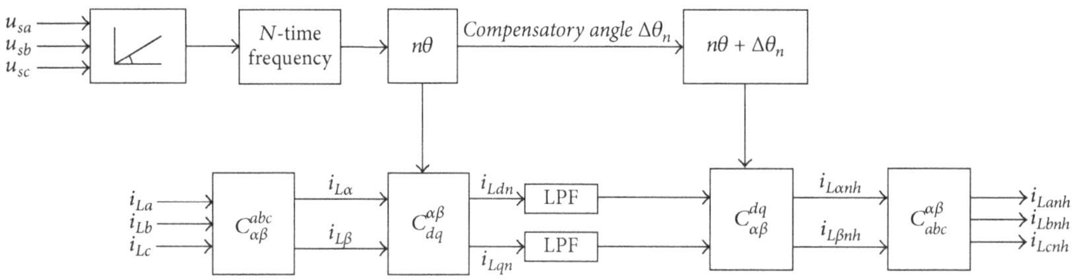

FIGURE 3: The basic principle of SRF with compensatory angle.

3. SRF-Based Random Harmonic Detection and Compensation

3.1. Stability of DC Side Capacitor Voltage. Stabling the DC side capacitor voltage is an important premise to effectively compensate the harmonic currents generated by load. The basic idea (of stabling the DC side capacitor voltage) is to absorb the fundamental active power in the grid by treating the active power filter as the current source device. The low-pass filter in the traditional harmonic detection algorithm is carried out in fundamental frequency SRF. When the load changes suddenly, the time delay during hall current detection of load side is usually neglected. There is a time delay in the harmonic compensating reference calculation process of LPF to extract the current, so the LPF is unable to trace the changes in input. If the load changes suddenly during the process of harmonics compensating, although the APF can instantly sense this change, some fundamental components will still be remained in harmonic detection algorithm. So for the time period from the beginning of mutation in load to the end of time delay in LPF, APF actually absorbs active power from the grid. These active components raise the DC side capacitor voltage, and stronger load current or change causes more obvious fluctuation in the DC side capacitor voltage.

The harmonic compensating reference in the random harmonic detection algorithm does not come from the result of subtracting the LPF output from the load current, so the compensating reference, when load changes do not contain fundamental component that largely fluctuate the DC side capacitor voltage, and the voltage can be easily maintained.

3.2. Random Harmonic Detection and Compensation. As stated above, when the load suddenly changes, the influence from random harmonic detection and compensation device on the DC side capacitor voltage is much smaller than that from a full compensation shown in Figure 2. The random harmonic detection method with preset compensatory angle is commonly used and given in Figure 3.

The application of the random harmonic detection algorithm in Figure 3, however, is limited by large computation. This algorithm will be updated in this paper to reduce computation load and realize quick random detection and compensation of harmonics. The updated algorithm can also ensure the stability of DC side capacitor voltage.

In two-phase stationary reference frame, the vector of the load current is composed of the vectors of fundamental and harmonic currents. The $(6n + 1)$th harmonic current vectors are in the same direction with the fundamental current vector and are defined as positive sequence components; correspondingly, the $(6n - 1)$th harmonic current vectors are in the opposite direction to the fundamental current vector and are defined as negative sequence components. The rotational angular velocity of a pair of adjacent harmonic current of positive and negative sequences relative to the fundamental vector is $6n\omega$ in two-phase stationary reference frame; therefore, the fundamental, positive, and negative sequences of the load currents under a stationary reference frame can be, respectively, described as follows:

$$i_{L\alpha 1} = \sqrt{\frac{3}{2}} I_1 \sin(\omega t),$$
$$i_{L\beta} = -\frac{\sqrt{3}}{2} I_1 \cos(\omega t),$$

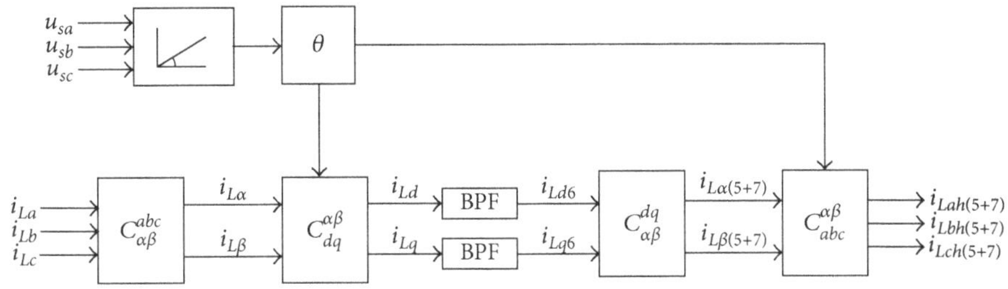

FIGURE 4: The schematic diagram of new SRF-based algorithm for harmonic detection.

$$
\begin{aligned}
i_{L\alpha h}^{+} &= \sqrt{\frac{3}{2}} I_n \sin(n\omega t), \quad n = 6k+1, \ k = 1,2,\ldots, \\
i_{L\beta h}^{+} &= -\frac{\sqrt{3}}{2} I_n \cos(n\omega t), \quad n = 6k+1, \ k = 1,2,\ldots, \\
i_{L\alpha h}^{-} &= \sqrt{\frac{3}{2}} I_n \sin(n\omega t), \quad n = 6k-1, \ k = 1,2,\ldots, \\
i_{L\beta h}^{-} &= \frac{\sqrt{3}}{2} I_n \cos(n\omega t), \quad n = 6k-1, \ k = 1,2,\ldots.
\end{aligned}
\tag{1}
$$

Transfer them with the expression in (2), and then fundamental and harmonic components in fundamental frequency SRF can be obtained, as shown in (3):

$$
\begin{bmatrix} i_{Ld} \\ i_{Lq} \end{bmatrix} = C_{dq}^{\alpha\beta} \begin{bmatrix} i_{L\alpha} \\ i_{L\beta} \end{bmatrix} = \begin{bmatrix} \cos\theta & \sin\theta \\ -\sin\theta & \cos\theta \end{bmatrix} \begin{bmatrix} i_{L\alpha} \\ i_{L\beta} \end{bmatrix}, \tag{2}
$$

$$
\begin{aligned}
i_{Ld} &= \sqrt{\frac{3}{2}} \left[I_1 + \sum_{k=1,2,\ldots} \left(-I_{6k-1} + I_{6k+1}\right) \cos(6k\omega t) \right], \\
i_{Lq} &= \sqrt{\frac{3}{2}} \left[I_1 + \sum_{k=1,2,\ldots} \left(I_{6k-1} + I_{6k+1}\right) \sin(6k\omega t) \right],
\end{aligned}
\tag{3}
$$

where I_1 is the amplitude value of the fundamental current and I_n of the nth harmonic current. The physical meaning of (3) is that $(6k \pm 1)$th harmonics in stationary reference frame turn into $6k$th harmonic in fundamental frequency SRF.

The sum of adjacent positive and negative sequence harmonic current vectors in stationary reference frame is transferred into that of harmonic current vectors with the same rotational speed in fundamental frequency SRF. On the basis of SRF, two different harmonics $(6k \pm 1)$ in stationary reference frame can be abstracted through a pair of band-pass filter (BPF) $(6k)$ in fundamental frequency SRF. For example, the schematic diagram detecting 5th harmonic negative and 7th harmonic positive sequence components of a rectifier load is shown in Figure 4, in which the center frequency of second-order BPF is 300 Hz.

Extended from the SRF algorithm to detect harmonic, this new algorithm is able to control the power factor as well, so the reactive power compensation can be achieved. Meanwhile, the number of second-order BPFs and the computation of CPU are also reduced.

TABLE 2: Simulation parameters of all digital harmonic devices.

Parameters	Value
Grid voltage	0.4 KV
DC side voltage	0.9 KV
DC side capacitor voltage	6800 μF
Output impedance	0.6 mH, 0.01 Ω
Load resistance	5 Ω

4. Simulation Analysis

4.1. Random Harmonic Detection and Compensation. From the above analysis, it is possible to detect harmonics by designing multigroup of BPFs with center frequencies of $6k$, $k = 1, 2, 3, \ldots$ in the fundamental frequency SRF. Each group of BPF can extract positive and negative harmonic sequence harmonics with different frequency at the same time.

To verify the correctness of the proposed algorithm, a model of harmonic suppression is built in MATLAB (Simulink), of which simulation parameters are shown in Table 2. Current waveform of the rectifier bridge with pure resistive load is shown in Figure 5.

The d, q components of harmonic current with the order less than 31 are shown in Figures 6 and 7, respectively. The compensated current and its spectrum are given in Figure 8.

According to the current waveform and harmonic distortion rate after compensation, harmonics of all frequency are well compensated by the APF with the proposed method.

4.2. Waveform Comparison under Sudden Change of Load. For sudden changes in load, simulate and compare the effects of full compensation and random compensation. The nonlinear load consists of a bridge rectifier and a pure resistance, and the load current is suddenly doubled at 0.6 s and then halved at 0.75 s, as shown in Figure 9.

Figures 10 and 11 show the corresponded grid currents for full compensation and random compensation, respectively. Figures 12 and 13 are waveforms for DC side capacitor voltages under these two kinds of compensations.

These simulation results indicate that the SRF-based algorithm for harmonic detection and compensation proposed in

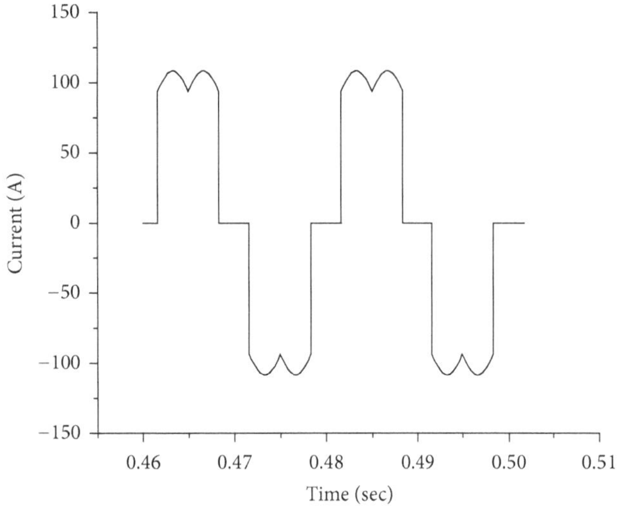

Figure 5: Current waveform of the rectifier bridge with pure resistive load.

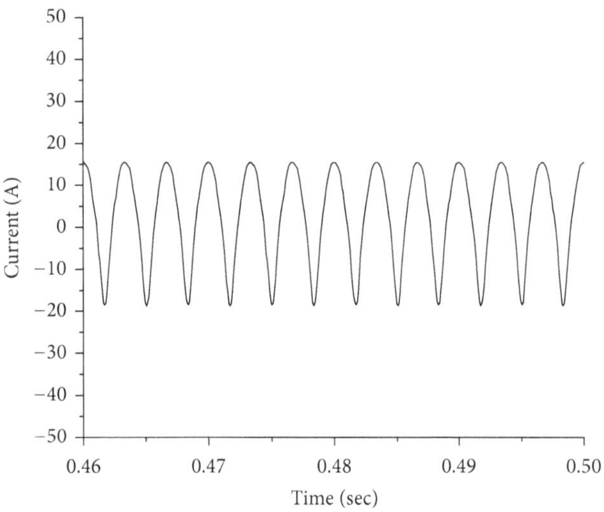

Figure 6: The d component of harmonic current.

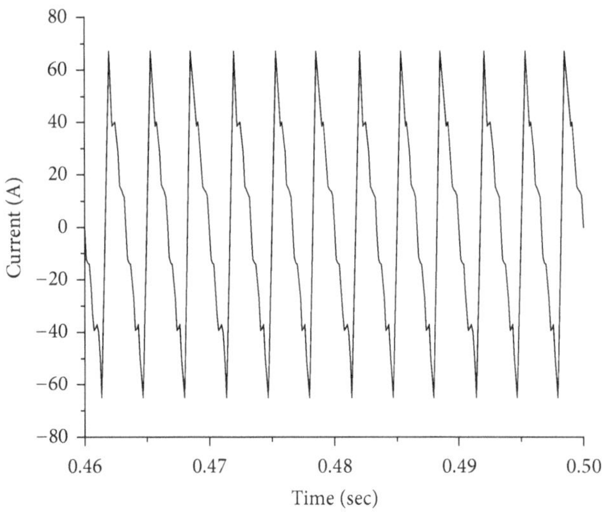

Figure 7: The q component of harmonic current.

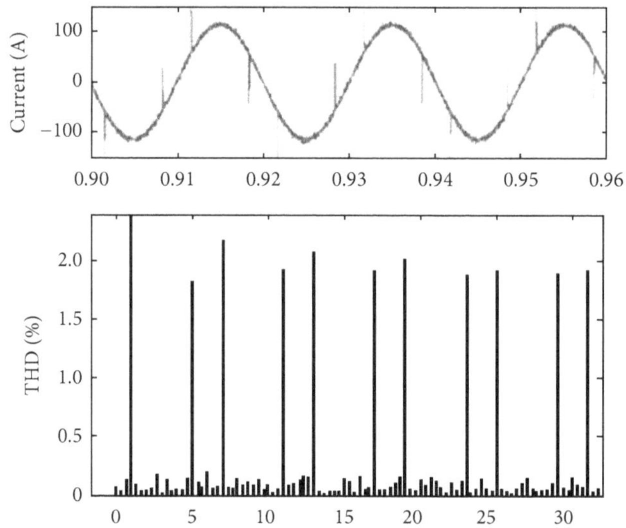

Figure 8: Compensated current waveform and spectral waveform.

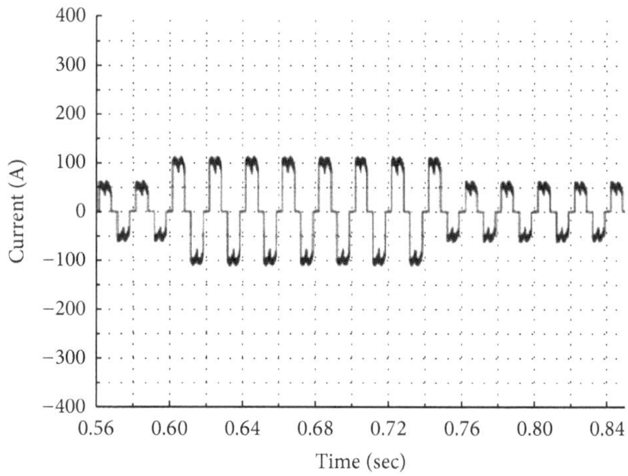

Figure 9: The load current waveform.

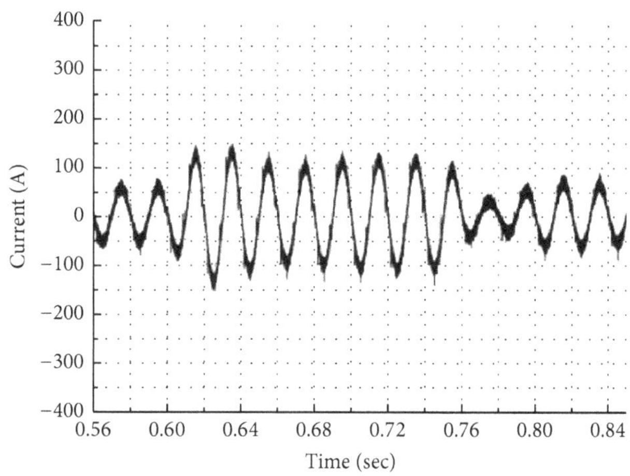

Figure 10: The grid current waveform under full compensation.

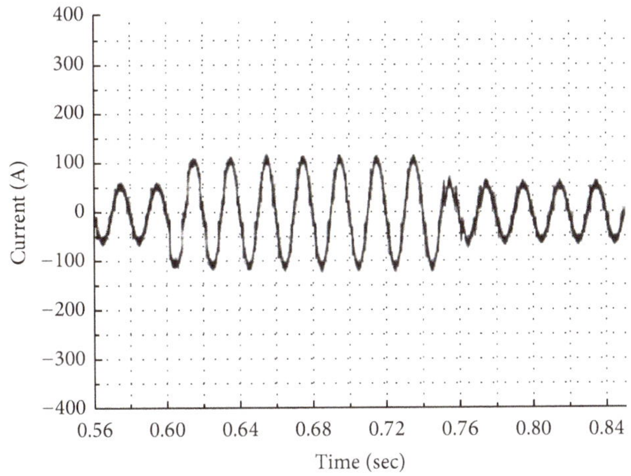

FIGURE 11: The grid current waveform under random compensation.

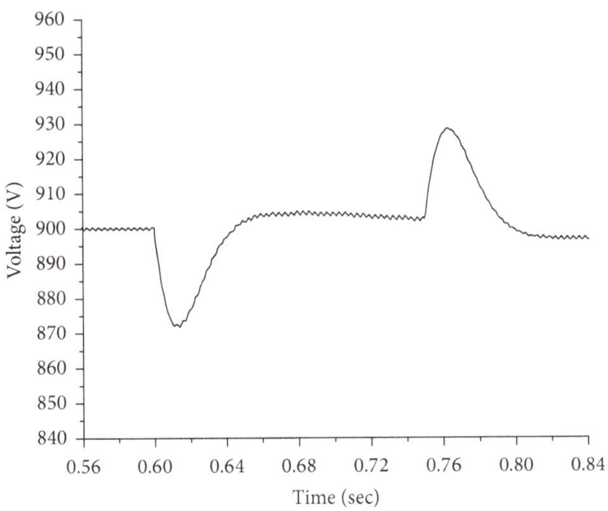

FIGURE 12: The capacitor voltage waveform under full compensation.

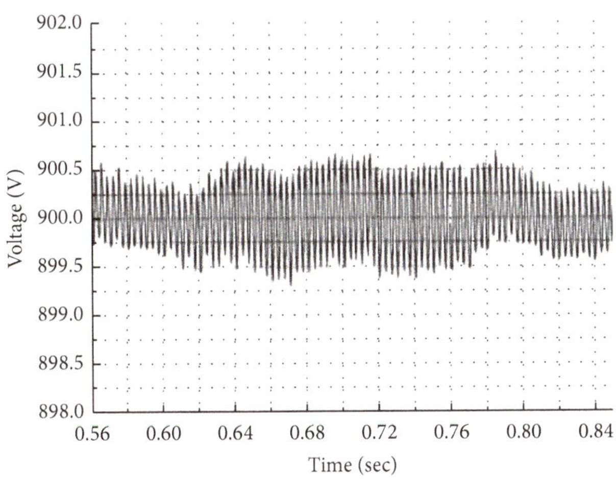

FIGURE 13: The capacitor voltage waveform under random compensation.

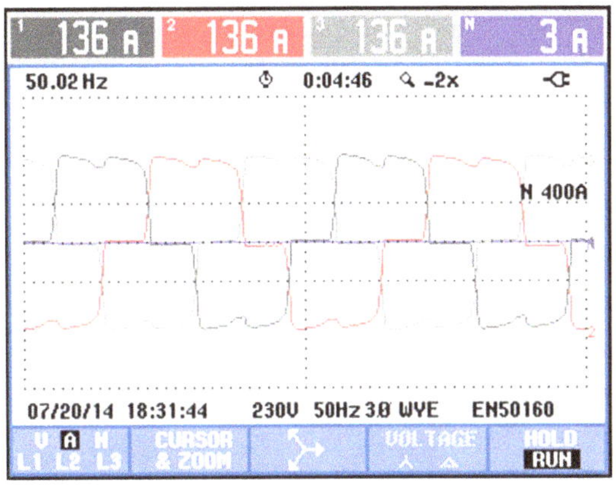

FIGURE 14: Current fluctuation curve before compensation.

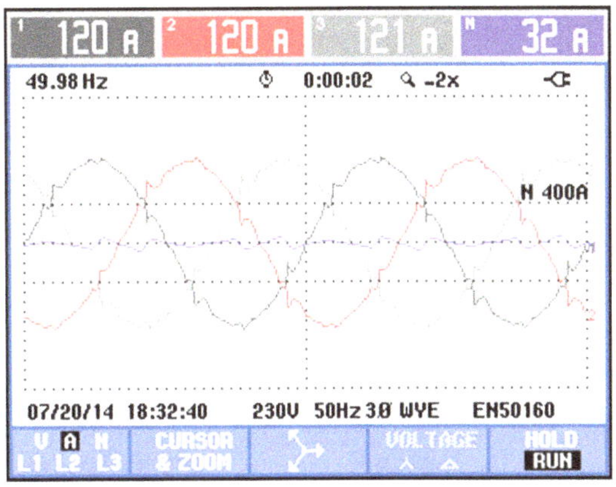

FIGURE 15: Current fluctuation curve after compensation.

this paper has good filtering effect to the harmonic current generated by simulation. When the load current changes suddenly, the random harmonic compensation algorithm does not import fundamental components, so it will not affect the DC side capacitor voltage.

5. Experimental Verification

The algorithm of random harmonic detection and compensation proposed in this paper is employed in a full digital controller for APF to verify its effect. The load is composed of a three-phase control rectifier and a pure resistance set to 3 Ω. A power quality analyzer of model FLUKE 434 was selected as measurement instrument. The current waveforms of three-phase grid before and after compensation are shown in Figures 14 and 15, respectively, and their harmonic contents are shown in Figures 16 and 17.

The heavily distorted current is restored to normal waveform after harmonic detection and compensation, and the effect of active power filter with our algorithm is significant. The experimental results show that the SRF-based method for harmonic detection and compensation can be well used to detect and compensate harmonics of specified order, and the DC side capacitor voltage can also be stable even when the

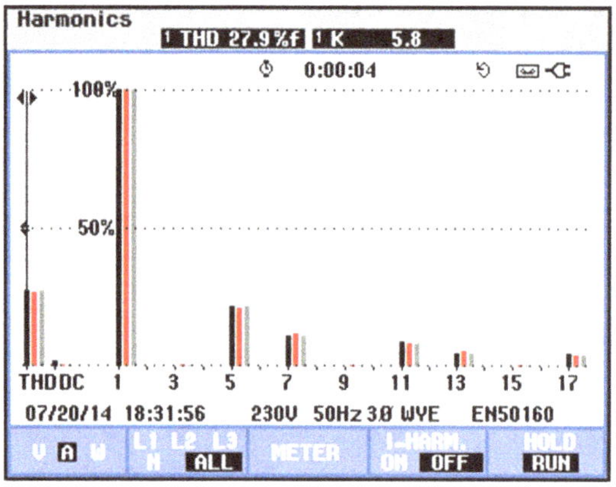

FIGURE 16: Current harmonic contents before compensation.

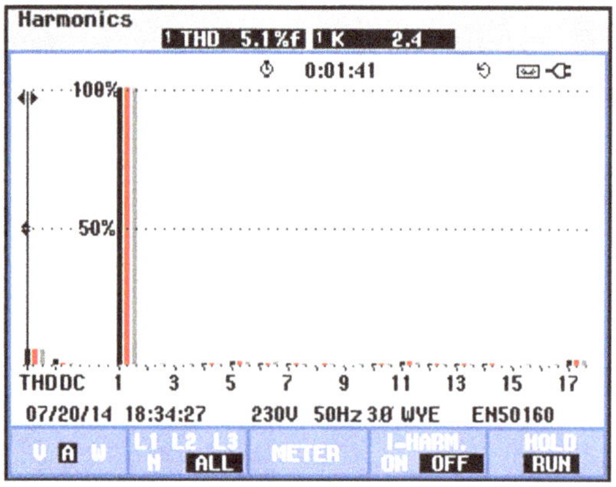

FIGURE 17: Current harmonic contents after compensation.

load changes suddenly. This is the key why the APF can filter efficiently.

6. Conclusion

The paper proposes a new SRF-based algorithm for random harmonic detection and compensation, and its mathematical principle is also deduced. Simulation and experiment show that the method is correct and practical. The conclusion can be drawn as follows.

Compared with traditional SRF harmonic detection algorithms, the computation is significantly reduced. When this new algorithm is applied to APF system, the operation burden of the digital controller can be reduced effectively and the control speed of the overall system is improved.

Avoiding obtaining the harmonic command from the subtraction of load current and output of the low-pass filter, this new algorithm does not affect the fluctuation of DC side capacitor voltage.

Conflicts of Interest

The authors declare that they have no conflicts of interest.

Acknowledgments

The authors would like to acknowledge financial support from National Key Research and Development Plan (no. 2016YFB0900204).

References

[1] B. Singh, S. Gairola, B. N. Singh, A. Chandra, and K. Al-Haddad, "Multipulse ac-dc converters for improving power quality: a review," *IEEE Transactions on Power Electronics*, vol. 23, no. 1, pp. 260–281, 2008.

[2] K. Z. Shuai, L. An, W. Zhu J et al., "Study on complex control of injection type hybrid active power filter," *Proceedings of the Chinese Society of Electrical Engineering*, vol. 28, no. 15, pp. 84–91, 2008.

[3] P. Dang, T. Ellinger, and J. Petzoldt, "Dynamic interaction analysis of APF systems," *IEEE Transactions on Industrial Electronics*, vol. 61, no. 9, pp. 4467–4473, 2014.

[4] L. an, *Harmonic Control and Reactive Power Compensation Technology and Equipment*, China power press, Beijing, China, 2006.

[5] X. Xia, A. Luo, and S. Luo, "Application of high-capacity active power filter with single injection circuit," *Electric Power Automation Equipment*, vol. 30, no. 4, pp. 115–118, 2010.

[6] M. S. Hamad, M. I. Masoud, K. H. Ahmed, and B. W. Williams, "A shunt active power filter for a medium-voltage 12-pulse current source converter using open loop control compensation," *IEEE Transactions on Industrial Electronics*, vol. 61, no. 11, pp. 5840–5850, 2014.

[7] C. Gao, X. Jiang, and L. I. Yongdong, "A novel voltage self-balance diode-clamped cascaded multilevel active power filter," *Automation of Electric Power Systems*, vol. 36, no. 12, pp. 94–100, 2012.

[8] X. Cui, C. Lam, W. Choi, and M. Wong, "Active part minimum capacity analysis of three-phase four-wire hybrid active power filters," *Automation of Electric Power Systems*, vol. 37, no. 3, pp. 122–128, 2013.

[9] C. Zhu, H. Yuan, and X. Guo, "Design and realization of harmonic extraction of active power filter based on DSP," *Low Voltage Apparatus*, 2012.

[10] A. Prasai and D. M. Divan, "Control of dynamic capacitor," *IEEE Transactions on Industry Applications*, vol. 47, no. 1, pp. 161–168, 2011.

[11] Q. Liu, Y. Deng, and X. He, "A novel AC-AC shunt active power filter without large energy storage elements," in *Proceedings of the 14th European Conference on Power Electronics and Applications (EPE '11)*, pp. 1-9, IEEE, September 2011.

[12] C. Li, Y. Deng, Z. Lv, W. Li, X. He, and Y. Wang, "Virtual quadrature source-based sinusoidal modulation applied to high-frequency link converter enabling arbitrary direct AC-AC power conversion," *IEEE Transactions on Power Electronics*, vol. 29, no. 8, pp. 4195–4208, 2014.

[13] M. El-Habrouk, M. K. Darwish, and P. Mehta, "Active power filters: a review," *Electric Power Applications, IEE Proceedings*, vol. 147, no. 5, pp. 403–413, 2000.

[14] C. Yu, "Harmonic detection in electric power system based on wavelet multi-resolution analysis," in *Proceedings of the International Conference on Computer Science and Software Engineering (CSSE '08)*, pp. 1204–1207, December 2008.

[15] H. C. Lin, "Intelligent neural network-based fast power system harmonic detection," *IEEE Transactions on Industrial Electronics*, vol. 54, no. 1, pp. 43–52, 2007.

[16] M. Cirrincione, M. Pucci, G. Vitale, and A. Miraoui, "Current harmonic compensation by a single-phase shunt active power filter controlled by adaptive neural filtering," *IEEE Transactions on Industrial Electronics*, vol. 56, no. 8, pp. 3128–3143, 2009.

[17] F. Zhang, Z. Geng, and W. Yuan, "The algorithm of interpolating windowed FFT for harmonic analysis of electric power system," *IEEE Transactions on Power Delivery*, vol. 16, no. 2, pp. 160–164, 2001.

[18] M. K. Darwish, "Design and implementation of a modified fourier analysis harmonic current computation technique for power active filters using DSPs," *IEEE. Electrical Power Application*, vol. 148, no. 1, pp. 21–28, 2001.

[19] W. M. Grady, A. H. Noyola, and M. J. Samotyj, "Survey of active power line conditioning methodologies," *IEEE Transactions on Power Delivery*, vol. 5, no. 3, pp. 1536–1542, 1990.

[20] L.-Q. Tong, Z.-M. Qian, and F.-Z. Peng, "Synchronous reference frame harmonic detection modeling and digital realization," *Proceedings of the Chinese Society of Electrical Engineering*, vol. 29, no. 19, pp. 111–117, 2009.

[21] N. F. A. Rahman, M. K. Hamzah, S. Z. M. Noor, and A. S. A. Hasim, "Single-phase hybrid active power filter using single switch parallel active filter and simple passive filter," in *Proceedings of the International Conference on Power Electronics and Drive Systems (PEDS '09)*, pp. 40–45, IEEE, Taipei, Taiwan, January 2009.

[22] L. Asiminoaei, F. Blaabjerg, and S. Hansen, "Evaluation of harmonic detection methods for active power filter applications," in *Proceedings of the 20th Annual IEEE Applied Power Electronics Conference and Exposition (APEC '05)*, pp. 635–641, March 2005.

[23] P. Zheng, P. Wang, Y. Sui, C. Tong, F. Wu, and T. Li, "Near-five-vector SVPWM algorithm for five-phase six-leg inverters under unbalanced load conditions," *Journal of Power Electronics*, vol. 14, no. 1, pp. 61–73, 2014.

[24] C.-S. Lam, M.-C. Wong, W.-H. Choi, X.-X. Cui, H.-M. Mei, and J.-Z. Liu, "Design and performance of an adaptive low-DC-voltage-controlled LC-hybrid active power filter with a neutral inductor in three-phase four-wire power systems," *IEEE Transactions on Industrial Electronics*, vol. 61, no. 6, pp. 2635–2647, 2014.

[25] L. Asiminoael, F. Blaabjerg, and S. Hansen, "Detection is key—harmonic detection methods for active power filter applications," *IEEE Industry Applications Magazine*, vol. 13, no. 4, pp. 22–33, 2007.

[26] Y. F. Wang and Y. W. Li, "An overview of grid fundamental and harmonic components detection techniques," in *Proceedings of the 5th Annual IEEE Energy Conversion Congress and Exhibition (ECCE '13)*, pp. 5185–5192, September 2013.

[27] C. Li, Y. Deng, H. Peng, X. He, W. Li, and Y. Wang, "Partial power conversion device without large electrolytic capacitors for power flow control and voltage compensation," *IEEE Transactions on Power Electronics*, vol. 27, no. 12, pp. 4847–4857, 2012.

[28] H. Akagi, E. H. Watanabe, and M. Aredes, "Instantaneous Power Theory and Applications to Power Conditioning," *Instantaneous Power Theory and Applications to Power Conditioning*, pp. 1–379, 2006.

7

A Novel Optimization of Plug-In Electric Vehicles Charging and Discharging Behaviors in Electrical Distribution Grid

Haiming Fu,[1] **Yinghua Han**,[2] **Jinkuan Wang,**[1] **and Qiang Zhao**[2]

[1]*College of Information Science and Engineering, Northeastern University, Shenyang, Liaoning, China*
[2]*Engineering Optimization & Smart Antenna Institute, Northeastern University at Qinhuangdao, Qinhuangdao, Hebei, China*

Correspondence should be addressed to Yinghua Han; yhhan723@126.com

Academic Editor: Luigi Piegari

In most countries, the problems of energy and environment are becoming worse. To deal with the environmental impacts and the dependence on fossil energy, many solutions were proposed. Plug-in electric vehicles (PEVs) is one of the best technique among these solutions. However, the large number of PEVs connected to the power grid simultaneously might increase power fluctuation or even cause the electricity shortage and thus affecting the typical use of the basic load. To cope with this issue and inspire PEV users coordinating with scheduling results, an algorithm was proposed to ensure the power transmission safety of branches and maximize the economic benefits. Considering the cost of both PEV owners and the power grid, a two-phase model of optimizing PEVs charging and discharging behaviors was built. According to the traveling purpose of PEV owners and the current electricity price, in the first phase, a novel model which defines each PEV's charging or discharging status was established. The number of PEVs' charging and discharging in each charging station can be obtained. Considering the constraints on the power transportation of branch, in the second phase, we built a mathematical model to maximize the benefit of both power grid and PEV owners. The genetic algorithm was used to optimize the charging and discharging power of PEVs. Simulation results show that the optimization method proposed in this paper has a better performance on the daily power curve compared with the uncoordinated PEVs charging.

1. Introduction

Nowadays, the problems of air pollution and the fossil fuel shortage have attracted public concern in many countries. As a potential solution to protect the environment and reduce the dependence on traditional fossil fuel, plug-in electric vehicles (PEVs) play an especially important role in the development of raising energy efficiency and alternative to traditional fuel-based automobiles [1–3]. When large-scale PEVs are connected to an electric distribution network simultaneously, some issues including the location of PEVs charging or discharging at a given time, the power fluctuation, active power loss, the state of charge (SOC) of PEVs, and the degree of PEV owners' satisfactory should be concerned [4, 5]. Considering that PEVs can be regarded as both distributed energy source and load that can be scheduled, vehicle-to-grid (V2G) system was proposed by researchers. In V2G system, power can flow between the power grid system and PEVs. Meanwhile, PEVs can reduce the operation cost of power grid and relieve daily power curve. However, uncoordinated charging and discharging behaviors of PEVs may cause circuit damage, power congestion, electricity shortage, and other grid problems [6, 7]. Therefore, the optimization of PEVs charging and discharging is necessary.

There are many existing studies on the optimal scheduling of PEVs' charging and discharging behaviors. A new mathematical model was established in the literature [5], to increase the degree of satisfaction of PEV owners and decrease the power grid's operation cost. Based on the Monte Carlo simulation algorithm, the charging load curves of different types of PEVs were established in the literature [8, 9]. Considering the economic benefit of power grid and PEV owners, a bilayer system was designed in [9]. It verified that the electricity price can adjust the time and the location of PEVs charging and discharging. In [10, 11], an aggregator was used to exchange information between power grid and PEV owners through VANWT, Wi-Fi, or cellular network. Several studies established optimization model aiming at minimizing the peak-to-valley difference and using the genetic algorithm to optimize the users' charging time in valley electricity price period [12, 13]. The literature [14] introduced V2G robust energy-scheduling problems, and proposed an algorithm to adjust the trade-off between economic performance and reliable operation. In [15], an algorithm was proposed, which not only can efficiently shift the peak load and reduce the total operation cost, but also can provide great flexibility in adjusting the trade-off between economic performance and reliable operation. In [16], a combined control and communication approach considering distributed features and vehicle preferences was proposed. The literature [17] used the flexibility and schedulability of PEVs to relieve the power fluctuation in power distribution network. To solve the optimal dispatch problem of renewable generators and PEVs, Liu et al. [18] developed a novel algorithm to solve the optimal dispatch problem of renewable generators and PEVs. Most PEVs and power grid problems in V2G system are multiagent problem, and a novel method to solve the multiagent problems was used by [19].

Most of the current researches have focused solely on the management of the distribution grid or economic benefits. However, few researches so far put the two problems together and rarely the literature refers to the PEVs' traveling purpose. This paper proposed a novel optimization method of PEVs charging and discharging based on the previous methods; meanwhile, the benefit of power grid and PEV users, the traveling purpose of PEV users, and the initial SOC of each PEV were taken into account. To reduce the impact of PEV-centralized charging on the branch in power grid, the main contribution of this paper is considering both the benefit of the power grid and PEV users to ensure the stable operation of the power grid and satisfy the traveling purpose of PEV users. We dominate this problem as the *Max Power Grid and PEV Users Profit* problem. Considering the initial SOC of PEVs arriving at charging stations, the traveling purpose of PEV users, and the load power without PEVs plugging in at this time, this paper proposed a novel charging and discharging model for PEV users to determine the PEV users' charging and discharging status and time. Meanwhile, the charging and discharging power of PEVs at a given time is optimized to reduce the power fluctuation of distribution network. Daily basic load power curve is shown in Figure 1 [1].

The rest of the parts of this paper are organized as follows: In Section 2, considering about the current electricity price, the traveling purpose of PEV users, and the SOC_{ini} of each PEV, the number of PEVs charging and discharging is defined. According to the Max Power Grid and PEV Users Profit problem and the number of PEVs charging and discharging that is obtained from Section 2, Section 3 builds a mathematical model to optimize PEVs' charging and discharging power. This paper used genetic algorithm (GA) to calculate the optimal result. Section 4 gives the specific algorithm flow. Section 5 shows the simulation results of the mathematical model which is built in the Section 3. Finally, the conclusion of this paper and future work are given in the Section 6.

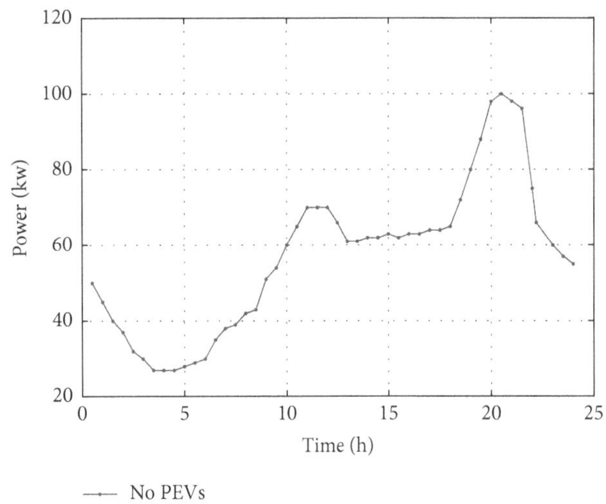

FIGURE 1: Daily power curve without PEVs.

2. The Model of PEVs Charging and Discharging

2.1. The Model of PEVs Connected to the Power Grid and Initial SOC. This paper analyzed the private PEV-centralized charging and discharging behaviors in the workdays mainly. According to the surveys, the annual average mileage of a vehicle in China is about 15,000~20,000 km and the average daily mileage is 40 km. The peak hours of private PEVs to go to work are from 7:00 to 8:00 and the charging time is during the working hours. The peak hours of private PEVs to come home from work are from 17:00 to 19:00 and the charging time is from 19:00 to 6:00 the next day. So, a PEV can connect to the power grid twice a day, and the initial SOC obeys the normal distribution $N(0.6, 0.01)$ [8]. Considering the traveling purpose of PEV users, the SOC after PEV charging or discharging must be higher than 0.6.

In this part, we assumed that each PEV's charging power is constantly equal to 3.12 kW, the start time of charging obeys the normal distribution $N(9, 0.25)$, $N(19, 1.25)$, respectively, [5] and the corresponding probability of PEVs connected to the power grid is 0.5 and 0.7, respectively [8]. The Monte Carlo simulation method is used to extract the initial SOC, the charging start time, and the charging finish time of PEVs. Then, we can obtain the result of load calculation under the constraint conditions of meeting the charging time requirements [8]. The calculation model of PEVs charging load based on the Monte Carlo simulation method is shown in Figure 2. We assumed that the maximum capacity in each charging station is 10 PEVs.

2.2. The Cases of PEVs Charging and Discharging. To achieve the peak shaving of the power grid and to ensure the power transportation of each branch smooth, considering the bidirectional flow of energy between PEVs and power grid, the charging and discharging behaviors of PEVs can be coordinated to restrain the power fluctuation of power distribution network.

In this article, we improved the interests of PEV users to stimulate PEV users join in the optimal scheduling of power

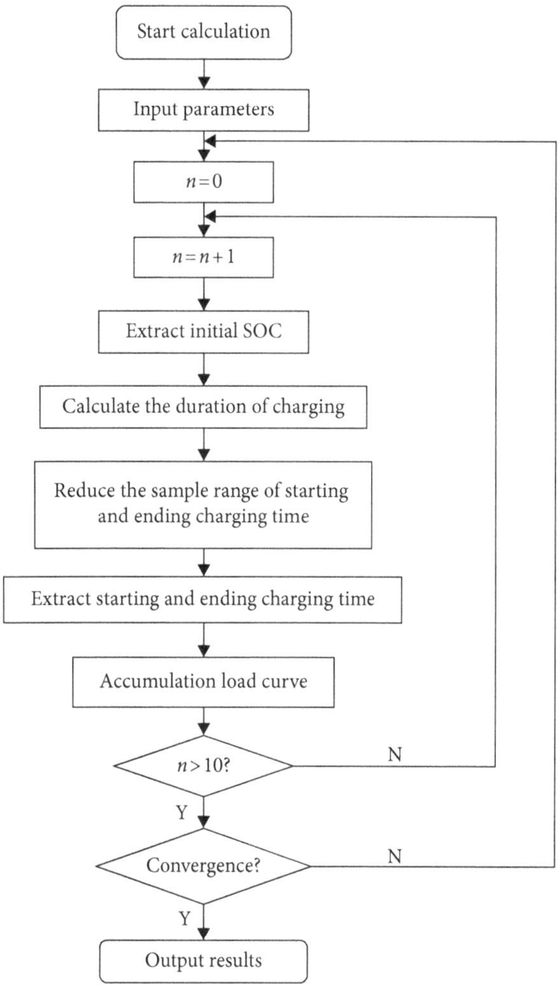

FIGURE 2: The flowchart of PEVs charging based on the Monte Carlo simulation method.

distribution network. The electricity price at a given time interval is formulated thanks to Figure 1; the principle of formulation is that when the basic load power is higher, the electricity price is higher and when the basic load power is lower, the electricity price is lower, simulating the PEV users charging at the low-load time period and discharging at the high-load time period on the basis of satisfying the traveling purpose.

Due to the different electricity prices and to make calculation easier, 24 hours a day can be divided into 3 parts: low-price period, normal-price period, and high-price period, and the time interval is 24:00–10:00, 10:00–18:00, and 18:00–24:00, respectively. According to the initial SOC of PEV users and the different time period, the optimization system defines the charging or discharging status for each PEV user, which connected to the power grid. Just take one PEV as example, and all of the PEVs charging load could get after adding. The situations of a PEV charging or discharging are divided into the following four types:

Case 1. During the high-price period ($S=10$). When the $SOC_{ini} \leq 0.6$ PEV charges to its $SOC=0.6$, if it is still in the high-price period, PEV would charge again when the low-price period starts.

Case 2. During the high-price period ($S=10$). When the $SOC_{ini} > 0.6$ PEV discharges to its $SOC=0.6$, if it is still in the high-price period, PEV would charge when the low-price period starts.

Case 3. During the normal-price period ($S=01$). When the $SOC_{ini} \leq 0.8$ PEV charges to its $SOC=0.8$, if it is still during the normal price period or turning to the high-price period, PEV would charge again when the low-price period starts.

Case 4. During the low-price period ($S=00$). PEV charges to its $SOC=1$.

The algorithm flowchart of calculating the number of PEVs charging and discharging in each time interval (0.5 h) is shown in Figure 3. $N_{c,t}$ and $N_{d,t}$ represent the number of PEVs charging and discharging at the given time. We assumed that each node of the distribution network can accommodate at most 10 PEVs, the PEVs' charge and discharge power is constantly equal to 3.12 kW, the maximum capacity of each PEV battery is 24 kWh [8], and the parameter inputted include the initial SOC, arrival time, and departing time of PEVs. According to the initial SOC of PEVs and the level of electricity price when PEV users connected to power grid, the charging or discharging behaviors of PEVs could be coordinated on the basis of satisfying the propose of PEVs traveling. The SOC of PEVs is updated after each time interval, and the number of PEVs charging and discharging is exported.

3. The Mathematical Model that Optimized PEVs' Charging and Discharging Power

In this section, we established a mathematical model that the goal is to maximize the benefit of PEV users and minimize the fluctuation of the daily load curve in distribution network; meanwhile, the PEVs' charging and discharging behaviors do not affect the normal work of basic loads and the power transmission constraints of each branch as the constraint conditions. The power fluctuation of distribution network is reduced, and the stable power transportation of branch is guaranteed through the optimization of charging and discharging power of PEVs.

3.1. The Objective Function. To raise the power grid companies' profit, the power fluctuation needs to be reduced. Therefore, reducing the power fluctuation is an objective. So, the optimization objective function of power fluctuation g_1 is denoted as

$$g_1 = \min \sum_{t=0}^{T} \left(P_{c,t} * N_{c,t} + P_{Load,t} - P_{d,t} * N_{d,t} - \overline{P} \right)^2, \quad (1)$$

where $P_{c,t}$ and $P_{d,t}$represent the charging and discharging power of PEV users at time t, respectively, $N_{c,t}$ and $N_{d,t}$ represent the number of PEV users' charging and discharging at time t, respectively, $P_{Load,t}$ represents the basic load power at time t, and $\overline{P}$ represents the average load power during a day without PEVs. $P_{Load,t}$ and $\overline{P}$ can be got by gathering the historical data.

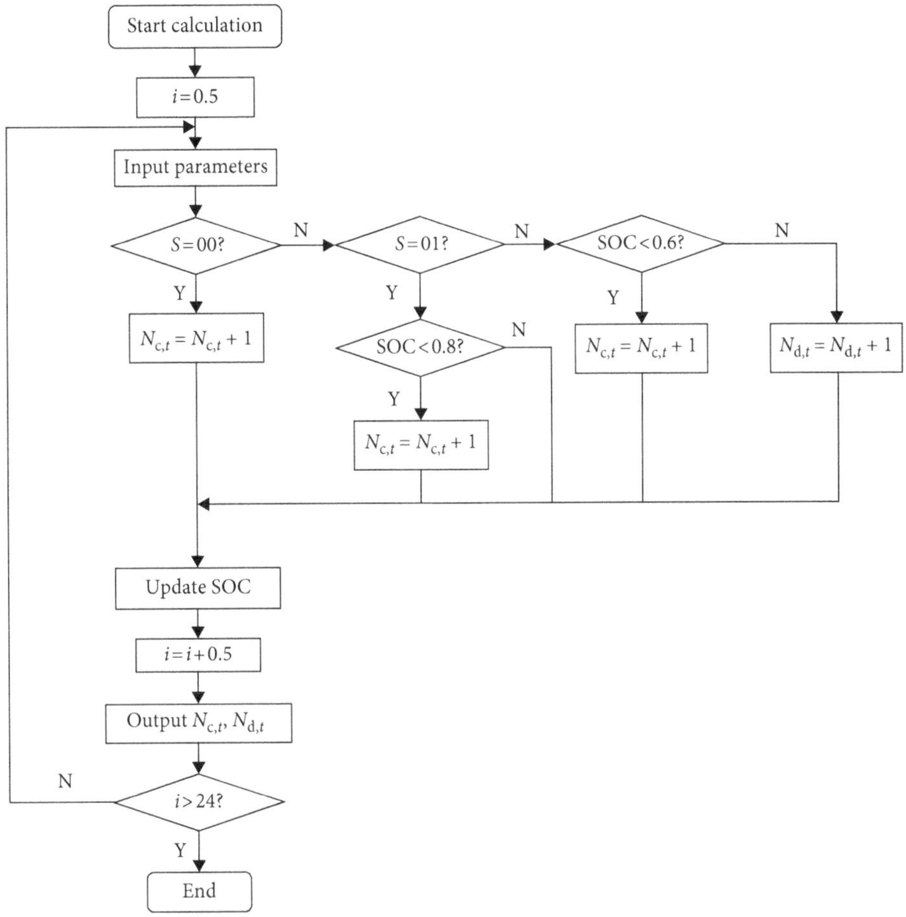

Figure 3: The flowchart of the number of PEVs charging and discharging.

To raise the PEV users' profit, the optimization objective function g_2 is denoted as

$$g_2 = \min \rho_{c,t} * N_{c,t} * P_{c,t} * \Delta t - \rho_{d,t} * N_{d,t} * P_{d,t} * \Delta t, \tag{2}$$

where $\rho_{c,t}$ and $\rho_{d,t}$ represent the charging and discharging electricity price at time t and Δt represents time interval (0.5 h).

We put all the objective functions together, so the method can be regarded as a multiobjective optimization problem. Because coupling does not exist between the objective functions in this paper, the problem can be translated into a single-objective optimization problem by weighted linear combination in the following equation:

$$\min f = \alpha_1 \cdot g_1 + \alpha_2 \cdot g_2, \tag{3}$$

where the α_1 and α_2 is the weight of g_1 and g_2, respectively, g_2 represents the benefit of PEV owners , and g_1 represents the benefit of power grid where it is more important than g_2 because of the demand of reality. Therefore, we set $\alpha_1 = 0.6$, $\alpha_2 = 0.4$.

3.2. Constraints. The following constrains in the mathematical model must be met at each time interval Δt, and the definitions of corresponding parameters are shown in Table 1.

Table 1: The parameters of constraints.

Symbol	Definition
$\mathrm{SOC}_{\mathrm{ini},h,t}$	Initial SOC of PEV h when it arriving at the charging station at time t
$\mathrm{SOC}_{\mathrm{fin},h,t}$	Final SOC of PEV h when it leaving the charging station at time t
$C_{\max}$	The maximum capacity of PEV's battery
$E_{\mathrm{total},t}$	Total energy in distribution network at time t
$E_{\mathrm{load},t}$	Total energy used by basic load at time t
$P_{l,\max}$	Maximum active power transmission of line l
$N^{\mathrm{S}}_{\max}$	Maximum accommodation capacity of the station S
$P_{\mathrm{c,min}}$	Minimum charging power of PEV
$P_{\mathrm{c},max}$	Maximum charging power of PEV
$P_{\mathrm{d,min}}$	Minimum discharging power of PEV
$P_{\mathrm{d},max}$	Maximum discharging power of PEV
$N^{\mathrm{S}}_{\mathrm{p},t}$	The number of PEVs neither charging or discharging at the charging station S at time t
$N^{\mathrm{S}}_{\mathrm{c},t}$	The number of PEVs charging at the charging station S at time t
$N^{\mathrm{S}}_{\mathrm{d},t}$	The number of PEVs discharging at the charging station S at time t

$$\left(\mathrm{SOC}_{\mathrm{fin},h,t} - \mathrm{SOC}_{\mathrm{ini},h,t}\right) * C_{\max} * N_{\mathrm{c},t} \le E_{\mathrm{total},t} - E_{\mathrm{load},t}, \tag{4}$$

$$P_{\mathrm{Load},t} + P_{\mathrm{c},t} * N_{\mathrm{c},t} - P_{\mathrm{d},t} * N_{\mathrm{d},t} \le P_{l,\max}, \tag{5}$$

$$P_{c,\min} \le P_{c,t} \le P_{c,\max}, \quad (6)$$

$$P_{d,\min} \le P_{d,t} \le P_{d,\max}, \quad (7)$$

$$0 \le N^{S}_{c,t} \le N^{S}_{\max}, \quad (8)$$

$$0 \le N^{S}_{d,t} \le N^{S}_{\max}, \quad (9)$$

$$0 \le N^{S}_{p,t} \le N^{S}_{\max}, \quad (10)$$

$$0 \le N^{S}_{c,t} + N^{S}_{d,t} + N^{S}_{p,t} \le N^{S}_{\max}. \quad (11)$$

Constraint (4) ensures the total PEV users' charging energy cannot exceed the surplus energy of distribution network, where the maximum capacity of PEV's battery ($C_{\max}$) is identically equal to 24 kWh. Constraint (5) indicates that the limit of active power transportation of each branch, which ensures the branch safety at power distribution network would not be influenced by PEVs' charging and discharging behaviors. The constraints (6) and (7) represent the scope of charging and discharging power of PEVs, respectively. The PEVs would be broken if charging or discharging power exceeds the maximum power of PEVs; meanwhile, the excessive charging and discharging power would increase the fluctuation of the power grid. However, if the PEVs' charging and discharging power is lower than $P_{c,\min}$ and $P_{d,\min}$, the waiting time would become so long that the owners' degree of satisfaction would decrease. Constraints (8)–(11) represent that the total number of PEVs cannot exceed the maximum accommodation capacity of charging stations, which is to reduce the PEV owner's waiting time.

4. Power Optimization Algorithm

The key of the power optimization algorithm for PEVs charging and discharging is how to deal with the optimal power flow model, which is a nonlinear, nonconvex, large-scale, static optimization problem with both continuous and discrete control variables. Genetic algorithm is applied to find optimal result, the algorithm considered the parameter ($P_{c,t}$, $P_{d,t}$) as optimal variables. The optimal charging and discharging power of PEVs and the daily power curve of distribution network can be obtained under the PEV users' responsiveness δ.

We can obtain the number of PEVs charging and discharging in each charging station at the given time from Section 3. The total PEVs charging power in a charging station equals the sum of charging power of each PEV in the charging station, which is denoted as Equation (12). Similarly, the total PEVs discharging power is denoted as Equation (13). According to the optimization objective functions in the Section 3, choose the excellent parameter ($P_{c,t}$, $P_{d,t}$).

$$P_{c,\text{total}} = P_{c,t} * N_{c,t}, \quad (12)$$

$$P_{d,\text{total}} = P_{d,t} * N_{d,t}. \quad (13)$$

The main process of algorithm is as follows:

(1) Initialize grid load parameters and PEVs load parameters, which include PEV users' responsiveness, basic load power without PEVs connecting to the power grid in each time interval, average load power during one day without PEVs, the maximum battery capacity of PEVs, initial SOC, arriving time, leaving time, and average energy consumption.

(2) Calculate the number of PEVs charging and discharging at the current time interval.

(3) Generate 100 initial populations of EV charging power and discharging power ($P_{c,t}$, $P_{d,t}$) randomly.

(4) Calculate the constraint value of populations, which include equality constraints and inequality constraints, and eliminate the populations that unsatisfied the constrains of the mathematical model. Then, calculate the interests of PEV users charging and discharging and the PEVs charging and discharging power at the current time interval according to the initial populations that satisfied constrains. Based on the basic load power at the current time, PEVs charging and discharging power after coordinating and the average basic load power during one day, the power fluctuation of distribution network can be calculated.

(5) According to the power fluctuation of distribution network and the charging and discharging interests of PEV users, calculate and sort the fitness value. Select and retain excellent individuals through the roulette rule to ensure that the superior genes are retained.

(6) The surviving genes were crossed and mutated to generate offspring populations, where the probability of crossover and mutation is 0.7 and 0.2, respectively.

(7) Repeat the iterative optimization, until reaching the maximum genetic algebra, and then output the optimal result ($P_{c,t}$, $P_{d,t}$) .

(8) Calculate PEVs charging and discharging power in each charging station ($P_{c,\text{total}}$, $P_{d,\text{total}}$).

(9) Output the daily load power curve that before and after optimization of distribution network.

The optimization algorithm flowchart of charging and discharging power is shown in Figure 4.

5. Simulation Results

According to the optimization strategy that proposed in this paper, simulation results can be shown by MATLAB.

Just consider the charging behavior of PEVs and without any coordination, a branch of daily load power curve shown is in Figure 5. As can be seen from the figure, the peak-to-valley difference of a branch increased due to the behavior of PEVs uncoordinated charging, especially between 19:00 and 21:00 hours, the load power become too high, which might impact on the power transportation or even damage the branch.

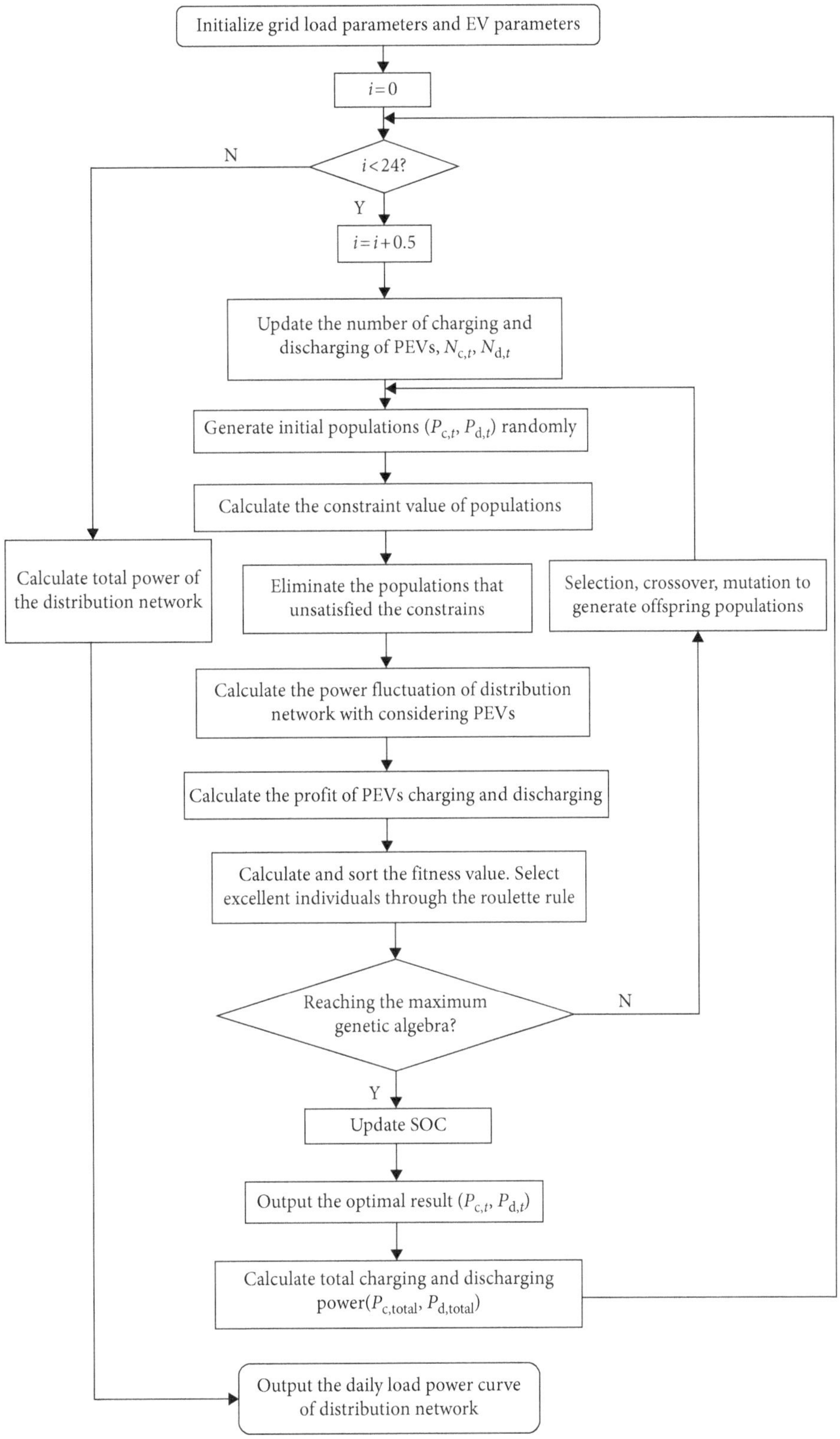

FIGURE 4: The optimization algorithm flowchart of charging and discharging power.

The load power curve after price optimized coordination is shown in Figure 6. Compared with the load power curve of distribution power grid that uncoordinated charging, peak and off-peak difference of the load power curve is decreased thanks to the optimization strategy of electricity price and thus ensure the stable operation of each branch. However, as shown, the power fluctuation is large and PEVs' charging and discharging power needs to be optimized for each time period in a day.

According to the power optimization in Section 3, the simulation results of one branch and the whole power are

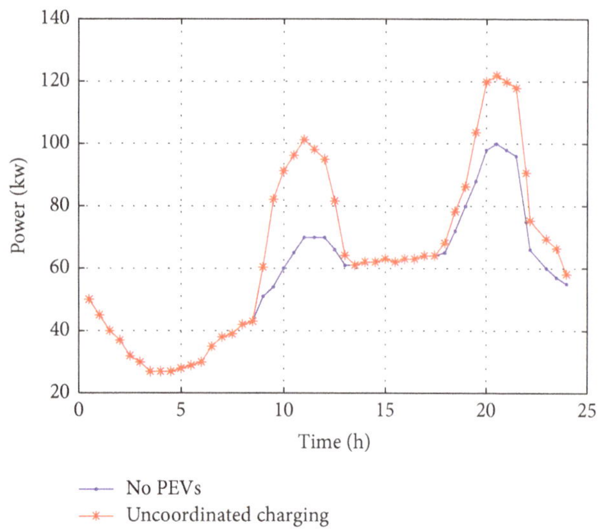

FIGURE 5: The daily load curve of a branch.

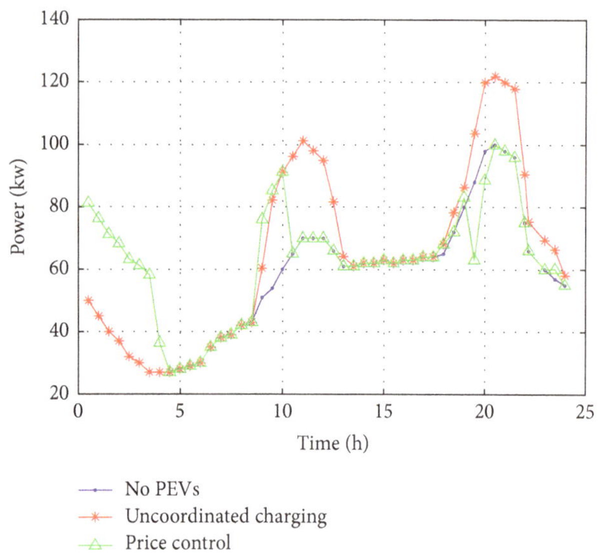

FIGURE 6: The load power curve after price-optimized coordination.

shown in Figures 7 and 8. As can be seen from Figure 7, the peak and off-peak difference in power curve of one branch would increase if PEVs are connected to the power grid while without charging coordination, especially during the 18:00–22:00 hours, which might lead to the electricity shortage of basic load or even damage the circuit. Considering V2G system, the peak and off-peak difference in power curve of one branch can be reduced through the electricity price control and the optimization of charging and discharging power, which ensures the normal operation of branch.

This paper used an IEEE 33-bus distribution network to simulate the whole distribution network, and the specific data can be obtained by [20]. Figure 8 shows the load curve comparison chart of three kinds of cases: basic load without considering the PEVs, uncoordinated PEVs charging and

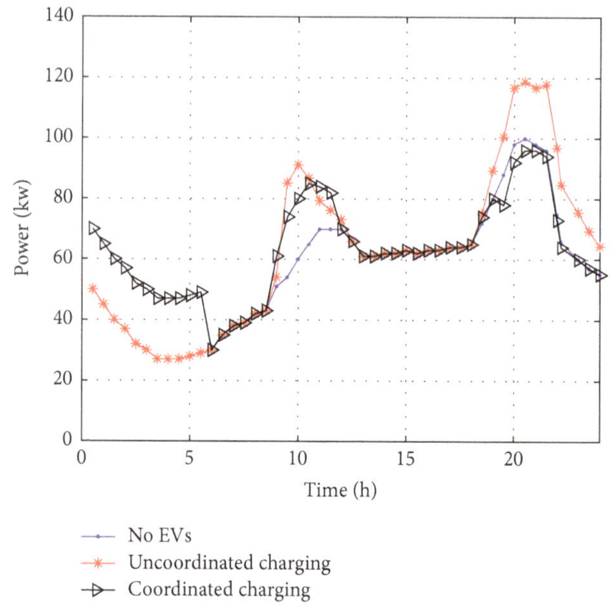

FIGURE 7: Simulation result of one branch.

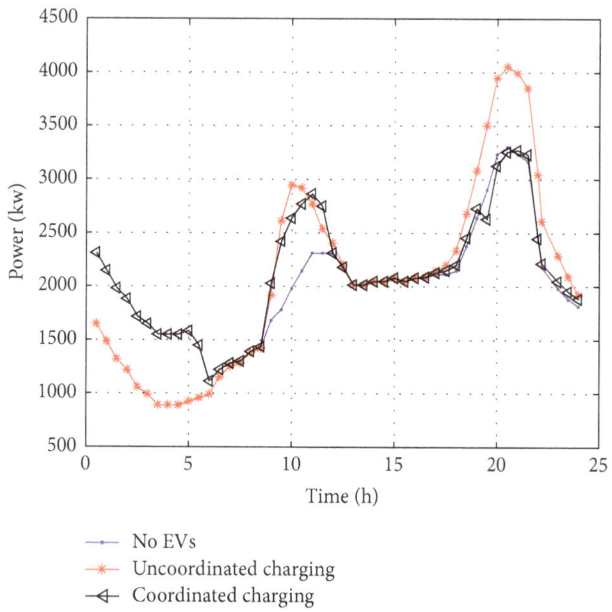

FIGURE 8: Simulation result of the whole distribution network.

discharging behaviors, PEVs joined in the optimal dispatch of distribution network. As can be seen from Figure 8, the optimization strategy of charging and discharging behaviors of PEVs has a better property on the peak and off-peak difference; meanwhile, daily load power curve is smoother compared with uncoordinated charging and no PEVs.

6. Conclusions

In this paper, we have relieved the intensifying power fluctuation of power grid caused by a large-scale PEVs connecting to the power grid simultaneously. To this aim, we have investigated optimization of the PEVs charging and discharging behaviors to balance the profit of power grid and PEV owners. A novel two-phase optimization method is proposed for PEVs charging and discharging. The first phase is about optimization of electricity price. According to the historical daily power curve, the status of electricity price is defined based on the principle of high load and high price. This phase ensures the maximum profit of both the power grid and PEV owners, and the PEV owners' traveling purpose is taken into account at the same time. The second phase is about optimization of PEVs' charging and discharging power. This phase aims to reduce the fluctuation and ensures the power transmission safety of distribution network. Compared with the power curve of uncoordinated charging and non-PEVs, simulation results have shown that the optimization strategy proposed in this paper can reduce the peak and off-peak difference of distribution network and protect each branch of power transportation which is not damaged obviously.

Conflicts of Interest

The authors declare that there are no conflicts of interest exist.

Acknowledgments

This work was supported by National Key Research and Development Program of China (2016YFB0901900), Natural Science Foundation of Hebei Province of China (F2017501107), Open Research Fund from the State Key Laboratory of Rolling and Automation, Northeastern University (2017RALKFKT003), and Technology Support Foundation of Northeastern University at Qinhuangdao (no. XNK201603).

References

[1] R. Yu, W. Zhong, S. Xie, C. Yuen, S. Gjessing, and Y. Zhang, "Balancing power demand through EV mobility in vehicle-to-grid mobile energy networks," *IEEE Transactions on Industrial Informatics*, vol. 12, no. 1, pp. 79–90, 2016.

[2] Z. Xia, X. Wang, X. Sun, and Q. Wang, "A secure and dynamic multi-keyword ranked search scheme over encrypted cloud data," *IEEE Transactions on Parallel & Distributed Systems*, vol. 27, no. 2, pp. 340–352, 2016.

[3] P. Yi, T. Zhu, B. Jiang, R. Jin, and B. Wang, "Deploying energy router in an energy internet based electric vehicles," *IEEE Transactions on Vehicular Technology*, vol. 65, no. 6, pp. 4714–4725, 2016.

[4] J Yang, L. He, and S. Fu, "An improved PSO-based charging strategy of electric vehicles in electrical distribution grid," *Applied Energy*, vol. 128, no. 3, pp. 82–92, 2014.

[5] L. He, J. Yang, J. Yan, Y. Tang, and H. He, "A bi-layer optimization based temporal and spatial scheduling for large-scale electric vehicles," *Applied Energy*, vol. 168, pp. 179–192, 2016.

[6] K. Ma, L. Xie, and P. R. Kumar, "A layered architecture for EV charging stations based on time scale decomposition," in *Proceedings of IEEE International Conference on Smart Grid Communications*, pp. 674–679, IEEE, Piscataway, NJ, USA, November 2014.

[7] S. V. Chakraborty, S. K. Shukla, and J. Thorp, "A hierarchical networked micro-simulator to study grid-integration of renewables and electric vehicles," in *Proceedings of Modeling and Simulation of Cyber-Physical Energy Systems*, pp. 1–6, IEEE, Piscataway, NJ, USA, 2013.

[8] L. Bedogni, L. Bononi, A. D'Elia, M. Di Felice, M. Di Nicola, and T. S. Cinotti, "Driving without anxiety: a route planner service with range prediction for the electric vehicles," in *Proceedings of International Conference on Connected Vehicles and Expo*, pp. 199–206, IEEE, Piscataway, NJ, USA, November 2014.

[9] S. Shrestha and T. M. Hansen, "Spatial-temporal stochasticity of electric vehicles in an integrated traffic and power system," in *Proceedings of IEEE International Conference on Electro Information Technology*, pp. 0227–0232, IEEE, Piscataway, NJ, USA, May 2016.

[10] J. C. Mukherjee and A. Gupta, "Distributed charge scheduling of plug-in electric vehicles using inter-aggregator collaboration," *IEEE Transactions on Smart Grid*, vol. 8, no. 1, pp. 331–341, 2017.

[11] L. Bedogni, L. Bononi, M. D. Felice et al., "An integrated simulation framework to model electric vehicle operations and services," *IEEE Transactions on Vehicular Technology*, vol. 65, no. 8, pp. 5900–5917, 2016.

[12] S. Y. Ge, L. Huang, and H. Liu, "Optimization of peak-valley TOU power price time-period in ordered charging mode of electric vehicle," *Power System Protection & Control*, vol. 40, no. 10, pp. 1–5, 2012.

[13] W. Wei, F. Liu, and S. Mei, "Charging strategies of EV aggregator under renewable generation and congestion: a normalized nash equilibrium approach," *IEEE Transactions on Smart Grid*, vol. 7, no. 3, pp. 1630–1641, 2016.

[14] Z. Zhou, C. Sun, R. Shi, Z. Chang, S. Zhou, and Y. Li, "Robust energy scheduling in vehicle-to-grid networks," *IEEE Network the Magazine of Global Internetworking*, vol. 31, no. 2, pp. 30–37, 2017.

[15] K. Wang, L. Gu, X. He et al., "Distributed energy management for vehicle-to-grid networks," *IEEE Network*, vol. 31, no. 2, pp. 22–28, 2017.

[16] L. Wang, S. Sharkh, and A. J. Chipperfield, "Optimal decentralized coordination of electric vehicles and renewable generators in a distribution network using A∗ search," *International Journal of Electrical Power & Energy Systems*, vol. 98, pp. 474–487, 2018.

[17] A. Dutta and S. Debbarma, "Frequency regulation in deregulated market Using Vehicle-to-Grid Services in Residential Distribution Network," *IEEE Systems Journal*, vol. 99, pp. 1–9, 2017.

[18] D. Liu, Y. Wang, and Y. Shen, "Electric vehicle charging and discharging coordination on distribution network using multi-objective particle swarm optimization and fuzzy decision making," *Energies*, vol. 9, no. 3, p. 186, 2016.

[19] H. N. D. Melo, J. P. F. Trovao, P. G. Pereirinha et al., "A controllable bidirectional battery charger for electric vehicles with vehicle to grid capability," *IEEE Transactions on Vehicular Technology*, vol. 67, no. 1, pp. 114–123, 2018.

[20] M. E. Baran and F. F. Wu, "Network reconfiguration in distribution systems for loss reduction and load balancing," *IEEE Transactions on Power Delivery*, vol. 4, no. 2, pp. 1401–1407, 1989.

8

A Novel Hybrid T-Type Three-Level Inverter based on SVPWM for PV Application

Ayiguzhali Tuluhong,[1,2] Weiqing Wang,[1] Yongdong Li,[2] and Lie Xu[2]

[1]*Electrical Engineering, Xinjiang University, Urumqi, Xinjiang 830046, China*
[2]*State Key Laboratory of Power System, Department of Electrical Engineering, Tsinghua University, Beijing 100084, China*

Correspondence should be addressed to Weiqing Wang; wwq59@xju.edu.cn

Academic Editor: Yongheng Yang

We describe several, recently reported, new topologies and compare them with each other, in order to find out the optimal multilevel grid-connected inverter topology. Then, we classify these topologies according to the basic unit which predecessors proposed. Eventually, we propose the hybrid T-type inverter topology structure, which is composed of two best basic units. This structure takes full advantage of the two components, to reduce the harmonic content and the power loss of the converter and improve the conversion efficiency of the system. At the same time, the space vector pulse width modulation (SVPWM) method is used to simulate the proposed topology in the MATLAB/SIMULINK platform, while the loss of each semiconductor switch is calculated using MELCOSIM software. The results show that the proposed structure is superior to the most widely used topology, i.e., the diode clamped and the T-type three-level circuits.

1. Introduction

Nowadays, the contradictions between the consumption pattern of the traditional petrochemical energy and the economic development and environmental protection are becoming more and more prominent. People gradually realize the importance of taking sustainable development road, vigorous developing, and utilizing of the renewable energies. Among the renewable energies, the solar energy represents the largest and the most commonly distributed resource. The photovoltaic power generation technology using the solar cells effectively absorbs the solar energy and changes it into electricity. The grid-connected inverter is the key component and important equipment in a photovoltaic grid-connected system.

In the design of general inverters, synthetically considering the cost-effective factors, insulated-gate bipolar transistor (IGBT) represents the most employed device. However, due to the nonlinearity of the IGBT's conduction voltage drop, it does not significantly increase with the increase of current, thus ensuring that the inverter still presents a relatively low loss and high efficiency at the maximum load condition. However, since the European efficiency is mainly related to the efficiency of the inverter at different light-load, the aforementioned characteristics of the IGBT represent the disadvantage of the photovoltaic grid-connected inverter. For light loads, the turn-on voltage drop of the IGBT does not significantly reduce, which in turn reduces the European efficiency of the inverter. In contrast, due to linear conduction voltage drop of the MOSFET, it provides lower turn-on voltage drop for light loads. Considering the excellent dynamic characteristics and high frequency work ability, MOSFET becomes the first choice for photovoltaic inverting [1].

Three-level inverter has been widely used in the middle and high voltage large capacity AC speed regulating fields, since its output has higher power quality, lower harmonic contents, better electromagnetic compatibility, lower switching losses, and other advantages. However, it still suffers from some key problems, including the simplification of the three-level algorithm, neutral point voltage control in the overmodulation region, and the stability of the system at high voltage. In view of the above problems, this paper studies

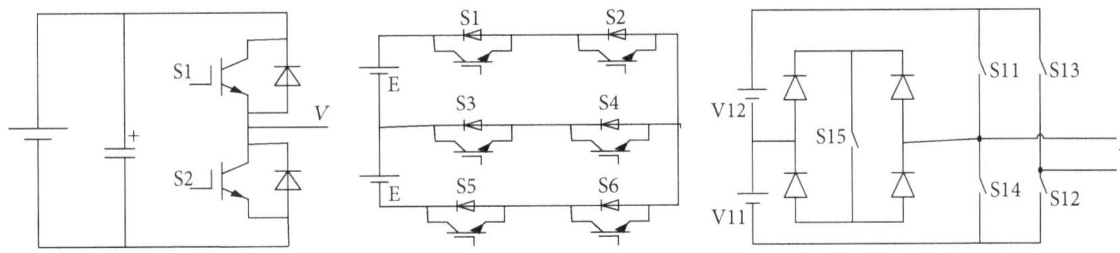

(a) Basic unit A (b) Basic unit B (c) Basic unit C

FIGURE 1: Basic unit of the inverter.

the structure and principle of the three-level inverter, the control of the neutral point voltage of the capacitor, and the realization of the SVPWM algorithm.

In this paper, we study novel T-type inverter topology in PV system using SVPWM control algorithm. The structure is organized as follows: Section 2.1 introduces basic cells of the new multilevel PV inverters and classifies them. Section 2.2 presents and compares new types of multilevel inverters. Section 2.3 analyzes and compares switch losses and conversion efficiency of diode clamped T-type and proposed hybrid T-type. Section 2.4 details SVPWM control algorithm. Section 3 gives simulation results. Finally, Section 4 concludes the paper.

2. Materials and Methods

2.1. Basic Unit. Inverter basic unit refers to the minimum component that meets the demands of the topological multilevel grid-connected inverter. Odeh [2] proposed one basic concept of the basic unit of multilevel grid-connected inverter.

As shown in Figure 1(a), after achieving the basic unit, it may be extended in form of series-parallel and parallel-series connections to present higher voltage levels. Odeh [3] analyzes the connection between the topologies of the multilevel grid-connected inverters and proposes a regular pattern that is followed when general multilevel grid-connected inverter topology simplifies to other multilevel grid-connected inverter topology. For example, the main switches must be all maintained, all switching devices must be deleted symmetrically, diodes and capacitors must be distributed symmetrically, and so on. With the development of power electronic devices, in order to unify the pressure drop of the power electronic switch transistors, a more practical basic unit is put forward.

Figure 1(b) shows the stacked commutation cells or three-pole cells. As the employed conduction strategy, one of the two transistors in the outer leg is in a frequent switching state only during half-cycle, while the other transistor is switched only once during the fundamental wave period, which greatly reduces the switching loss. This conduction strategy also avoids the voltage-equalization problem caused by the simultaneous turn-on and turn-off of the series devices. Figure 1(c) shows the structure of a multilevel inverter basic unit reported by Draxe et al. [4]. This inverter is formed using five switches with antiparallel diodes, where S11 to S14 are arranged as the traditional CHB inverter, while, however, S15 is added to increase the output voltage levels by selecting the appropriate voltage source. This structure produces higher voltage levels with minimum number of switches by optimizing the circuit layout and reducing the gate drive circuitry.

Recently reported inverter topologies aim to reduce the number of the power electronic devices to improve the conversion efficiency. Figure 2 shows the classification of the inverters based on their topologies. As we see from this figure, the largest portion of the proposed inverter topologies formed by the basic unit A, due to its relatively simple construction. The topologies composed of basic units B and C have not yet appeared because of their complex basic constitution unit.

The basic unit of the inverter consists of a DC power supply and a pair of switches. By using the same basic unit, taking Figure 1(a), for example, in the form of series-parallel combinations a new circuit topology, as shown in Figure 3, called single phase H-bridge topology, is obtained. Or by series-parallel combinations between different basic units also a new one is got. For instance, three parallel basic units A and one basic unit B constitute the three-level T-type inverter topology, as shown in Figure 4. Different multilevel inverters can be obtained by multiseries and multiparallel connections of multiple basic units. And the construction of the other topologies in Figure 2 can be deduced by the same analogy.

2.2. New Type Inverter Topology. The work in [5] proposed the split capacitor H-bridge (SC-HB) inverter topology. Adding a simple DC-DC converter, this topology overcomes the capacitance-voltage balance problem, while reducing the leakage current, with improved efficiency. The work in [2] presented the improved cascaded H-bridge topology which consists of a half-bridge level latching circuit and a main inverter H-bridge. Thus, it reduces the number of the devices, switching loss, and harmonic content. An asymmetric cascaded H-bridge topology, with different, and proportional, DC voltage source values is reported in [4]. This reduces the capacity and number of the devices, as well as the costs. A sine-wave pulse width modulation (SPWM) three-phase multilevel inverter topology may be achieved by inserting two auxiliary switches in each phase bridge to change the basic two-level to three-level inverter to synthesize higher levels [3]. In comparison with traditional diode clamps, flying

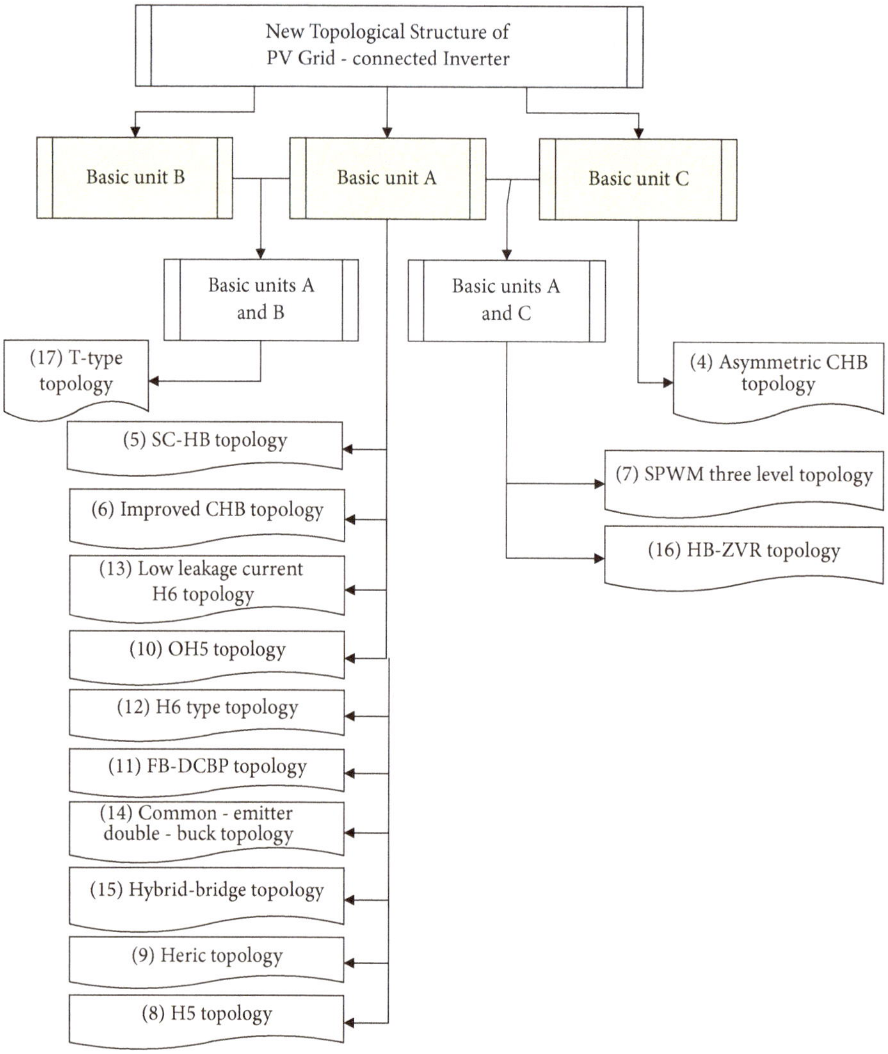

FIGURE 2: Per-phase circuit configuration of the conventional multilevel inverter.

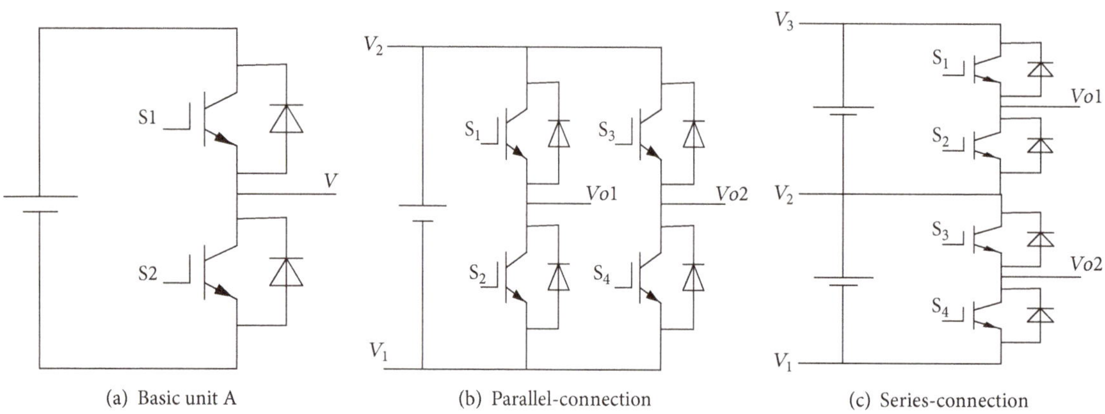

FIGURE 3: Parallel-series-connection topology of the same basic units.

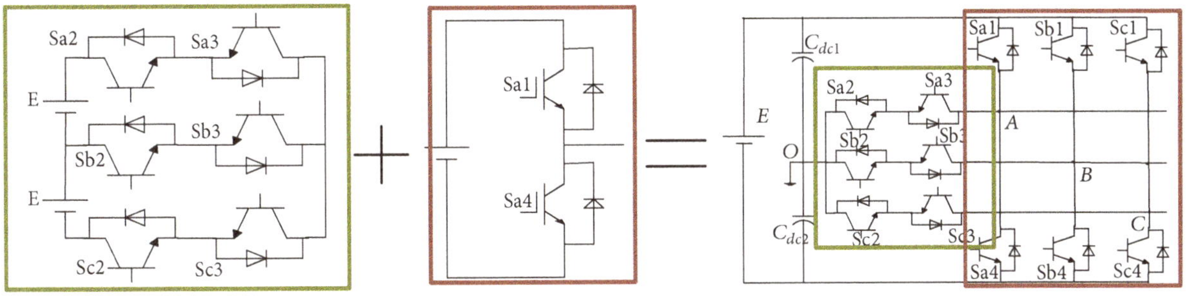

FIGURE 4: Parallel-series-connection topology of the different basic units.

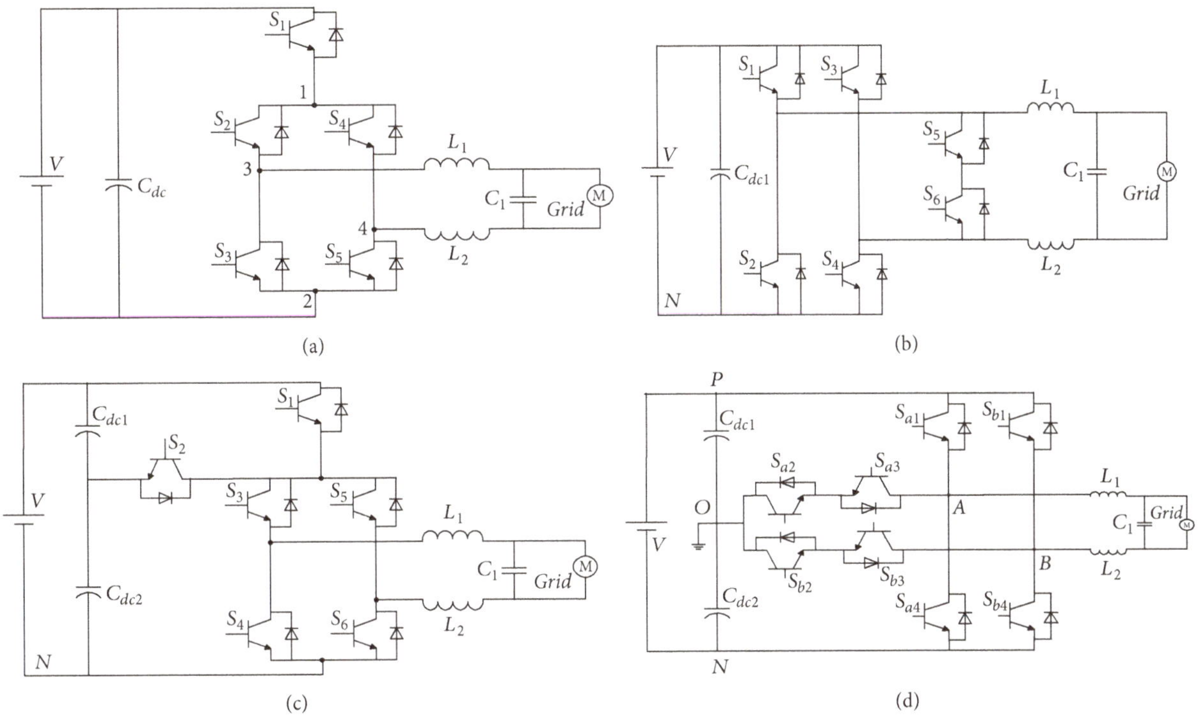

FIGURE 5: Some new topology structure. (a) H5 topology, (b) Heric topology, (c) OH5 topology, and (d) T-type topology.

capacitors, and cascaded H-bridge inverters, this topology uses only a DC power supply and lower power electronics providing the same number of the levels.

Figure 5(a) shows the H5 topology [6], composed of a freewheeling circuit by adding a high frequency switch in the positive end of the DC input side, where the efficiency may reach up to 98.1%. However, it suffers from big loss and heat and unbalanced thermal stress. Figure 5(b) shows the structure of the Heric topology [7], where a new freewheeling circuit is added, which itself is composed of a set of bidirectional switch branches on the AC side, based on the traditional full-bridge inverter topology. The efficiency of this topology may reach more than 98%. The OH5 topology is shown in Figure 5(c) [8], which uses a switch and capacitor to form the bidirectional clamping circuit on the basis of the H5 topology, which greatly inhibits the leakage current, and improves the conversion efficiency. However, in such a structure, the clamping circuit is constrained by the dead time of the switch, and a large on-state loss is presented. The work in [9] proposes the FB-DCBP topology, which achieves the complete elimination of the leakage current by adding two controllable switch and clamp diodes in the DC side to format the clamping circuit. The leakage current suppression capability of this topology is stronger than that of the H5 and Heric topologies but suffers from large on-state loss and high number of devices, providing the maximum efficiency of 97.4%.

H6-type topology is proposed in [10], and it is a variant of the Heric topology. It does not require setting dead time between the switches of the same bridge arm and does not pass through the body diode of the switch. Therefore, all switches may be formed by MOSFET, but the topology requires extra two freewheeling diodes, so the cost is increased. Moreover, the efficiency is lower than the Heric topology, and the efficiency is up to 97.6%. The low-leakage current H6 topology is reported in [11], which is a compromise between the H5 and Heric topologies. A new freewheeling circuit is formed based on the H5 topology by adding a switch to make the number of switches through two half-frequency cycles of the network current at power transfer mode not the same, yielding reduced on-state loss. The efficiency of the H6 topology is lower than that of the Heric topology and higher than that of H5 topology. The common mode is inferior to H5 topology and superior to Heric topology. A common emitter double BUCK topology may be considered [12], composed of four controllable switches and two diodes. The leakage current of this structure is basically zero, and there is no high frequency dead zone between the switches. In addition, this structure presents no break-through phenomenon, high reliability, and small switching loss, where the efficiency may reach more than 98.5%. However, the topological magnetic component utilization is low and cannot output reactive current.

The hybrid bridge topology proposed in [13] consists of six controllable switches and two freewheeling diodes. This topology moved the switch on the AC side of the Heric topology to the middle of the bridge arm A, similar to H6-type topology. HB-ZVR (H-bridge zero-voltage state rectifier) topology is proposed in [14], introducing an AC bypass circuit, which itself is composed of a IGBT and a group of diode rectifiers, which is clamped to the midpoint of the two capacitors on the DC bus, to achieve low common mode current and high efficiency inverter topology. The HB-ZVR topology solves the problem that bidirectional switches S5 and S6 of the Heric topology cannot turn on, and this topology will always find a way in the bidirectional switch. Figure 5(d) shows the T-type inverter topology [15], formed by a set of switches Sa3/Sa4 as a bidirectional switch to achieve the main switch Sal/Sa2 clamping function. It uses the clamp diodes or clamp bit capacitance to improve the midpoint clamping circuit, reducing the number of devices and uneven distribution of the loss. In the topological selection, T-type three-level circuit leverages many advantages of the nonisolation technology as well as the multilevel technology. Therefore, it is very suitable for the photovoltaic grid-connected power generation occasions; however, it is required to effectively suppress the circuit leakage current and system efficiency.

By comparing the above different new topologies, we see that the proposed multilevel inverter topology optimizes the performances of the inverter by adding auxiliary/clamping circuit, using hybrid switch or asymmetric structure.

Figure 6 shows the topology of the hybrid T-type inverter, which is on the basis of T-type structure and composed of nine MOSFET switches, i.e., Sa2, Sb2, Sc2, . . ., Sa4, Sb4, Sc4. We choose IGBT for Sal, Sbl, and Scl, since the reverse recovery ability of the body diode in the field effect transistor is poor; therefore they act in low frequencies. The high frequency MOSFET semiconductor switches, i.e., Sa2, Sb2, Sc2, . . ., Sa4, Sb4, Sc4, provide good switching characteristics and low on-resistance. Moreover, due to the low speed characteristics of the built-in diode of MOSFETs, MOSFET cannot be used in the upper bridge arm. We take advantage of the two devices, to reduce the harmonic content and the power loss of the converter and improve the conversion efficiency of the system.

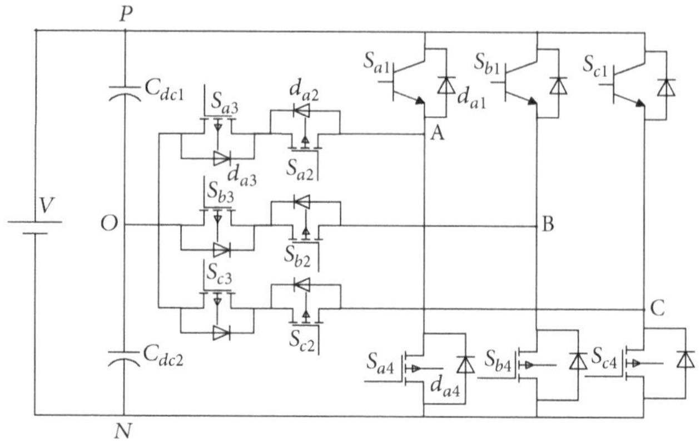

FIGURE 6: The proposed hybrid T-type inverter.

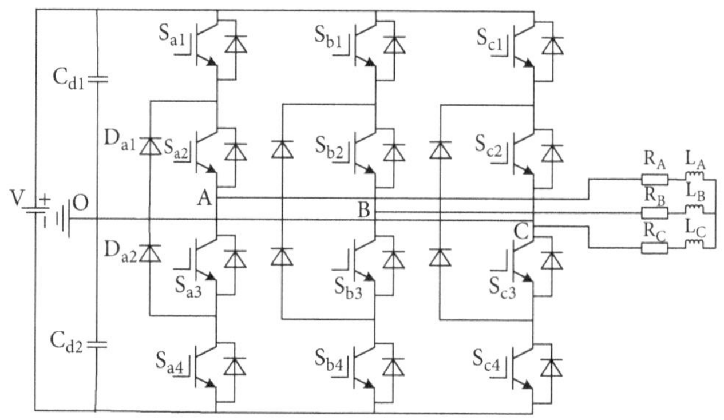

FIGURE 7: Diode clamped three-level inverter.

2.3. Loss Analysis of the Proposed Topology. The high power inverter works in high voltage and large current situations, where various losses caused by the opening device are relatively large. Besides, optimization of the switching characteristics of the power electronic devices yields bigger conduction losses of the system, especially for soft switching technology, where the switching loss of the power electronic devices is reduced, and the source of the power loss is converted to the conduction loss [16]. Therefore, it is a key step to accurately calculate the state loss for the design of grid-connected inverter systems. Take A phase of Figures 6 and 7 as an example, we compare the various losses of diode clamped (Figure 7) and hybrid T-type (Figure 6) inverter losses and their conversion efficiency. We assume that the output current of the grid-connected inverter is an ideal sine wave; the output voltage of the inverter integrates the conduction period in the period T. Then, the conduction losses per device may be expressed as

$$P_{S_{a1}} = \frac{1}{2\pi}\left[\int_0^{\pi-\theta} d(\omega t)\, U_{\text{com}}(i)\left(I_{\text{com}}\sin(\omega t)\, d(\omega t)\right)\right] \quad (1)$$

where $P_{S_{a1}}$ is the conduction loss of the per device, $d(\omega t)$ is the duty cycle, $U_{\text{com}}(i)$ represents the conduction voltage drop of the IGBT, I_{com} denotes the conduction current peak of the IGBT, and ω is the angle speed. The integral interval from 0 to π-θ is a chopper phase of the semiconductor switch Sal in a fundamental period. The conduction loss of the antiparallel diode in the IGBT reads

$$P_{da1} = \frac{1}{2\pi}\int_{\pi-\theta}^{\pi} (-m * \sin(\omega t + \phi)) \left[U_{od} I_{com} \sin(\omega t) + R_{od} (I_{com} \sin(\omega t))^2\right] d(\omega t) \quad (2)$$

where P_{da1} is the conduction loss of the antiparallel diode in the IGBT, m is the modulation index, Φ is the output power factor angles, and R_{od} and U_{od} are the conduction pressure drop constants of the antiparallel diode.

When IGBT/MOSFET is turned off, the current flows from the clamped diodes and then reads the following clamped diode conduction loss:

$$\begin{aligned} P_{Da1} = {} & \frac{1}{2\pi}\int_{0}^{\pi-\theta} (1 - m * \sin(\omega t + \phi)) \\ & \cdot \left[U_{dt} I_{com} \sin(\omega t) + R_{dt} (I_{com} \sin(\omega t))^2\right] d(\omega t) + \frac{1}{2\pi} \\ & \cdot \int_{0}^{\pi-\theta} (1 + m * \sin(\omega t + \phi)) \left[U_{dt} I_{com} \sin(\omega t) + R_{dt} (I_{com} \sin(\omega t))^2\right] d(\omega t) \end{aligned} \quad (3)$$

where U_{dt} and R_{dt} are the conduction pressure drop constants of the clamped diode.

We considered the DC side voltage of 700 V, carrier frequency fc of 5 kHz, gate resistance of 1.65 ohm, output frequency equal to 60 Hz, modulation rate of unity, and power factor of 0.8. We used MELCOSIM software to simulate the loss. Table 1 presents the calculated loss of different components.

In Table 1, the power losses of Sa1 and Sa4 as well as Sa2 and Sa3 IGBT switches are identical. The four antiparallel diode power dissipations, da1, da2, da3, and da4, are also identical. Moreover, clamping diodes, Da1 and Da2, have the same loss. Table 2 lists the overall loss of the two topologies, which may be calculated based on the number of components contained in two topologies. This table also presents the total power of the inverter, which is 10 kW, and the conversion efficiency of each topology.

The conversion efficiency is lower than 90% because of the higher switching frequency. We chose the frequency of 10 Khz MOSFET, and the switching frequency is relatively high; at the same time the switching loss itself is higher than at power frequency, so the conversion efficiency is lower than 90%.

Table 3 shows the comparison between the NP and T-type three-level inverters.

2.4. SVPWM Control Method. The theoretical basis of the SVPWM is the mean equivalent principle, that is, combining the fundamental voltage vectors in a switching cycle to make the average value equal to the given voltage vector. At a certain moment, the voltage vector rotates into a certain region, which may be achieved by a different combination over time of two adjacent nonzero vectors and zero vectors that make up this region. The action time of the two vectors is repeatedly applied in one sampling period, so that it controls the action time of each voltage vector. This rotates the voltage space vector in accordance with the circular trajectory and approaching ideal flux circle through the actual magnetic flux, generated by different switching states of the inverter. Then, it determines the inverter switch state by the comparison, at the end form of the PWM waveform.

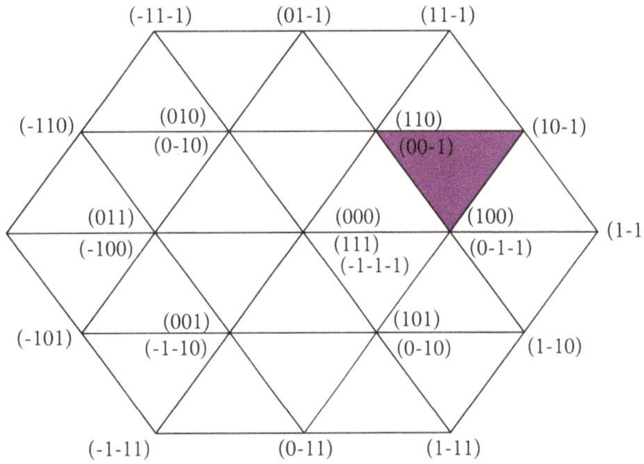

FIGURE 8: Three-level space vector diagram.

SVPWM presents lower total harmonic distortion (THD) compared to other control strategies, and it may refine the steady and dynamic state performances of the PV grid-connected system, simultaneously. Meanwhile, output results of the inverters with SVPWM control strategy provide better power quality than that of the inverters with other control strategies. The SVPWM control strategy with three-phase three-level voltage source inverter is effective and feasible [17]. The detailed equations of the SVPWM strategies based on the proposed topology are as follows.

In this paper we used decomposition hexagon method.

Idea: decompose the multilevel space vector into a combination of multiple two-level space vectors to achieve greatly simplified PWM calculation method.

The three-level space vector diagram is shown in Figure 8. Any reference vector must fall within a small triangle. The vertex of this triangle is the basic voltage vector that composes this reference vector.

The reference vector falling into the shadow in Figure 8 can be decomposed into an offset vector and a two-level vector, as shown in Figure 9.

Steps

2.4.1. Sector Judgments. Based on Clark's transformation, the normalized output vectors transformed from abc to $\alpha\beta$ reference frame may be expressed as

$$\begin{pmatrix} V_\alpha \\ V_\beta \\ V_o \end{pmatrix} = \begin{pmatrix} \frac{2}{3} & -\frac{1}{3} & -\frac{1}{3} \\ 0 & \frac{1}{\sqrt{3}} & -\frac{1}{\sqrt{3}} \\ 1 & 1 & 1 \end{pmatrix} \begin{pmatrix} V_A \\ V_B \\ V_C \end{pmatrix} \quad (4)$$

TABLE 1: The loss of each device in hybrid T-type inverter.

Switch type	device	Loss (W) Conduction loss	Switching loss	Total loss
IGBT	S_{a1}	90.45	26.46	116.92
	S_{a2}	31.78	1.03	32.82
Freewheeling diode	d_{a1}	40.59	0.47	41.07
Clamped diode	D_{a1}	2.23	9.93	12.17

TABLE 2: The power loss and conversion efficiency of two topologies.

component	The number of DC	Loss (W) IGBT	diode	Total loss	conversion efficiency
NPC	1	998.88	310.72	1309.6	0.8690
Hybrid T-type	1	898.44	319.44	1217.88	0.8782

TABLE 3: Similarities and differences of diode clamped and T-type three-level circuit topologies.

Compared items	Diode clamped three-level inverter	T-type three-level inverter
Switch pressure	Sa1~Sa4: 0.5 UPV	Sa1/Sa2: Upv; Sa3/Sa4: 0.5 Upv
Commutation path	Long commutation path and short commutation path	All paths are consistent
Efficiency	The higher switching frequency (>16 kHZ) increases efficiency	The lower switching frequency (<16 kHZ) increases efficiency
Modulation strategy	Traditional control strategy of current PI loop	Traditional control strategy of current PI loop
Number of components	4 switches plus 2 diodes	4 switches
Drive power	4 groups	3 groups

TABLE 4: The relations between N and the sector where V_{ref} is located.

N	1	2	3	4	5	6
sector	II	VI	I	IV	III	V

where V_α, V_β, V_o are voltages on the two-phase stationary coordinate system and V_A, V_B, V_C are voltages on the three-phase stationary coordinate system.

We define $N = A+2B+4C$, and the value of N determines in which sector the reference vector V_{ref} is located. So the relations between N and the corresponding sector are shown in Table 4.

2.4.2. Basic Vector Dwell Time Calculation. We assume V_{ref} is located at sector I, which yields the following equation:

$$V_\alpha T_s = T_1 |V_1| + T_2 |V_2| \cos \frac{\pi}{3}$$
$$V_\beta T_s = T_2 |V_2| \sin \frac{\pi}{3} \quad (5)$$
$$T_s = T_1 + T_2 + T_{0,7}$$

where T_s is the sampling period; T_1, T_2, and T_0 represent the dwell times of the basic vectors V_1, V_2, and V_0, respectively.

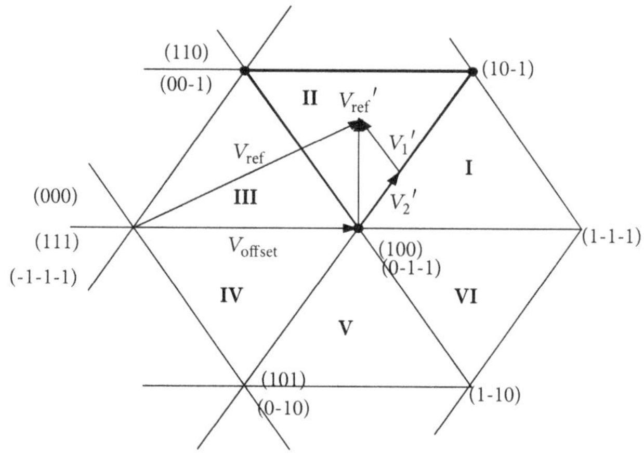

FIGURE 9: Decomposed two-level space vector diagram.

According to the above expressions, we can obtain the following equation:

$$T_1 = \frac{\sqrt{3}T_S}{2V_{dc}} \left(\sqrt{3}V_\alpha - V_\beta\right)$$
$$T_2 = \frac{\sqrt{3}T_s V_\beta}{V_{dc}} \quad (6)$$
$$T_{0,7} = T_s - T_1 - T_2$$

TABLE 5: The relations between vector dwelling time and its sector.

sector	I	II	III	IV	V	VI
T_1	Z	Y	-Z	-X	X	-Y
T_2	Y	-X	X	Z	-Y	-Z

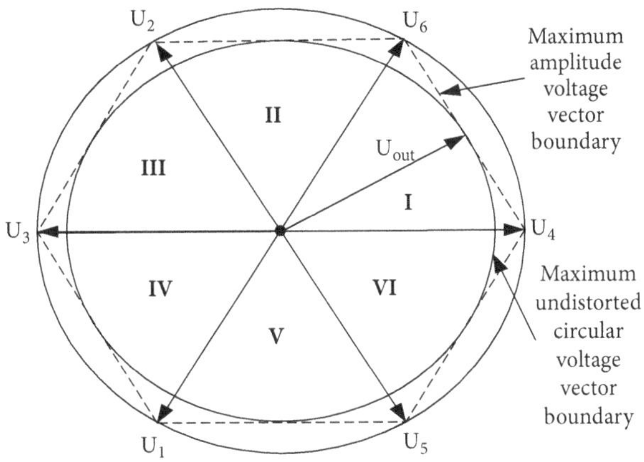

FIGURE 10: Voltage vector amplitude boundary in SVPWM mode.

Following the same principle, the dwell time of each vector, we achieve V_{ref} of different sectors. To facilitate the solution, we may define it as (7). The value of T_1 and T_2 at different sectors could be set according to Table 5.

$$X = \sqrt{3}\frac{T_S}{V_{dc}}V_\beta$$
$$Y = \frac{T_s}{V_{dc}}\left(\frac{\sqrt{3}}{2}V_\beta + \frac{\sqrt{3}}{2}V_\alpha\right) \tag{7}$$
$$Z = \frac{T_s}{V_{dc}}\left(\frac{\sqrt{3}}{2}V_\beta - \frac{\sqrt{3}}{2}V_\alpha\right)$$

2.4.3. Vector Switching Point Calculation. When the synthesized voltage vector endpoint falls between the regular hexagon and the circumscribed circle, as shown in Figure 10, the overmodulation has occurred and the output voltage will be distorted. So we use a proportional scaling algorithm to control the overmodulation. The vector dwell time that first occurs in each sector is defined as T_{Nx}, and the vector dwell time that occurs after is defined as T_{Ny}. When $T_x + T_y \leqslant T_{\mathrm{NPWM}}$, the vector endpoint is within the regular hexagon and no overmodulation occurs. When $T_{\mathrm{Nx}} + T_{\mathrm{Ny}} > T_{\mathrm{NPWM}}$, the vector endpoint is beyond the regular hexagon and overmodulation occurs. The output waveform will be seriously distorted and must take the following measures.

Suppose that the nonzero vector dwell time is T'_{Nx}, T'_{Ny}, when the endpoint of the voltage vector trace is pulled back to the inscribed circle of the regular hexagon, and then there is a proportional relationship:

$$\frac{T'_{Nx}}{T_{Nx}} = \frac{T'_{Ny}}{T_{Ny}} \tag{8}$$

Therefore, T'_{Nx}, T'_{Ny}, T_{N0}, T_{N7} can be obtained by the following formula:

$$T'_{Nx} = \frac{T_{Nx}}{T_{Nx} + T_{Ny}}T_{\mathrm{NPWM}}$$
$$T'_{Ny} = \frac{T_{Ny}}{T_{Nx} + T_{Ny}}T_{\mathrm{NPWM}} \tag{9}$$
$$T_0 = T_7 = 0$$

According to the above process, the action time of two adjacent voltage space vectors and zero-voltage vectors in each sector can be obtained. The operation relationship is shown in Figure 11 when U_{ref} is in sector I. After the U_{ref} sector and the corresponding effective voltage vector are determined, according to the PWM modulation principle, the value of each corresponding comparator is calculated, and the operation relationship is as follows:

$$t_{\mathrm{aon}} = \frac{\left(T_s - T_x - T_y\right)}{2}$$
$$t_{\mathrm{bon}} = t_{\mathrm{aon}} + T_x \tag{10}$$
$$t_{\mathrm{con}} = t_{\mathrm{bon}} + T_y$$

Other sectors follow the same above principle.

Here, T_{cm1}, T_{cm2}, T_{cm3} denote the transistor's switching time. And the relation between sector switching point and its appropriate sector is tabulated in Table 6.

3. Results and Discussion

This study put forward a novel hybrid T-type inverter topology which is composed of basic units A and B on the basis of previous research studies. We established a three-phase three-level hybrid T-type photovoltaic grid-connected inverter topology model, which is shown in Figure 12, using MATLAB platform. Considering the A-phase bridge leg, for example, it consists of one half-bridge IGBT, one half-bridge MOSFET, and two neutral point MOSFETs. Switches Sal and Sa2 work in the mutual intermittent state, and switches Sa3 and Sa4 only work near the current zero-crossing point with high frequency. This topology is rather suited for the photovoltaic nonisolated AC system applications.

The topological structure is on the basis of T-type structure, changing the nine switches into MOSFET, i.e., Sa2, Sb2, Sc2 ··· Sa4, Sb4, and Sc4. We choose IGBT for Sal, Sbl, and Scl; since the reverse recovery ability of the body diode in the field effect transistor is poor, therefore they act in low frequencies. The high frequency MOSFET semiconductor switches, i.e., Sa2, Sb2, Sc2, . . ., Sa4, Sb4, Sc4, provide good switching characteristics and low on-resistance. Moreover, due to the low speed characteristics of the built-in diode of MOSFETs, MOSFET cannot be used in the upper bridge arm. We take advantage of the two devices, to reduce the harmonic content and the power loss of the converter and improve the conversion efficiency of the system.

Regarding the topology selection, the three-level circuit combines the advantages of the nonisolation and multilevel

TABLE 6: The relationship between the vector switching point and its corresponding sector.

sector	I	II	III	IV	V	VI
T_a	T_{bon}	T_{aon}	T_{aon}	T_{con}	T_{con}	T_{bon}
T_b	T_{aon}	T_{con}	T_{bon}	T_{bon}	T_{aon}	T_{con}
T_c	T_{con}	T_{bon}	T_{con}	T_{aon}	T_{bon}	T_{aon}

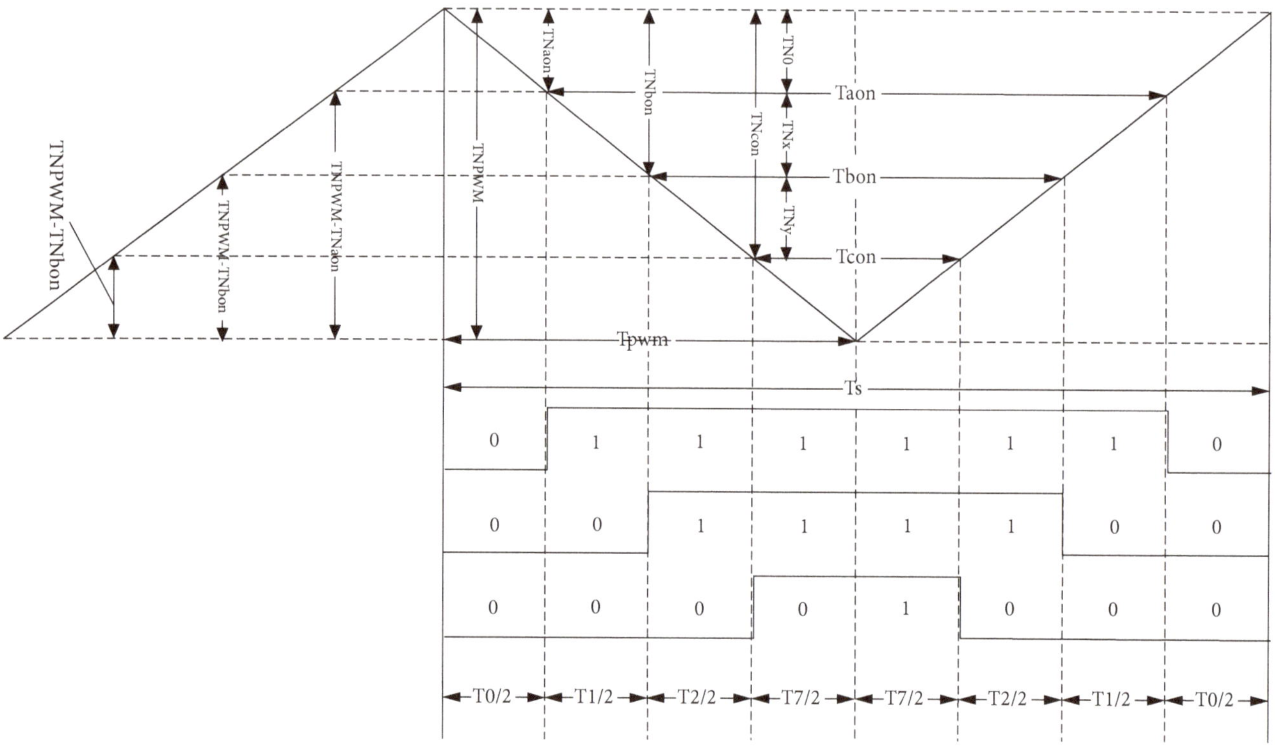

FIGURE 11: Operation relationship when U_{ref} is in sector I.

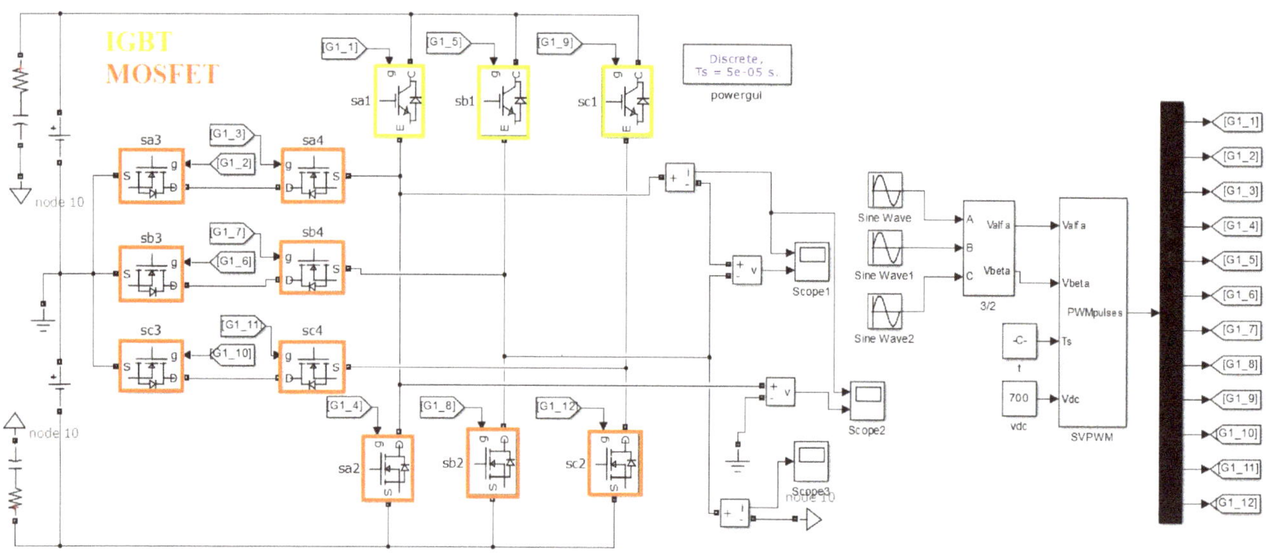

FIGURE 12: The simulation model of the SVPWM controlled novel T-type three-phase three-level inverter.

TABLE 7: The classification of the influence of the switching state of the short and medium vectors on the direction of the neutral point current.

Positive short vector switching status	io	Negative short vector switching status	io	Medium vector switching status	io
100	ia	211	- ia	210	ib
221	ic	110	- ic	120	ia
010	ib	121	- ib	021	ic
122	ia	011	- ia	012	ib
001	ic	112	- ic	102	ia
212	ib	101	- ib	201	ic

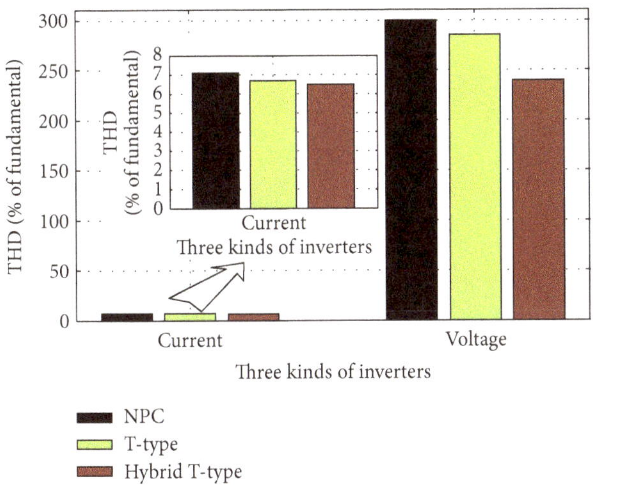

FIGURE 13: The current and voltage THD comparison of three types of inverters.

technologies, which is very suitable for the photovoltaic grid-connected power generation.

System simulation parameters are as follows: The DC-link voltage is 700 V, with the frequency of 50 Hz, the value of the support capacitor reads 3300 uF, with the AC-side inductance of 3 mH, the AC-side resistance is 0.1 Ω, and the switching frequency is 10 kHz. The value of the stray capacitance is 10e−6 F, and the small resistor is connected in parallel with the stray capacitance of $R = 10e - 6$ ohm.

Table 7 shows the neutral point current when the positive and negative short vector and medium vector act. It can be seen that positive and negative short vector and medium vectors will cause neutral voltage fluctuations.

The neutral point voltage control method is based on the SVPWM. According to the influence of the medium and short vectors on the neutral point voltage offset and by selecting the appropriate switching state and the most suitable switching sequence for the neutral point voltage, the midpoint voltage offset during each control cycle has been minimized.

So selecting the excellent transistor sequence is really important. To solve this problem we run through all the transistors in every possible combination and permutation on the condition of the following switch rules.

(1) The switching states of the devices in every bridge leg are independent. (2) The switching states of any two adjacent switches of each bridge leg are complementary (e.g., if Sa1 is turned on, then Sa2 must be switched off). (3) According to the principle of complementarity, if the switching state of any of the devices in the same bridge leg is determined, the state of the other switching devices of that bridge leg can also be confirmed [18].

The influence of harmonic current on the power grid is greater than the harmonic voltage and it is the fundamental cause of most of the problems, and the neutral voltage fluctuations are proportional to the amount of the harmonic current. So, as can be seen from Table 8, giving comprehensive consideration, choose the following optimal transistor sequence: Sa1-Sa3-Sa4-Sa2-Sb1-Sb3-Sb4-Sb2-Sc1-Sc3-Sc4-Sc2.

The control method is achieved by the SVPWM with 12 trigger pulses. Switching sequence reads Sa1-Sa3-Sa4-Sa2-Sb1-Sb3-Sb4-Sb2-Sc1-Sc3-Sc4-Sc2.

Figure 13 shows the output voltage and current harmonic content of the three types of inverter topology structure, using SVPWM control method. From Figure 13 we could see that current and phase voltage THD (total harmonic distortion) of the NPC inverter is the biggest, the T-type inverter is the middle one, and the proposed hybrid T-type is the smallest one. Meanwhile, the amplitude of the three kinds of inverter current harmonics is not obvious; however, the voltage THD results of the three kinds of inverter are sharply comparable.

The topology and control strategy of the two circuits are the same, except the devices used. The T-type topology consists of 12 IGBTs, while the hybrid T-type topology consists of 9 MOSFETs and 3 IGBTs. And the off-delay time and dead time of the IGBT are longer than those of the MOSFET.

Although the proportion of dead time is often very small relative to one switching cycle, the dead zone will make the three-phase control system deviate from the ideal mathematical model. When the switching frequency becomes higher, the dead-zone effect will gradually accumulate, and when it accumulates to a certain degree, it will distort the AC-side voltage waveform, which will affect the waveform quality of the input current. It will also cause fluctuations in the DC voltage, which degrades the accuracy of the entire system. In addition, the dead time can cause common mode voltage waveform distortion, resulting in higher harmonics [19]. So the THD of the novel T-type inverter is the lowest.

From Figure 14 we see that the common mode current of the proposed novel T-type inverter is smaller than the previous T-type inverter topology. In conclusion the proposed

TABLE 8: Comparison of output parameters under various switching sequence (simulation time: 0.1 s).

Sequence of the transistors	Power factor	Current harmonics%	voltage harmonics%
Sal- Sa2- Sbl- Sb2- Scl- Sc2- Sa3- Sa4- Sb3- Sb4- Sc3- Sc4	0.7209	27.27	158.99
Sal- Sa2- Sbl- Sb2- Scl- Sc2- Sa4- Sa3- Sb4- Sb3- Sc4- Sc3	0.7263	21.73	236.75
Sal- Sc2- Sbl- Sa2- Scl- Sb2- Sa3- Sa4- Sb3- Sb4- Sc3- Sc4	0.647	11.61	149.13
Sal- Sc2- Sbl- Sa2- Scl- Sb2- Sa4- Sa3- Sb4- Sb3- Sc4- Sc3	0.6158	7.81	224.95
Sal- Sa2- Sa3- Sa4- Sbl- Sb2- Sb3- Sb4- Scl- Sc2- Sc3- Sc4	0.5799	0.63	156.03
Sal- Sa2- Sa4- Sa3- Sbl- Sb2- Sb4- Sb3- Scl- Sc2- Sc4- Sc3	0.447	0.92	230.51
Sal- Sa3- Sa4- Sa2- Sbl- Sb3- Sb4- Sb2- Scl- Sc3- Sc4- Sc2	0.7132	6.61	248.44
Sal- Sa4- Sa3- Sa2- Sbl- Sb4- Sb3- Sb2- Scl- Sc4- Sc3- Sc2	0.6225	6.91	158.98

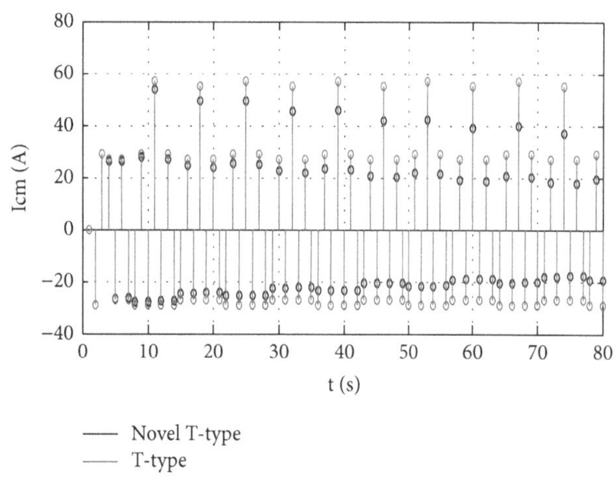

FIGURE 14: The obtained CM-current of the novel and previous T-type inverters.

hybrid T-type inverter has priority compared to the NPC and T-type inverter.

The instantaneous common mode voltage and ground leakage current could be given by the following equation:

$$u_{cm} = \frac{u_{AN} + u_{BN} + u_{CN}}{3}$$
$$i_{cm} = C\frac{d_{u_{cm}}}{dt} \quad (11)$$

where u_{AN}, u_{BN}, and u_{CN} are the pulse voltages between the branch midpoint and the dc bus minus terminal, respectively

It can be seen from the equation that the common mode current i_{cm} is proportional to the change rate of the common mode voltage u_{cm}.

According to the above interpretation it can be got that the dead time of the proposed topology is the shortest because of the smallest number of IGBTs compared with the other two topologies, so that the common mode voltage and current are also the smallest.

4. Conclusions

We introduced a variety of new multilevel photovoltaic grid-connected inverter topologies, compared them with each other, and classified them according to the basic unit which predecessors proposed. We proposed the hybrid T-type inverter topology, which is composed of two best basic units. This structure makes full advantages of the two devices, which leads to reducing the harmonic content and power loss of the converter and improving the conversion efficiency of the system. By comparing the new topology, the new multilevel inverter topology is formed by the basic constituent elements to increase the auxiliary/embedded circuit or with a hybrid switch or with asymmetric structure to optimize the inverter performance. We use the space vector pulse width modulation (SVPWM) method to build the SIMULINK models of the proposed hybrid T-type and NPC three-level inverter in the MATLAB software and compare them with each other. The simulation results show that the proposed topology presents lower output harmonics than the NPC topology. We then provided further loss analysis for these two topologies, using the MELCOSIM software to calculate the loss of different components in each topology as well as the total power loss. The outcome is that the loss of the T-type topology is significantly lower than that of the NPC topology, and the conversion efficiency is higher than that of the NPC topology. Moreover the simulation results show that the common mode current of the proposed novel T-type inverter is smaller than the previous T-type inverter topology.

The proposed inverter topology has a certain practicality and economy to meet the actual needs.

Conflicts of Interest

All the authors declare that there are no conflicts of interest regarding the publication of this paper.

Acknowledgments

The work was supported by the National Natural Science Foundation of China (Grants no. 51667020 and no. 61364010), the Outstanding Doctor Graduate Student Innovation Project of Xinjiang University (no. XJUBSCX-2016018), and Graduate Student Innovation Project of Xinjiang Uygur Autonomous Region (no. XJGRI2017025).

References

[1] F. Michael and E. Temesi, "High Efficient Topologies for Next Generation Solar Inverter," *Inverter World*, vol. 03, pp. 31–36, 2009.

[2] C. I. Odeh, "A cascaded multi-level inverter topology with improved modulation scheme," *Electric Power Components and Systems*, vol. 42, no. 7, pp. 768–777, 2014.

[3] C. I. Odeh, "Sinusoidal pulse-width modulated three-phase multi-level inverter topology," *Electric Power Components and Systems*, vol. 43, no. 1, pp. 1–9, 2015.

[4] K. P. Draxe, M. S. B. Ranjana, and K. M. Pandav, "A cascaded asymmetric multilevel inverter with minimum number of switches for solar applications," in *Proceedings of the 2014 Power and Energy Systems Conference: Towards Sustainable Energy, PESTSE 2014*, 15, 13 pages, ind, March 2014.

[5] M. Azri, N. A. Rahim, and W. A. Halim, "A highly efficient single-phase transformerless H-bridge inverter for reducing leakage ground current in photovoltaic grid-connected system," *Electric Power Components and Systems*, vol. 43, no. 8-10, pp. 928–938, 2015.

[6] H. Schmidt, C. Siedle, and J. Ketterer, "Inverter for transforming a DC voltage into an AC current or an AC voltage," Tech. Rep. EP1369985(A2), Germany, 2003-12-10.

[7] D. Jovcic and K. Ahmed, "Method of converting a direct current voltage form a source of direct current voltage, more specifically from a photovoltaic source of direct current voltage, into a alternating current voltage," Tech. Rep. US7411802(B2), United States, 2008.

[8] H. Xiao, S. Xie, Y. Chen, and R. Huang, "An optimized transformerless photovoltaic grid-connected inverter," *IEEE Transactions on Industrial Electronics*, vol. 58, no. 5, pp. 1887–1895, 2011.

[9] R. González, J. López, P. Sanchis, and L. Marroyo, "Transformerless inverter for single-phase photovoltaic systems," *IEEE Transactions on Power Electronics*, vol. 22, no. 2, pp. 693–697, 2007.

[10] W. Yu, J.-S. Lai, H. Qian, and C. Hutchens, "High-efficiency MOSFET inverter with H6-type configuration for photovoltaic nonisolated AC-module applications," *IEEE Transactions on Power Electronics*, vol. 26, no. 4, pp. 1253–1260, 2011.

[11] L. Zhang, K. Sun, and L. L. Feng, "H6 non-isolated full bridge grid-connected PV inverters with low leakage currents," in *Proceedings of the CSEE 32*, vol. 32.15, pp. 1–7, 2012.

[12] S. V. Araújo, P. Zacharias, and R. Mallwitz, "Highly efficient single-phase transformerless inverters for grid-connected photovoltaic systems," *IEEE Transactions on Industrial Electronics*, vol. 57, no. 9, pp. 3118–3128, 2010.

[13] X. Zhang, L. L. Sun, and P. Xu, "Research on the restraining of common mode current in the single-phase transformerless grid-connected PV system," *Acta Energiae Solaris Sinica 30.9*, vol. 30, no. 9, pp. 1202–1208, 2009.

[14] T. Kerekes, R. Teodorescu, P. Rodríguez, G. Vázquez, and E. Aldabas, "A New high-efficiency single-phase transformerless PV inverter topology," *IEEE Transactions on Industrial Electronics*, vol. 58, no. 1, pp. 184–191, 2011.

[15] U.-M. Choi, F. Blaabjerg, and K.-B. Lee, "Reliability improvement of a T-type three-level inverter with fault-tolerant control strategy," *IEEE Transactions on Power Electronics*, vol. 30, no. 5, pp. 2660–2673, 2015.

[16] C. Yang and Y. Liu, "The contrast of conduction dissipation of T-type 3-level inverter and diode-clamped 3-level inverter," *Autocontrol 02*, vol. 02, pp. 61–66, 2015.

[17] W. Kun, Y. Xiaojie, W. Chenchen, and Z. Minglei, "An equivalent dual three-phase SVPWM realization of the modified 24-sector SVPWM strategy for asymmetrical dual stator induction machine," in *Proceedings of the 2016 IEEE Energy Conversion Congress and Exposition, ECCE 2016*, Milwaukee, WI, USA, September 2016.

[18] T. Mingting, *Research on Three-Phase T-Type Three-Level Non-Isolated Grid-Connected Inverter*, Hefei University of Technology, 2013.

[19] J.-W. Choi and S.-K. Sul, "Inverter output voltage synthesis using novel dead time compensation," *IEEE Transactions on Power Electronics*, vol. 11, no. 2, pp. 221–227, 1996.

9

Modelling and Automated Implementation of Optimal Power Saving Strategies in Coarse-Grained Reconfigurable Architectures

Francesca Palumbo,[1] Tiziana Fanni,[2] Carlo Sau,[2] Paolo Meloni,[2] and Luigi Raffo[2]

[1]*POLCOMING, Information Engineering Unit, University of Sassari, Sassari, Italy*
[2]*Department of Electrical and Electronic Engineering (DIEE), University of Cagliari, Cagliari, Italy*

Correspondence should be addressed to Tiziana Fanni; tiziana.fanni@diee.unica.it

Academic Editor: Wen B. Jone

This paper focuses on how to efficiently reduce power consumption in coarse-grained reconfigurable designs, to allow their effective adoption in heterogeneous architectures supporting and accelerating complex and highly variable multifunctional applications. We propose a design flow for this kind of architectures that, besides their automatic customization, is also capable of determining their optimal power management support. Power and clock gating implementation costs are estimated in advance, before their physical implementation, on the basis of the functional, technological, and architectural parameters of the baseline design. Experimental results, on 90 and 45 nm CMOS technologies, demonstrate that the proposed approach guides the designer towards optimal implementation.

1. Introduction

Electronic devices on the market rely on the execution of computation-intensive applications on complex heterogeneous systems. Coarse-grained reconfigurable (CGR) platforms combine the high performance levels provided by Application Specific Integrated Circuit (ASIC) designs with an increased flexibility, allowing the execution of a larger set of applications over the same substrate [1, 2]. However, in the dark silicon era, due to the limited available power budget, a gap exists between the number of transistors that can be placed within a die and the number that can be actually active during execution [3, 4]. Therefore, systems are also required to be energy efficient and CGR designs must integrate specific power management techniques of the functional logic regions constituting them.

Several effective techniques for power monitoring [5] and reducing [6] have been presented at the state of the art. Among them, voltage/frequency scaling [7, 8] and power shut-off schemes [9, 10] can be extremely beneficial. However, their integration requires manual intervention of the designer, resulting in a complex, error prone, and time consuming process. While commercial synthesizers [11] allow the automatic implementation of low overhead saving strategies at the gate level, such as fine-grained clock gating, they only provide implementation-level instruments to apply more complex strategies, like power gating. In the CGR systems field, the Multi-Dataflow Composer tool (MDC), combining the dataflow-based system specification approach with the coarse-grained reconfigurable design paradigm, is capable of automatically generating run-time reconfigurable multifunctional systems, featuring flexibility and area minimization [12]. MDC was originally meant to address reconfigurable codec implementations and was conceived to be exploited within MPEG Reconfigurable Video Coding (MPEG-RVC) studies. However, it was successfully adopted also in different resource and power-constrained scenarios [13, 14], where only microprogrammed solutions have been used so far, either exploiting single-core digital signal processors [15] or custom multicore embedded processors [16]. MDC design suite is composed of different extensions. The work presented in this paper is related to MDC power management extension, which has been previously addressed in [17, 18]. MDC tool identifies in the generated CGR system the minimum set of disjointed functionally homogeneous logic areas of the system, called logic regions. These latter are exploited to automatically implement dynamic power management strategies, applying indistinctly to all of them either clock [17] or power gating methodologies [18].

The work we are presenting in this paper intends to propose a power modelling methodology and to improve the MDC power management extension by integrating such methodology within its automated flow. We introduce in this paper an algorithm that analyses the identified logic regions and, on the basis of one single synthesis and a minimal set of simulations (one for each scenario of the multifunctional problem), is capable of optimally characterizing the power management support. This flow, in a separate manner for each logic region of the CGR design, is capable of assessing both clock and power gating management costs and of determining which is the optimal power saving strategy (if any) prior to any physical system implementation. The algorithm is based on detailed static and dynamic power consumption models that take into account functional, architectural, and technological parameters to define the potential overhead and benefits of the considered solutions. As a future perspective, besides its application within the MPEG-RVC scenario, the proposed modelling strategy may also be extended to support other complex autonomous computing systems [19], where the number of involved resources may change at runtime.

The rest of this paper is organized as follows. Section 2 reports the background of this work. Section 3 describes the operation of MDC: in particular, Section 3.1 focuses on the base operation, while Section 3.2 details the proposed power estimation models and their integration within the tool. Section 4 presents the designs under test used to validate the proposed approach: Section 4.1 involves a FFT use case targeting a 90 nm CMOS technology, while Section 4.2 presents the experimental results conducted to assess the enhanced MDC flow on a zoom coprocessor (targeting both 90 nm and 45 nm CMOS technologies). Finally, Section 4.4 details the benefits of the proposed models, before concluding with some final remarks in Section 5.

2. Background

This work presents a model of power consumption for CGR systems. The model is capable of estimating the system power dissipation since the early stages of the design flow and it has been integrated within an automated flow that decides which power saving technique, between clock gating and power gating, has to be applied to each portion of the design.

This section provides an overview of the state of the art in the field of power-aware optimization and, in more detail, on the main aspects involved in the proposed approach. Section 2.1 introduces the distinctive features of CGR systems and the main architectural trends for such a kind of devices. Section 2.2 deals with power issues in digital system, with a particular emphasis on modelling strategies and design automation.

2.1. Coarse-Grained Reconfigurable Systems. Reconfigurable architectures are usually conceived as collections of functional units (FUs) whose functionality and connections can be configured at run time, to adapt them to different applications or operating modes. Such systems can be classified according to the granularity of the FUs. Fine-grained approaches, typically exploiting FPGA devices as underlying technology, involve bit-level FUs, resulting in a higher flexibility but requiring long configuration time (due to the configuration bitstream size). Coarse-grained reconfigurability, on the other hand, provides word-level FUs, thus providing less flexibility while guaranteeing faster configuration phases. CGR architectures are usually exploited to design flexible ASICs, making them capable of switching among a finite set of functionality. In such design cases, high efficiency in terms of area obstruction of the designed system can be easily obtained, but not all the resources are evenly involved in the computation. Dedicated power management techniques are needed to reduce the overhead, in terms of power consumption, related to resources that are not involved in each operating mode [17]. CGR architectures already demonstrated being suitable to address application scenarios that require flexibility along with strong area, power, and execution efficiency [13, 20].

One of the main issues of CGR architectures is their complex mapping and programming [21, 22]. Several works tried to automate the mapping of applications and computational kernels onto CGR and multicore systems [23–25]. The mapping problem requires specific knowledge of the considered kernels that usually have to be identified and specified by means of hardware description languages. The mapping effort is directly proportional to the number of involved kernels [26]. Recently, dataflow models demonstrated to be very useful in this scenario [27, 28]. Dataflows describe programs through a graph whose nodes are processing elements (*actor*) linked by point-to-point unidirectional channels managed according to a FIFO protocol. Actors encapsulate their own state and communicate only through atomic packets of data (*tokens*). Due to their intrinsic modularity, dataflows favour hardware and software components definition and reuse. Furthermore, they are natively capable of highlighting the intrinsic parallelism of the specified applications. The Multi-Dataflow Composer (MDC) tool, adopted within the presented work, relies on dataflows (RVC-CAL formalism by MPEG is currently supported) to solve the CGR mapping problem. It exploits the characteristics of such kind of models to provide several advanced features (e.g., power management [17, 18] or coprocessing units automatic generation [29]).

2.2. Power Management. Power consumption in digital devices is composed mainly of two different contributions: dynamic and static. The former is due to capacitance charging/discharging when logic transitions occur (i.e., switching activity). The latter is due to leakage currents and it is consumed also when no circuit activity is present. Modern designers need to consider both terms when conceiving smart management strategies. Several techniques (clock gating, multifrequency, operand isolation, multithreshold, multisupply libraries, power gating, etc.) exist and, in some cases, they are automatically implemented by commercial synthesis/place-and-route tools. In custom computing systems, some advanced design tools support the designers in the application-driven customization of the hardware architectures [30, 31]. However, generally speaking, invasive techniques (requiring insertion of additional logic and target technology support (such as the availability of dedicated cells and processes on the implementation stack)) still need tools to be guided with significant manual effort by the designer.

Clock gating is an example of quite noninvasive technique. It may reduce the dynamic power consumption due to the clock tree and to sequential logic up to the 40% [32]. It consists in shutting off the clock of the unused synchronous logic, by means of simple *AND* gates. Clock gating has been deeply automated and it is available on most of the commercial synthesizers. In the MPEG-RVC community, recent studies [33] presented an extension of a High-Level Synthesis tool, Xronos, to selectively switch off clock signal for parts of the circuit that are idle due to stalls in the pipeline, to reduce power consumption. Moreover, as mentioned, the MDC tool has the capability of identifying, by means of a graph-based analysis of the input dataflow specifications, independent circuitry regions. These logic regions can be clock gated to dynamically adapt power consumption when switching between different functionalities [17, 18]. From the technical point of view, in ASIC designs *AND* gates can be used directly on the clock to disable it, while in FPGA designs the clock network cannot be modified by the insertion of any custom logic and dedicated cells are required (Xilinx boards, e.g., are equipped with dedicated blocks (*BUFGs*), whose outputs can drive distinct regions of logic powering down different design portions (when enabled)). Clock gating can be applied at different granularities: fine-grained approaches act on single registers, whereas coarse-grained ones are referred, as in [17, 18, 33], to a set of resources. Commercial synthesizers normally can automate only fine-grained strategies.

Power gating is quite invasive. The main idea behind it is as follows: if a specific portion of the design is not used in a given computation mode, then it can be completely switched-off by means of a sleep transistor. This technique, as the clock gating one, is applicable at different granularities: fine-grained approaches require driving a different sleep transistor for every cell in the system, while coarse-grained ones, again, operate on a set of resources instantiating one sleep transistor to drive different cells connected to a shared power network. MDC, as discussed in [18], supports also automatic power gating for CGR architectures. Each identified logic region in the CGR system is implemented (no matter of its nature or characteristics) as a different power domain (PD) that, in order to be managed, requires to insert and drive the following resources:

(i) The *sleep transistor* between the gated region and the main power supply to switch on/off the derived power supply

(ii) The *isolation logic* between the gated region and normally-on cells to avoid the transmission of spurious signals in input to the normally-on cells

(iii) The *state retention logic* to maintain, where needed, the internal state of the gated region

MDC, besides defining the power gated design netlist, provides also the automatic definition of the power format file. This file specifies the shut-off, isolation, and state retention rules (if you have 10 PD you are required to define $3 * 10 = 30$ interfaces), along with their respective enable signals ($[1 *$ shut-off $+ 1 *$ isol. $+ 2 *$ reten.$] * 10 = 40$ signals) and a dedicated Power Controller to properly drive them. The power format file automatically created by MDC is compliant with the Silicon Integration Initiative's Common Power Format (CPF), whose definition is driven mainly by developers using Cadence [34] tools.

2.2.1. Modelling. To the best of our knowledge, literature does not treat the problem of modelling power gating and clock gating costs in CGR designs. Some approaches only partially address the issue. For example, [35] focusses on low-power techniques and power modelling for FPGAs. In [36], only clock gating is taken into account: different power states (on the basis of the clock enable signals) are defined and their consumption is characterized by low-level Power Analysis results. [37] focusses on estimating the leakage reduction for power gating and reverse body bias.

The CASPER simulator for shared memory many-core processors [38] includes precharacterized libraries containing power dissipation models of different hardware components, enabling accurate power estimation at a high-level exploration stage. In particular the authors implement Chip-wide Dynamic Voltage, Frequency Scaling, and Performance Aware Core-Specific Frequency Scaling. The FALPEM framework [39] provides power estimations at preregister transfer level (RTL) stage, specifically targeting the power consumed by clock network and interconnect, but power and clock gating costs are not defined. Other approaches perform an estimation that considers different components. Li et al. [40] propose an architecture-level integrated power, area, and timing modelling framework for multicore systems, which evaluates system building blocks (CPU, buses, etc.) for different technology nodes, providing also power gating support. Finally, the work presented in [41] focuses on on-chip networks.

3. Design Suite for Coarse-Grained Power-Aware Systems

This section discusses the proposed technique for modelling the power consumption of a CGR system when clock gating or power gating are applied. These models, combined in a selection algorithm, can be exploited for developing an automated design flow for power-efficient CGR systems, where the optimal saving strategy is selected for each identified working set of resources.

In this work, we have embedded these models and the algorithm in the Multi-Dataflow Composer (MDC) tool, a framework capable of CGR systems characterization. MDC provides a comprehensive design suite automating several development tasks of the synthesis and development of CGR systems, within design flows targeting both FPGA [29] and ASIC. The tool provides extensive support to dynamic power management [17, 18], addressing power-constrained design cases scenarios, completely automating implementation and control of clock gating and power gating strategies in the final CGR platform. Nevertheless, such techniques are not currently addressed in a hybrid manner, the users must choose the approach to be used in the design without an a priori automated analysis process. Such an unsupported selection may easily lead to suboptimal implementations on the final platform.

In the following, Section 3.1 provides an overview of the MDC baseline functionality and of the current power management support. Section 3.2 discusses the proposed power models and automated selection algorithm identifying optimal power management strategy in CGR systems. In both sections, step-by-step examples are presented to clarify the methodology.

3.1. The Multi-Dataflow Composer Tool. MDC automates generation and management of CGR systems, facing the complex mapping of multiple applications onto a single reconfigurable architecture. It automates the mapping process and guarantees the minimization of hardware resources, allowing for significant area/energy savings [12, 42]. In literature, this problem is known as *datapath merging* and it deals with the combination of a set of input datapaths, described by means of graphs, onto a single reconfigurable datapath. It aims at maximally sharing (among the different input graphs) both processing nodes and connections.

MDC is naturally compliant with the RVC-CAL formalism and natively supports Dataflow Process Network (DPN) models as input. Currently, it is interfaced with the Open RVC-CAL Compiler, Orcc [43]. The Orcc front-end is responsible for parsing, one-by-one, the high-level DPN specifications of the different datapaths that MDC will merge within the CGR system. Please note that MDC can be interfaced with other graph parsers, so that it will be able to be easily adapted to any other dataflow-based modelling environment.

As depicted by Figure 1, the MDC baseline flow involves three main phases:

(1) The input *DPNs parsing*, performed by the Orcc front-end, translates the RVC-CAL specifications into Java Intermediate Representations (IRs), which are basically directed graphs.

(2) The *datapath merging*, performed by the MDC front-end, combines the IRs into a reconfigurable IR, inserting (where necessary) special switching actors (responsible for properly distributing the token flow among the different merged DPNs) keeping trace of the system programmability through a dedicated *Configuration Table* (TAB in Figure 1).

(3) The *hardware platform generation*, performed by the MDC back-end, leads to the creation of the RTL that describes the CGR system itself, where each actor of the reconfigurable IR is mapped onto a different hardware FU. In this phase, the hardware communication protocol and the *HDL (Hardware Description Language) components library* (providing the RTL descriptions of the required FUs, manually or automatically generated) are provided as input to the tool.

At the hardware level, reconfiguration takes place in a single clock cycle. It is achieved through low overhead switching elements (SBoxes) that allow the sharing of common resources among different input DPNs. SBoxes are simple combinatorial multiplexers and demultiplexers, whose configuration is stored into dedicated Look-Up tables that, according to the *Configuration Table*, compute the selectors necessary for the correct data forwarding in order to implement the requested functionality.

3.1.1. Automated Power Management. Dealing with reconfigurable architectures, and in particular with CGR systems, the power consumption has to be carefully taken into consideration. Such a kind of systems is affected by resource redundancy, mainly due to the FUs that are not shared among different functionalities. Thus, when a certain functionality is executed, part of the design (not involved in the computation) is in an idle state and can uselessly consume precious power. Fortunately the unused resources, for each implemented functionality, depend on the input specifications and, therefore, are fixed at design-time.

Given these considerations, then, a CGR system can be characterized by a set of disjointed *logic regions* (LRs), grouping the resources that are always active/inactive at the same time. The MDC power management extension is capable of automatically identifying LRs. It performs the *LRs identification* at a high-level of abstraction, on the reconfigurable IR, by exploiting the intrinsic modularity of the dataflow graphs. Once the LRs have been identified, the MDC dynamic power manager automatically applies, according to the user selection, either clock gating [17] or power gating [18] on the resulting CGR hardware platform. The identification of the minimal number of LRs is guaranteed, to minimize the power overhead of the extra logic needed to implement the selected power saving strategies. An overview of the power management extension is provided by Figure 1.

For each input DPN G_i, the currently available algorithm determines the set V_i', which contains all the resources of the reconfigurable IR activated by G_i. These are the original sets of LRs that represent the starting point for the algorithm to find the final LRs by iteratively comparing two V_i' sets at a time, determining their possible overlapping. If overlapping is found, its resources are removed by the two considered sets and a new V_j' (corresponding to a new LR) involving these shared resources is issued. V_j' groups resources that are shared among different input DPNs, while the remaining resources in the two V_i' will uniquely belong to the originally considered DPNs. This compare and split identification process guarantees that the number of LRs found by the MDC dynamic power manager is the minimum achievable one.

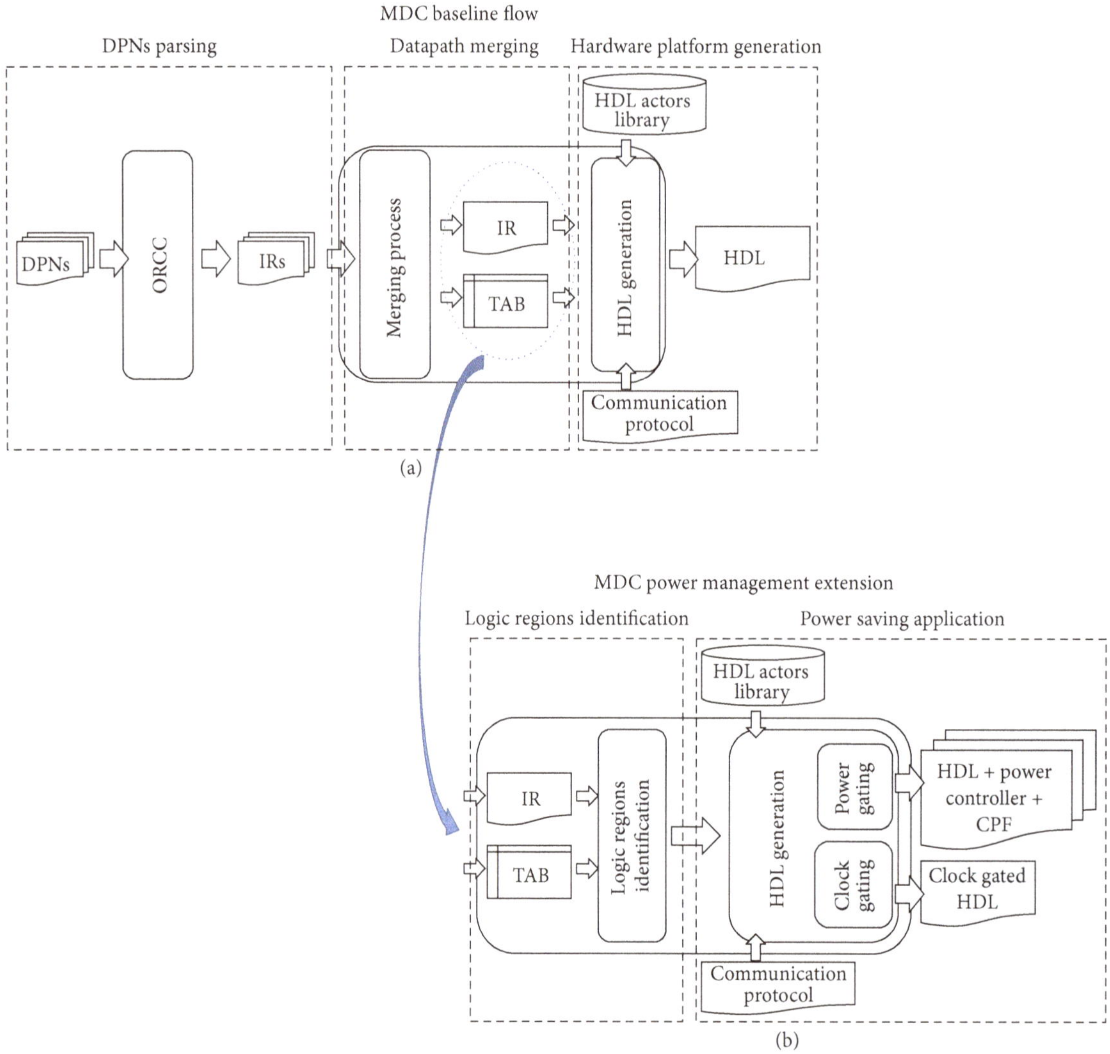

FIGURE 1: MDC design suite: baseline flow (a) and corresponding power extension (b).

If this number is still too high for the considered target platform, as it can happen if FPGAs are the target devices (in FPGA devices the number of hardware blocks that can drive the different LRs is limited; e.g., 32 BUFG units are available in Xilinx boards for clock management purposes), a *LRs merging* process has to be applied. MDC users are required to specify the target technology and the maximum number of implementable LRs. This latter is compared with the number of LRs determined by the compare and split identification process and, if necessary, the *LRs merging* process is applied. Two LRs at a time are unified (details on how to merge different LR sets can be found in [17], where two merging strategies (a power-aware one and a number-aware one) are presented) until the constraint fixed by the user is met. This process leads to a suboptimal system implementation: each DPN, while activating its corresponding LRs, may also activate some resources that do not contribute to its computation, leading to extra unnecessary power consumption.

MDC power management extension, during the HDL generation phase, provides also the implementation of the chosen power management strategy upon the identified LRs. It blindly applies the selected strategy to all the identified LRs, without any warranty on the approach effectiveness. Clock gating acts only on the dynamic contribute of the power consumption and requires a minimum logic overhead on the final platform. Indeed, the simplest implementation is achieved by means of one *AND* gate for each LR plus one unique *Enable Generator* to properly set the enable signals of the *AND* gates according to the desired functionality. On the contrary, power gating is able to reduce both power contributions, static and dynamic, by shutting off the power supply of the region. However, it is quite more invasive, since it requires one *power switch* for each LR, one *state retention* cell for each Flip-Flop whose state has to be kept also when the corresponding LR is off, and one *isolation* cell for each bit-wise wire that goes from a disabled LR to an enabled one. A different clock gating cell (again an *AND* gate) is required for each LR, according to the switching-off protocol for the proper operation of the retention cells (details on the power gating switch on/off protocol can be found in [11]). Furthermore,

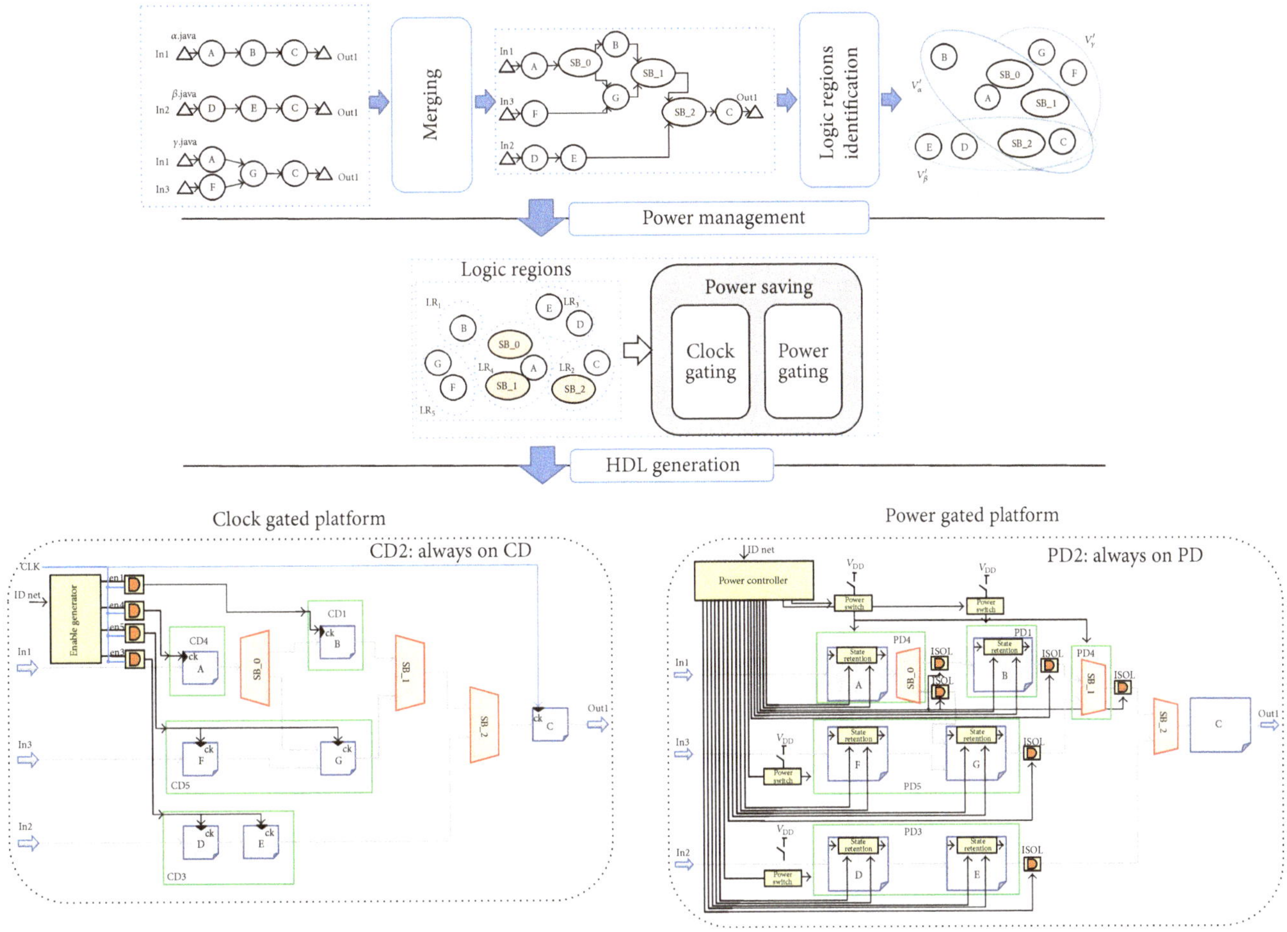

FIGURE 2: Step-by-step example of the MDC baseline and dynamic power management features. Saving strategies are blindly applied by the dynamic power manager on each identified LR.

one *Power Controller* block (involving a different finite state machine for each LR) is needed to properly drive the inserted *power switch*, *state retention* and *isolation* cells.

3.1.2. Step-by-Step Example. In order to clarify the features provided by MDC baseline functionality and its related dynamic power manager extension, this section describes a step-by-step example of the whole flow. Three different input functionalities, labelled α, β, and γ, are considered and modelled as DPNs. As first step of the baseline MDC functionality, the DPNs are parsed by the Orcc font-end and translated into Java IR graphs. Figure 2 depicts an overview of the whole flow starting from these input IRs (α*.java*, β*.java*, and γ*.java*). At this level, MDC combines the dataflows into a reconfigurable IR inserting the SBox actors (SB in Figure 2). Three SBoxes are required to share actor A between α and γ and actor C among all the three functionalities.

Once the reconfigurable IR has been derived, the dynamic power manager can identify the corresponding LRs. The starting V_i' sets are

(i) $V_\alpha' = \{A, B, C, \text{SB\_0}, \text{SB\_1}, \text{SB\_2}\}$;

(ii) $V_\beta' = \{D, E, C, \text{SB\_2}\}$;

(iii) $V_\gamma' = \{A, F, G, C, \text{SB\_0}, \text{SB\_1}, \text{SB\_2}\}$.

The compare and split identification process produces five different LRs:

(i) $\text{LR}_1 = \{B\}$.

(ii) $\text{LR}_2 = \{C, \text{SB\_2}\}$.

(iii) $\text{LR}_3 = \{D, E\}$.

(iv) $\text{LR}_4 = \{A, \text{SB\_0}, \text{SB\_1}\}$.

(v) $\text{LR}_5 = \{F, G\}$.

LR_4 and LR_2 involve shared resources, being activated, respectively, by α and γ and by α, β, and γ, while the remaining LRs involve nonshared resources. LR_1 is activated only by α, LR_3 only by β, and LR_5 only by γ.

At this point, the selected power saving strategy is applied during the CGR system HDL generation. Figure 2 shows both the final designs resulting from the application of clock gating and power gating.

The clock gated platform is shown in the bottom left corner of Figure 2. In this case, the identified LRs become the Clock Domains (CDs) of the resulting architecture, meaning that the involved actors are all driven by the same gated clock. SBoxes are not included in any CD since they are fully combinatorial modules. It can be noticed by Figure 2 that the clock gating overhead is limited to four *AND* gates (LR_2 is activated by all the implemented functionalities and does not need to be turned off) and one *Enable Generator* that properly assigns the clock enable values.

The power gated platform is depicted in the bottom right corner of Figure 2. LRs define the architecture power domains (PDs) where, in this case, also SBoxes are taken into consideration. Power gating turns off the whole PD power supply and it has effect also on combinatorial blocks. Figure 2 clearly shows that the logic overhead of power gating is larger than the clock gating one. In the power gated platform *power switches*, *state retention* cells, *isolation* cells, and one *Power Controller* are inserted. Please note that clock gating cells are not reported for simplicity. Again the logic necessary to switch off LR_2 is avoided since this region is an always on one (being activated by all the input DPNs).

3.2. Automated Power Management with Hybrid Clock and Power Gating. To overcome the limits of a blindly applied unique power management strategy, we propose, in this paper, a power estimation flow capable of (1) characterizing, at a high-level of abstraction, the LRs identified by the MDC power extension, and of (2) autonomously applying the optimal power reduction technique for each LR. Power and clock gating overhead are estimated, based on LRs characteristics, before any physical implementation. This strategy is meant for ASIC technologies, which allow hybrid power and clock gating support over the same CGR design.

The estimation is based on two sets of models that determine the static and dynamic consumptions of each LR when clock gating or power gating are applied. The proposed models are derived after a single logic synthesis of the baseline CGR system generated by MDC, carried out with commercial synthesis tools from the analysis of the power reports obtained after netlist simulation. Such a synthesis constitutes the only implementation effort required for the designer, besides the characterization of technique-specific blocks such as the *Enable Generator* or the *Power Controller*. Models are technology-dependent since they include parameters that are characteristic of the chosen target technology library, as it will be discussed in Section 3.2.4.

3.2.1. Power Gating: Static Power Consumption Model. Static power can be estimated on the basis of the leakage contributes provided, for each cell, by the targeted ASIC library. Given any hardware FU (uniquely corresponding to an actor of the reconfigurable IR) its static power can be obtained by summing up the single contributions of the adopted cells. The static power consumption term is tightly related to the LR area: the more cells are included in the considered region, the more is its corresponding static dissipation.

The proposed model for the static power consumption is defined as follows:

$$\begin{aligned} P_{lkg}(LR_i) &= P_{lkgON}(LR_i) + \text{Ext\_Over}_{lkg}(LR_i) \\ &= \sum_{\text{actors} \in LR_i} \Big[P_{lkg}(\text{cmb}) + P_{lkg}(\text{RC}) * \#\text{rtn} \\ &\quad + P_{lkg}(\text{reg}) * \frac{\#\text{reg} - \#\text{rtn}}{\#\text{reg}} \Big] * \text{Ti}_{ON} + \big[P_{lkg}(\text{ISO}_{ON}) \\ &\quad * \text{Ti}_{ON} + P_{lkg}(\text{ISO}_{OFF}) * \text{Ti}_{OFF} \big] * \#\text{iso} \\ &\quad + \big[P_{lkg}(\text{Contr}_{ON}) * \text{Ti}_{ON} + P_{lkg}(\text{Contr}_{OFF}) \\ &\quad * \text{Ti}_{OFF} \big] + \big[P_{lkg}(\text{CG}_{ON}) * \text{Ti}_{ON} + P_{lkg}(\text{CG}_{OFF}) \\ &\quad * \text{Ti}_{OFF} \big]. \end{aligned} \tag{1}$$

Dealing with a prospective power gating implementation, the static estimation (1) for each LR involves two terms: $P_{lkgON}(LR_i)$ corresponds to the static consumption when the LR is active and $\text{Ext\_Over}_{lkg}(LR_i)$ refers to the power overhead due to the additional power gating logic. This second term does not consider the *power switch* overhead, since it is not included in the prelayout netlist. Power gating prevents, by definition, any static dissipation on the LR when disabled; therefore, (1) does not present any $P_{lkgOFF}(LR_i)$.

$P_{lkgON}(LR_i)$ is obtained as the multiplication of the LR activation time Ti_{ON} and the sum of leakage power of the involved actors, considering separately combinatorial and sequential logic. The former, $P_{lkg}(\text{cmb})$, is equal to the leakage of the combinatorial cells within the considered LR. The latter is related to the number of registers (#reg) within the LR and their need (according to the implemented functionality) of preserving or not their status, by means of *state retention* cells, when the region is inactive. Then it involves, in turn, two terms. The first one refers to the registers whose state can be lost and it is estimated on the basis of the static consumption of the sequential cells ($P_{lkg}(\text{reg})$), as an average on the number of registers that are not retained. The second one refers to the retention cells and it is estimated starting from the number of registers whose state has to be maintained (#rtn) multiplied by the leakage of a single *state retention* cell ($P_{lkg}(\text{RC})$), whose value is retrieved from the target ASIC library.

$\text{Ext\_Over}_{lkg}(LR_i)$ is composed of three terms: the first one is related to the *isolation* cells (#iso), the second one to the *Power Controller*, and the third one to the clock gating cell (the power gating switch off protocol requires applying clock gating at the region level, before retaining the registers value). Note that, unlike P_{lkgON}, for the three abovementioned terms, Ext_Over_{lkg} characterizes the LR static consumption in both its on and off states. In the on state, the model accounts for the static consumption in the on state (e.g., $P_{lkg}(\text{ISO}_{ON})$) multiplied by the activation time Ti_{ON} and by the overall number of cells within the LR (e.g., #iso). In the off state, the model accounts for the static consumption in the off state (e.g., $P_{lkg}(\text{ISO}_{OFF})$) multiplied by the inactive time Ti_{OFF} and by the overall number of cells within the LR (e.g., #iso). Please note that there is just one *Power Controller* for all the LRs and one clock gating cell per LR, but an a priori characterization phase would be required to the designer, since their consumption values cannot be retrieved directly from any ASIC library.

3.2.2. Power Gating: Dynamic Power Consumption Model. Estimating the dynamic power is more complex than estimating the static one, since this term strongly depends on the nodes switching activity. Frequently, commercial tools (e.g., Cadence Encounter Digital Implementation System) consider dynamic power as composed of two main terms, as depicted by (2): a *net* contribution due to the power dissipated throughout the wires linking the cells, and an *internal* contribution due to the dissipation occurring inside the cells [44]:

$$\begin{aligned} P_{\text{dyn}} &= P_{\text{net}} + P_{\text{int}} \\ &= \frac{1}{2} f V_{DD}^2 \sum_{\text{net}_j} C_{\text{load}_j} \text{SW}_j + f \sum_{\text{cell}_i} P_i \text{SW}_i. \end{aligned} \tag{2}$$

The operating frequency, f, influences both terms. P_{net} accounts for the load capacitance of each net_j (bearing a specific capacitance C_{load_j}) and the related switching activity (SW_j), whereas P_{int} depends on the power per MHz dissipated by each cell (P_i) and the related switching activity (SW_i).

Currently, the developed model is able to estimate only the P_{int} contribution that can be expressed for each single LR as follows:

$$\begin{aligned} P_{\text{int}}(\text{LR}_i) &= P_{\text{intON}}(\text{LR}_i) + \text{Ext\_Over}_{\text{int}}(\text{LR}_i) \\ &= \sum_{\text{actors} \in \text{LR}_i} \left[P_{\text{int}}(\text{cmb}) + P_{\text{int}}(\text{RC}) * \#\text{rtn} \right. \\ &\quad \left. + P_{\text{int}}(\text{reg}) * \frac{\#\text{reg} - \#\text{rtn}}{\#\text{reg}} \right] * \text{Ti}_{\text{ON}} + \left[P_{\text{int}}(\text{ISO}_{\text{ON}}) \right. \\ &\quad \left. * \text{Ti}_{\text{ON}} + P_{\text{int}}(\text{ISO}_{\text{OFF}}) * \text{Ti}_{\text{OFF}} \right] * \#\text{iso} \\ &\quad + \left[P_{\text{int}}(\text{Contr}_{\text{ON}}) * \text{Ti}_{\text{ON}} + P_{\text{int}}(\text{Contr}_{\text{OFF}}) \right. \\ &\quad \left. * \text{Ti}_{\text{OFF}} \right] + \left[P_{\text{int}}(\text{CG}_{\text{ON}}) * \text{Ti}_{\text{ON}} + P_{\text{int}}(\text{CG}_{\text{OFF}}) \right. \\ &\quad \left. * \text{Ti}_{\text{OFF}} \right]. \end{aligned} \tag{3}$$

Considering a prospective power gating implementation, the different parts of (3) basically reflect the ones of (1). The main difference among the static power model and the dynamic one is that this latter requires accurate data in terms of nodes switching activity. For this reason, the netlist of the baseline CGR system is not sufficient to retrieve accurate values from the power reports and one different simulation of the netlist for every implemented functionality is required. Thus, dynamic power model takes into consideration the real system switching activity provided by the hardware simulations.

The P_{net} term of (2) is not currently addressed in our model. Nevertheless, as demonstrated further on in this work (please see Tables 8, 9, 12, and 14) neglecting this term seems not to affect the optimal identification of the region to be gated. We have already planned to extend the proposed methodology to include the nets contribution in a near future.

3.2.3. Clock Gating: Static and Dynamic Power Consumption Models. Clock gating static and dynamic models are less complicated than the power gating ones, since clock gating requires a very low logic overhead and it positively acts only on the dynamic dissipation. Equations (4) and (5) report the models adopted, respectively, for the static power estimation and for the dynamic power one, referring to a clock gated design:

$$\begin{aligned} P_{\text{lkg}}(\text{LR}_i) &= P_{\text{lkg}}(\text{LR}_i) + \text{Ext\_Over}_{\text{lkg}}(\text{LR}_i) \\ &= \sum_{\text{actors} \in \text{LR}_i} \left[P_{\text{lkg}}(\text{cmb}) + P_{\text{lkg}}(\text{reg}) \right] \\ &\quad + \left[P_{\text{lkg}}(\text{Enab}_{\text{ON}}) * \text{Ti}_{\text{ON}} + P_{\text{lkg}}(\text{Enab}_{\text{OFF}}) * \text{Ti}_{\text{OFF}} \right] \\ &\quad + \left[P_{\text{lkg}}(\text{CG}_{\text{ON}}) * \text{Ti}_{\text{ON}} + P_{\text{lkg}}(\text{CG}_{\text{OFF}}) * \text{Ti}_{\text{OFF}} \right], \end{aligned} \tag{4}$$

$$\begin{aligned} P_{\text{int}}(\text{LR}_i) &= P_{\text{int}}(\text{LR}_i) + \text{Ext\_Over}_{\text{int}}(\text{LR}_i) \\ &= P_{\text{int}}(\text{combLR}_i) + P_{\text{intON}}(\text{seqLR}_i) \\ &\quad + \text{Ext\_Over}_{\text{int}}(\text{LR}_i) \\ &= \sum_{\text{actors} \in \text{LR}_i} \left[P_{\text{int}}(\text{cmb}) + P_{\text{int}}(\text{reg}) * \text{Ti}_{\text{ON}} \right] \\ &\quad + \left[P_{\text{int}}(\text{Enab}_{\text{ON}}) * \text{Ti}_{\text{ON}} + P_{\text{int}}(\text{Enab}_{\text{OFF}}) * \text{Ti}_{\text{OFF}} \right] \\ &\quad + \left[P_{\text{int}}(\text{CG}_{\text{ON}}) * \text{Ti}_{\text{ON}} + P_{\text{int}}(\text{CG}_{\text{OFF}}) * \text{Ti}_{\text{OFF}} \right]. \end{aligned} \tag{5}$$

At the logic region level, (4) considers always the combinatorial and sequential contributions for both the on or off states, since clock gating does not affect the system leakage, whereas (5) considers always the combinatorial part for both the on or off states, since combinatorial logic cannot benefit from clock gating and the sequential contribution only during the LR active time. The overhead, $\text{Ext\_Over}_{\text{int}}(\text{LR}_i)$, is given by the clock gating cell and the *Enable Generator*. Please remember that implementing clock gating management at a coarse-grained level, just one clock gating cell per LR has to be inserted within the system. Equation (5) is pretty much the same as (4), a part from the fact that, dealing with the dynamic model, clock gating effects are estimated by omitting the contribute of sequential logic when the LR is off.

3.2.4. Parameters Discussion. The proposed models are determined by the intrinsic features of the LRs. In particular, they consider the following:

(i) *Architectural Parameters.* LRs composition determines the amount of involved combinatorial and sequential cells.

(ii) *Functional Parameters.* LRs behaviour defines the region activation time and if its status has to be preserved or not.

(iii) *Technological Parameters.* Target technology has an impact on the ratio between dynamic and static power (as it will be demonstrated in Section 4.2) and on the different cells characterization.

Table 1 reports, for each parameter considered in (1), (3), (4), and (5), their classification. A deeper explanation about $P_{lkg/int}(\text{cmb})$ and $P_{lkg/int}(\text{reg})$ is necessary. They are not associated with any specific parameters class, indeed they depend on type and number of involved cells composing the considered LR and also on the system switching activity (especially for the internal contribute). These values are gathered by the reports of the baseline CGR system netlist, assuming that the amount and type of cells composing the FUs do not change as power saving strategies are applied (except for the retained registers).

3.2.5. Step-by-Step Example. The example proposed in Figure 2 shows the merging process of three DPNs in a CGR system, where five LRs have been identified. Equations (1), (3), (4), and (5) can be applied to all of them. However, as already discussed, LR_2 can be discarded: being common to all the input DPNs, it is always active and does not require to be switched-off. As defined in the previous section, the parameters reported in Table 2 are extracted by the reference technology library or characterized by synthesis trials (see the definition provided in Table 1). The power consumption values in Table 3 have been extracted by the synthesis reports of the baseline (with no power saving applied) CGR platform and required three hardware simulations (one for each input kernel). These simulations are necessary to correctly estimate the internal power consumption of the different LRs, taking into account the real switching activity of the design. In practice, power values are determined as an average of those obtained according to the different switching activity profiles.

Starting from the data in Tables 3 and 2, here follows the detailed equations characterization for LR_5, which include actors F and G.

When power gating is applied, the static power consumption of LR_5 is derived according (1), as follows:

$$\begin{aligned} P_{lkg}(LR_5) &= \Bigg[\bigg(P_{lkg}(\text{comb}_F) + P_{lkg}(\text{RC}) * \#\text{rtn}F \\ &+ P_{lkg}(\text{reg}_F) * \frac{\#\text{reg}F - \#\text{rtn}F}{\#\text{reg}F}\bigg) + \bigg(P_{lkg}(\text{comb}_G) \\ &+ P_{lkg}(\text{RC}) * \#\text{rtn}G + P_{lkg}(\text{reg}_G) \\ &* \frac{\#\text{reg}G - \#\text{rtn}G}{\#\text{reg}G}\bigg)\Bigg] * T5_{ON} + \big[P_{lkg}(\text{ISO}_{ON}) \\ &* T5_{ON} + P_{lkg}(\text{ISO}_{OFF}) * T5_{OFF}\big] * \#\text{iso}_5 \\ &+ \big[P_{lkg}(\text{Contr}_{ON}) * T5_{ON} + P_{lkg}(\text{Contr}_{OFF}) \\ &* T5_{OFF}\big] + \big[P_{lkg}(\text{CG}_{ON}) * T5_{ON} + P_{lkg}(\text{CG}_{OFF}) \\ &* T5_{OFF}\big] = [(213 + 17.15 * 128) + (273 + 17.15 \\ &* 64 + 1385 * 0.5)] * 0.3 + [4.27 * 0.3 + 1.39 \\ &* 0.7] * 32 + [95.44 * 0.3 + 88.63 * 0.7] + [5.77 \\ &* 0.3 + 4.71 * 0.7] = 1509.219. \end{aligned} \tag{6}$$

The internal power consumption is given by (3):

$$\begin{aligned} P_{int}(LR_5) &= \Bigg[\bigg(P_{int}(\text{comb}_F) + P_{int}(\text{RC}) * \#\text{rtn}F \\ &+ P_{int}(\text{reg}_F) * \frac{\#\text{reg}F - \#\text{rtn}F}{\#\text{reg}F}\bigg) + \bigg(P_{int}(\text{comb}_G) \\ &+ P_{int}(\text{RC}) * \#\text{rtn}G + P_{int}(\text{reg}_G) \\ &* \frac{\#\text{reg}G - \#\text{rtn}G}{\#\text{reg}G}\bigg)\Bigg] * T5_{ON} + \big[P_{int}(\text{ISO}_{ON}) \\ &* T5_{ON} + P_{int}(\text{ISO}_{OFF}) * T5_{OFF}\big] * \#\text{iso}_5 \\ &+ \big[P_{int}(\text{Contr}_{ON}) * T5_{ON} + P_{int}(\text{Contr}_{OFF}) \\ &* T5_{OFF}\big] + \big[P_{int}(\text{CG}_{ON}) * T5_{ON} + P_{int}(\text{CG}_{OFF}) \\ &* T5_{OFF}\big] = [(537 + 383.25 * 128) + (363 + 383.25 \\ &* 64 + 44068 * 0.5)] * 0.3 + [2.7 * 0.3 + 0 * 0.7] \\ &* 32 + [1449 * 0.3 + 1488 * 0.7] + [169 * 0.3 \\ &+ 292 * 0.7] = 30217.72. \end{aligned} \tag{7}$$

When clock gating is considered, (4) and (5) are computed as follows:

$$\begin{aligned} P_{lkg}(LR_5) &= \big[\big(P_{lkg}(\text{comb}_F) + P_{lkg}(\text{reg}_F)\big) \\ &+ \big(P_{lkg}(\text{comb}_G) + P_{lkg}(\text{reg}_G)\big)\big] + \big[P_{lkg}(\text{Enab}_{ON}) \\ &* T5_{ON} + P_{lkg}(\text{Enab}_{OFF}) * T5_{OFF}\big] + \big[P_{lkg}(\text{CG}_{ON}) \\ &* T5_{ON} + P_{lkg}(\text{CG}_{OFF}) * T5_{OFF}\big] = [(213 + 1232) \\ &+ (273 + 1385)] + [84.51 * 0.3 + 76.54 * 0.7] \\ &+ [5.77 * 0.3 + 4.71 * 0.7] = 3186.96, \\ P_{int}(LR_5) &= [(P_{int}(\text{comb}_F) + P_{int}(\text{reg}_F) * T5_{ON}) \\ &+ (P_{int}(\text{comb}_G) + P_{int}(\text{reg}_G) * T5_{ON})] \\ &+ [P_{int}(\text{Enab}_{ON}) * T5_{ON} + P_{int}(\text{Enab}_{OFF}) \\ &* T5_{OFF}] + [P_{int}(\text{CG}_{ON}) * T5_{ON} + P_{int}(\text{CG}_{OFF}) \\ &* T5_{OFF}] = [(537 + 22489 * 0.3) \\ &+ (363 + 44068 * 0.3)] + [1351 * 0.3 + 1320 * 0.7] \\ &+ [169 * 0.3 + 292 * 0.7] = 22451.8. \end{aligned} \tag{8}$$

Table 4 summarizes all the values achieved applying the proposed static and dynamic models to all the different logic regions.

3.2.6. Hybrid Clock and Power Gating Support and Integration in MDC. The discussed models (described in (1), (3), (4), and (5)) have been integrated in the MDC design flow, in order to implement a fully automated power management

TABLE 1: Parameter classification. Table reports for each considered parameter, typology (architectural, functional, and technological), description, and extraction method.

Parameter	Arch.	Funct.	Tech.	Description	Extraction
#reg	x			Number of sequential cells in the considered LR	Provided by the synthesis reports
#iso	x			Number of estimated isolation cells in the design	Obtained by counting the number of wires that connect the different LR among each other in the dataflow model
$\mathrm{Ti}_{\mathrm{ON}}$		x		Activation time	Found by means of a high-level (directly on the dataflow) profiling of the targeted scenario
$\mathrm{Ti}_{\mathrm{OFF}}$		x		Off time	
#rtn		x		Number of retention cells in the design	Strictly related to the LR functionality and is chosen by the designer
$P_{\mathrm{lkg/int}}(\mathrm{RC})$			x	Power estimation of retention and isolation cells	Determined by the selected synthesis process and can be obtained without any implementation run
$P_{\mathrm{lkg/int}}(\mathrm{ISO}_{\mathrm{ON}})$			x		
$P_{\mathrm{lkg/int}}(\mathrm{ISO}_{\mathrm{OFF}})$			x		
$P_{\mathrm{lkg/int}}(\mathrm{Contr}_{\mathrm{ON}})$			x	Power estimation of PG and CG controller when the LR is on or off	It requires being characterized, according to the target technology, with dedicated synthesis trials
$P_{\mathrm{lkg/int}}(\mathrm{Contr}_{\mathrm{OFF}})$			x		
$P_{\mathrm{lkg/int}}(\mathrm{Enab}_{\mathrm{ON}})$			x		
$P_{\mathrm{lkg/int}}(\mathrm{Enab}_{\mathrm{OFF}})$			x		
$P_{\mathrm{lkg/int}}(\mathrm{cmb})$	x	x	x	Power consumed by combinatorial and sequential cells within the considered LR	Gathered by the reports of the baseline CGR system netlist
$P_{\mathrm{lkg/int}}(\mathrm{reg})$	x	x	x		

TABLE 2: Contributions of static and internal power consumption extracted by the reference technology library or characterized by synthesis trials.

Parameters	lkg power [nW]	int power [nW]
$Enab_{ON}$	84.51	1351
$Enab_{OFF}$	76.54	1320
$Contr_{ON}$	95.44	1449
$Contr_{OFF}$	88.63	1488
CG_{ON}	5.77	169
CG_{OFF}	4.71	292
ISO_{On}	4.27	2.7
ISO_{OFF}	1.39	0
$P(RC)$	17.15	383.25

```
PG_set is empty;
CG_set is empty;
foreach LR_i in set LRs do
    evaluate_area(LR_i, area_th)
end
function: evaluate_area(LR_i, area_th):
calculate_LR_i_area;
if area_LR > area_th then
    evaluate_PG(LR_i);
else
    evaluate_CG(LR_i);
end
function: evaluate_PG(LR_i):
estimate_PG_total_overhead;
    if PG_total_overhead < 0 then
        estimate_CG_total_overhead;
        if PG_total_overhead < CG_total_overhead then
            add LR_i to PG_set;
        else
            add LR_i to CG_set;
        end
    else
        evaluate_CG(LR_i);
end
evaluate_CG(LR_i);
function: evaluate_CG(LR_i):
estimate_CG_total_overhead;
if CG_total_overhead < 0 then
    add LR_i to CG_set;
end
```

ALGORITHM 1: Automatic power saving strategy selection for CGR systems.

strategy. Designers are guided towards the optimal solution for each LR, rather than choosing a one-fit-to-all switching-off technique for all of them.

This automated selection flow is implemented as reported in Algorithm 1. For each LR, identified by the MDC power management extension, Algorithm 1 executes the following steps, embodied by different functions.

(1) Area Thresholding (See evaluate_area Function). As previously discussed, power gating is a quite invasive technique, requiring a lot of extra logic to be inserted in the nonswitchable always on domain. Thus, for small LRs, we can assume that it will not bring any benefit, so that power gating is not to be considered for implementation. Indeed, clock gating may still be beneficial, due to its very small additional logic amount.

(2) Power Gating Overhead Estimation (See evaluate_PG Function in Algorithm 1). Power gating cost is estimated in order to find out if it can lead to power saving or not. The prospective power and clock gating implementations are compared on the basis of their overall consumption. Equation (1) is applied and summed to (3), if there is not total power saving the algorithm goes to the *clock gating overhead estimation*. On the contrary, if there is saving it has to be compared with the sum of (4) and (5) to determine whether the current LR may benefit from power gating (despite its larger overhead) or from clock gating.

(3) Clock Gating Overhead Estimation (See evaluate_CG Function in Algorithm 1). Clock gating cost is estimated to investigate the possibility of achieving power saving with this technique. If the LR clock gating achievable saving does not counterbalance its implementation costs, the LR is discarded. This means that when the MDC back-end generates the RTL description of the CGR system, the LR logic is included in the always on domain. On the contrary, if the clock gating leads to an overall saving in terms of total power, the LR will be clock gated during the implementation.

The output of Algorithm 1 is the classification of the LRs, stating which one should be power gated (see PG_set in Algorithm 1), which one should be clock gated (see CG_set in Algorithm 1), and which ones should be included in the always on domain.

Figure 3 provides an overview of the modified design flow. As it can be noticed, the MDC tool and its power management extension are directly interfaced with the logic synthesizer. Algorithm 1 is implemented within the *Power Analysis* block. MDC baseline tool provides the HDL description of the plain CGR system and all the scripts to perform the synthesis of the CGR design and all the different hardware simulations (one for each input DPNs), as required by the proposed power estimation models. The power reports are then fed back to the MDC power management extension and parsed within the *Power Analysis* to execute Algorithm 1. The LRs classification (see LR class in Figure 3), generated by the *Power Analysis* block, is used by the *CG/PG HDL Generation* block to automatically define the hybrid, clock and power gating, power management support for the given CGR design.

Summarizing, this flow, with respect to what is discussed in Section 3.1, does not require designers to opt for a specific power management technique. On the basis of the proposed power estimation models and by linking MDC with a logic synthesis tool, the presented flow is capable of overcoming the limit of providing a one-fit-to-all solution. Each LR, in a CGR design, is supported (where necessary) with the optimal power management technique.

TABLE 3: Parameter and power consumption of each LR, extracted by the synthesis reports of the baseline CGR platform.

Logic region	Kernel	T_{ON}	Actors	#iso	#reg	#rtn
$LR_1 LR_1$	α	0.1	B	32	514	24
LR_3	β	0.6	D, E	32	8	8
LR_4	α, γ	0.4	A, SB_0, SB_1	96	256	64
LR_5	γ	0.3	F, G	32	265	192
Actor			Power [nW]			
	lkg seq.	int seq	lkg comb.	int comb.	#reg	#rtn
B	801	104987	121411	3916599	512	24
D	48	1104	51	319	4	4
E	56	1437	53	198	4	4
A	3264	89238	0	0	256	64
SB_0	0	0	307	409	—	—
SB_1	0	0	225	350	—	—
F	1232	22489	213	537	128	128
G	1385	44068	273	363	128	64

TABLE 4: Resulting power consumption of the different LRs when the proposed models are applied.

Logic region	lkg PG [nW]	int PG [nW]	lkg CG [nW]	int CG [nW]
LR_1	1240.69	404358.71	122294.15	3928700.60
LR_3	342.67	3884.44	294.67	3599
LR_4	2062.11	38705.08	3880.86	5598.6
LR_5	1509.58	30712.72	3186.96	22451.8

3.2.7. Step-by-Step Example. In this section, a step-by-step example of the application of Algorithm 1 is presented, considering the same example proposed in Figure 2. In that case, MDC *LRs identification* led to determining five LRs and the user-specified power management technique is blindly applied to all of them except LR_2. This region is used by all the input DPNs; thus, it is never disabled and does not require any power management support. In the following step-by-step example, shown in Figure 4, the threshold on the area ($area_{th}$) is set to 5%.

(i) LR_1 is processed by invoking evaluate_area(LR_1, 5).

(a) Its area is calculated: $area_{LR_1}$ = 52% of total area.

(b) $area_{LR_1}$ > $area_{th}$, so that a prospective power gating implementation on LR_1 is taken into consideration by invoking evaluate_PG(LR_1).

(1) The static and the dynamic overheads are estimated, respectively, applying (1) and (3). The power gating overhead on the overall consumption is calculated by subtracting the power consumption of the LR when PG is applied, to the power consumption of the LR in the baseline design, the result is then divided by the total power consumption of the baseline design in order to estimate the total percentage power variation. The power variation when PG is applied to region LR_1 is −86.45%. Since this value is negative, power gating may be convenient if its total saving is larger than the clock gating one.

(2) Equation (4) is calculated and summed up to (5) to determine clock gating overhead on the overall consumption, which is −2.15%.

(3) Power gating is more beneficial than clock gating determining, overall, a larger power saving. Thus, LR_1 is added to PG_set.

(ii) LR_3 is processed by invoking evaluate_area(LR_3, 5).

(a) Its area is calculated: $area_{LR_3}$ = 0.4% of total area.

(b) $area_{LR_3}$ < $area_{th}$, so that a prospective clock gating implementation on LR_3 is considered straight away by invoking evaluate_CG(LR_3).

(1) Equations (4) and (5) are evaluated to determine clock gating overhead on the overall consumption: CG_{over} = +0.01%.

(2) Clock gating is not beneficial since its overhead is positive. Thus LR_3 is discarded and no power management policy will be applied to it.

(iii) LR_4 is processed by invoking evaluate_area(LR_4, 5).

(a) Its area is calculated: $area_{LR_4}$ = 7% of total area.

(b) $area_{LR_4}$ > $area_{th}$, so that a prospective power gating implementation on LR_4 is taken into consideration by invoking evaluate_PG(LR_4).

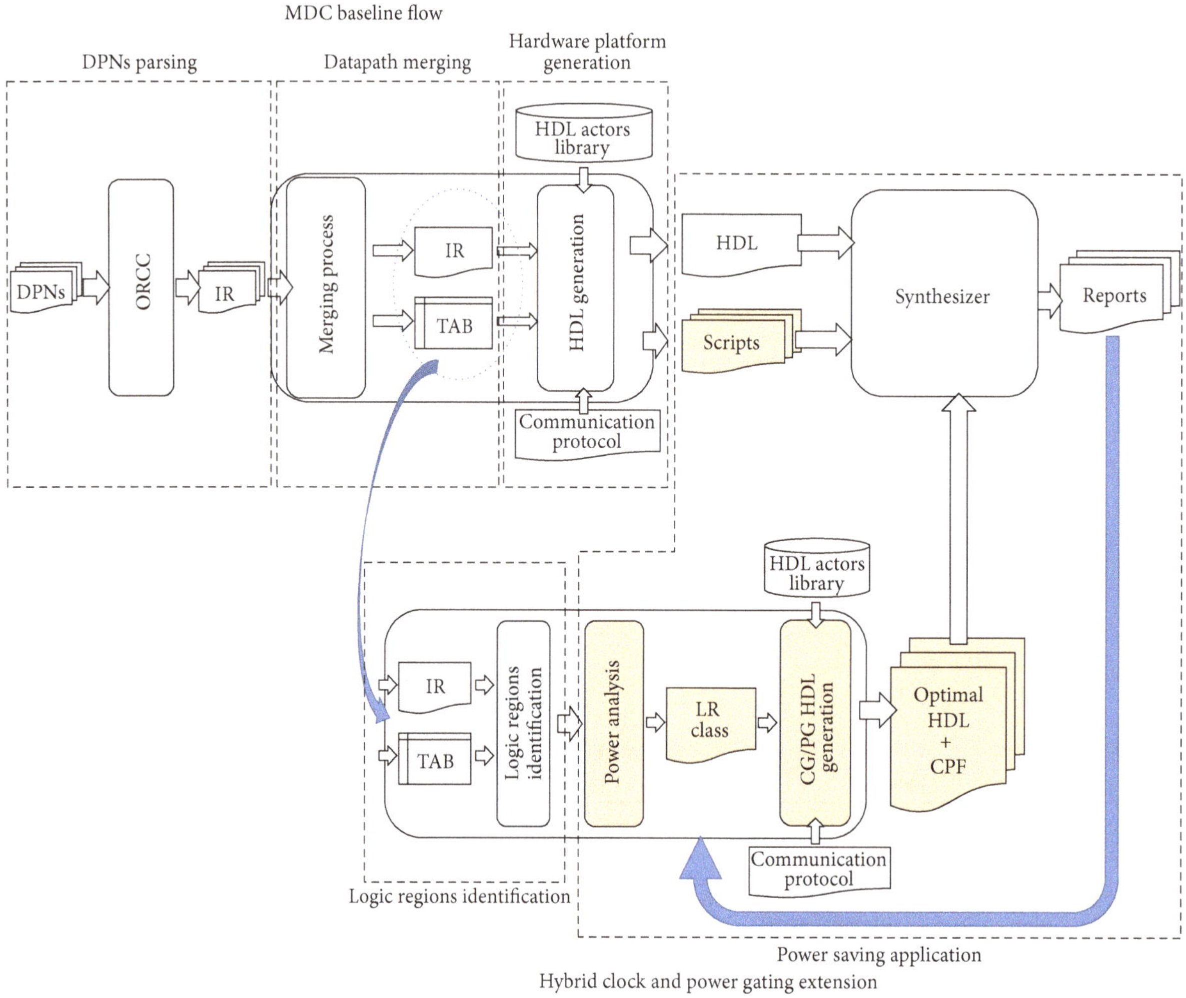

FIGURE 3: Enhanced MDC design suite: integration of the automated hybrid, clock, and power gated support.

(1) The static and the dynamic overheads are estimated, respectively, applying (1) and (3). The power gating overhead on the overall consumption is −1.2%. Since this value is negative, power gating may be convenient if its total saving is larger than the clock gating one.

(2) Equation (4) is calculated and summed up to (5) to determine clock gating overhead on the overall consumption, which is −2.0%.

(3) Clock gating is more beneficial than power gating determining, overall, a larger power saving. Thus, LR_4 is added to CG_set.

(iv) Finally, LR_5 is processed by invoking evaluate_area(LR_5, 5).

(a) Its area is calculated: $area_{LR_5}$ =15% of total area.

(b) $area_{LR_5}$ > $area_{th}$, so that a prospective power gating implementation on LR_5 is taken into consideration by invoking evaluate_PG(LR_5).

(1) The static and the dynamic overheads are estimated, respectively, applying (1) and (3). The power gating overhead on the overall consumption is −0.89%. Since this value is negative, power gating may be convenient if its total saving is larger than the clock gating one.

(2) Equations (4) and (5) are evaluated to determine clock gating overhead on the overall consumption, which is −1.04%.

(3) Clock gating is more beneficial than power gating determining, overall, a larger power saving. Thus, LR_5 is added to CG_set.

The resulting hardware design with the hybrid application of clock gating and power gating is shown in Figure 5. Comparing this design with the two reported in Figure 2, we can notice as now the power gating is applied only to region LR_1 (called PD1 in the figure), while the clock gating is applied to regions LR_4 and LR_5 (called CD4 and CD5 in the figure); the SBoxes SB_0 and SB_1 included in region LR_4 are purely combinatorial, so they are not affected by the application

LRs	Area%	Base (nW)	PG (nW) (over%)	CG (nW) (over%)
LR_1	52%	4143798	416765 (−86.45%)	4050955 (−2.15%)
LR_3	0.4%	3266	4227 (+0.02%)	3894 (+0.01%)
LR_4	7%	93793	40767 (−1.2%)	9479 (−2.0%)
LR_5	15%	70560	32222 (−0.89%)	25639 (−1.04%)

FIGURE 4: Step-by-step example of the enhanced MDC power extension implementing Algorithm 1.

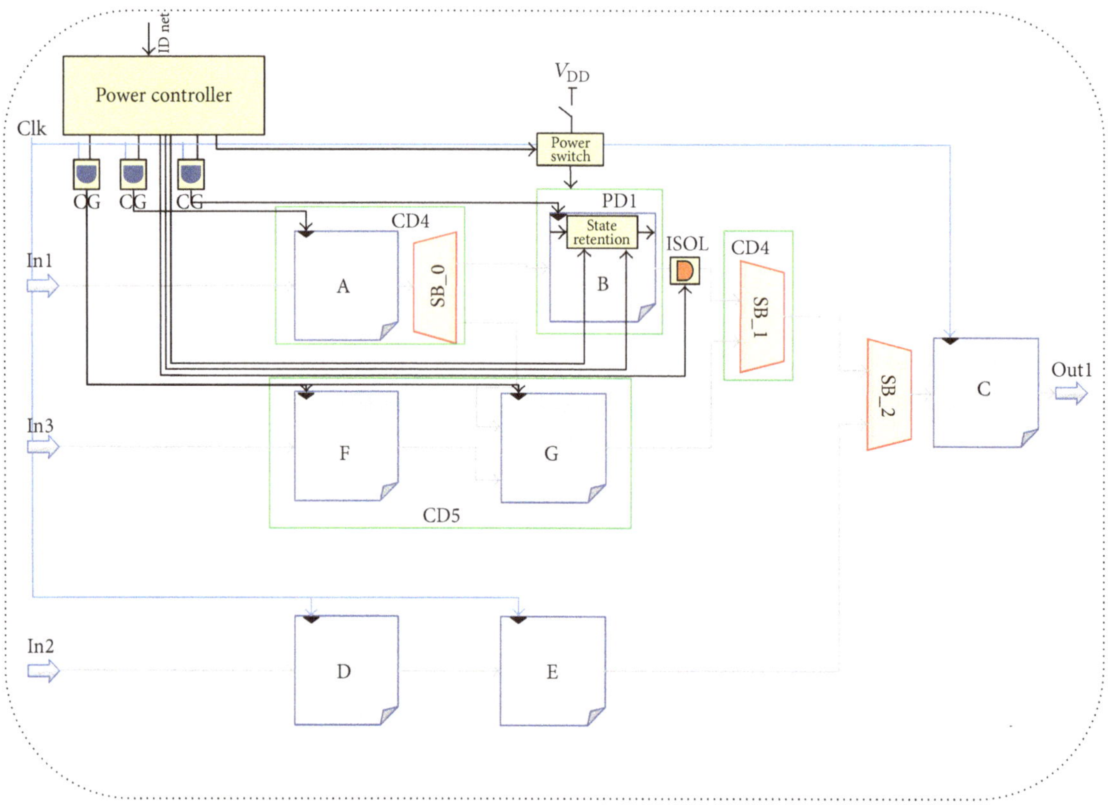

FIGURE 5: Enhanced MDC design suite: hardware platform with hybrid application of clock gating and power gating methodologies.

of the clock gating. All the remaining logic, which includes region LR_3, is always on.

4. Assessments

In order to assess the proposed power estimation flow and the effectiveness of the hybrid clock and power gating management, in this section we discuss two use cases, which are completely different in terms of both behaviour and resulting power consumption contributions. The first one deals with a simple FFT algorithm implemented on a 90 nm CMOS technology and it has been mainly adopted to evaluate in detail the proposed flow. The second one presents a more complex scenario. An image coprocessing unit, accelerating a zoom application, has been implemented both on a 90 nm and on a 45 nm CMOS technology in order to access the robustness of the proposed flow with different technology parameters.

In the following, Section 4.1 discusses the evaluation phase involving the FFT use case, while Section 4.2 describes the experimental results conducted to validate the approach on the zoom application. Finally, Section 4.4 details the benefits of adopting the proposed models and their correlated flow.

4.1. Evaluation Phase. This section deeply discusses the results obtained considering the FFT use case targeting a 90 nm CMOS technology.

4.1.1. Fast Fourier Transform Algorithm. Fast Fourier Transform (FFT) is an optimised algorithm for the Discrete Fourier Transform (DFT) calculation. It is widely adopted in several applications, ranging from the solving of differential equations to the digital signal processing. We refer to the original DFT equation:

$$X_k = \sum_{n=0}^{N} x_n e^{-i2\pi k(n/N)}, \quad k = 0, 1, \ldots, N-1. \tag{9}$$

The FFT algorithm that has been adopted for this use case has been proposed by Cooley and Tukey [45]. It aims at speeding up the calculation of a given size N DFT by considering smaller DFTs of size r, called radix. To obtain the whole original DFT, M stages of size r DFTs are required, where $N = r^M$. Small DFTs have to be multiplied by the so-called twiddle factors, according to the decimation in time variant of the algorithm. When the radix $r = 2$, the DTFs take the name of *butterflies* by their block scheme. The equations describing a butterfly are

$$\begin{aligned} X_0 &= x_0 + x_1 w_n^k, \\ X_1 &= x_0 - x_1 w_n^k, \end{aligned} \tag{10}$$

where X_0 and X_1 are the outputs, while x_0 and x_1 are the corresponding inputs. w_n^k are the twiddle factors, defined as

$$w_n^k = e^{-i2\pi k(k/N)}, \tag{11}$$

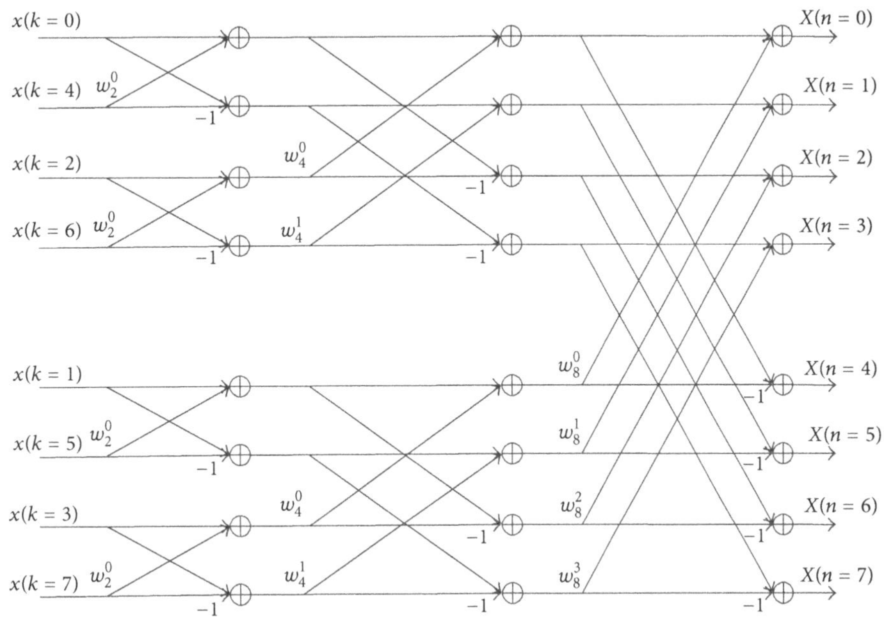
FIGURE 6: FFT use case: original design with 12 radix-2 butterflies for an FFT of size 8. Twiddle factors w_n^k are calculated according to (11).

where k and n are integers depending on the butterfly position in the FFT.

The adopted use case involves a radix-2 FFT of size 8, as depicted in Figure 6, obtained by means of three stages involving four butterflies each, meaning 12 butterflies overall ($r = 2$, $M = 3$, $N = r^M = 2^3 = 8$). Stages have been pipelined to keep the system critical path short. The baseline 12 butterflies design then requires three clock periods for the outputs elaboration.

From the baseline 12 butterfly design, several variants have been derived through the decrease of the involved butterflies number. In such a way, the available resources of the design must be multiplexed in time and reused. Therefore, the overall computation latency increases and the throughput becomes lower. In particular, four size 8 FFT configurations are considered:

(i) 12b is the baseline 12 butterflies FFT design, taking 3 clock periods to finalize the transform.

(ii) 4b involves 4 butterflies for an overall execution latency of 6 cycles.

(iii) 2b involves 2 butterflies for an overall execution latency of 12 cycles.

(iv) 1b involves 1 single butterfly for an overall execution latency of 24 cycles.

4.1.2. FFT CGR System Implementation. The abovementioned configurations have been modelled as dataflow networks and the corresponding CGR system has been assembled with MDC. The activation percentage, resource utilization, and power consumption of each FFT variant are shown in Table 5. In general, the higher the number of butterflies is, the more the corresponding area and dissipation are.

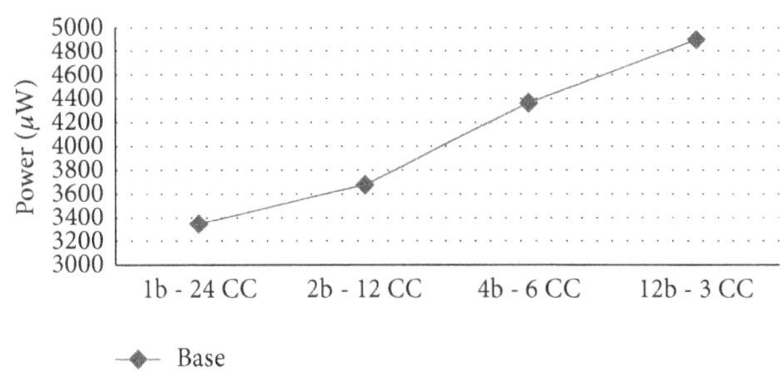

FIGURE 7: FFT use case: latency versus power consumption trade-off for the 4 different 8-size FFT configurations.

The main purpose of the resulting CGR system is to enable several trade-off levels between power dissipation and throughput, as illustrated in Figure 7. Such a system is capable of dynamically switching among the different configurations, fitting to external environment requests. For instance, in a battery operated environment, when the battery level becomes lower than a given threshold, some throughput can be waived to consume less power.

MDC identifies 8 different LRs in the CGR system. The LRs activated by each FFT variant are listed in Table 5, while their characteristics are reported in Table 6. In this table, given any LR, its activation time (T_{ON}) has been obtained summing up the activation times of the FFT configurations activating the same region (provided in Table 5). For example, LR_2 is activated by 1b, 2b, and 4b. Its T_{ON} is 0.67, which is the sum of $T_{\mathrm{ON}}(1\mathrm{b}) = 0.42$, $T_{\mathrm{ON}}(2\mathrm{b}) = 0.21$, and $T_{\mathrm{ON}}(4\mathrm{b}) = 0.04$.

4.1.3. Power Modelling and Hybrid Power Management Assessment. As explained in Section 3.2, the proposed flow requires

Table 5: FFT use case: features of the different configurations integrated on the CGR design. Data refer to a 90 nm CMOS target technology.

FFT configuration	T_{ON}	LRs	Area%	Static [nW]	Internal [nW]
1b	0.42	(2, 6, 8)	12.86	1750407	1307259
2b	0.21	(2, 5, 7, 8)	20.81	1752106	1539855
4b	0.04	(2, 3, 4, 7)	35.28	1757485	2020953
12b	0.33	(1, 3, 7, 8)	98.03	1776056	2411257

Table 6: FFT use case: logic regions architectural and functional characteristics.

	LR_1	LR_2	LR_3	LR_4	LR_5	LR_6	LR_7	LR_8
T_{ON}	0.33	0.67	0.37	0.04	0.21	0.42	0.58	0.96
#iso(e)	1024	1112	512	2324	1948	1756	256	5259
#rtn	1024	518	256	0	0	0	128	512
#reg	1024	518	256	0	0	0	128	512
#iso(r)	1024	1032	512	2306	1926	1740	256	4934

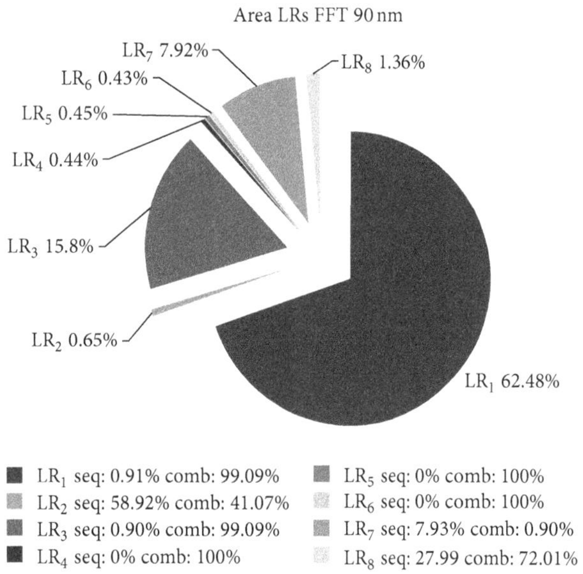

Figure 8: FFT use case: area percentage per LR.

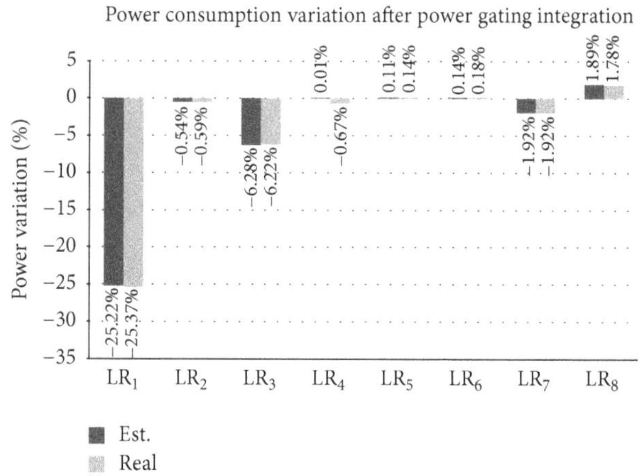

Figure 9: FFT use case: comparison between the estimated and real power variation due to the power gating integration.

a preliminary synthesis of the baseline (without any implemented power saving strategy) CGR system. From the synthesized design, it is possible to retrieve area occupancy and logic composition (combinatorial and sequential contributes) of the 8 LRs, as depicted in Figure 8. The area is given in terms of percentage with respect to the overall system area. The biggest region is LR_1; it occupies more than 60% of the whole system area. It is the region that mainly impacts power consumption. Furthermore, it is quite entirely combinatorial (99.18%), so that power gating should be a very suitable strategy for this LR. By Figure 8 it is possible to notice that LR_2, LR_4, LR_5, LR_6, and LR_8 are extremely small. For all these regions, the proposed power modelling strategy can be extremely beneficial to investigate if power saving techniques may lead or not to an effective power saving. Figure 8 also suggests that LR_4, LR_5, and LR_6 cannot benefit from clock gating, since they are fully combinatorial.

In order to evaluate the proposed power model, Figures 9 and 10 compare the estimated and measured (retrieved from the postsynthesis reports) overhead, respectively, due to power gating and clock gating. In both cases, the reported power refers to both the static and internal contributions, as taken into consideration by the power model. The remaining term, the net one, is neglected. The error of neglecting the net contribution is discussed in Section 4.1.4. The proposed power models are generally able to accurately approximate the power saving strategy overhead. As expected, LR_1 is the region with the highest power saving, regardless of the considered strategy (please note that a negative overhead implies a saving in power). It is interesting to notice that LR_8, despite being one of the smallest regions, does not provide any saving if power gated, but it can achieve a little power reduction when clock gated.

The static and internal power estimations obtained by applying (1) and (3), for a prospective power gating implementation, and (4) and (5), considering a possible clock gating implementation, are shown in Table 7. Please notice

TABLE 7: FFT use case: detailed static and dynamic saving due to power (PG) and clock gating (CG).

LR	PG overhead%		CG overhead%	
	Static	Internal	Static	Internal
LR_1	−40.87	−9.47	0.00	−11.33
LR_2	+0.23	−1.32	0.00	−2.85
LR_3	−10.44	−2.11	0.00	−2.59
LR_4	−0.08	0.10	—	—
LR_5	0.08	0.13	—	—
LR_6	0.11	0.17		
LR_7	−3.44	−0.39	0.00	−0.79
LR_8	1.42	2.35	0.00	−0.25

TABLE 8: FFT use case: *power gating overhead estimation* step accuracy.

LR	PG overhead%			Overhead error%			Net
	Est.	Real	Err.%	Iso	Rtn	Contr.	Err.%
LR_1	−25.22	−25.37	0.59	5.18	4.64	0.75	10.3
LR_3	−6.28	−6.22	1.07	16.34	2.97	0.76	12.6
LR_7	−1.92	−1.92	0.24	9.24	1.08	0.83	13.6

TABLE 9: FFT use case: *clock gating overhead estimation* step accuracy.

LR	CG overhead%			Overhead	Net
	Est.	Real	Err.%	CG_{cell} Err.%	Err.%
LR_1	−5.73	−5.77	0.06	0.83	0.18
LR_2	−1.42	−1.42	0.12	1.03	0.40
LR_3	−1.34	−1.38	0.05	0.84	0.21
LR_7	−0.39	−0.39	0.05	0.96	0.22
LR_8	−0.12	−0.12	0.29	1.08	0.47

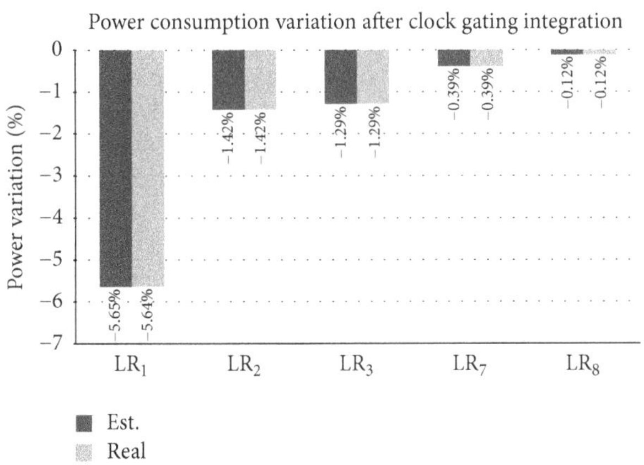

FIGURE 10: FFT use case: comparison between the estimated and real power variation due to the clock gating integration.

that clock gating static overhead is not appreciable, since one single clock gating cell is required per LR.

Algorithm 1 (see Section 3.2.6) implies a preliminary *area thresholding* step. Two different thresholds have been considered for the algorithm evaluation:

(i) DAT_5%: Threshold set to 5%. Regions with area above the 5% are LR_1, LR_3, and LR_7, so that the *power gating overhead estimation* step is performed for each of them. All the considered regions lead to an overall saving (negative overhead) larger than those achievable with a prospective clock gating implementation; therefore, they are selected as eligible regions for power gating. *Clock gating overhead estimation* is performed on all the remaining subthreshold regions. The regions capable of providing saving, when clock gated, are LR_2 and LR_8, since LR_4, LR_5, and LR_6 are fully combinatorial. Thus, clock gating will be implemented only on LR_2 and LR_8.

(ii) DAT_10%: Threshold set to 10%. Only LR_1 and LR_3 are above the area threshold and, as occurred also for DAT_5%, they both achieve power saving if implemented with power gating strategies. In this second case, the *clock gating overhead estimation* step is performed also on LR_7, which results in a negative overhead. Then, the regions to be clock gated are LR_2, LR_7, and LR_8, while LR_4, LR_5, and LR_6 are again discarded.

To access the proposed flow, five designs have been assembled:

(i) Base: the baseline CGR design without any power saving

(ii) PG_full: the CGR design, where power gating is applied blindly to all the regions

(iii) CG_full: the CGR design, where clock gating is applied blindly to all the regions

(iv) DAT_5%: derived with the proposed automated flow capable of hybrid power and clock gating support, setting the Area Threshold to 5% in Algorithm 1

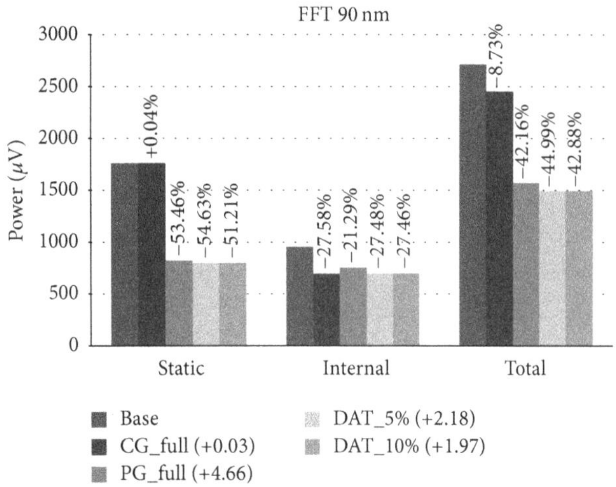

Figure 11: FFT use case: comparison between the base design and the four gated designs. Legend shows, in brackets, the power management area overhead for each design with respect to the Base one.

(v) DAT_10%: derived with the proposed automated flow capable of hybrid power and clock gating support, setting the Area Threshold to 10% in Algorithm 1

These designs have been synthesized with Cadence RTL Compiler, targeting the same 90 nm CMOS technology adopted to synthesise and simulate the baseline CGR design, whose results have fed the *Power Analysis* block of the proposed enhanced power management flow to assemble DAT_5% and DAT_10%.

The power consumption (internal, static, and total) of these designs is depicted in Figure 11. The reported data correspond to the real power consumed by the synthesised designs. Since LR_1 occupies more than the 60% of the design area and it is mainly combinatorial, little differences among the entirely power gated design (PG_full) and the hybrid clock and power gating ones (DAT_5% and DAT_10%) are visible. Nevertheless, DAT_5% achieves the largest power saving (−45.12%) among all the designs, validating the proposed hybrid and selective management with respect to a one-fit-to-all solution. CG_full, capable of diminishing only the dynamic power consumption, is the worst design among those applying power management.

The area overhead of the implemented power management strategies, reported in the legend of Figure 11, proves that the proposed hybrid management leads to less area hungry designs than the entirely power gated one. In fact, DAT_5% and DAT_10% present half of the area overhead of PG_full. CG_full data confirms that clock gating has a very little impact on the baseline design, presenting a negligible area overhead (two orders of magnitude smaller than DAT_5% and DAT_10%).

For the sake of completeness, in Figure 12 the trade-off levels between power and latency (and, in turn, throughput) are illustrated for all the considered designs. The trade-off curves demonstrate that power management strategies are generally extremely beneficial within a CGR scenario.

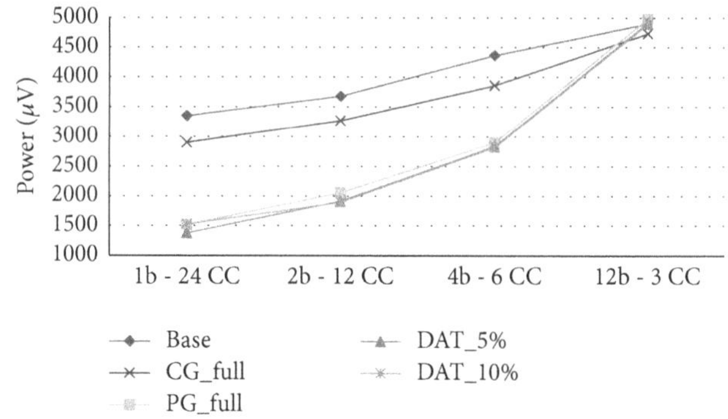

Figure 12: FFT use case: atency versus power consumption trade-off for the 4 different 8-size FFT configurations, when gated designs are adopted.

4.1.4. Accuracy and Errors. The accuracy of the proposed power modelling approach is assessed by Tables 8 and 9, respectively, considering the *power gating overhead estimation* step and the *clock gating overhead estimation* step of Algorithm 1. These tables, in each row, report the estimation errors with respect to the real consumption of the baseline CGR system, where the given power saving strategy (i.e., power gating in Table 8 and clock gating in Table 9) is applied only on the LR specified in the first column.

Looking at Table 8, power overhead estimations per LR demonstrate to be very accurate, leading to errors that are always below 1.1%. Also errors related to the estimation of state retention and Power Controller overhead are quite low (resp., below 5% and 1%). The isolation cells overhead estimation is less precise, resulting in an error of 16.36% for PD3, due to the fact that the static and internal values of $P(ISO_{ON})$ and $P(ISO_{OFF})$ are characterized as average values, the same for each LR. Nevertheless, this error has no visible impact on the total estimation one that is 1.07%.

Table 9 depicts an overview of the estimations accuracy for the clock gating overhead. In this case, estimations are even more accurate. The error on the clock gating overhead is always below the 0.3%. The error due to the clock gating cells overhead is very limited too, being always under the 1.1%.

Tables 8 and 9 report also the errors caused by omitting the power net term (column net % (Err.)) in (3) and (5). This error is obtained by comparing the estimated overhead (not comprehensive of the net contribute) with the real measured overhead comprehensive of the net term, as extracted by the synthesis reports. The net error is higher in the power gating overhead estimation (13.6% for LR_7) with respect to the clock gating one (at maximum 0.47% for LR_8), since power gating requires more additional logic to be implemented.

4.2. Validation Phase. In order to validate the proposed approach, a second use case has been assessed targeting the same 90 nm technology used for the FFT use case and a smaller 45 nm library. The reconfigurable computing core of an image coprocessing unit, accelerating a zoom application, has been assembled. Its characteristics are discussed in Section 4.2.1, while Sections 4.2.2 and 4.2.3 analyse the achieved results.

Table 10: Zoom coprocessor use case: computational kernels distinctive features.

Kernel	# actors	T_{ON}	Functionality	LRs
abs	1	0.03	Absolute value calculation	(4)
chgb	7	0.33	Bilevel/grayscale block checking	(5, 6, 7, 12, 13)
cubic	10	0.06	Linear combination calculation	(1, 5, 9, 10, 13)
cubic_conv	6	0.09	Cubic filter convolution	(5, 7, 8, 9)
median	9	0.06	Median calculation	(1, 3, 5, 11, 13)
min_max	1	0.01	Maximum/minimum finding	(11)
sbwlabel	17	0.42	Edge block checking	(2, 4, 5, 6, 13)

Table 11: Zoom coprocessor at 90 nm CMOS technology: characterization of the hybrid, clock, and power gated designs, achieved with the proposed automated flow. DAT_5%: area threshold 5%. DAT_10%: area threshold 10%. NA stands for not assigned and includes those LRs that placed in the always on domain.

Design	>Th	PG_set	CG_set	NA
DAT_5%	$LR_1, LR_2, LR_3, LR_6, LR_8, LR_{10}, LR_{12}, LR_{13}$	$LR_1, LR_2, LR_3, LR_8, LR_{10}, LR_{12}$	$LR_4, LR_6, LR_7, LR_9, LR_{11}, LR_{13}$	LR_5
DAT_10%	LR_1, LR_8	LR_1, LR_8	$LR_2, LR_3, LR_4, LR_6, LR_7, LR_9, LR_{10}, LR_{11}, LR_{12}, LR_{13}$	LR_5

4.2.1. CGR System of the Zoom Coprocessor. The zoom application is meant to scale an image according to the given zooming factor. Missing pixels of the zoomed image are calculated by adaptively interpolating the neighbouring values. The original application is a C program. It has been executed and profiled on a general purpose machine in order to identify the most computational intensive portions of the code (computational kernels) with the intention of accelerating them on a CGR hardware accelerator. Seven computational kernels have been identified and modelled as dataflow networks. Table 10 summarizes the kernels composition (in terms of number of dataflow actors), activation time, and main functionality. Then, these dataflow kernels have been combined by MDC to obtain a multidataflow specification, constituting the computing core of the CGR accelerator in charge of accelerating the zoom application.

Thirteen LRs are identified on the CGR zoom coprocessor. The main difference between this scenario and the FFT one is that in the zoom coprocessor it is not necessary to retain the status of any kernel when switching among them. This means that, applying power gating, no retention cells are needed in the identified regions.

4.2.2. Zoom Coprocessor Validation Results at 90 nm CMOS Technology. This section is meant to provide the discussion of the achieved results in the zoom coprocessor scenario using the same 90 nm CMOS technology adopted for the FFT CGR designs assessment. From the implementation point of view, we will discuss the same designs we considered for the FFT use case, defined as follows:

(i) Base: the baseline CGR design without any power saving

(ii) PG_full: the CGR design, where all the 13 LRs are power gated

(iii) CG_full: the CGR design, where all the 13 LRs are clock gated

(iv) DAT_5%: the CGR design, where hybrid power and clock gating support are implemented by means of the proposed flow, setting the Area Threshold to 5% in Algorithm 1

(v) DAT_10%: the CGR design, where hybrid power and clock gating support are implemented by means of the proposed flow, setting the Area Threshold to 10% in Algorithm 1

Please refer to Table 11 for the composition of DAT_5% and DAT_10%.

Figure 13 depicts static, internal, and total power consumption for each considered design. In this case, the differences among CG_full, PG_full, DAT_5%, and DAT_10% are not so evident. The reason is that, in this scenario, the dynamic power consumption (due to the internal power) is considerably higher than the static one. As you can see in the reported histograms, on average, there are approximately more than two orders of magnitude of difference. Power gating and clock gating demonstrate to be equally capable of cutting down the internal power consumption.LR_5 is the only region that Algorithm 1 completely discards by any form of power management, both in the DAT_5% design and in the DAT_10% one. It is fully combinatorial; therefore, clock gating does not provide any positive effect on it. Nevertheless, it is so small (0.65% of the whole system area) that, if power gated, it cannot provide any substantial benefit. A closer observation of the histograms confirms what we already got for the FFT: despite the similar trend for all the designs, which lead to more than the 62% of power saving, DAT_5% consumes less than any other (62.61% of saving), while the CG_full design is the less beneficial (62.29% of saving).

Focusing on the static histograms, as expected, the CG_full design introduces a small overhead with respect to

TABLE 12: Zoom coprocessor at 90 nm CMOS technology: *power gating overhead estimation* step and clock gating overhead estimation step accuracy.

LR	PG saving%			Net	CG saving%			Net
	Est.	Real	Err.%	Err.%	Est.	Real	Err.%	Err.%
LR_1	−6.322	−6.331	0.15	0.18	−6.307	−6.313	0.09	0.17
LR_2	−23.464	−23.463	0.00	1.26	−23.369	−23.373	0.02	1.67
LR_3	−6.537	−6.544	0.10	1.65	−6.507	−6.514	0.10	1.59
LR_4	—	—	—	—	−0.958	−0.964	0.68	1.34
LR_5	—	—	—	—	—	—	—	—
LR_6	—	—	—	—	−0.870	−0.878	0.81	1.22
LR_7	—	—	—	—	−1.033	−1.039	0.63	0.15
LR_8	−7.062	−7.058	0.05	2.02	−6.987	−6.986	0.01	1.83
LR_9	—	—	—	—	−2.436	−2.443	0.27	1.55
LR_{10}	−5.498	−5.503	0.10	1.65	−5.474	−5.480	0.11	1.57
LR_{11}	—	—	—	—	−3.329	−3.285	1.32	3.67
LR_{12}	−4.278	−4.288	0.23	1.83	−4.267	−4.273	0.15	1.86
LR_{13}	—	—	—	—	−0.649	−0.656	1.01	1.13

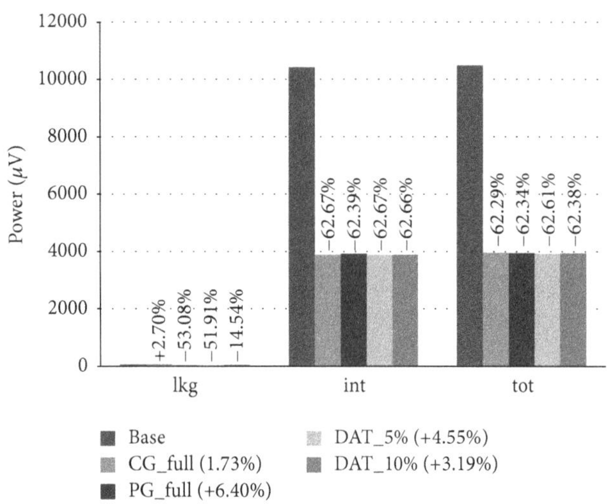

FIGURE 13: Zoom coprocessor at 90 nm CMOS technology: comparison between the base design and the four gated designs. Legend shows, in brackets, the power management area overhead for each design with respect to the Base one.

base. That is due to the 12 (one for each region but LR_5) clock gating cells introduced in the always on domain of this design, which never contribute to save any static power consumption. When power gating is applied, there is always a benefit in terms of static power consumption: DAT_5% saving is slightly higher than the PG_full one, both being over 51%; DAT_10% is still beneficial, but its saving is limited to the 15%. Please note that the difference between DAT_5% and DAT_10% (in terms of static consumption) demonstrates that, in the *area thresholding* step of the proposed Algorithm 1, it is better to opt for small area threshold values to achieve higher saving results.

In terms of area occupancy, reported in the legend of Figure 13, the PG_full design is the one with the largest overhead, +6.4%. DAT_5%, which is the most beneficial in terms of power, shows a slightly smaller overall overhead, +4.55% of the whole system area. The CG_full is again the less invasive one, leading just to +1.73% of area overhead. Summarizing, DAT_5% constitutes the optimal solution for the zoom coprocessor scenario, considering a 90 nm technology. DAT_10%, which is less beneficial than DAT_5% in saving static power consumption, is a better solution than a fully power gated design, presenting basically the same power saving (−62.38% for DAT_10% versus −62.29% for PG_full) but a smaller area overhead (+3.19% for DAT_10% versus +6.4% for PG_full).

Table 12 reports the estimation error of the proposed automated hybrid power management design flow, when the power saving percentages for the considered domains, respectively, considering power gating (*power gating overhead estimation*) and clock gating (*clock gating overhead estimation*), are evaluated. *PG saving*% errors are always below 0.3% and *CG saving*% ones do not exceed 1.5%. These data confirm the accuracy of the proposed models, as in the FFT use case. Table 12, for both power and clock gating implementations, depicts also the error of neglecting the net term in the dynamic power consumption. Again, as in the previously discussed scenario, models are not affected by this simplification.

4.2.3. Zoom Coprocessor Validation Results at 45 nm CMOS Technology. In order to provide a robust validation of the proposed approach, we decided to assess the same zoom coprocessor designs on a different technology targeting a 45 nm CMOS technology. The idea is assessing what changes when the ratio between the static and the dynamic consumption is varied. Here follows the list of the implemented designs:

(i) Base: the same as in the 90 nm synthesis trial

(ii) PG_full: the same as in the 90 nm synthesis trial

(iii) CG_full: the same as in the 90 nm synthesis trial

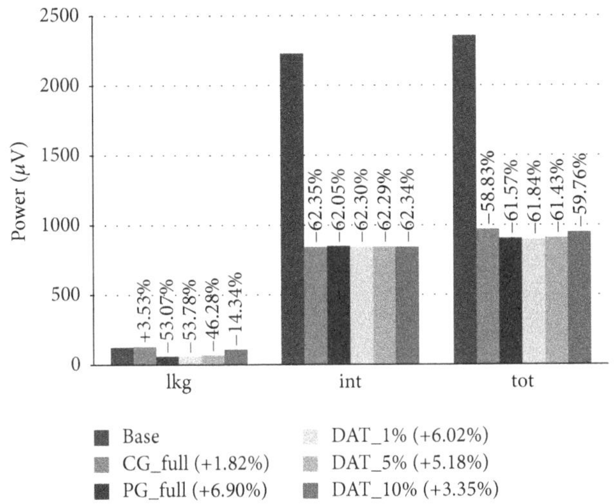

Figure 14: Zoom Co-Processor at 45 nm CMOS technology: Comparison between the base design and the five gated designs. Legend shows, in brackets, the power management area overhead for each design with respect to the Base one.

(iv) DAT_1%: the CGR design, where hybrid power and clock gating support are implemented by means of the proposed flow, setting the Area Threshold to 1% in Algorithm 1

(v) DAT_5%: the same as in the 90 nm synthesis trial

(vi) DAT_10%: the same as in the 90 nm synthesis trial

Targeting a smaller technology and having already established that the 10% area threshold leads to power results comparable to those of the fully power gated design, we have decided to add in this second trial an additional design: DAT_1%. Setting the area threshold to 1%, quite all the LRs are considered for a prospective power gating implementation. Please refer to Table 13 for the composition of DAT_1%, DAT_5%, and DAT_10%.

The power consumption in terms of static, internal, and total contributions is illustrated in Figure 14. The dynamic power consumption is still higher than the static one, determining the overall trend of the total power. However, with the scaling of the channel length, the ratio among internal and static power on average has decreased from a factor of 100 to approximately 10. In this second trial, the influence of the static power consumption is partially reflected on the total one. Technology scaling and the different static versus dynamic power ratio are such that PG_full is capable of providing better overall saving results than DAT_5% and DAT_10%. At 45 nm technology, designers are required to select a very low area threshold in Algorithm 1 to achieve really optimal results. DAT_1%, which basically excludes from power gating only the 3 LRs, saves up to 61.84% of total base power, and represents the optimal design solution for the zoom coprocessor in this second synthesis trial. Please note also that, lowering down the area threshold, the area of the optimal design and that of the fully power gated one are pretty similar.

We can conclude that as technology gets smaller the *area thresholding* step of the proposed algorithm is less beneficial still, in the automated flow, its presence makes the overall process more robust, avoiding useless iterations on not convenient by construction designs when the technology are not so constrained or the ratio among static and dynamic consumption is larger.

The accuracy of the proposed models, targeting the 45 nm CMOS technology, is reported in Table 14, which contain both power gating and clock gating estimation errors. The models, even neglecting the net contribution in the discussed equations, are extremely accurate (the error never exceeds 3.70%).

4.3. Power Switch Overhead. The sleep transistors are inserted in the design during the place-and-route process and their overhead is strictly use case dependent since it is related to the aspect ratio of the macro and to the style of power routing that is selected in the target design. Since the proposed power estimation model is based on the synthesis of the design, the contribution of these cells is not considered yet.

The insertion of header/footer switches (we used header ones in this work as reference) determines two kinds of power overhead: (1) a leakage-related overhead; (2) the dynamic power dissipated during sleep and wake-up transition. Another overhead that has to be taken into consideration is the time necessary to wake up the power domain. For the proper operation of the power gating methodology, the gating logic has to be enabled/disabled according to a switch on/off protocol [11] that requires 4 clock cycles for each transition. Thus, when a kernel is off, at least 4 clock cycles are necessary before it is switched on and then it can receive new input data.

The leakage-related contribution is fixed by the technology library and it is always present regardless of the ON/OFF state of the power domain. It could be inserted in (1) as $P_{lkg}(SW) * \#switches$ where $P_{lkg}(SW)$ is the static power consumption of the considered power switch, as reported in the technology library, and #switches is the number of power switches inserted in the power domain. In the library that we took as reference, a header switch has the same leakage power of a 64-bit wide register, and one column of N switches can be used to switch down N horizontal virtual power stripes of the power grid. If we assume to use only one vertical real power stripe to supply power to N horizontal virtual power stripes of a power domain, only one column of N power switches is inserted. In this case, gating the virtual power supply easily saves enough leakage power to counterbalance the leakage dissipation of the inserted switch.

The dynamic power contribution is only relevant when intervals between successive kernel switches are in the order of tens of cycles (Hu et al. [46]). When the computation of the kernels last tens of cycles also the wake-up time is not relevant. Thus, in designs with low switching rates, these two overhead contributions could be neglected.

The FFT use case is a really simple design, used only for the development of the power estimation model and, as reported in Section 4.1, its kernels are far from lasting tens of cycles. The zoom application adopted for the validation

TABLE 13: Zoom coprocessor at 45 nm CMOS technology: characterization of the hybrid, clock, and power gated designs, achieved with the proposed automated flow. DAT 1%: area threshold 1%. DAT 5%: area threshold 5%. DAT_10%: area threshold 10%. NA stands for not assigned and includes those LRs that are placed in the always on domain.

Design	>Th	PG_set	CG_set	NA
DAT_1%	LR_1, LR_2, LR_3, LR_4,LR_6, LR_7, LR_8, LR_9, LR_{10}, LR_{11}, LR_{12}, LR_{13}	LR_1, LR_2, LR_3, LR_4, LR_7, LR_8, LR_9, LR_{10}, LR_{11}, LR_{12}	LR_6, LR_{13}	LR_5
DAT_5%	LR_1, LR_2, LR_3, LR_6, LR_8, LR_{10}, LR_{12}, LR_{13}	LR_1, LR_2, LR_3, LR_8, LR_{10}, LR_{12}	LR_4, LR_6, LR_7, LR_9, LR_{11}, LR_{13}	LR_5
DAT_10%	LR_1, LR_8	LR_1, LR_8	LR_2, LR_3, LR_4, LR_6, LR_7, LR_9, LR_{10}, LR_{11}, LR_{12}, LR_{13}	LR_5

TABLE 14: Zoom coprocessor at 45 nm CMOS technology: *power gating overhead estimation* step and *clock gating overhead estimation* step accuracy.

LR	PG saving%			Net	CG saving%			Net
	Est.	Real	Err.%	Err.%	Est.	Real	Err.%	Err.%
LR_1	−5.686	−5.699	0.23	0.47	−5.507	−5.501	0.13	0.53
LR_2	−22.990	−22.982	0.03	1.02	−22.063	−22.029	0.15	0.81
LR_3	−6.569	−6.566	0.04	1.13	−6.273	−6.262	0.18	0.91
LR_4	−0.946	−0.944	0.15	1.82	−0.919	−0.917	0.22	1.52
LR_5	—	—	—	—	—	—	—	—
LR_6	−0.747	−0.748	0.25	0.25	−0.853	−0.833	0.26	1.59
LR_7	−0.955	−0.957	0.17	1.36	−0.935	−0.933	0.21	1.49
LR_8	−7.464	−7.460	0.06	1.36	−6.724	−6.714	0.16	0.66
LR_9	−2.475	−2.482	0.25	1.10	−2.350	−2.345	0.18	1.09
LR_{10}	−5.473	−5.470	0.04	1.09	−5.270	−5.261	0.17	0.91
LR_{11}	−3.422	−3.361	1.79	2.58	−3.257	−3.205	1.63	2.04
LR_{12}	−4.327	−4.330	0.08	1.19	−4.176	−4.169	0.17	0.98
LR_{13}	−0.587	−0.580	1.25	3.70	−0.623	−0.621	0.28	1.85

phase of the proposed model is a real use case but it is a small size design, where the execution of the fastest kernels lasts 24 clock cycles. Then, the wake-up time overhead is, in the worst case, almost 17% of the total execution time. If we consider a bigger and more complex real use case, such as interpolation filtering for motion compensation in High Efficiency Video Coding [47], we would achieve the condition for neglecting the dynamic power consumption of the sleep transistors and the wake-up time overhead. This application involves 2-dimensional filters working on subblocks of pictures belonging to the same video sequence. The smallest block, corresponding to the fastest execution time, has 8 × 8 pixels. In this case, 160 overall cycles are required to filter the whole block and the wake-up time overhead is just 2.5% of the total execution time.

TABLE 15: FFT use case at 90 nm CMOS technology: *power gating overhead estimation* step accuracy, using reports generated without the real switching activity.

LR	PG overhead%		
	Est.	Real	Err.%
LR_1	−30.54	−26.10	17.07
LR_2	−0.19	−1,99	90.22
LR_3	−21.83	−6.95	214.30
LR_4	−0.12	−0.67	80.97
LR_5	−0.06	−0.59	90.11
LR_6	−0.07	0.56	86.60
LR_7	−15.39	−2.64	482.93
LR_8	−0.38	−0.02	1941.81

4.4. Advantages of the Proposed Approach. Considering a CGR system implementing N different functionalities and partitioned into k different LRs, the proposed selection algorithm, based on the power models embodied in (1), (3), (4), and (5), requires the synthesis of the baseline CGR design (without any power saving strategy applied) and N hardware simulations, each one running a different functionality (i.e., executing the different DPNs provided as input to the MDC tool). The hardware simulations are needed since the real switching activity is essential for correct dynamic power estimation. Table 15, targeting the FFT scenario and the power gating implementation, depicts the estimated and real power overhead when estimations are performed adopting the default synthesis reports (without taking into account the real switching activity). The estimation errors are extremely high when the switching activity is neglected; therefore, the proposed models are not capable of properly determining which LRs would actually benefit of power gating.

In order to understand the advantages of the proposed approach, let us compute the effort needed to determine the optimal saving strategy for each region if our flow is not adopted. It is required to

(1) synthesize the baseline design without any power management support;

(2) synthesize one power gated design and one clock gated design for each LR;

(3) perform N different hardware simulations for the baseline design, to retrieve the real switching activity of the system;

(4) perform N different hardware simulations for each power gated and clock gated design, to retrieve their real switching activity;

(5) compare each power gated design and clock gated design, in the different operating conditions, with respect to the synthesized baseline CGR design.

Our flow requires only point 1 and point 3. In numbers it corresponds to one single synthesis and N hardware simulations. On the contrary, not using our approach, $2 * k + 1$ synthesis

(k for power gating evaluation, k for clock gating evaluation plus the baseline one) and $N*(2*k+1)$ hardware simulations are necessary. The only simplification that may be done, even without adopting the proposed approach, is when a given region is fully combinatorial. This would save the effort related to its perspective clock gating evaluation.

Dealing with the presented use cases, for the FFT (Section 4.1) there are $N = 4$ different functionalities and $k = 8$ LRs. Among these latter 3 are fully combinatorial. The proposed approach required one synthesis and 4 hardware simulations, rather than 14 synthesis (8 power gated LRs, 5 clock gated LRs and the baseline design) and 56 hardware simulations (4 for each synthesized design). Considering the zoom coprocessor (Section 4.2), $N = 7$ and $k = 13$, with only 1 fully combinatorial LR. The proposed approach required one synthesis and 7 hardware simulations, rather than 26 synthesis (13 power gated, 12 clock gated and the baseline designs) and 182 hardware simulations.

5. Conclusions

In this paper, we addressed the problem of power management in coarse-grained reconfigurable (CGR) systems. Such systems are as suitable to accelerate multifunctional applications as, potentially, energy inefficient. In fact, on a CGR substrate, while a particular task is executed, the resources not involved in the computation may potentially waste precious power if not properly managed. On top of that, these architectures are also characterized by an intrinsic design difficulty: mapping several applications on the same substrate, customizing the datapath, is not so straightforward and requires a deep knowledge of the target applications. Dataflow models of computation turned out to be very efficient for the development of automated mapping problems. In our studies, we have exploited a dataflow-based approach to define a complete design suite for multifunctional CGR systems: the Multi-Dataflow Composer (MDC) tool. Besides automatically managing dataflow to hardware systems composition, MDC also supports the automated characterization of power and clock gated platforms.

The proposed work makes some steps further both in the MPEG-RVC field, which the MDC tool belongs to, and in the definition of optimal power management strategies for CGR designs. In this paper, we have presented an automated methodology capable of estimating, prior to any physical implementation, the effectiveness and costs that power gating or clock gating would have when implemented on top of the functional logic regions constituting a CGR system. This methodology is based on static and dynamic power estimation models that, in a separate manner for each logic region in the CGR design, are capable of determining the overhead of clock gating and power gating on the basis of the functional, technological, and architectural parameters of the baseline CGR system. These models and the corresponding estimation algorithm are applicable in any CGR scenario and are currently integrated in the MDC tool, improving its basic functionality. In fact, MDC was normally applying a one-fit-to-all user-specified power reduction technique, either clock or power gating, without any warranty of its effectiveness on the different identified regions.

By considering two different scenarios and adopting different ASIC technologies, our assessments proved that the enhanced MDC flow is capable of guiding the designers towards the definition of the optimal power management support. It is more efficient than the previous, blindly applied, methodology and the proposed models turned out to be extremely accurate. Finally, as demonstrated in Section 4.4, the new flow drastically reduces the number of designs to be synthesized and simulated, leading to saving both designer effort and computational time.

Competing Interests

The authors declare that they have no competing interests.

Acknowledgments

Tiziana Fanni is grateful to Sardinia Regional Government for supporting her Ph.D. scholarship (P.O.R. F.S.E., European Social Fund 2007–2013, Axis IV Human Resources).

References

[1] R. Hartenstein, "A decade of reconfigurable computing: a visionary retrospective," in *Proceedings of the Design, Automation and Test in Europe Conference and Exhibition (DATE '01)*, pp. 642–649, March 2001.

[2] P. Meloni, G. Tuveri, L. Raffo et al., "System adaptivity and fault-tolerance in NoC-based MPSoCs: the MADNESS project approach," in *Proceedings of the 15th Euromicro Conference on Digital System Design (DSD '12)*, pp. 517–524, 2012.

[3] H. Esmaeilzadeh, E. Blem, R. St. Amant, K. Sankaralingam, and D. Burger, "Dark silicon and the end of multicore scaling," in *Proceedings of the International Symposium on Computer Architecture*, pp. 365–376, San Jose, Calif, USA, 2011.

[4] M. B. Taylor, "Is dark silicon useful? Harnessing the four horsemen of the coming dark silicon apocalypse," in *Proceedings of the 49th Annual Design Automation Conference (DAC '12)*, pp. 1131–1136, San Francisco, Calif, USA, June 2012.

[5] F. Oboril, J. Ewert, and M. B. Tahoori, "High-resolution online power monitoring for modern microprocessors," in *Proceedings of the Conference on Design, Automation and Test in Europe*, pp. 265–268, March 2015.

[6] Synopsys, "Advanced low power techniques," http://goo.gl/2FqEjf.

[7] S. Herbert and D. Marculescu, "Analysis of dynamic voltage/frequency scaling in chip-multiprocessors," in *Proceedings of the International Symposium on Low Power Electronics and Design (ISLPED '07)*, pp. 38–43, August 2007.

[8] S. Eyerman and L. Eeckhout, "Fine-grained DVFS using on-chip regulators," *Transactions on Architecture and Code Optimization*, vol. 8, no. 1, article 1, 2011.

[9] M. Arora, S. Manne, Y. Eckert, I. Paul, N. Jayasena, and D. M. Tullsen, "A comparison of core power gating strategies implemented in modern hardware," in *Proceedings of the Conference on Measurement and Modeling of Computer Systems*, pp. 559–560, 2014.

[10] B. Jeff, "Advances in big.LITTLE technology for power and energy savings," ARM White Paper, 2012.

[11] Power Forward Initiative, *A Practical Guide to Low Power Design*, 2009.

[12] F. Palumbo, N. Carta, D. Pani, P. Meloni, and L. Raffo, "The multi-dataflow composer tool: generation of on-the-fly reconfigurable platforms," *Journal of Real-Time Image Processing*, vol. 9, no. 1, pp. 233–249, 2014.

[13] N. Carta, C. Sau, F. Palumbo, D. Pani, and L. Raffo, "A coarse-grained reconfigurable wavelet denoiser exploiting the multi-dataflow composer tool," in *Proceedings of the 7th Conference on Design and Architectures for Signal and Image Processing (DASIP '13)*, pp. 141–148, October 2013.

[14] N. Carta, C. Sau, D. Pani, F. Palumbo, and L. Raffo, "A coarse-grained reconfigurable approach for low-power spike sorting architectures," in *Proceedings of the 6th International IEEE EMBS Conference on Neural Engineering (NER '13)*, pp. 439–442, San Diego, Calif, USA, November 2013.

[15] D. Pani, F. Usai, L. Citi, and L. Raffo, "Real-time processing of tfLIFE neural signals on embedded DSP platforms: a case study," in *Proceedings of the 5th International IEEE/EMBS Conference on Neural Engineering (NER '11)*, pp. 44–47, Cancun, Mexico, April 2011.

[16] N. Carta, P. Meloni, G. Tuveri, D. Pani, and L. Raffo, "A custom MPSoC architecture with integrated power management for real-time neural signal decoding," *IEEE Journal on Emerging and Selected Topics in Circuits and Systems*, vol. 4, no. 2, pp. 230–241, 2014.

[17] F. Palumbo, C. Sau, and L. Raffo, "Coarse-grained reconfiguration: dataflow-based power management," *IET Computers and Digital Techniques*, vol. 9, no. 1, pp. 36–48, 2015.

[18] F. Palumbo, T. Fanni, C. Sau, and P. Meloni, "Power-awarness in coarse-grained reconfigurable multi-functional architectures: a dataflow based strategy," *Journal of Signal Processing Systems*, 2016.

[19] D. Pani and L. Raffo, "Self-coordinated on-chip parallel computing: a swarm intelligence approach," in *Parallel and Distributed Computational Intelligence*, pp. 91–112, Springer, Berlin, Germany, 2010.

[20] M. Yan, Z. Yang, L. Liu, and S. Li, "ProDFA: accelerating domain applications with a coarse-grained runtime reconfigurable architecture," in *Proceedings of the 18th IEEE International Conference on Parallel and Distributed Systems (ICPADS '12)*, pp. 834–839, December 2012.

[21] S. M. Carta, D. Pani, and L. Raffo, "Reconfigurable coprocessor for multimedia application domain," *Journal of VLSI Signal Processing Systems for Signal, Image and Video Technology*, vol. 44, no. 1-2, pp. 135–152, 2006.

[22] V. V. Kumar and J. Lach, "Highly flexible multimode digital signal processing systems using adaptable components and controllers," *EURASIP Journal on Applied Signal Processing*, vol. 2006, no. 1, Article ID 079595, pp. 1–9, 2006.

[23] B. Mei, S. Vernalde, D. Verkest, H. De Man, and R. Lauwereins, "Exploiting loop-level parallelism on coarse-grained reconfigurable architectures using modulo scheduling," in *Proceedings of the Design, Automation and Test in Europe Conference and Exhibition (DATE '03)*, pp. 296–301, March 2003.

[24] F. Palumbo, D. Pani, L. Raffo, and S. Secchi, "A surface tension and coalescence model for dynamic distributed resources allocation in massively parallel processors on-chip," in *Nature Inspired Cooperative Strategies for Optimization (NICSO 2007)*, vol. 129 of *Studies in Computational Intelligence*, pp. 335–345, Springer, Berlin, Germany, 2007.

[25] G. Ansaloni, K. Tanimura, L. Pozzi, and N. Dutt, "Integrated kernel partitioning and scheduling for coarse-grained reconfigurable arrays," *IEEE Transactions on Computer-Aided Design of Integrated Circuits and Systems*, vol. 31, no. 12, pp. 1803–1816, 2012.

[26] C. C. de Souza, A. M. Lima, G. Araujo, and N. B. Moreano, "The datapath merging problem in reconfigurable systems: complexity, dual bounds and heuristic evaluation," *ACM Journal of Experimental Algorithmics*, vol. 10, no. 2.2, Article ID 1180613, 2005.

[27] R. Giorgi and A. Scionti, "A scalable thread scheduling co-processor based on data-flow principles," *Future Generation Computer Systems*, vol. 53, pp. 100–108, 2015.

[28] L. Verdoscia, R. Vaccaro, and R. Giorgi, "A clockless computing system based on the static dataflow paradigm," in *Proceedings of the 4th Workshop on Data-Flow Execution Models for Extreme Scale Computing (DFM '14)*, pp. 30–37, Edmonton, Canada, August 2014.

[29] C. Sau, L. Fanni, P. Meloni, L. Raffo, and F. Palumbo, "Reconfigurable coprocessors synthesis in the MPEG-RVC domain," in *Proceedings of the International Conference on ReConFigurable Computing and FPGAs (ReConFig '15)*, pp. 1–8, IEEE, Riviera Maya, Mexico, December 2015.

[30] L. Jozwiak, M. Lindwer, R. Corvino et al., "ASAM: automatic architecture synthesis and application mapping," in *Proceedings of the 15th Euromicro Conference on Digital System Design (DSD '12)*, pp. 216–225, September 2012.

[31] L. Jozwiak, M. Lindwer, R. Corvino et al., "ASAM: automatic architecture synthesis and application mapping," *Microprocessors and Microsystems*, vol. 37, no. 8, pp. 1002–1019, 2013.

[32] Y. Zhang, J. Roivainen, and A. Mammela, "Clock-gating in FPGAs: a novel and comparative evaluation," in *Proceedings of the 9th Euromicro Conference on Digital System Design: Architectures, Methods and Tools*, p. 590, 584, Dubrovnik, Croatia, August-September 2006.

[33] E. Bezati, S. Casale-Brunet, M. Mattavelli, and J. W. Janneck, "Coarse grain clock gating of streaming applications in programmable logic implementations," in *Proceedings of the Electronic System Level Synthesis Conference*, pp. 1–6, 2014.

[34] Silicon Integration Initiative, Si2 Common Power Format SpecificationTM-Version 2.1, 2014.

[35] M. Shafique, L. Bauer, and J. Henkel, "Adaptive energy management for dynamically reconfigurable processors," *IEEE Transactions on Computer-Aided Design of Integrated Circuits and Systems*, vol. 33, no. 1, pp. 50–63, 2014.

[36] J. Yi and J. Kim, "Power modeling for digital circuits with clock gating," *IEICE Electronics Express*, vol. 12, no. 24, Article ID 20150817, 2015.

[37] H. Xu, R. Vemuri, and W.-B. Jone, "Run-time active leakage reduction by power gating and reverse body biasing: an energy view," in *Proceedings of the 26th IEEE International Conference on Computer Design (ICCD '08)*, pp. 618–625, IEEE, October 2008.

[38] K. Datta, A. Mukherjee, G. Cao et al., "CASPER: embedding power estimation and hardware-controlled power management in a cycle-accurate micro-architecture simulation platform for many-core multi-threading heterogeneous processors," *Journal of Low Power Electronics and Applications*, vol. 2, no. 1, pp. 30–68, 2012.

[39] A. Chhabra, H. Rawat, M. Jain et al., "FALPEM: framework for architectural-level power estimation and optimization for large memory sub-systems," *IEEE Transactions on Computer-Aided Design of Integrated Circuits and Systems*, vol. 34, no. 7, pp. 1138–1142, 2015.

[40] S. Li, J. H. Ahn, R. D. Strong, J. B. Brockman, D. M. Tullsen, and N. P. Jouppi, "The McPAT framework for multicore and many-core architectures: simultaneously modeling power, area, and timing," *ACM Transactions on Architecture and Code Optimization*, vol. 10, no. 1, article 5, pp. 1–29, 2013.

[41] D. Zoni and W. Fornaciari, "Modeling DVFS and power-gating actuators for cycle-accurate NoC-based simulators," *ACM Journal on Emerging Technologies in Computing Systems*, vol. 12, no. 3, article 27, pp. 1–24, 2015.

[42] C. Sau, L. Raffo, F. Palumbo, E. Bezati, S. Casale-Brunet, and M. Mattavelli, "Automated design flow for coarse-grained reconfigurable platforms: an RVC-CAL multi-standard decoder use-case," in *Proceedings of the 14th International Conference on Embedded Computer Systems: Architectures, Modeling and Simulation (SAMOS '14)*, pp. 59–66, Agios Konstantinos, Greece, July 2014.

[43] Open RVC-CAL Compiler, http://orcc.sourceforge.net/.

[44] Cadence, Low Power in Encounter RRTL Compiler Product Version 14.1, July 2014.

[45] J. W. Cooley and J. W. Tukey, "An algorithm for the machine calculation of complex Fourier series," *Mathematics of Computation*, vol. 19, no. 90, pp. 297–301, 1965.

[46] Z. Hu, A. Buyuktosunoglu, V. Srinivasan, V. Zyuban, H. Jacobson, and P. Bose, "Microarchitectural techniques for power gating of execution units," in *Proceedings of the International Symposium on Low Power Electronics and Design (ISLPED '04)*, pp. 32–37, IEEE, August 2004.

[47] C. M. Diniz, M. Shafique, S. Bampi, and J. Henkel, "A reconfigurable hardware architecture for fractional pixel interpolation in high efficiency video coding," *IEEE Transactions on Computer-Aided Design of Integrated Circuits and Systems*, vol. 34, no. 2, pp. 238–251, 2015.

10

HTLS Conductors: A Way to Optimize RES Generation and to Improve the Competitiveness of the Electrical Market

M. G. Ippolito,[1] F. Massaro ,[1] and C. Cassaro[2]

[1]*Department of Energy, Information Engineering and Mathematical Models, DEIM, University of Palermo, Viale delle Scienze, Building n.9, 90128 Palermo, Italy*
[2]*Terna Rete Italia S.p.A., Direzione Territoriale Centro Sud, Via Castellana 195, Palermo, Italy*

Correspondence should be addressed to F. Massaro; fabio.massaro@unipa.it

Academic Editor: Renato Procopio

Stringent environmental constraints make more difficult to identify new energy corridors and build new power lines. The increase in the generation of electricity from renewable energy sources (RESs) makes the operation of electrical systems increasingly difficult in some areas. The transmission system operators (TSOs), in Italy Terna, are forced to exploit the existing overhead transmission lines, increasing the possibility of dispatching energy, in particular RES, optimizing the transmission capacity. Therefore, after a brief presentation of the Sicilian electricity system and a brief description of high-temperature low-sag (HTLS) conductors, a case study is presented. It is shown how it is possible to optimize the dispatching from renewable sources and, finally, it is demonstrated how the use of HTLS conductor can contribute, together with other actions taken by the TSO, to the decrease of the zonal price of electricity, to the creation of a market with fewer constraints and to new operating conditions that increase the resilience of the electricity system in case of disturbances.

1. Introduction

The 2007–2013 Energy Saving Interregional Operative Program (POI Energia), with a financial allocation of 1,071 billion euro, financed in Italy 1,887 projects of public administrations and companies of the Convergence Regions (Calabria, Campania, Puglia, and Sicily) (Figure 1). The investments made with the resources of the program have concerned energy efficiency and the production of energy from renewable sources, investment support, upgrading of the network, carrying out studies, and assessing the potential for energy development [1].

Within the overall development policy framework outlined above, the integration between growth and environmental protection is also confirmed by the founding principles of the new European energy policy which aims at the following:

(i) To achieve a true internal energy market

(ii) To accelerate the transition to a low carbon economy, by acting on the development of renewable sources, on the diversification of the mix of sources, and on research in the field of energy technologies able to reduce emissions from energy production

(iii) To adopt a multisector impact energy efficiency plan, with the proposal of a new international agreement for the achievement of common quantitative targets by 2020

The energy-climate strategy at the European level is based on a package of measures aimed, on the one hand, at combating climate change through the reduction of greenhouse gas emissions and, on the other, at reducing the dependence on energy imports and the increase of prices; in this context, the production of energy from renewable sources plays a fundamental role in achieving these objectives. The European Union has recently launched a series of measures that clearly outline the path to be followed from now to 2020 to drastically reduce the effects of energy consumption on the climate; energy and environmental

Italia
Structural funds 2007–2013: convergence objective

Objective	Boundaries
Convergence regions	National
Phasing-out regions	NUTS 2

FIGURE 1: POI Energia 2007–2013—convergence regions in Italy.

policy at the community level has been strengthened by the decision of the European Council of 9 March 2007, which pursues the integration of energy and environmental policies, setting different targets for 2020, among which they appear relevant, for the purposes of this program:

(i) A 20% penetration of renewable energy sources on primary energy consumption (including 10% of biofuels)

(ii) A 20% reduction in primary energy consumption compared with the current trend

(iii) A 20% reduction in greenhouse gas emissions compared with 1990

So far, the interventions on the electricity grid, which by their nature require long implementation times, have mostly been planned by the operators as a consequence of the evolution of electricity demand and supply and have often followed the same administrative procedures for plant authorization. Instead, conditions must be created for the intrinsic development of the electricity infrastructure, which ensures the conditions for a rational use of the territory and for a reliable operation of the electricity grid. With this in mind, measure 2.4 of the POI Energia aims to introduce an incentive effect towards network operators to allow for the anticipation of direct investments in the evolution of the transmission network, not depending on the individual connection requests, but on the basis of a territorial planning by the regions. The investments for this project, perfectly in line with the purposes of this Public Notice, are not envisaged in Terna's 2014–2018 business plan and therefore would not be realized in the times and in the manner indicated in the absence of the public financial intervention. The allocation for the Sicily region for interventions on power lines amounts to a total of about €23,000,000.

The growing installation of generation from renewable energy sources (RES) and the strict environmental restrictions for the construction of new power lines cause difficulties in the operation of the electricity system. As described in [2–6], the Italian Transmission System Operator (TSO), Terna S.p.A., is optimizing the use of existing transmission lines. Starting from the studies carried out in [7], this document demonstrates how the use of high-temperature low-sag (HTLS) conductors is a possibility to allow the optimization of the use of current energy corridors and also a greater integration of these sources (RES) on the electrical system. After a brief description of some of the technologies of high-temperature low-sag conductors existing in the market today, the authors will evaluate the new range at the thermal limit through the use of the mathematical model proposed by Schurig and Frick. Afterwards, thanks to POI Energia investments, a real case study of reconductoring is analysed checking the benefits in terms of transport capacity, increasing the N-1 security criterion, electricity market benefits, etc. Since reconductoring has affected many 150 kV lines on which wind farms are connected, an analysis of the zonal price of electricity in Sicily in last years will be carried out to demonstrate how this structural intervention, together with others [8], has allowed better integration of RES and therefore a decrease in zonal prices and in HHI index (Herfindahl–Hirschman index).

2. The Power Grid in Sicily

The Sicilian high-voltage electrical system has three voltage levels: 400 kV, 220 kV, and 150 kV. It is connected to the European mainland by a double circuit of 400 kV submarine AC cables. The 400 kV (overhead) electric system consists of a few lines, the 220 kV lines form a ring along the Sicilian coast, and the 150 kV subtransmission system is discreetly meshed. The existing production units in Sicily are large power units (CCGT) and renewable energy sources (RES), such as hydroelectric, PV, and wind farms [6]. In recent years, thanks to the incentive policies developed by the Italian government, numerous wind and photovoltaic plants have been built; today, the installed power in wind farms in Sicily is more than 2 GW, whereas the photovoltaic power is more than 1.5 GW. The average electricity demand, in Sicily, is estimated at 2.1 GW. A model of the Sicilian electrical system was created by authors in a previous paper [7] in order to carry out studies and simulations. The main objective of these simulations was to determine the values of the power flows of each power line. The model of the Sicilian high-voltage network consists of 336 nodes, 346 lines, 90 transformers, 66 synchronous machines, and numerous RES production units. The simulations carried out in [7] have concerned standard operating configurations, relating to winter and summer conditions. Both of these simulated conditions showed the violations of transport capacity limits in some 150 kV overhead lines. The electric system at 220 kV and 400 kV instead has never shown problems of overload (good margins of loadability) and therefore the study will focus on the 150 kV power grid.

3. HTLS Conductors

Usually, the electric transmission lines use bimetallic conductors (ACSR-aluminum conductor steel reinforced). This type of conductor consists of a steel core characterized by a high mechanical strength and external layers of different aluminum wires wound spirally on the core. Today, the market offers different reconductoring solutions with heat-resistant conductors and low linear thermal elongation (called high-temperature low-sag conductor). These features offer, besides the possibility of being able to operate the line up to temperatures even over 180°C, even to be able to use the same existing energy corridors without substantial changes to the structure of the power line. The main HTLS conductors are as follows: TACSR, GTASCR, ZTACIR, and ACSS [9]. These conductors can operate at temperatures between 150°C and 250°C without changing the mechanical and chemical properties. The TACSR conductor (conductor made of heat-resistant aluminum alloy in reinforced steel) has steel wires in the core and TAL wires (aluminum-zirconium alloy which has stable mechanical properties up to 150°C) around it (Figure 2).

GTACSR (gap-type thermal aluminum conductor steel reinforced) conductor presents a small gap between steel core and aluminum outer layers in order to apply strain only on the steel. The gap is filled with heat-resistant grease (filler) to decrease friction between core and outer layers and to prevent water penetration (Figure 3).

ZTACIR (super thermal aluminum conductor invar reinforced) conductor consists of steel-invar galvanized alloy core (ACI core) and ZTAL wires in the outer layers. It does not present annealing phenomena up to 210°C (Figure 4).

ACSS (aluminum conductor steel supported) conductor presents steel wires on the core and aluminum wires in the outer layers subjected to an annealing process. There are two different types: "atandard round strand ACSS" (aluminum rope with circular cross-sectional wires) or "trapezoidal aluminum wire ACSS" (aluminum rope with trapezoidal cross section's wires) as shown in Figure 5.

The conductor most used in electric lines in Italy at 150 kV is the ACSR (aluminum conductor steel reinforced) with a diameter of 22.8 mm. For the purposes of this paper, among the high-temperature and low-sag (HTLS) conductors, the following types of conductors have been considered with the same (or similar) diameter:

(i) G(Z)TACSR (gap-type thermal resistance aluminum alloy conductor), diameter 22.6 mm

(ii) ZTACIR (invar reinforced thermal-resistant aluminum alloy conductor), diameter 22.75 mm

(iii) ACSS (aluminum conductor steel supported), diameter 20.9 mm

The HTLS conductors used, described in [9, 10], are both electrically and geometrically very similar to the conventional ACSR conductor. The main differences are the resistance to high temperatures of the aluminum alloy which constitutes the outer layer and the low value of linear thermal elongation. The advantage deriving from the use of HTLS conductors is that therefore they can withstand a higher temperature (up to 150°C or 210°C) and present lower sag (Figure 6), guaranteeing at the time a certainly higher transport capacity [11–19].

Table 1 compares the electrical and mechanical characteristics of the conductors under study. For the three HTLS conductors, the coefficient of linear expansion is referred to the core.

The authors highlight that other types of HTLS conductors have been developed in recent years. Particularly, conductors with composite core have experienced a high expansion in certain countries and are commercially available. The authors are studying these last types of conductors for future investigation, but they will not be included in the comparison of this paper.

4. Ampacity Calculation

The study, as mentioned, concerned the electrical lines of the Sicilian territory at 150 kV which were overloaded in some operating conditions according to [7] and therefore do not allow a correct dispatching of renewable sources (in particular wind farms).

More precisely, in [7] the following were carried out:

(i) Simulations of the electrical system through the help of Neplan® software and identification of network elements (in particular power lines) in overload that limit the possibility of freely dispatching renewable sources

(ii) Calculation of the maximum conductor temperature

(iii) Comparison between the maximum temperature of different HTLS conductors and the relative sag

The simulation results in standard winter and summer conditions showed that the 400 kV and 220 kV systems do not have overload problems; the only power lines that present these problems were some of those at 150 kV. These lines create problems of local congestion and therefore difficulties in dispatching the RES. In the following, a hypothesis of reconductoring of overloaded lines will be carried out by replacing the traditional ACSR conductor with the HTLS conductors previously described, showing a possibility of operating the electrical system with greater flexibility.

For the calculation of the maximum transport capacity at the thermal limit of the lines, the mathematical model Schurig and Frick was used [20]:

$$P_j = P_r + P_c - P_s, \quad (1)$$

where P_j = Joule losses; P_r = radiative losses; P_c = power dissipated by convection; and P_s = power absorbed by solar radiation.

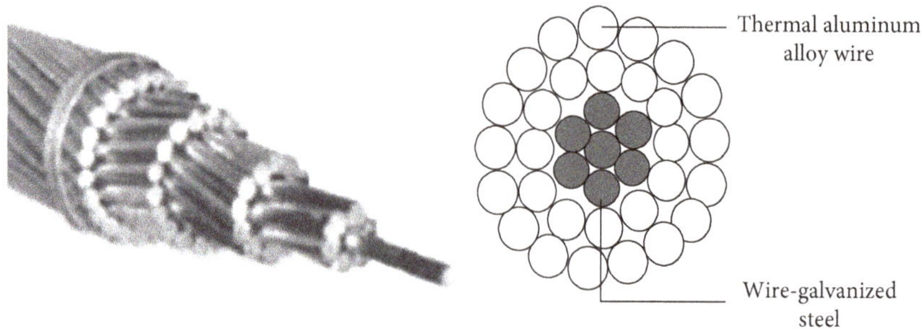

FIGURE 2: TACSR conductor.

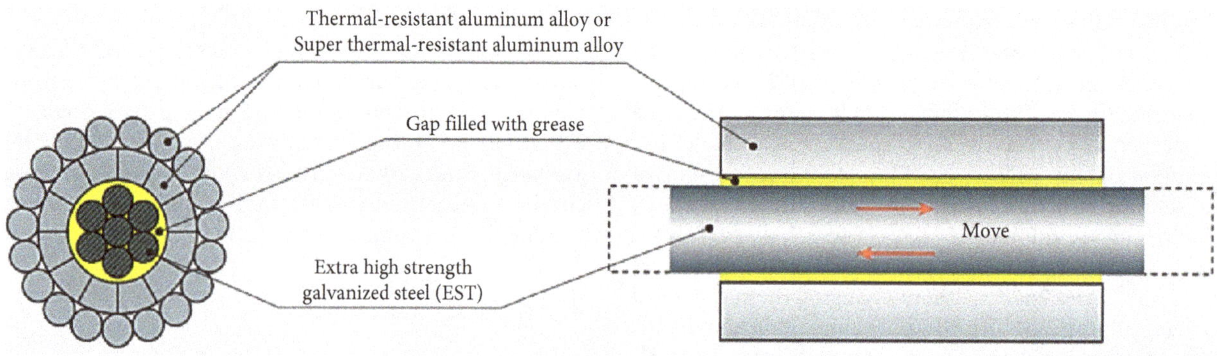

FIGURE 3: Gap-type conductor.

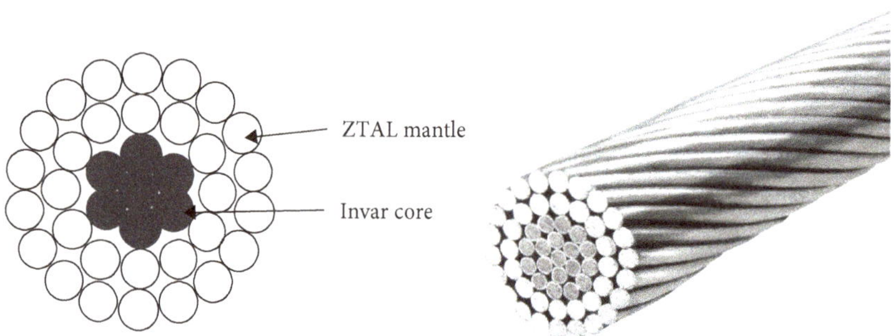

FIGURE 4: ZTACIR conductor.

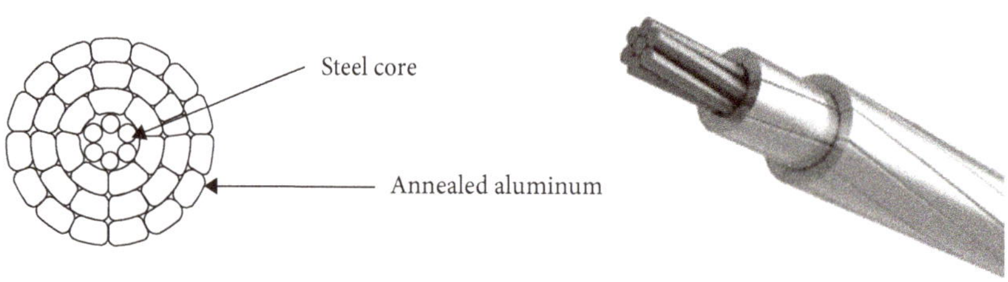

FIGURE 5: ACSS/TW conductor.

Current is estimated using the following equation [21]:

$$I = 5,6 \cdot \frac{\sqrt{(P_r + P_c - P_s) \cdot 10^4 \cdot D}}{R_T}, \quad (2)$$

where D = diameter of conductor (cm) and R_T = resistance of the conductor (Ohm/km).

The authors remark that the resistance varies a lot with the temperature (particularly important for temperatures above 75°C), and skin and magnetic effects may be considerable at high current densities.

The electrical connections that presented overloads are located in different areas of Sicily. In this paper, authors consider a case study located in the east part of the island; in order

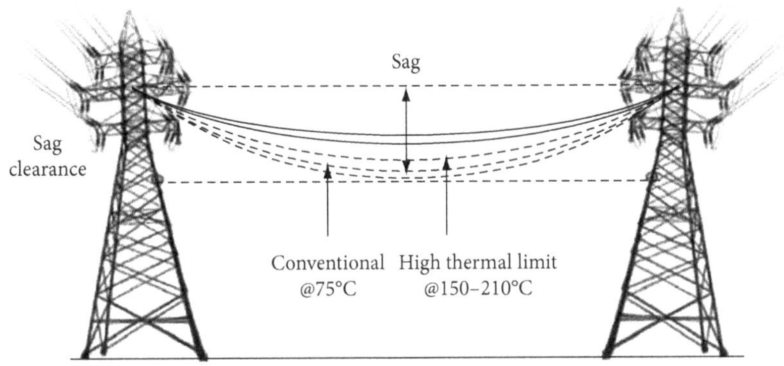

Figure 6: Comparison of sag for ACSR and HTLS temperature limit.

Table 1: Electrical and mechanical characteristics of HTLS conductor.

	Conductor			
	ACSR	G(Z)TACSR	ZTACIR	ACSS
Area (mm^2)	307.7	308.4	306.94	307.7
Rated Tensile strength (kgf)	9944	10962	10066	9901
Rdc at 20°C (ohm/km)	0.109	0.111	0.11068	0.1077
Weight (kg/km)	1068	1098	1083	1068
Coeff. of linear expansion (10^{-6}/°C)	18.9	11.5	4.7	11.5

to study the heat exchange between the environment and the conductor more precisely, the meteorological conditions have been identified considering 10-year historical series extracted from meteorological stations close to the considered lines.

The following Table 2 shows the ambient conditions taken into consideration for the line under investigation. The data that were used in Equation (2) are, in a precautionary way, the worst of the series, that is, those that return the lowest current values.

The maximum current value that can cross the electrical conductors has been evaluated both in winter and in summer. The calculation of the maximum transport capacity was carried out for the traditional ACSR conductor and subsequently for the HTLS conductors with which the reconductoring of the lines was hypothesized. The Table 3 shows the maximum permissible current values for ACSR and HTLS conductors according to the area of the Sicilian territory where the overloaded electric line is located.

These values have been compared with those of power lines with overload problems, and the results of the simulation (after reconductoring) show a considerable improvement in the dispatching of RES in the 150 kV electricity grid.

The authors remark that the overhead lines are often limited by sag, and, therefore, the sag-temperature performance of each conductor has to be considered to determine their maximum temperature. In the following case study, the TSO consider lines with sufficient “buffer” and so not limited by sag.

5. Case Study

As said, the allocation for the Sicily region for interventions (POI Energia) on power lines amounts to a total of about €23,000,000.00, and National High Voltage Transmission Network (in Sicily) has been subject of numerous interventions aimed at optimizing the transport capacity on some important 150 kV lines (Figure 7).

The Sicilian electrical system, as well as 400 kV and 220 kV interconnections, is equipped with a 150 kV network which also performs a transport function rather than subtransmission.

In particular, with high demand and the absence of distributed generation (PV on the MV and LV network) or in conditions of high wind power generation in the HV lines, the 150 kV network must be able to carry the power flows to the substation in the first case and to transmit the wind generation (which almost insists on the 150 kV network) in the second case. In addition, the 150 kV network must have characteristics such as to exceed the security operational criterion N-1 in the event of tripping of a section of the 220 kV ring in double circuit.

In addition to these operating conditions, a new N-1 security criterion condition was added in June 2016, linked to the high import power flows (maximum 1200 MW) that

Table 2: East Sicily ambient conditions.

	Winter	Summer
Ambient temperature (°C)	20	33
Wind speed (km/h)	2	2
Coefficient of solar radiation (W/m^2)	1029.49	1029.49
Coefficient ε emissivity of the conductor	0.7	0.7
Sky conditions	Clean	Clean
Irradiation	Yes	Yes

Table 3: East Sicily HTLS conductors ampacity.

Zone: East Sicily		Maximum allowable conductor temperature (°C)	Maximum allowable current (A)	
Type of Conductor	Diameter (mm)		Winter	Summer
ACSR	22.8	55	390	250
ZTACIR	22.75	180	630	580
ACSS	20.9	250	910	820
G(Z)TACSR	22.6	210	1080	1040

FIGURE 7: 150 kV lines object of reconductoring.

the Sicilian electrical system normally receives from the continental network in case of tripping of one of the 400 kV mainland interconnections.

In all the scenarios described above, the 150 kV network of the Sicilian electrical system has to be able to continue operating without overload.

In this context, Terna replaced, for some lines which go under overload and here presented as a case study, ACSR conductor with a diameter of 22.8 mm (aluminum conductor steel reinforced) with high-temperature low-sag conductors ZTACIR with a diameter of 22.75 mm.

The following figure (Figure 8) reports a schematization of the section conductor, the geometrical data, and the main mechanical characteristics.

The outer layer is made of an aluminum-zirconium alloy with a thermal limit higher than that of traditional aluminum (ZTAL: super thermal-resistant aluminum alloy) and this means the possibility of optimizing the transport of current. A core consisting of an aluminum-coated Fe-Ni alloy (ACI: alluminum clad invar) characterized by a reduced coefficient of thermal expansion that allows to keep sags similar to those of traditional conductors, even if they operate at higher temperatures. The ZTACIR (diameter mm 22.75) conductor is designed for the replacement of the ACSR analogue diameter mm 22.8 but has a similar current range to that of diameter mm 31.5 ACSR conductor used for 220 kV lines.

Other positive consequences in the reconductoring of some 150 kV links of the Sicilian electric system are as follows:

(i) Reduction of the limitations of wind generation due to local overload on 150 kV
(ii) Possibility to connect new production plants on the reconductored 150 kV lines
(iii) More flexibility for the programmable internal production resources
(iv) Reduction/elimination of some operating restrictions with more flexibility in operation and increase in quality in supplying power

In order to provide an example for the case under investigation, in Figure 9, it is compared the trend of the current recorded for the line under study in a period of a month characterized by the presence of strong wind and therefore a high value of energy production from wind farms. The current flow in the windy period with ACSR is, on average (147 A), lower since the TSO, for operational security, was forced to redispatch production from RES, sometimes not accepting it on the system. The presence of the ZTACIR conductor, on the other hand, allows a greater transits (average value of 242 A), thanks to the possibility of accepting all the wind generation available at the moment with clear environmental and economic benefits for the community (e.g., decrease in the zonal price of electricity and CO_2 emissions). The comparison in terms of average values highlights that, thanks to HTLS conductors, the line can transport (on average) an higher value of current and then of energy (the increase is estimated more than 50%).

In order to verify the positive effects of the substitution activities of the conductors on the Sicilian electrical system, the present paper also considers an analysis of the statistics of GME (Italian energy market operator) on the zonal prices of the electricity market in Sicily from 2014 to 2016 (the 2017 report is not yet available) [22] and on the HHI index (Herfindahl–Hirschman index), also related to the region under study. This is an aggregate market index that measures the degree of concentration and dispersion of the volumes offered and/or sold by market participants. The HHI is calculated, for each hour and for each macro zone, as the

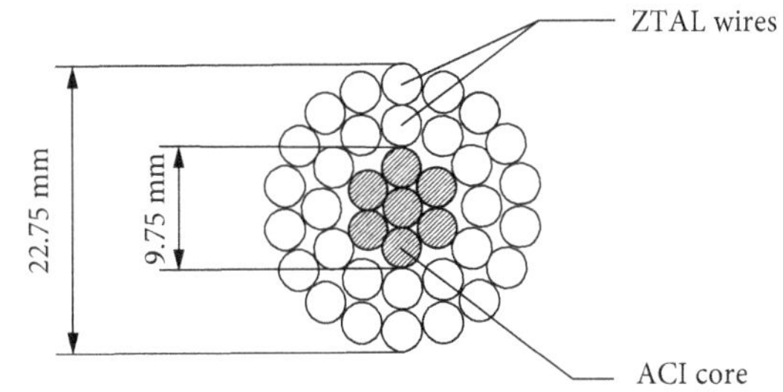

Technical data

Conductor code	-	249-AT3/58-ACI20SA
Standard	-	UX LC3914
Stranding	-	7x3.25+30x3.25
Diameter of Alu Clad Fe-Ni wire	mm	3.25 +/- 1.5%
Diameter of ACI core	mm	9.75 +/- 1.5 %
Diameter of ZTAL aluminium wire (AT3)	mm	3.25 +/- 1.0 %
Diameter of complete conductor	mm	22.75 +/- 1.0 %
Outer layer clockwise		RIGHT
ACI core cross-sectional area	mm^2	58.07
Total ZTAL aluminium cross-sectional area	mm^2	248.87
Total cross-sectional area of compl. conductor	mm^2	306.94
Unit weight of ACI	kg/km	355
Unit weight of ZTAL aluminium	kg/km	688
Unit weight of complete conductor	kg/km	1083 +/-2 %
Calculated breaking force	daN	9782
Nominal transition temperature	°C	119±5
Modulus of elasticity for ACI	kN/mm^2	138.50
Modulus of elasticity for conductor	kN/mm^2	72
Coefficient of linear expansion for ACI	1/°Cx10^{-6}	4.7
Coefficient of linear expansion for conductor	1/°Cx10^{-6}	16.4
DC resistance at 20°C (max.)	ohm/km	0.11068
Allowable continuous operating temperature	°C	180
Allowable operating temperature temporarily	°C	210

MATERIALS:
- Thermal resistant aluminium alloy: AT3 (ZTAL: Super Thermal Resistant Aluminum Alloy) according to EN 62004
- High temperature Fe-Ni alloy: ACI (Aluminum Clad Invar) EN61232 20SA + customer specification

FIGURE 8: Geometrical data and the main mechanical characteristics of ZTACIR.

sum of the shares of the volumes sold (or offered) on the market by the market operators is multiplied by 100 and squared. The value of the HHI can vary from 0 (perfect competition) to 10,000 points (monopoly). The HHI is calculated on an hourly basis and then aggregated in monthly simple averages. Figure 10 shows the trend of the zonal price of electricity in Sicily, whereas Figure 11 shows the trend of the HHI index. It is possible to see how after

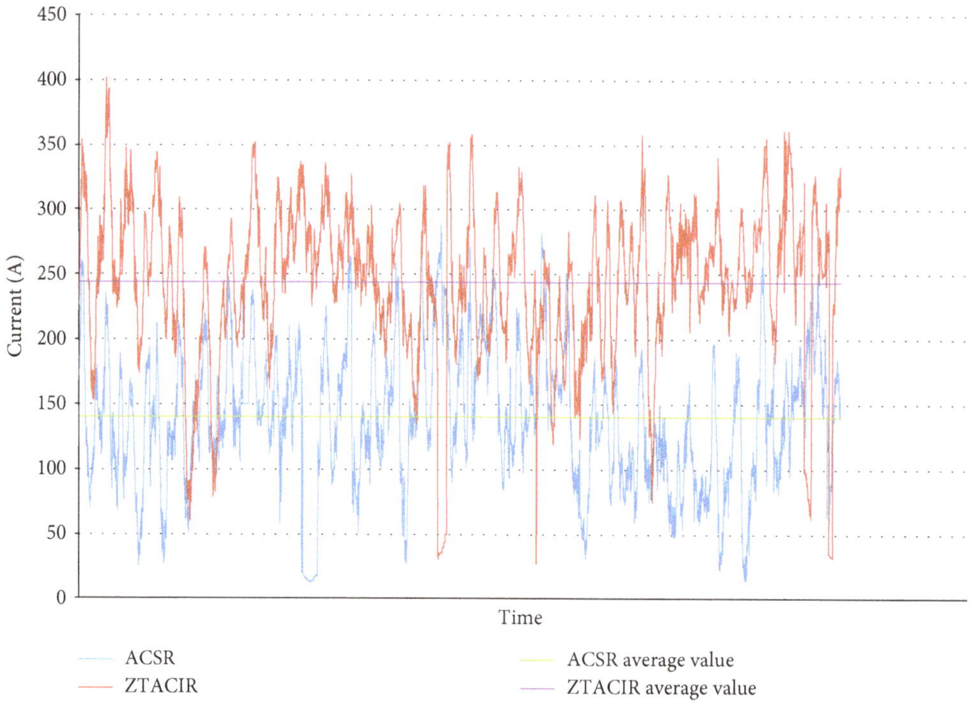

FIGURE 9: Comparison between ACSR and ZTACIR current flow and relative average values.

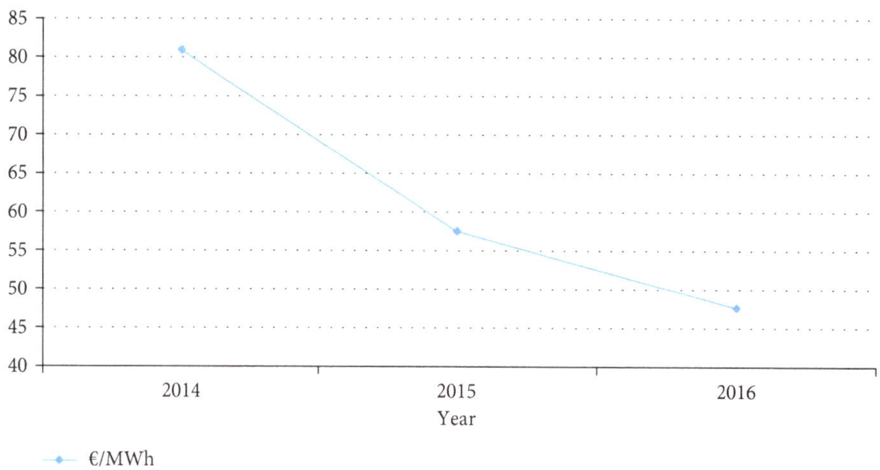

FIGURE 10: Sicilian electricity (zonal) price (period 2014–2016).

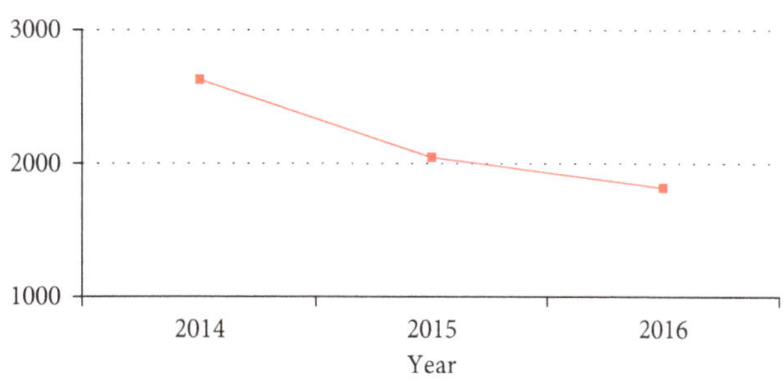

FIGURE 11: Sicilian HHI value (period 2014–2016).

2014, the year in which reconductoring operations were carried out on the 150 kV network, the zonal price of Sicily decreased significantly from €82/MWh at a price of around €47/MWh recorded in 2016. The Herfindahl–Hirschman index (HHI), which characterizes the competitiveness of a market, also recorded a significant decrease, thus demonstrating greater flexibility of the system. It should be noted that the reconductoring cited is one of the actions implemented by TERNA in Sicily, including the doubling of the connection to the mainland, the connection with the island of Malta [7] and the use of DTR (dynamic thermal rating systems) [23].

The authors highlight that the PUN has decreased throughout Italy, but in percentage terms, Sicily and other regions of southern Italy have had a more marked decrease. This is due to a greater penetration of the RESs on the electricity system that has allowed the energy price to be mitigated.

6. Conclusions

In order to obtain a flexible operation of the electricity system, with particular reference to the possibility of dispatching the power generated by renewable sources without constraints and to mitigate the effects of the unpredictability of the determination of the price in the Sicilian market, it is advisable to use HTLS conductors. Terna, thanks to POI Energia 2007–2013, invested ingent capitals on four regions of the south (including Sicily) for the reinforcement of electricity grids with the aim of making smart grid (HTLS and DTR-dynamic thermal systems) and encouraging the integration of RES for decreasing the zonal price of electricity and limiting the emissions of CO_2. This paper, after a brief presentation of the Sicilian electricity system and a brief description of high-temperature low-sag conductors (HTLS), presents a case study of reconductoring. The analysis demonstrates how the current flow in the windy period with ACSR is, on average (147 A), lower since the TSO, for operational security, was forced to redispatch production from RES, sometimes not accepting it on the system. The presence of the ZTACIR conductor, on the other hand, allows a greater transits (average value of 242 A), thanks to the possibility of accepting all the wind generation available at the moment with clear environmental and economic benefits for the community (e.g., decrease in the zonal price of electricity and CO_2 emissions). The comparison, in terms of average values, highlights that HTLS conductors can transport a higher value of current and then of energy (the increase is estimated more than 50%).

In order to verify also the positive effects of the substitution activities of the conductors on the Sicilian electrical system (the reconductoring was completed at the end of 2014), the present paper also considers an analysis of the statistics of GME (Italian energy market operator) on the zonal prices of the electricity market in Sicily from 2014 to 2016 (the 2017 report is not yet available) and on the HHI index (Herfindahl–Hirschman index), also related to the region under study. The authors highlight that the PUN has decreased throughout Italy, but in percentage terms Sicily and other regions of southern Italy have had a more marked decrease. This is due to a greater penetration of the RESs on the electricity system that has allowed the energy price to be mitigated. It is shown how it is possible to optimize the dispatching from renewable sources and, finally, it is demonstrated how, the use of HTLS conductor, can contribute, together with other actions taken by the TSO, to the decrease of the zonal price of electricity, to the creation of a market with fewer constraints.

Conflicts of Interest

The authors declare that they have no conflicts of interest.

References

[1] 2.4 POI Energia 2007–2013 Misure, "Interventi per il potenziamento e l'adeguamento delle reti di trasporto ai fini della diffusione delle fonti rinnovabili e della piccola e micro cogenerazione," 2016, http://www.sviluppoeconomico.gov.it/images/stories/normativa/TERNA2.PDF.

[2] V. Cosentino, S. Favuzza, G. Graditi et al., "Smart renewable generation for an islanded system. Technical and economic issues of future scenarios," *Energy*, vol. 39, no. 1, pp. 196–204, 2012.

[3] E. M. Carlini, S. Favuzza, S. E. Giangreco, F. Massaro, and C. Quaciari, "Uprating an overhead line: Italian TSO applications for integration of RES," in *Proceedings of ICCEP*, Alghero, Italy, June 2013, ISBN 978-1-4673-4430-2.

[4] E. M. Carlini, S. Favuzza, S. E. Giangreco, F. Massaro, and C. Quaciari, "Uprating an overhead line: Italian TSO applications to increase system N-1 security," in *Proceedings of 2nd International Conference on Renewable Energy Research and Applications, ICRERA 2013*, Madrid, Spain, October 2013, ISBN: 978-1-4799-1464-7.

[5] E. M. Carlini, F. Massaro, and C. Quaciari, "Methodologies to uprate an overhead line. Italian TSO case study," *Journal of Electrical Systems*, vol. 9, no. 4, pp. 422–439, 2013.

[6] E. M. Carlini, S. Favuzza, M. G. Ippolito, A. Madonia, F. Massaro, and C. Quaciari, "Impact of non-programmable renewable sources in Day-Ahead Market in Sicily: reasons to make more flexible and smarter the national transmission grid," in *Proceedings of International Conference IEEE PES ISGT 2014*, Istanbul, Turkey, October 2014.

[7] G. Filippone, M. G. Ippolito, F. Massaro, and A. Puccio, "On the roadmap to Supergrid in Sicily: LIDAR technology and HTLS conductors for uprating the 150 kV lines," in *Proceedings of International Conference IEEE PES ISGT 2014*, Istanbul, Turkey, October 2014.

[8] M. G. Ippolito, S. Favuzza, F. Massaro, L. Mineo, and C. Cassaro, "New high voltage interconnections with islands in the Mediterranean Sea: Malta and Sicily: analysis of the effects on renewable energy sources integration and benefits for the electricity market," *Energies*, vol. 11, no. 4, p. 838, 2018.

[9] CIGRE WG B2.12, *Conductors for the Uprating of Overhead Lines, ELECTRA*, vol. 213, CIGRE, Paris, France, 2004, ISSN: 1286-1146.

[10] M. J. Tunstall, S. P. Hoffmann, N. S. Derbyshire, and M. J. Pyke, "Maximizing the ratings of national grid's existing transmission lines using high temperature, low sag conductor," in *Proceedings of CIGRE 2000*, Paris, France, August 2000.

[11] Hitachi Cable Ltd., *Introduction of Hitachi-Invar Reinforced Aluminium-Conductors*, TD 80-12006, Hitachi Cable, Tokyo, Japan, 1999.

[12] The Furukawa Electric Co. Ltd., *Thermo-Resistance Aluminium-Alloy Conductors Invar Reinforced*, Furukawa Electric Co. Ltd., Japan, n. TD 89-035 E, 2003.

[13] F. R. Thrash, "ACSS/TW–an improved conductor for upgrading existing lines or new construction," in *Proceedings of IEEE Transmission I Distribution Conference (99CH36333)*, vol. 2, pp. 852–857, New Orleans, LA, USA, 1999.

[14] S. Kotaka, H. Itou, T. Matsuura, and H. Morikawa, "Applications of gap-type small-sag conductors for overhead transmission lines," *SEI Technical Review*, vol. 50, 2000.

[15] ASTM B856-18, *Standard Specification for Concentric-Lay-Stranded Aluminium Conductors: Coated Steel Supported (ACSS)*, ASTM, West Conshohocken, PA, USA, 2018.

[16] A. J. M. Zamora, P. Eguia, R. Criado, C. Alonso, J. Iglesias, and J. R. Saenz, "High-temperature conductors: a solution in the uprating of overhead transmission lines," in *Proceedings of Power Tech Conference*, Porto, Portugal, 2001.
[17] A. A. P. Silva and J. M. B. Bezerra, "Applicability and limitations of ampacity models for HTLS conductors," *Electric Power Systems Research*, vol. 93, pp. 61–66, 2012.
[18] J. Heckenbergerova and J. Hosek, "Dynamic thermal rating of power transmission lines related to wind energy integration," in *Proceedings of 11th International Conference on Environment and Electrical Engineering (EEEIC)*, pp. 798–801, Venice, Italy, May 2012.
[19] J. Hosek, "Dynamic thermal rating of power transmission lines and renewable resources," in *Proceedings of ES1002 Workshop*, pp. 1–3, Paris, France, March 2011.
[20] IEEE STD 738–2012, *IEEE Standard for Calculating the Current-Temperature of Bare Overhead Conductors*, IEEE Power Engineering Society, Piscataway, NJ, USA, 2012.
[21] O. R. Schurig and C. W. Frick, "Heating and current-carrying capacity of bare conductors for outdoor service," *General Electric Review*, vol. 33, no. 3, pp. 141–157, 1930.
[22] http://www.mercatoelettrico.org/It/Default.aspx, 2018.
[23] M. G. Ippolito, F. Massaro, G. Zizzo, G. Filippone, and A. Puccio, "Methodologies for the exploitation of existing energy corridors: GIS analysis and DTR applications," *Energies*, vol. 11, p. 979, 2018.

11

Extreme Learning Machine and Moving Least Square Regression based Solar Panel Vision Inspection

Heng Liu,[1] Caihong Zhang,[2] and Dongdong Huang[1]

[1]*School of Computer Science and Technology, Anhui University of Technology, Maxiang Road, Ma'anshan 243032, China*
[2]*BOCOM Smart Network Technologies Inc., Shanghai 200433, China*

Correspondence should be addressed to Heng Liu; hengliusky@aliyun.com

Academic Editor: Lei Zhang

In recent years, learning based machine intelligence has aroused a lot of attention across science and engineering. Particularly in the field of automatic industry inspection, the machine learning based vision inspection plays a more and more important role in defect identification and feature extraction. Through learning from image samples, many features of industry objects, such as shapes, positions, and orientations angles, can be obtained and then can be well utilized to determine whether there is defect or not. However, the robustness and the quickness are not easily achieved in such inspection way. In this work, for solar panel vision inspection, we present an extreme learning machine (ELM) and moving least square regression based approach to identify solder joint defect and detect the panel position. Firstly, histogram peaks distribution (HPD) and fractional calculus are applied for image preprocessing. Then an ELM-based defective solder joints identification is discussed in detail. Finally, moving least square regression (MLSR) algorithm is introduced for solar panel position determination. Experimental results and comparisons show that the proposed ELM and MLSR based inspection method is efficient not only in detection accuracy but also in processing speed.

1. Introduction

Over the past decade, machine vision has been widely applied in industrial inspection, especially in electric and electronic industrial inspection [1–3]. Superior to manual inspection, the benefit of machine vision detection is that no direct contact is required between the inspector and the target. Therefore, automatic nondestructive inspection can be achieved in machine vision detection system.

Since the accuracy of traditional vision inspection depends much on the effect and the stability of image processing technique, the traditional inspection way will lose its' functionality when facing the changing environment or the altered imaging condition. Therefore, not only processing the images of industry objects but also learning the features (shapes, positions, and orientations) from image samples is required. Actually, with the fast development of computer vision, recently, there have appeared some works [4, 5] for machine learning based vision inspection.

In modern photovoltaic electronic industry, for the solder joint of solar PCB is so fine, the vision based solder joint defect identification is challenging and thus needs to be investigated in depth [6, 7]. In addition, during the manufacture of solar panel, there is a strong demand for automatic position determination of the solder panel [8, 9]. In this work, in the face of the two major challenges, we present a set of defect identification and position determination methods.

So far, Capson and Eng [10] have used a tiered-color illumination approach for solder joint's inspection. Matsuyama et al. [11] also design a system for automated solder joint inspection. In addition, Chiu and Perng [12] utilize the reflection area of solder joint to describe the image features, and Tae-Hyeon et al. [13] develop a vision inspection system for the classification of solder joint. Moreover, unlike previous methods, Zhang and Liu in [14] propose utilizing the principle of ELM [15] with least square fitting to achieve vision detection on solder joint and solar panel. In this work, we extend and improve the previous work [14] in

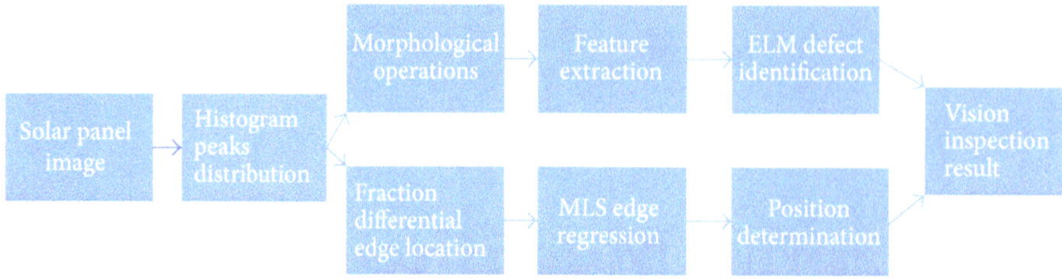

FIGURE 1: The proposed scheme of solder joint defect detection and solar panel position determination.

three aspects: histogram peaks distribution (HPD) [16] and fractional calculus [17–20] are adopted to improve the effect of image preprocessing and initial edge extraction; instead of least square fitting, MLSR is adopted to enhance the accuracy of edge estimation of the solar panels; more vision inspection experiments and comparisons of more solar PCB images are implemented.

Specifically, for solder joint defect identification, we take histogram peaks distribution (HPD) [16] with morphological operation to extract image features and then apply extreme learning machine (ELM) [21–24] to classify the input images. As for solar panel position determination, accurate edges location is prerequisite. To achieve this, we extract the initial edge points by a novel signal processing method, fractional calculus [16, 22–24], and then utilize MLSR algorithm [25, 26] to refine initial edges. Once the final edges of solder panel are acquired, the position parameters, position shift, deflection angle, and edge lengths, are easily determined. The overall proposed scheme of solder joint defect detection and solar panel position determination is illustrated in Figure 1.

The paper is organized as follows. Section 2 discusses the related techniques of image preprocessing and feature extraction used for defect identification and position detection. Section 3 describes how to use ELM to recognize solder joint defect in detail. Section 4 presents utilizing MLSR algorithm to achieve solder panel position determination. The abundant experimental results and comparisons are given in Section 5. Section 6 makes a short conclusion and future research prospective.

2. Image Preprocessing and Feature Extraction

In this section, we firstly make a detailed discussion on using HPD and morphological operation to preprocess the images of solder joint. And then we introduce fractional differential to extract initial edges of solar panel. Finally, we also elaborate on the specific feature extraction method used for solder joint defect classification.

2.1. Image Preprocessing Based on HPD and Morphological Operation. In machine vision inspection system, image preprocessing, including noise removing, image binarization, and edge detection, is the basic guarantee for following detection. However, due to illumination or view angle changing and noise intervention, such image preprocessing task under complex environment is always challenging. Fortunately, in solar panel vision detection system, the illumination and the view angle are under control. And imaging noise can be suppressed in advance by camera calibration beforehand. Thus the preprocessing effect of solar panel images can be guaranteed.

Image thresholding is an effective technique for image binarization. Due to the unevenness of the lighting distribution, threshold finding is not simple and straightforward. How to acquire an appropriate threshold becomes the key issue for image binarization. Here, we take HPD [16] to find an accurate threshold for solder panel image binarization. HPD threshold search can handle several histogram distributions, which are illustrated in Figure 2.

Morphological operation can obtain the backbone of the image object which can unify a host of techniques such as noise suppression, shape analysis, skeletonizing, and other related topics [27]. After thresholding, the basic morphology operations, dilation and erosion, are usually used to process binary images. In addition, the other two important morphological operations, closing and opening, are also typically used after thresholding. The closing operation smooths the contours of the image to eliminate narrow and long gaps or remove holes or fill contours. The open operation is used to eliminate tiny image objects or separate objects connected to narrow areas.

For solder joint images, we firstly convert image color space from *RGB* to *Ycbcr*. Then in Y-intensity subspace, we utilize HPD algorithm to obtain global threshold to binarize the image. Then, we use morphological closing and opening operations to get the final binary image. The processing results are shown in Figure 3.

2.2. Initial Edge Extraction via Fractional Gaussian Differential. There are two different backgrounds in solder panel images. One is red background and the other is gray background. The solar panel image under red background is used to calculate position shift and deflection angle against the reference position, while gray background image can be used to extract the inner standard lines for measuring edge length of solar panel. For preprocessing, we also perform color space conversion and select *cr* subspace for red background image, while for gray one we take *G* subspace from *RGB* space. Then, in these color subspaces, we all use HPD for image binarization.

After obtaining the binary image, we extract the initial edge of the solar panel by the new fractional calculus [17, 20]. Fractional calculus has been a powerful image enhancement

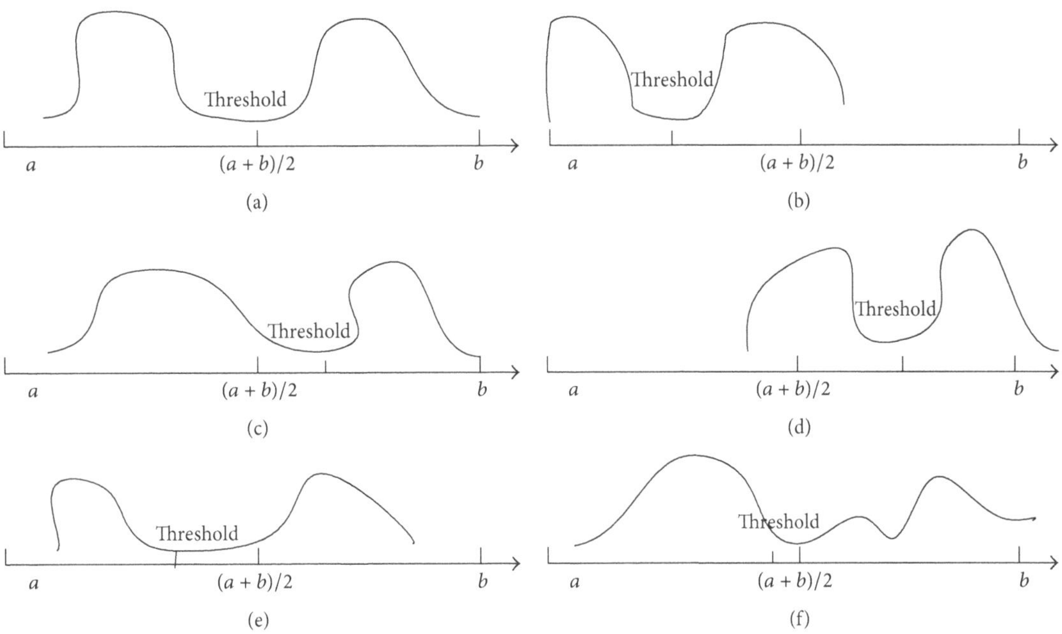

FIGURE 2: HPD threshold search sketch map.

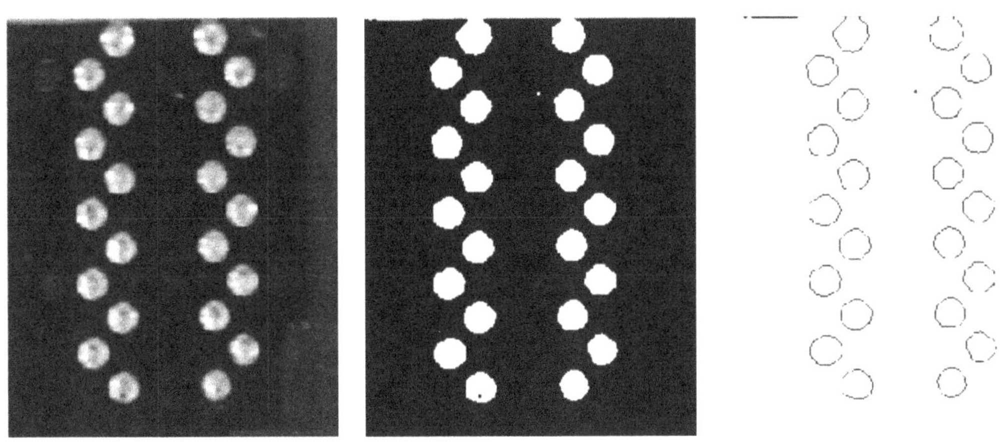

(a) (b) (c)

FIGURE 3: Solder joint image preprocessing: (a) original image; (b) HPD image binarization; (c) final morphological processing result.

technique in recent years. By adjusting the fractional order of the Gaussian difference, an accurate and antinoise edge extraction effect can be obtained [16, 17].

There are three ways to define fractional calculus, namely, the Grünwald-Letnikov (G-L) definition, the Riemann-Liouville definition, and the Capotu definition. The G-L definition comes from the classic definition of the calculus from integer-order continuous function and generalizes the integer calculus into fractional order. The G-L definition is described as [16]

$$ {}_{a}^{G}D_{t}^{v} f(x) = \lim_{h\to 0, nh=t-a} h^{-v} \sum_{r=0}^{n} \begin{bmatrix} -v \\ r \end{bmatrix} f(x-rh). \quad (1) $$

Thus, fractional partial differential can be defined as (v is fractional differential order)

$$ \begin{aligned} \frac{\partial^{v} F(x,y)}{\partial x^{v}} &= F(x,y) + (-v) F(x-1,y) \\ &\quad + \frac{(-v)(-v+1)}{2} F(x-2,y) \\ &\quad + \frac{(-v)(-v+1)(-v+2)}{6} F(x-3,y), \\ \frac{\partial^{v} F(x,y)}{\partial y^{v}} &= F(x,y) + (-v) F(x,y-1) \end{aligned} $$

$$+\frac{(-v)(-v+1)}{2}F(x,y-2)$$
$$+\frac{(-v)(-v+1)(-v+2)}{6}F(x,y-3). \tag{2}$$

Based on the above two equations, if combined with Gaussian kernel $g(x,y) = (1/2\pi\sigma^2)e^{-((x^2+y^2)/2\sigma^2)}$, the fractional Gaussian filtering function can be naturally deduced as

$$\begin{aligned} G_x^v &= g(x,y)\Big[1+(-v)e^{(2x+1)/2\sigma^2} \\ &+ \frac{(-v)(-v+1)}{2}e^{(4x+4)/2\sigma^2} \\ &+ \frac{(-v)(-v+1)(-v+2)}{6}e^{(6x+9)/2\sigma^2}\Big], \\ G_y^v &= g(x,y)\Big[1+(-v)e^{(2y+1)/2\sigma^2} \\ &+ \frac{(-v)(-v+1)}{2}e^{(4y+4)/2\sigma^2} \\ &+ \frac{(-v)(-v+1)(-v+2)}{6}e^{(6y+9)/2\sigma^2}\Big]. \end{aligned} \tag{3}$$

Then if we set the size of filtering window $w = 3$, the variance $\sigma = 1$, and fractional order $v = 1.4$, the fractional Gaussian filtering template [19] can be acquired as Tables 1 and 2.

When the Gaussian fractional template is used to extract the initial edge of the solar panel image, the threshold and maximum suppression are also performed as conventional edge detectors do but without performing neighborhood connections. The solar panel image preprocessing results are shown in Figures 4 and 5.

2.3. Solder Joint Image Feature Extraction. Once solder joint image is binarized, the image features for defect identification can be extracted. It should be noted that the extracted image features are capable of reflecting the quality of the solder joint; that is, the image features extracted from the good solder point image and the defective one should be different. According to the principle of machine vision inspection [27], since the area and gravity of the image can indicate whether the shape in the image is uniform or not and the anisotropy and the moment of inertia can reflect the symmetry and overall distribution of the image, through (4)–(8), we calculate the area, the gravity (x_g, y_g), the anisotropy alpha, and the moment of inertia (l_x, l_y) of the binary image as the extracted image features.

For area, this feature can count all nonzero pixels by

$$\text{area} = \sum_i (p_i == 1). \tag{4}$$

Table 1: G_x fractional Gaussian template.

0.0406	0.0669	0.0406
0.0183	0.0301	0.0183
−0.0595	−0.0980	−0.0595

Table 2: G_y fractional Gaussian template.

0.0406	0.0669	0.0406
0.0183	0.0301	0.0183
−0.0595	−0.0980	−0.0595

Gravity can reflect the uniformity quality of solder joint.

$$\begin{aligned} x_g &= \frac{\sum_{(i,j)\in S} iI(i,j)}{\sum_{(i,j)\in S} I(i,j)} = \frac{\sum_{(v(i,j)=1)\in S} i}{\text{sum}(v(i,j)==1)}, \\ y_g &= \frac{\sum_{(i,j)\in S} jI(i,j)}{\sum_{(i,j)\in S} I(i,j)} = \frac{\sum_{(v(i,j)=1)\in S} j}{\text{sum}(v(i,j)==1)}. \end{aligned} \tag{5}$$

And anisotropy reflects the symmetry quality of solder joint. Before calculating the anisotropy, we should get the $p+q$ center moment in image region S:

$$u_{p,q} = \frac{\sum_{(i,j)\in S} (i-x_g)^p (j-y_g)^q I(i,j)}{\sum_{(i,j)\in S} I(i,j)}. \tag{6}$$

Then the anisotropy can be acquired by

$$\text{alpha} = \frac{\sqrt{2(u_{20}+u_{02}+4u_{11}^2)+\sqrt{(u_{20}-u_{02})^2}}}{\sqrt{2(u_{20}+u_{02}+4u_{11}^2)-\sqrt{(u_{20}-u_{02})^2}}}. \tag{7}$$

Moreover, the moment of inertia can be utilized to describe the whole imaging distribution of solder joint:

$$\begin{aligned} l_x &= \frac{\sum_{(i,j)\in S} i^2 I(i,j)}{x_g^2}, \\ l_y &= \frac{\sum_{(i,j)\in S} j^2 I(i,j)}{y_g^2}. \end{aligned} \tag{8}$$

These six image features $[\text{area}, x_g, y_g, \text{alpha}, l_x, l_y]$ are representative and can describe the quality of solder joint at different views. Some examples of solder joint features extraction are shown in Figure 6.

3. ELM-Based Solder Joint Defect Detection Method

ELM is always treated as a single-hidden layer feedforward neural network (SLFN) [21, 22]. For N labeled samples (x_i, t_i), where $x_i = [x_{i1}, x_{i2}, \ldots, x_{in}]^T \in R^n$ and $t_i = [t_{i1}, t_{i2}, \ldots, t_{im}]^T \in R^m$, suppose there are $\widetilde{N}$ hidden neurons with

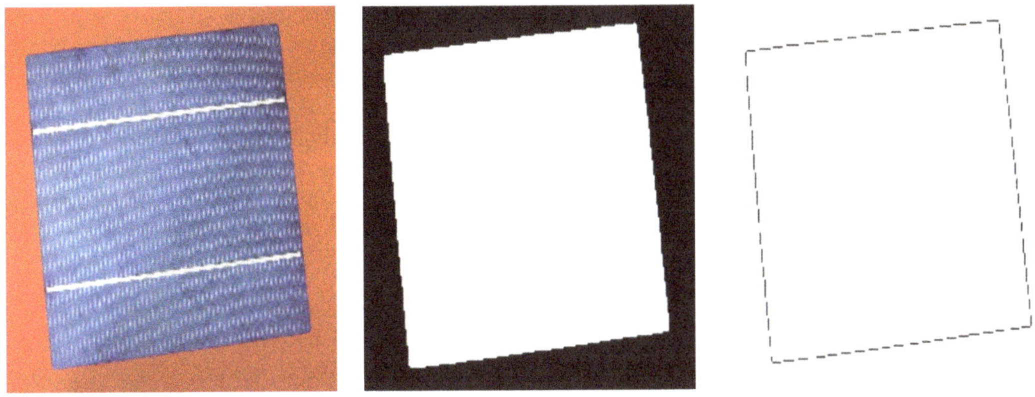

(a) (b) (c)

FIGURE 4: Red background solder panel image processing: (a) original image; (b) HPD thresholding; (c) initial edge extracted by fractional Gaussian template.

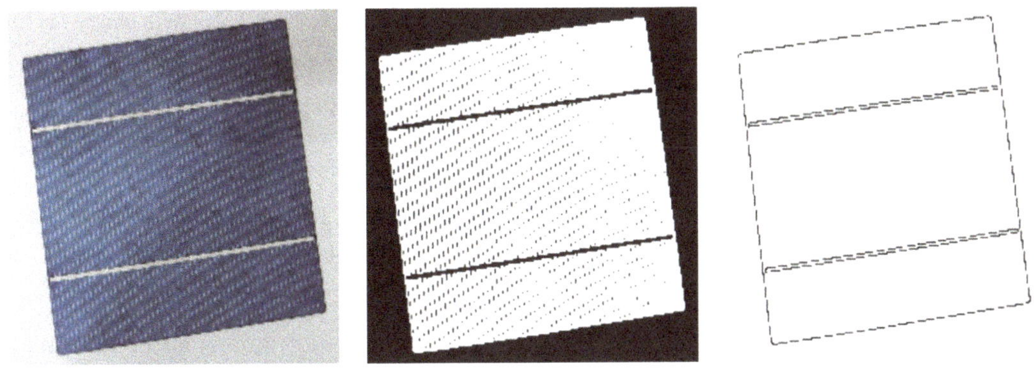

(a) (b) (c)

FIGURE 5: Gray background solder panel image processing: (a) original image; (b) HPD thresholding; (c) initial edge extracted by fractional Gaussian template.

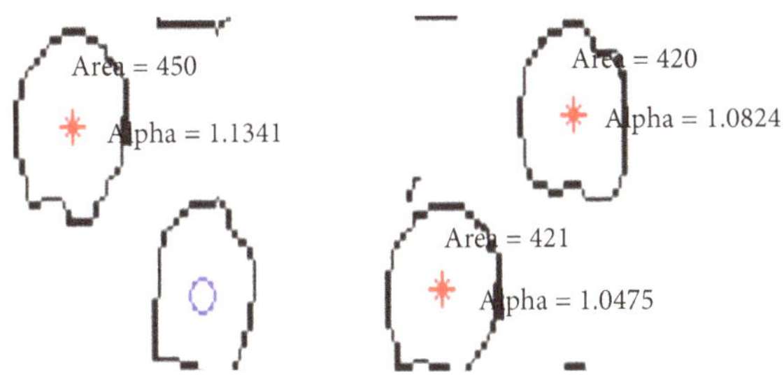

FIGURE 6: Features extraction examples. Both "$*$" and "o" indicate the gravity coordinate (x_g, y_g).

activation function $g(x)$; the standard SLFN can approximate these N samples with zero error, meaning that there exist β_i, w_i, and b_i such that

$$\sum_{i=1}^{\widetilde{N}} \beta_i g\left(w_i * x_j + b_i\right) = t_j, \quad j = 1, \dots, N, \tag{9}$$

where $w_i = [w_{i1}, w_{i2}, \dots, w_{in}]^T$ is the weight vector connecting the ith hidden neuron and the input neurons, $\beta_i = [\beta_{i1}, \beta_{i2}, \dots, \beta_{im}]^T$ is the weight vector connecting the ith hidden neuron and the output neurons, and b_i is the threshold of ith hidden neuron.

Then if we use matrix form, (9) can be written in short as

$$H\beta = T, \tag{10}$$

where

$$H = \begin{bmatrix} g\left(w_1 * x_1 + b_1\right) & \cdots & g\left(w_{\widetilde{N}} * x_1 + b_{\widetilde{N}}\right) \\ \vdots & \cdots & \vdots \\ g\left(w_1 * x_N + b_1\right) & \cdots & g\left(w_{\widetilde{N}} * x_N + b_{\widetilde{N}}\right) \end{bmatrix}_{N \times \widetilde{N}}, \tag{11}$$

$$\beta = \begin{bmatrix} \beta_1^T \\ \vdots \\ \beta_{\widetilde{N}}^T \end{bmatrix}_{\widetilde{N} \times m},$$

$$T = \begin{bmatrix} t_1^T \\ \vdots \\ t_N^T \end{bmatrix}_{N\times m}. \quad (12)$$

H is called the hidden layer output matrix of neural network. In ELM, the input weights and the biases are randomly assigned, and then the output weight β can be approximated by

$$\widehat{\beta} = H^{+}T, \quad (13)$$

where H^{+} is the Moore-Penrose generalized inverse (pseudoinverse) [21] of the hidden layer output matrix H.

From mathematical analysis, it is easy to see that ELM can be trained efficiently, and its prediction is fast and robust. In practice, when we apply ELM for solder joint defect detection, we firstly extract the mentioned features $[\text{area}, x_g, y_g, \text{alpha}, l_x, l_y]$ from solder joint images and input them to ELM. And the output prediction value is −1 or 1. Here, the output "−1" indicates that the solder joint is defective, whereas the output "1" means that it is good one. Some defect identification examples are illustrated in Figure 8.

4. MLSR Based Edge Estimation Algorithm

Accurate and robust edge estimation plays a key role in determining the position of the solar panel. Since the initial edges extracted by fractional Gaussian differential are small line segments with noised fluctuations, they should be properly corrected and estimated to be a clear and accurate edge line. Based on moving least square regression (MLSR) [25, 26], we improve the least square fitting used in previous work [14] for more accurate and robust solder panel edge estimation.

If we treat the edge as a position function $f(x)$, then the accurate edge estimation becomes the position regression optimization problem when given initial line segments. Through the base function expansion, the edge regression function can be expressed as

$$f(x) = \sum_{m=1}^{M} \alpha_m p_m(x) = p^T(x)\alpha(x), \quad (14)$$

where $f(x_i)$ is position regression value for certain position x. $p(x) = [p_1(x), p_2(x), \ldots, p_M(x)]^T$ is the k-order complete monomial basis function, $\alpha(x) = [\alpha_1(x), \alpha_2(x), \ldots, \alpha_M(x)]^T$ is the coefficient of the basis function, and M is the number of basis functions. From the point of view of moving least square, the coefficient matrix $\alpha(x)$ should be determined by minimizing the weighted least square error measure $J(x)$ at x. $J(x)$ can be defined as

$$J(x) = \sum_{i=1}^{N} w(x - x_i)\left(p^T(x_i)\alpha(x_i) - f(x_i)\right)^2, \quad (15)$$

where x_i is the discrete point to be fitted within the influence domain of x. And $w(x - x_i)$ is weight function which defines the influence domain of x and attributes a weight to each discrete point depending on its position relative to x. The weight function is positive and the value should be increased with the decrease of the distance $\|x - x_i\|$ between x and x_i.

With the original extracted edge, we can fit an initial edge line by randomly choosing two pixels. Then, based on the distance to the initial edge of the rest pixels, we can determine whether we use them to update the edge line or discard them. The detailed MLSR algorithm is given as follows.

Step 1. Assume that the two initial edge pixels are (rx_1, ry_1) and (rx_2, ry_2). Denote the initial fitted line as $y_0(x)$; if $x_2 = x_1$, then $y_0(x) = x_1$; else $y_0(x) = ((y_2 - y_1)/(x_2 - x_1))x + b_0$, where $b_0 = y_1 - x_1(y_2 - y_1)/(x_2 - x_1)$.

Step 2. Compute the distance d between a rest edge pixel and the fitted edge line $y_0(x)$. If $d \le \theta$ (θ is the distance threshold), then the pixel can be regarded as a true edge pixel that can be used to update $y_0(x)$. Save such pixel position (x_i, y_i) in a matrix: $E(:, i) = \left[\begin{smallmatrix} x_i \\ y_i \end{smallmatrix}\right]$, $i = 1, 2, \ldots, n$, where i is the number of true edge pixels.

Step 3. Repeat Step 2 until all candidate edge pixels are processed; then we acquire the turn estimated edge pixels' position matrix E.

Step 4. According to the matrix E, use MLSR algorithm to update the initial edge line and acquire this turn edge line denoted as $y(x)$.

Step 5. If the iteration turn number reaches the setting, then exit and output the final refined edge line $y(x)$; else let $y_0(x) = y(x)$ and go to Step 2 to continue.

Once the accurate edge line of solar panel is acquired, we can calculate the position parameters position shift and deflection angle against reference position.

5. Experimental Results and Comparisons

5.1. Solder Joint Defect Detection. In our experiments, 30 PCB solder joint images are captured and each of them has 22 solder joints, which means that totally there are 660 solder joints. Some solder joint examples are shown in Figure 7. For ELM solder defect identification, we take 500 solder joints as the training samples and the remaining 160 ones are regarded as the testing samples.

Then we extract the six image features $[\text{area}, x_g, y_g, \text{alpha}, lx, ly]$ for each solder joint image and utilize ELM to determine whether the solder joint is defective or not. Some defect detection examples are shown in Figure 8.

Generally speaking, the good solder joint can be treated as a tiny circle that means the anisotropy alpha = 1. However, in practice, the value of alpha for good solder joint is sometimes with small fluctuation around 1. In our experiments, if alpha $\in [1, 1.25]$ and area $\in [200, 500]$, then the extracted region is identified as a good solder joint; otherwise it will be a defective solder joint or not a solder joint.

Given the training set $\Delta = [X, T]$, where $X = [x_1, x_2, \ldots, x_N]$, $x_i = [\text{area}, x_g, y_g, \text{alpha}, lx, ly]^T \in R^n$ and

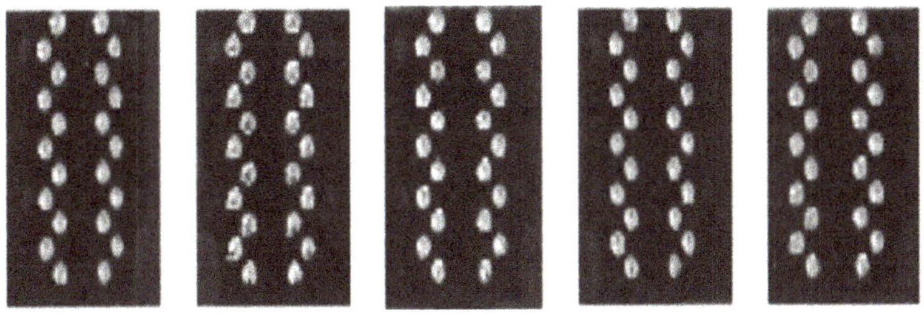

FIGURE 7: Some solder joint samples.

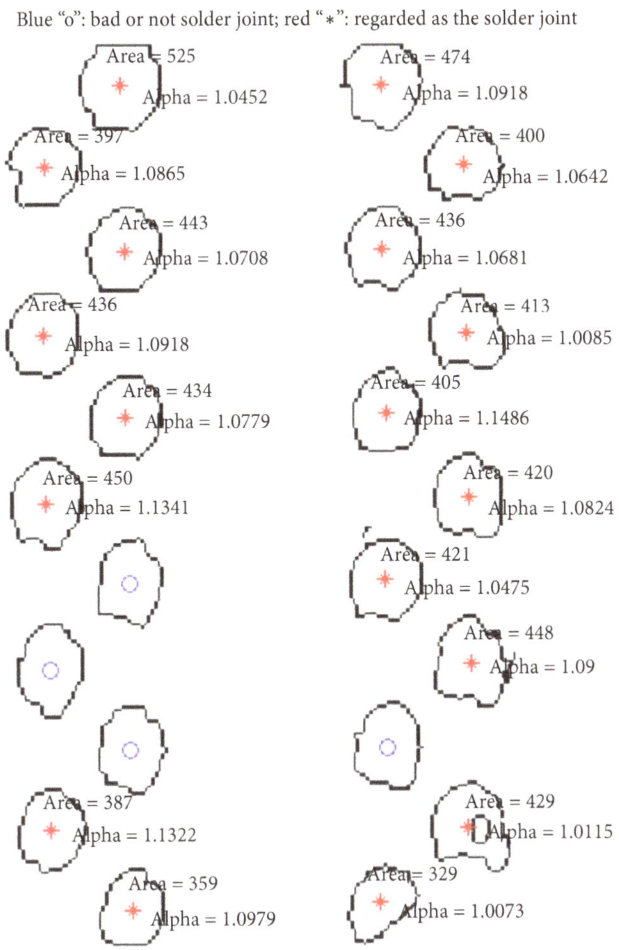

FIGURE 8: Some defect detection examples. Red "*" shows good solder joint and blue "o" indicates the defect one.

$T = [t_1, t_2, \ldots, t_m]$, $t_i \in \{-1, 1\}$. The specific procedure of the ELM solder joint defect identification is given as follows.

Step 1. Choose the activation function as $g(x) = \exp(-(x - (-1+2\delta))^2/0.3)$, where $\delta \in (-1, 1)$, and set the hidden neuron number as $\widetilde{N}$.

Step 2. Assign the input weights w_i and biases b_i, $i = 1, \ldots, \widetilde{N}$, randomly.

Step 3. Utilize the solder joint training images to train the network; that is, calculate the hidden layer output matrix H by (11).

Step 4. Estimate the output weight β: $\widehat{\beta} = H^{+}T$ by (13).

Step 5. After training, calculate the output T by (10), and then identify whether the input solder joint is defective or not.

For comparison, we also used our previous method [14] and SVM classifier to do solder joint defect detection. In our experiments, we conducted eight tests. The detection performance and the comparison results are shown in Table 3, and the consumption time is also shown in Table 3. As can be seen from the table, our method's average recognition rate is 98.6%; recognition speed is not up to 5 seconds (in Matlab 2014). Experimental results and comparison show that the proposed ELM-based defect detection scheme is effective in both recognition rate and processing speed.

5.2. Solar Panel Position Determination and Edge Estimation. Before determining the position of the solar panel, we should estimate its edge lines accurately under red and gray backgrounds. For red background solar panel image, we only need to acquire four external edge lines. Suppose that there are four vertexes, (rx_1, ry_1), (rx_2, ry_2), (rx_3, ry_3), and (rx_4, ry_4); the initial edge lines can then be fitted by the four vertexes. After using MLSR algorithm, the final edge lines estimation result is shown in Figure 9. The corresponding line parameters, slope and bias, are shown in Table 4. And the result of least square fitting (LSF) method in [14] is compared in Table 4. From the table, we can see that the position estimated by MLSR is more accurate than that of LSF. And we can also find that the estimated edge lines in solder image are not actually parallel, which may be caused by imaging distortion according to the machine vision principles [27].

For gray background solar panel image, it will need eight vertexes to determine two inner standard lines in the middle part and the four edge lines. Suppose that there are eight vertexes, (wx_1, wy_1), (wx_2, wy_2), (wx_3, wy_3), (wx_4, wy_4), (wx_5, wy_5), (wx_6, wy_6), (wx_7, wy_7), and (wx_8, wy_8); after using MLSR algorithm, the final refined edges are shown in Figure 10. Lengths of estimated edge lines are calculated and shown in Table 5. Then we can take the difference between the actual detected edge length and the given length to determine whether the solar panel is qualified or not. Similarly, LSF

TABLE 3: Solder joint defect detect performance.

Times	SVM Recognition rate (%)	[14] Recognition rate (%)	The work Recognition rate (%)	SVM Time cost (s)	[14] Time cost (s)	The work Time cost (s)
1	96.3	**99.0**	98.5	4.42	**4.05**	4.32
2	97.1	**98.3**	97.7	4.53	**4.11**	4.54
3	97.3	**98.1**	97.3	4.60	**4.10**	4.58
4	98.1	**99.8**	99.6	4.47	**4.10**	4.50
5	97.3	**98.5**	97.8	4.40	**4.00**	4.36
6	97.4	**98.4**	97.8	5.02	**4.43**	4.98
7	96.6	**97.2**	**97.2**	4.72	**4.20**	4.80
8	98.0	**99.4**	99.0	4.90	**4.30**	4.72
Average	97.3	**98.6**	98.1	4.63	**4.16**	4.60

TABLE 4: Comparisons of position parameters for acquired edges.

Edges	Slop		Bias		Angle difference	
	MLSR	LSF [14]	MLSR	LSF [14]	MLSR	LSF [14]
Top	**−0.1017**	−0.1471	**1.1775**	1.6235	**3.2397**	3.3435
Bottom	**−0.1199**	−0.1265	**8.3395**	8.3327		
Left	**9.9545**	9.7217	**−4.5644**	−4.7879	**5.2545**	5.3326
Right	**7.6886**	7.3248	**−56.9778**	−58.174		

TABLE 5: Length comparisons of estimated edges.

Length	The numbers of image pixels		Realistic lengths (mm)		MSE (mm)	
	The proposed MLSR	LSF [14]	The proposed MLSR	LSF [14]	The proposed MLSR	LSF [14]
Average top line	652.6	655.2	156.38	158.23	**0.439**	0.575
Average bottom line	650.1	650.5	155.78	155.39		
Top two lines' distance	164.8	166.3	39.49	41.43	N/A	
Bottom two lines' distance	161.5	160.2	38.70	39.84		

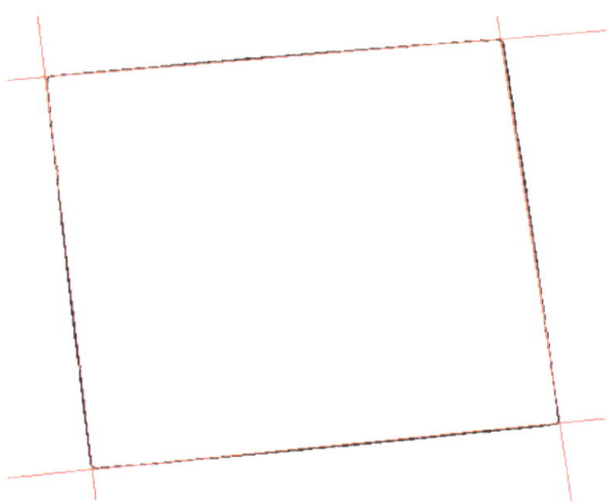

FIGURE 9: Edge estimation result under red background.

FIGURE 10: Edge estimation result under gray background.

results are also compared in Table 5. From the table, we can see that the mean square error (MSE) of MLSR estimated lines is less than 0.5 mm. This indicates that the performance of MLSR edge estimation is superior to previous one.

All experiments and comparisons under different background show that the proposed MLSR algorithm is efficient in solar panel position determination and edge lines estimation.

6. Conclusions

In this work, machine learning based methods for solder joint defect detection and solar panel position determination are explored in detail. After image preprocessing, six representative image features are extracted and ELM classification is used for defect identification. The experiments and comparisons show that the proposed feature extraction and ELM classification method are effective. Moreover, MLSR algorithm is proposed to estimate the edge lines of solar panel and then to determine the panels' position. The practical experiments show that the edge lines in different background can be robustly estimated. And then the position parameters can be accurately determined.

In the future, our further study will focus on using deep learning features for solder joint defect detection and developing more robust regression algorithm to overcome the influence of image distortion.

Conflicts of Interest

The authors declare that there are no conflicts of interest regarding the publication of this paper.

Acknowledgments

This work is partly supported by the Major and Key Project of Natural Science of Anhui Provincial Department of Education (Grant nos. KJ2015ZD09 and KJ2016A085). It is also supported by Anhui Provincial Natural Science Foundation (Grant no. 1608085MF129).

References

[1] E. N. Malamas, E. G. M. Petrakis, M. Zervakis, L. Petit, and J.-D. Legat, "A survey on industrial vision systems, applications and tools," *Image and Vision Computing*, vol. 21, no. 2, pp. 171–188, 2003.

[2] X. Deng, X. Ye, J. Fang, C. Lin, and L. Wang, "Surface defects inspection system based on machine vision," in *Proceedings of the International Conference on Electrical and Control Engineering (ICECE '10)*, pp. 2205–2208, China, June 2010.

[3] J. Y. Lee, C. H. Han, K. W. Ko, and S. Lee, "Development of vision system for defect inspection of electric parts in the tape and reel package," in *In Proceedings of the 16th International Conference on Control, Automation and Systems (ICCAS '16)*, pp. 437–439, Gyeongju, South Korea, October 2016.

[4] J.-K. Park, B.-K. Kwon, J.-H. Park, and D.-J. Kang, "Machine learning-based imaging system for surface defect inspection," *International Journal of Precision Engineering and Manufacturing—Green Technology*, vol. 3, no. 3, pp. 303–310, 2016.

[5] Q. Sun, J. Cai, and Z. Sun, "Detection of surface defects on steel strips based on singular value decomposition of digital image," *Mathematical Problems in Engineering*, vol. 2016, Article ID 5797654, 12 pages, 2016.

[6] G. Acciani, G. Brunetti, and G. Fornarelli, "Application of neural networks in optical inspection and classification of solder joints in surface mount technology," *IEEE Transactions on Industrial Informatics*, vol. 2, no. 3, pp. 200–209, 2006.

[7] X. Zhou, J. Zhou, G. Tian, and Y. Wang, "Research on defects inspection of solder balls based on eddy current pulsed thermography," *Sensors (Switzerland)*, vol. 15, no. 10, pp. 25882–25897, 2015.

[8] N. S. S. Mar, P. K. D. V. Yarlagadda, and C. Fookes, "Design and development of automatic visual inspection system for PCB manufacturing," *Robotics and Computer-Integrated Manufacturing*, vol. 27, no. 5, pp. 949–962, 2011.

[9] F. Zhang, I. Peterson, J. Vila-Comamala et al., "Translation position determination in ptychographic coherent diffraction imaging," *Optics Express*, vol. 21, no. 11, pp. 13592–13606, 2013.

[10] D. W. Capson and S.-K. Eng, "A tiered-color illumination approach for machine inspection of solder joints," *IEEE Transactions on Pattern Analysis and Machine Intelligence*, vol. 10, no. 3, pp. 387–393, 1988.

[11] Y. Matsuyama, T. Honda, H. Yamamura et al., "Automated solder joint inspection system using optical 3-D image detection," in *Proceedings of the 3rd IEEE Workshop on Applications of Computer Vision (WACV '96)*, Sarasota, Fla, USA, 1996.

[12] S.-N. Chiu and M.-H. Perng, "Reflection-area-based feature descriptor for solder joint inspection," *Machine Vision and Applications*, vol. 18, no. 2, pp. 95–106, 2007.

[13] K. Tae-Hyeon, C. Tai-Hoon, Y. S. Moon, and S. H. Park, "Visual inspection system for the classification of solder joints," *Pattern Recognition*, vol. 32, no. 4, pp. 565–575, 1999.

[14] C. Zhang and H. Liu, "The detection of solder joint defect and solar panel orientation based on ELM and robust least square fitting," in *proceedings of the Chinese Control and Decision Conference (CCDC '11)*, pp. 561–565, Mianyang, China, May 2011.

[15] G.-B. Huang and C.-K. Slew, "Extreme learning machine: RBF network case," in *Proceedings of the 8th International Conference on Control, Automation, Robotics and Vision (ICARCV '04)*, Kunming, China, December 2004.

[16] H. Liu, J. Yan, Z. Li, and H. Zhang, "Portrait beautification: a fast and robust approach," *Image and Vision Computing*, vol. 25, no. 9, pp. 1404–1413, 2007.

[17] B. T. Krishna, "Studies on fractional order differentiators and integrators: a survey," *Signal Processing*, vol. 91, no. 3, pp. 386–426, 2011.

[18] A. Lodhia, S. Sheffield, X. Sun, and S. S. Watson, "Fractional Gaussian fields: a survey," *Probability Surveys*, vol. 13, pp. 1–56, 2016.

[19] H. Qirui and K. Liu, "A novel edge detection operator based on fractional gaussian differential," 2014, http://www.paper.edu.cn/download/downPaper/201405-237.

[20] K. S. Miller and B. Ross, *An Introduction to the Fractional Calculus and Fractional Differential Equations*, A Wiley-Interscience Publication, John Wiley & Sons, New York, NY, USA, 1993.

[21] G.-B. Huang, H. Zhou, X. Ding, and R. Zhang, "Extreme learning machine for regression and multiclass classification," *IEEE Transactions on Systems, Man, and Cybernetics, Part B (Cybernetics)*, vol. 42, no. 2, pp. 513–529, 2012.

[22] J. Tang, C. Deng, and G.-B. Huang, "Extreme learning machine for multilayer perceptron," *IEEE Transactions on Neural Networks and Learning Systems*, vol. 27, no. 4, pp. 809–821, 2016.

[23] L. L. Kasun, Y. Yang, G.-B. Huang, and Z. Zhang, "Dimension reduction with extreme learning machine," *IEEE Transactions on Image Processing*, vol. 25, no. 8, pp. 3906–3918, 2016.

[24] G.-B. Huang, Z. Bai, L. L. C. Kasun, and C. M. Vong, "Local receptive fields based extreme learning machine," *IEEE Computational Intelligence Magazine*, vol. 10, no. 2, pp. 18–29, 2015.

[25] L. Zhang, T. Gu, J. Zhao, S. Ji, M. Hu, and X. Li, "An improved moving least squares method for curve and surface fitting," *Mathematical Problems in Engineering*, vol. 2013, Article ID 159694, 6 pages, 2013.

12

Multiobjective Optimization Approach for Coordinating Different DG from Distribution Network Operator

Limei Zhang,[1] **Honglei Yang**,[2] **Jing Lv**,[1] **Yongfu Liu**,[1] **and Wei Tang**[3]

[1]*College of Information Science & Technology, Hebei Agricultural University, Baoding, China*
[2]*China Electric Power Research Institute, Beijing, China*
[3]*College of Information and Electrical Engineering, China Agricultural University, Beijing, China*

Correspondence should be addressed to Limei Zhang; lmzhch@126.com

Guest Editor: Mustafa I. Fadhel

Integrating with analysis of uncertainties, this paper presented a multiobjective optimization approach for coordinating different DG from the perspective of Distribution Network Operator (DISOPER). Aiming to three uncertain factors including fuzzy variable, random variable, and interval variable, the information entropy and interval analysis methods are adopted to construct multistate models of multisource uncertainty. The information entropy method is to convert fuzzy variable into equivalent random variable. Interval analysis method is to transform random variables into interval variables by setting a confidence level. Then plenty of simulation analysis based on the small probability event and expectation are investigated to reduce the computational burden and eliminate invalid computation. Subsequently, multiobjective formulations based on multistate are built by analyzing systematical power loss, voltage quality, reliability, and environment change provide some reference for DISOPER in dealing with access of privately owned DG units. Furthermore, based on network topology analysis and modified nondominated sorting genetic algorithm (NSGA), a combinatorial optimization method is proposed to reduce search space and solve the constructed formulations efficiently. Simulations are carried out on IEEE 37-bus systems and results are presented and discussed.

1. Introduction

Environmental deterioration, energy shortages, and climatic anomaly compel many countries to take renewable energy as future alternatives [1]. Consequently, distributed generation (DG) such as wind turbines (WTs) and photovoltaic (PV) systems are considered the most promising technologies to meet electrical loads because of their outstanding advantages such as environmentally benign, clean, and inexhaustible [1, 2]. Constrained by traditional power grid, one main application of DG in China is DG access to distribution system for the purpose of giving full play to the function of interests' integration [3, 4]. However, different from conventional power sources, DG penetration has a significant influence on power system due to its stochastic nature [2, 3]. Previous research has shown that the effect of DG access is closely related to siting and size of DG [4]. Therefore, the appropriate planning of DG in distribution system plays a crucial role in fully exerting DG advantages as well as restraining its disadvantages [5–7].

Apparently, DG planning in distribution network is an extraordinarily complicated problem, which is attributed to not only the diversity and intermittent nature of DG but also the complex structure and abundant metadata information of distribution network. So far, loads of researches have been reported [7–23]. Earlier studies focused on certain situation because of restrictions imposed by the treatment of uncertainty. In Reference [8], Shuffled frog leaping algorithm is presented for the placement of DG in distribution systems to reduce the real power losses and cost of the DG. When constructing multiobjective formulas, influence of load models is considered in siting and sizing of DG [9]. In Reference [10], the authors optimized three objectives including maximum DG utilization, minimizing system loss, and minimum environmental pollution by improving Pareto evolutionary algorithm. From DG Owner's and Distribution Company's Viewpoints, multiobjective optimization is proposed from the operational aspects and economic analysis [11]. These models and calculation methods can be used for optimum design of controlled DG and good effect can be obtained.

However, due to the profound effect of uncertainty on optimization results, certain programming models and algorithms lack adaptability for dealing with planning problem with plenty of uncertainties, which is especially prominent in planning renewable energy like WTs and PV generation. Therefore, uncertainty modeling and optimization algorithm are focused on recently published documents [9–29]. In describing uncertainty, the probability density functions is usually employed to model load, wind power, PV power, electricity market price, and so on [12–22]. Also, few studies have explored fuzzy mathematics models to describe uncertainties like load [3, 20] and constraints for voltage profile, feeder current, and substation capacity [25]. Treatment techniques of uncertainty involved Monte Carlo simulation [22, 24], scenario analysis [19, 23, 25, 26], and multistate model constructed by applying probability density functions [17, 18, 22]. In objective formulations, most of them were analyzed from economic benefits [9, 27–30], environmental concerns [30, 31], and technical constraints [9, 27, 28]. In optimization solution, some intelligent algorithm [28] and uncertain programming theory [14] were introduced to adapt to planning problems with plenty of uncertainties. Based on chance-constrained programming, a planning model is presented integrated with random characteristic of load and renewable DG power [14]. In Reference [29], Monte Carlo simulation is employed to deal with uncertainties, and NSGA-II is applied to find Pareto optimal front of two minimum objectives including system total cost and technical risk. Considering technical constraint dissatisfaction, costs, and environmental emissions, stochastic dynamic multiobjective model is constructed in Reference [25], and it applied a binary particle swarm algorithm and a fuzzy satisfying method to obtain the optimal solution. A sample average approximation algorithm-based technique is proposed in Reference [32] to solve the optimal placement of DG units, which is formulated as a stochastic optimization problem.

All these researches offer valuable contributions, but they lack the following important features. For one thing, multisource uncertainty is rarely considered together in distribution network optimization studies. For another thing, access control mechanisms for privately owned DG units are overlooked in terms of DISOPER.

In fact, there has been all kind of available DG in the market. It will be an inevitable trend that multitype and multiresource DG access power grid simultaneously, which is dominated by different factors such as policy interventions, benefit actuation, and traditional resource restriction. The advent of more charging stations for electric cars, whose charging were noted for variation in time and space, gave rise to frequent load fluctuation [17, 19, 20]. The implications of DG and load change make it necessary for DISOPER to integrate considerations regarding multisource uncertainty into DG planning of the distribution network. For the treatment of uncertainty, Monte Carlo Simulation requires enormous computational burden and is unsatisfactory to deal with complex models [33]. When multistate models and scenario analysis techniques [17–19, 22, 23, 25, 26] are applied to cope with uncertainties, selection of state or scenario apparently influences on the accuracy of optimization results. In this paper, three uncertainties including fuzzy variable, random variable, and interval variable are analyzed by information entropy and interval analysis methods to construct multistate model of uncertainties. Simultaneously, small probability event and expectation for multistate models as well as proportion compensation method are investigated to obtain representative state which can reduce invalid computation, improve computational efficiency, and guarantee the accuracy of results.

According to distribution network planning studies, the objective functions involving active power losses, emission, and reliability have been presented in many researches. However, existing papers have the drawback of overlooking the private ownership feature of DG units. Under current rules, it does not state clearly that DISOPER can randomly reject privately owned DG units whose accesses have an adverse impact on distribution network operation. But as the operator of distribution network, DISOPER has the right to change connected location of privately owned DG units to assure safe operation of distribution network. In addition, DG access concerned to multifaceted problems such as technicality, economics, environment, and social characteristics, which potentially correlate to DISOPER's benefits. However, it is difficult for DISOPER to qualify some of them in monetary terms alone. Therefore, based on the proposed multistate model, this article developed multiobjective optimization formulations through analyzing power loss improvement, voltage quality, systematical reliability, and environment change so as to provide some reference for DISOPER in dealing with access of privately owned DG units.

Furthermore, it has been studied and concerned widely in solving multiobjective optimization problems of strong constrained condition in the fields of engineering. At present, nondominated sorting genetic algorithm (NSGA-II) has been recognized as one of the most promising algorithms, primarily because it has the feature of high convergence speed, low computational complexity and good global convergence property [34, 35]. Compared with contemporary constraint-handling strategy, application of NSGA-II is encouraged to solve more complex and real-world multiobjective optimization problems [34]. NSGA-II has also played an indispensable role in solving constrained multiobjective optimization of the electric industry [35–37]. Fast and elitist NSGA-II algorithm is employed to avoid artificially balanced solutions [35]. In Reference [36], NSGA-II is applied to deal with simultaneously determining optimal capacities of active and reactive power reserve. NSGA-II and fuzzy set theory is chosen to find the best compromise solution [37]. In Reference [38], the authors adopted a novel building energy/exergy simulation tool with multiobjective optimization capabilities based on NSGA-II for carrying out simultaneously analysis of building energy use and design's Net Present Value. NSGA-II algorithm is used to the design of a standalone HRES comprising WTs, PV panel and battery bank [39]. In this paper, we proposed a combinatorial algorithm integrating network topology analysis and modified NSGA in order to reduce search space and efficiently solve the constructed formulations. The purpose of network topology is to eliminate obvious impossible access location of DG and avoid entire network search. In NSGA-II, the modified contents refer to nondominated sorting operator, individual crowding distance operator, and genetic operators. The purpose is to obtain effective Pareto solutions, assure uniform distribution of solutions, accelerate convergence rate of algorithm, and avoid local optimal or missing Pareto optimal solutions.

The main contribution of this paper is as follows. Firstly, three uncertainties including fuzzy variable, random variable, and interval variable are analyzed by information entropy and interval analysis methods to construct multistate model. Simultaneously, small probability event and expectation for multistate models as well as proportion compensation method are investigated to obtain representative state. Subsequently, from the perspective of DISOPER of dealing with privately owned DG units, multiobjective formulations based on multistate are constructed by analyzing power loss improvement, voltage quality, systematical reliability, and environment change. Also a combinatorial algorithm integrating network topology analysis and modified nondominated sorting genetic algorithm (NSGA) is presented to reduce search space and efficiently solve the constructed formulations.

2. Descriptions and Treatment for Multisource Uncertainties

The involved uncertainties in DG planning are related to not only connotation (e.g., load variations and renewable DGs fluctuation) but also extension (e.g., electricity market change, policy and regulation adjustment, and availability of system facilities). For simplification and integrating with research contents, three uncertainties including random variable, fuzzy variable and interval variable, which are related to load variations and DG fluctuation, are portrayed and treated.

2.1. Descriptions for Load and DG Uncertainties. Uncertainties of load and DG have been widely studied. Here, frequently used load and electric vehicle charging stations [40] are considered, and RES-based DGs include wind turbines (WTs) and photovoltaic generators (PVs).

Integrating with existing researches, the frequently used load is modeled by using uniform distribution of interval and truncated Gaussian distribution [14–16], which is a normal distribution that restricts the range of variable value [41]. Simultaneously, taking into account the unavailability of historical data, here, electric vehicle charging stations are modeled in (1) according to quick charging mode [17, 19, 20]:

$$P = \frac{QV}{\eta}, \tag{1}$$

where η is the charging efficiency for charger, V is the Charging voltage for the electric vehicle, and Q is the current sum of all recharging battery. Here, assume that Q is a fuzzy variable with the triangular membership function.

Output power of WTs is directly related to wind speed, which is usually taken as Weibull distribution. Research for the intermittent and stochastic of WTs has been relatively mature [13, 22]. The expression of wind power can be found in Reference [22].

Similar to WTs, PVs is also related to natural factors, geographic envissronment, season variation, etc. Here, PVs output power is given according to the composition of Grid-connected PV generation system [3]:

$$P = \eta_{\mathrm{mod}} A r_{\mathrm{tilt}} \eta_{\mathrm{wr}} \eta_{\mathrm{pc}}, \tag{2}$$

where η_{mod} is the efficiency under the ambient temperature, A is the total area of illumination, r_{tilt} is solar radiation intensity of inclined plane, η_{wr} is the distribution efficiency, and η_{pc} is the efficiency of power regulating system, which has very strong diurnal patterns [3, 23, 42]. η_{mod} can be shown in the following equation:

$$\eta = \eta_0 \left[1 - \gamma_{\mathrm{T}}\left(T_p - T_{\mathrm{r}}\right)\right], \tag{3}$$

where T_{r} is reference temperature, whose value is 298 K. η_0 is the conversion efficiency of PVs at the reference temperature. γ_{T} is the temperature coefficient of solar cell, which generally is 0.005. T_p is the temperature at time p. From (2) and (3), uncertainty of PVs output power is concerned to solar radiation intensity and the temperature. It is well known that temperature is affected by sunlight and changed with different weather conditions. That is, the temperature changes quickly in sunny days while it changes slowly in cloudy or rainy weather [43]. Furthermore, real output of PVs is still relate to illumination inhomogeneity, dust, production deviance, installation error, etc. Apparently, it is not easy to express such change through random uncertainty model. For simplification and ensuring rationalization, here PVs output power is simplified into (4) according to (2):

$$P_{\mathrm{PV}} = A r_{\mathrm{tilt}} \eta_1, \tag{4}$$

where P_{PV} is real output and η_1 is the conversion efficiency related to weather conditions. Here, assume that η_1 is a fuzzy variable considering that there is an inevitable fuzziness in conversion efficiency of representing the weather conditions. And integrated with under sunny, cloudy, and rainy days given in [3], here η_1 is modeled as trapezoid membership function shown in the following equation:

$$\mu(\eta_1) = \begin{cases} \frac{10\eta_1}{3}, & 0 \le \eta_1 \le 0.3, \\ 1, & 0.3 \le \eta_1 \le 0.55, \\ \frac{14-20\eta_1}{3}, & 0.55 \le \eta_1 \le 0.7, \\ 0, & \text{others.} \end{cases} \quad (5)$$

2.2. Treatment for Multisource Uncertainties. The purpose of treatment for multisource uncertainties is to convert various uncertainties into multistate variables to improve computational efficiency and reduce invalid computation. Based on the concept of possibility-probability consistency, information entropy is adopted to transform fuzzy number to stochastic variable. Interval analysis is applied to convert random variables into limited discrete intervals by setting a certain confidence level, and meanwhile, composite state is obtained by analyzing small probability event and expectation.

Entropy is the best measure to uncertainty. Fuzzy entropy G_x measures fuzziness of uncertainty, while probability entropy H_y measures randomness of uncertainty [44]. Here, set G_x is equal to H_y, then fuzzy variable is converted into an equivalent random variable in normal distribution with mean u_{eq} and variance σ_{eq} shown in the following equations:

$$\sigma_{eq} = \frac{1}{\sqrt{2\pi}} e^{G_x - 0.5}, \qquad \mu_{eq} = \frac{\int_x x\mu(x)dx}{\int_x \mu(x)dx}, \quad (6)$$

where $\mu(x)$ is the fuzzy membership function of fuzzy variable and σ_{eq} is variance. And mean u_{eq} is obtained when membership degree takes 1 [45, 46]. For improving calculating efficiency and ensuring data integrity, transformation of random variables to interval variable includes the following:

(i) Determination of the upper and lower limits. The method is to calculate cumulative probability of random variable. The value should be close to 1 and never lower than a specified confidence level, which is often more than 90%.

(ii) Division of interval variable. Setting interval number m, interval is divided by regularly or irregularly step-length on the basis of actual situation. Subsequently, a binary state sequence with state and its probability is obtained. The ith state probability in interval $[x_i, x_{i+1}]$ can be obtained by

$$P_i(\omega_i) = \int_{x_i}^{x_{i+1}} f(x)\,dx = F(x_{i+1}) - F(x_i). \quad (7)$$

Simultaneously, the normalization processing is carried out by the proportion compensation method to assure that the sum of state probability is 1 [47], whose principle is to assign loss probability based on the proportion of interval probability. The revised form of state probability is shown in the following equation:

$$P_i' = P_i + \frac{P_i(1 - \sum_{i=1}^m P_i)}{\sum_{i=1}^m P_i}. \quad (8)$$

The composite state for multisource uncertainty is analyzed under the independence assumption among random variables. Assuming that the kth composite state is composed of three binary sequence X: $\{\omega_{Fi}^k, P'(\omega_{Fi}^k)\}$, Y: $\{\omega_{Rj}^k, P'(\omega_{Rj}^k)\}$, and Z: $\{\omega_{Iq}^k, P'(\omega_{Iq}^k)\}$ from three uncertainties, the number of composite state and the kth joint probability are obtained by

$$N = \prod_{i=1}^{m_F} N_{Fi} \prod_{j=1}^{m_R} N_{Rj} \prod_{q=1}^{m_I} N_{Iq}, \quad (9)$$

$$P_{X,Y,Z}(k) = P'(\omega_{Fi}^k) P'(\omega_{Rj}^k) P'(\omega_{Iq}^k), \quad (10)$$

where m and N are the number of variables and interval, respectively. Subscript F, R, and I are orderly on behalf of fuzzy, random, and interval variables.

From (9), the number of composite state increases exponentially with increase of interval number. From (10), the consequence of interval division is noticeable. Too small interval will increase tendency to small probabilities, while too large range will reduce accuracy of solution. To avoid these, plenty of simulations based on the small probability event and expectation are carried out besides setting a confidence level. The simulations are to analyze the number of small probability events and guarantee reasonable expectation of the given formulation, while the application of confidence level is to reduce the number of composite state. Similarly, the normalization is also carried out by (8).

3. Multiobjective Formulations

DG penetration has significant impacts on systematical loss, voltage, reliability, etc. And compared with conventional power, DG cost is higher. However, DISOPER cannot randomly reject privately owned DG units although its accesses have an adverse impact on distribution network operation under the condition that the technical parameter is in conformity with access standard. Therefore, sometimes DISOPER has to change connected location of privately owned DG units or sacrifice personal economic interests to assure safe operation of distribution network. From the viewpoint of economics and technology, here, four maximum objectives are constructed by analyzing reduction of power loss, enhancement of voltage quality, improvement of reliability, and change of environment to optimize the location and capacity of DG, thereby providing reference for DISOPER in coordinating access of privately owned DG units.

3.1. The Reduction of Power Loss. The power loss rate in (11), which is defined as the ratio of power loss reduction with DG and power loss without DG, is proposed to reflect impact of DG on power loss:

$$\mathrm{IP}_{\mathrm{loss}} = \frac{\sum_{s=1}^{N_1} P(s)P_{\mathrm{loss}}^{s} - \sum_{q=1}^{N_2} P(q)P_{\mathrm{loss}}^{q}}{\sum_{s=1}^{N_1} P(s)P_{\mathrm{loss}}^{s}}, \quad (11)$$

where N_1 and N_2 denote the number of composite state without DG and with DG, respectively. $P(q)$ and $P(s)$ are state probability at the qth and sth composite state. P_{loss}^{s} and P_{loss}^{q} are power loss at the qth and sth composite state.

3.2. Enhancement of Voltage Quality. Currently, besides the consumers' higher requirements for power supply because of the rapid development of modern power electronic devices, electric market reform has brought fierce competition among enterprises. To improve vitality and competitiveness, DISOPER has to reinforce system to better control over voltage variations [48, 49], which is closed linked to technical and potential economic interest. Therefore, the quantification of DG impact on voltage quality is very necessary. Here, the improvement of systematical voltage offset, which is defined as the ratio of nodal voltage offset before and after DG installation, is given in (12) to quantify DG enhance on voltage quality:

$$\mathrm{IV}_{\mathrm{enha}} = \frac{\sum_{s=1}^{N_1} P(s)\sum_{i\in\phi}\left(V_{\mathrm{wo}i}^{s} - V_{i}^{\mathrm{rated}}\right)^2}{\sum_{q=1}^{N_2} P(q)\sum_{i\in\phi}\left(V_{\mathrm{w}i}^{q} - V_{i}^{\mathrm{rated}}\right)^2}, \quad (12)$$

where subscripts wo and w are without and with DG, respectively; φ is the load node set; V_i and V_i^{rated} are the real voltage magnitude and the nominal voltage at the ith load node.

3.3. Improvement of Reliability. DG access can improve reliability of load in island if DG island operation is feasible after system fault. Here Ratio of Expected Energy Not Supplied Reduction shown in (13) is employed to measure DG impact on systematical reliability:

$$\mathrm{IR}_{\mathrm{imp}} = 1 - \frac{\sum_{q=1}^{N_2} P(q)\mathrm{EENS}_{\mathrm{w}}^{q}}{\sum_{s=1}^{N_1} P(s)\mathrm{EENS}_{\mathrm{wo}}^{s}}, \quad (13)$$

$$\mathrm{EENS} = \sum_{i\in\phi} \beta_i T_i L_i, \quad (14)$$

where EENS is expected energy not supplied; β_i is the ith load fault rate (number of faults per year); and T_i is the average durations of fault (hour per time). The equations for β_i and T_i are proposed in [48–50].

3.4. Improvement of Environment. Without a doubt, electricity industry is playing an inescapable role for environmental pollution. DG is also propelled to meet requirements of low carbon emissions. Here, only carbon emission is taken into account because the others are looked as being proportion to it. The environmental improvement is investigated by the rate of carbon emission reduction in the following equation:

$$\mathrm{IE}_{\mathrm{imp}} = 1 - \frac{\sum_{q=1}^{N_2} P(q)\left(E_{\mathrm{main}}^{q} C_{\mathrm{con}} + \sum_{i=1}^{N_{\mathrm{DG}}} E_{\mathrm{DG}i}^{q} C_{\mathrm{DG}i}\right)}{\sum_{s=1}^{N_1} P(s)E_{\mathrm{wo}}^{s} C_{\mathrm{con}}}, \quad (15)$$

where E_{wo} is the electricity from main grid without DG (kWh); E_{main} is the electricity from main grid with DG (kWh); E_{DG} is the electricity from DG; and C is the amount of carbon emission per unit electricity (kg/kWh). Subscript con is the conventional energy.

3.5. The Constraints. For safe and stable operation, the optimization problem is subject to power flow equations constraint and some inequation limits like node voltage, branch capacity, etc. These constraints are shown in the following equations:

$$\begin{cases} P_{\mathrm{main}}^{q} + \sum_{i=1}^{N_{\mathrm{DG}}} P_{\mathrm{DG}i}^{q} = \sum_{i\in\phi} P_{\mathrm{L}i}^{q} + P_{\mathrm{loss}}^{q}, & \\ \Pr\{V_i^{\min} \le V_i^{q} \le V_i^{\max} \mid i \in \phi\} \ge 1 - \varepsilon_V, & \\ \Pr\left\{S_{ij}^{q} \le S_{ij}^{\max} \mid i, j \in \phi\right\} \ge 1 - \varepsilon_S, & q = 1, 2, \ldots, N, \\ N_{\mathrm{DG}} \le N_{\mathrm{DG}}^{\max}, & \\ \sum_{i=1}^{N_{\mathrm{DG}}} P_{\mathrm{DG}i} \le k\% * \sum_{i\in\phi} P_{\mathrm{L}i}, & \end{cases} \quad (16)$$

where P_{main} is the power from the main grid; P_{DG} is DG power; P_{loss} is systematical losses; N are the number of operating state; $\Pr\{\cdot\}$ represents the probability of an event; $V_i^{\max}$ and $V_i^{\min}$ are the ith nodal upper and lower of voltage; S_{ij}^{q} and $S_{ij}^{\max}$ are real line capacity at the qth state and permitted maximum line capacity; N_{DG} and $N_{\mathrm{DG}}^{\max}$ are the real number and the allowable ceiling of DG location; ε_V and ε_S are confidence level of bus voltage and line capacity; $P_{\mathrm{L}i}$ and $P_{\mathrm{DG}i}$ are base load and DG power at node i; and $k\%$ is the penetration rate of DG.

4. Optimization Method Based on Network Topological Analysis and Modified NSGA-II

Besides the inherent complexities of power flow calculation in DG planning, the previously constructed multiobjective optimization formulations are obviously constrained, nonlinear, with mixed integer variables. Consequently, optimization computation is a tremendously complex task. In order to improve the computational efficiency without influencing the accuracy of solutions, the topological analysis and modified NSGA are adopted to carry on combinatorial optimization. The purpose of topological analysis is to reduce search space by determining DG candidate locations in accordance with the actual situation, while that of modified NSGA is to improve convergence speed as well as avoid falling into local best during optimizing the constructed formulations.

4.1. Determination of DG Candidate Locations Based on Network Topology Analysis. Previous researches showed that DG impacts on distribution network are related to the connected bus as well as all others buses, but the bigger influence occurs on the bus near DG [51]. And DG layouts should try to keep decentralized to limit potential of interfering with each other in undesired ways. In addition, DG location is restricted by geographical position and local natural resources. Therefore, here DG candidate locations area is proposed to determine one DG location when carrying out multiobjective optimization. The number of DG candidate locations areas is corresponding to the maximum number of DG installation locations. Each area, which determines DG candidate locations set, is obtained through the limited depth-breadth priority algorithm. A DG candidate locations set is obtained according to nodal electric distances, nodal load moment, and reliability requirement. The detailed process is described as following:

Step 1. Obtain the DG candidate location area. Based on the maximum number of DG installation location h, the distribution network is divided into k area. Partitioning method is to add the base load starting from the end of line according to the limited depth-breadth priority algorithm. The sum of all base loads in one area is approximately equal to the value of the total base load divided by h.

Step 2. Select DG candidate locations based on the principle of reducing the power loss. Generally, the bus far from substations produced more power line losses. Therefore, here the nodal electric distance d is applied to determine the DG candidate locations. The selected node i is obtained by

$$d_i \geq \frac{\max\{d(j) \mid j \in \phi\}}{3}, \tag{17}$$

where d is electric distance and j is the alternative bus.

Step 3. Select DG candidate locations based on the principle of improving the nodal voltage. It has been proved that the constant PQ DG can definitely boost up the nodal voltage, while PV DG also can achieve the same effect under the condition that the connected nodal voltage is lower than the voltage of PV DG [51]. Furthermore, the voltage loss is in correspondence with the nodal load moment. Therefore, through the descending order of the nodal load moment, the first half is selected as DG candidate location.

Step 4. Select DG candidate locations based on the reliability requirements. As mentioned previously, DG access can improve the reliability in island load. Thus, buses having high reliability requirements are chosen as candidate location.

Step 5. Form DG candidate locations sets by combining and abandoning candidate locations in view of the partitioning area. One candidate location set is corresponding to one area. The principle of abandon and combination includes the following: (a) to reserve all candidate locations based on reliability, (b) to abandon the candidate locations without locating in installation locations area, (c) to abandon inaccessible geographical location, and (d) to keep candidate locations coming from both step 2 and step 3.

4.2. DG Island Separation Based on Network Topology Analysis. As mentioned in [48], DG impact on reliability is mainly concerned to the island's load. Then improvement of reliability is closely associated with the formation of the island. Consequently, DG island separation has a significant influence on reliability. Starting from DG connected bus, this paper goes into distribution network's structure using breadth search, giving the island scope by the following equation:

$$P_{\mathrm{DG}i}^{s} - \sum_{j=1}^{N_{\mathrm{il}}} P_j^s \leq \gamma, \tag{18}$$

where N_{il} is the number of load points in island; P_j^s are the load value at the sth state; $P_{\mathrm{DG}i}^{s}$ is power output at the ith DG; and γ is the threshold.

The formation of the island is described as following:

(i) Select the connected bus of DG as the reference node of determining electrical distance, and then obtain the electrical distance of all loads.
(ii) Give a lateral stratification of loads in accordance with electrical distance. Loads on the same layer are ranked in order of importance, which is represented in weighted coefficients of load.
(iii) Perform breadth search starting from the innermost layer. When the total load in the island is closed to threshold, the loads on the outermost layer are chosen according to the weighted coefficients of load.

To understand it more clearly, take Figure 1 as an example. Assume that threshold γ is constant, and bus 4 is selected as DG location. By carrying on the width search, the next-linked buses 3, 5, 7, 8 are selected. If the sum of these load value is not satisfied with (18), buses 2, 9, 6 are considered. If they are all chosen, the sum will gone beyond the threshold γ. Therefore, according to weighted coefficients of loads at buses 2, 9, 6, bus 9 with larger weighty is chosen. Then the buses in island include 3, 5, 7, 8, 9.

4.3. Multiobjective Optimization Based on NSGA. NSGA-II is currently one of the most popular multiobjective evolutionary algorithms because of its rapidly converging rate. However, considering that NSGA is inclined to result in prematurely converging to local Pareto optimal front, here the proposed modified NSGA-II can not only manage a variety of decision variables such as DG type, DG capacity, and DG location, but also deal with the complicated constraints expediently. The modified content mainly refers to genetic algorithm (GA) implementation to improve the GA's local searching capacity, accelerate the convergence rate, and effectively prevent the premature convergence. About population code, decimal encoding scheme based on

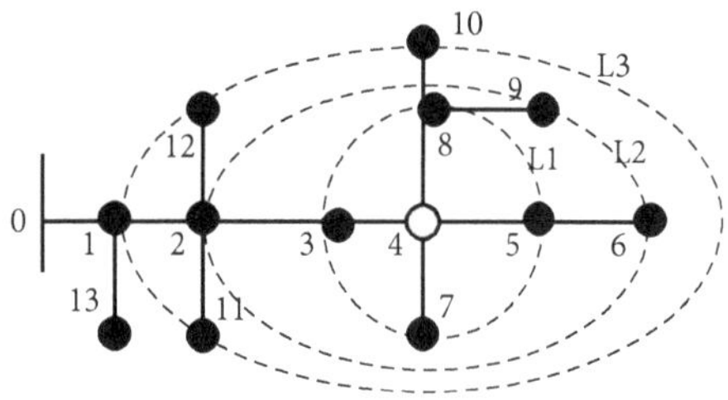

FIGURE 1: Schematic of island division.

segmented chromosome management is applied to adapt to characteristic of DG planning involving DG type, DG capacity, and DG location. That is, each solution is coded by using three segments, which orderly represent DG type, DG location, and DG size. Each segment is a vector, whose size is equal to maximum number of installation DG units [49]. In genetic manipulation, the selection, crossover, and mutation operator are properly collected besides the segmented point cross and mutation proposed in [49]. Figure 2 describes a coding example and the implementation details of the segmented point cross and mutation.

Selection operator is improved by using elitist preservation strategy and dual tournament selection to enhance the global convergence performance of algorithms and enlarge search space. Elitism preservation strategy is to copy a small proportion of the fittest candidates into the next generation to improve convergence and avoid loss of optimal solution. Dual tournament selection is to randomly pick up individuals from the population.

Crossover is preceded by correlation between two individuals to maintain population diversity and improve global search ability. The correlation is shown in the following equation [52]:

$$r(x_1, x_2) = \sum_{i=1}^{N_{\text{ch}}} x_1^i \cap x_2^i, \tag{19}$$

where $r(x_1, x_2)$ is the correlation between individuals x_1 and x_2; N_{ch} is the segment number of population; superscript i is the ith component of individuals x_1 or x_2; and $\cap$ is a binary operator, which is satisfied with the following equation:

$$x_1^i \cap x_2^i = \begin{cases} 0, & x_1^i = x_2^i, \\ 1, & x_1^i \neq x_2^i. \end{cases} \tag{20}$$

Obviously, the bigger the r is, the smaller the correlation is. Therefore, two individuals with bigger r are matched for crossover.

Nonuniform mutation operator is employed to improve local searching ability of GA and maintain diversity. Nonuniform mutation operator in Reference [39] can adaptively adjust the searching step size in genetic evolution.

In addition, application of segmented point cross and mutation maybe result in occurrence of individual beyond threshold due to diversity of heredity encoding. Therefore, according to [53], treatment of DG capacity border in (21) is used to maintain the crossed or mutated individual within normal range:

$$x_1^C = \begin{cases} k\% * \sum_{i \in \phi} P_{\text{L}i} - \sum_{i=1}^{N_{\text{DG}}-1} P_{\text{DG}i}, & \rho = 0, \\ 0, & \rho = 1, \end{cases} \tag{21}$$

where ρ is random variable of the Bernoulli distribution.

4.4. The Framework for Multiobjective Optimization. The proposed optimization method includes determination of DG candidate location sets and multiobjective optimization based on modified NSGA-II. Flowchart for multiobjective optimization is shown in Figure 3.

5. Numerical Example

5.1. Example for Multistate Model of Multisource Uncertainty. As stated in Section 2, fuzzy variables are transferred into random variables with normal distribution. Therefore, based on research contents, the discretization of the variable with standardized normal distribution is presented in details. Subsequently, multistate models for output power of PV power and output power of wind power. The composite state for PV power and a load are given.

Here, multistate models of a standardized normal random variable are investigated, and then others can be extended to the demanded range. For a standardized normal random variable, the probability in interval $[-3, 3]$ is over 9.7%. So, the upper and lower bounds are specified as 3 and -3. When interval number is equal to 10, the schematic diagram is shown in Figure 4, which is obtained in accordance with the relationship between standardized normal distribution and general normal distribution. The multistate models corresponding to general normal distribution are obtained. Setting random variable $L_1 \sim N_{11}$ (0.5, 2), the state value and its probability is shown in Table 1.

Multistate models of wind power, whether its wind speed is taken as Weibull distribution or interval variable, are similar with that of load because they are all treated as stochastic variable here. In addition, WTGs has a requirement of cut-in wind speed, rated wind speed, and cut-out wind speed, so the lower and upper of wind discretization is confirmed in cut-in wind speed and cut-out wind speed. Supposing that $V_i = 4\,(\text{m/s})$, $V_r = 14\,(\text{m/s})$, and $V_o = 25\,(\text{m/s})$, the shape parameter and the scale parameter of wind are 2 and 8 m/s, interval number is 10, the rated power P_r is 45 kW, and then the binary state of wind power are {{0, 0.2213}, {4.725, 0.2197}, {12.175, 0.2094}, {23.625, 0.1591}, {33.075, 0.1001}, {42.525, 0.0531}, {45, 0.0374}}.

Different from wind power, output of PVs is concerned to fuzzy variable η_1. The parameter of fuzzy variable η_1 refers to Equation (5). Integrated with previously proposed method, the mean and variance of the equivalent normal distribution transformed from fuzzy variable are 0.4537 and 0.104. Binary sequences of fuzzy variable η_1 are {{0.1729, 0.0069}, {0.2353, 0.0278}, {0.2977, 0.0794}, {0.3601, 0.1596}, {0.4225, 0.2264}, {0.4849, 0.2264}, {0.5473, 0.1596}, {0.6097, 0.0794}, {0.6721, 0.0278}, {0.7345, 0.0069}}.

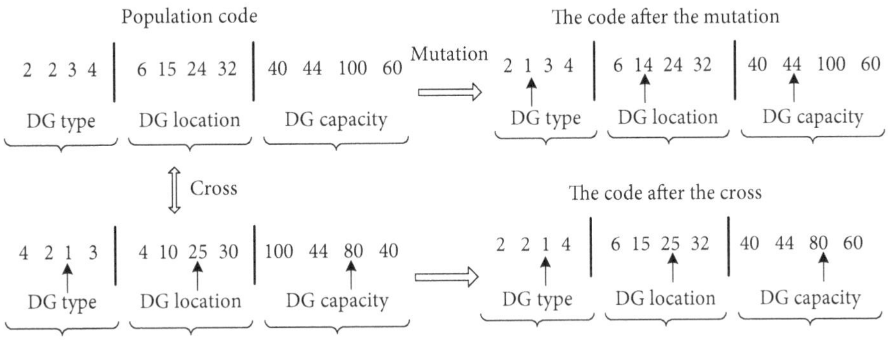

Figure 2: The segmented point cross and mutation.

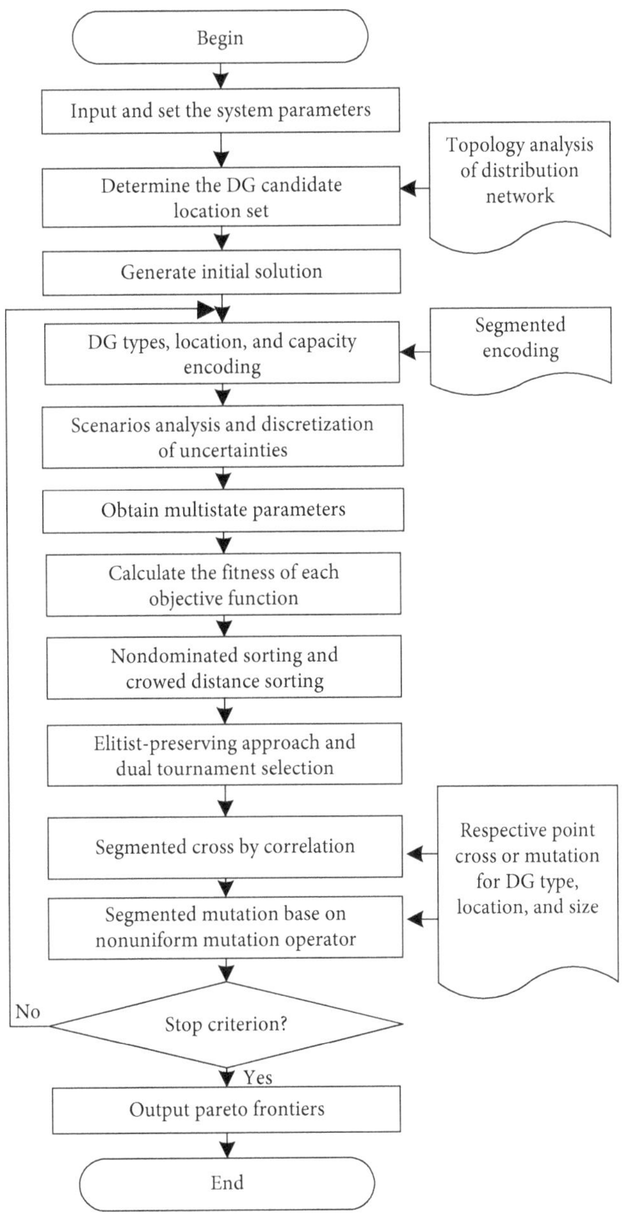

Figure 3: Flowchart for multiobjective optimization.

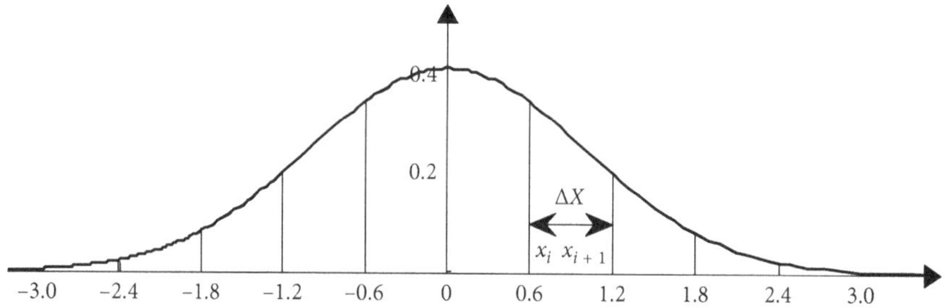

Figure 4: Schematic diagram for discretization of standardized normal distribution.

Table 1: State variables and probability of $N(0, 1)$ and $N_{l1}(0.5, 2)$.

State for $N(0, 1)$	State for $N_{l1}(0.5, 2)$	Probability	Corrected probability
−2.7	−4.9	0.0068	0.0068
−2.1	−3.7	0.0277	0.0278
−1.5	−2.5	0.0791	0.0794
−0.9	−1.3	0.1592	0.1596
−0.3	−0.1	0.2257	0.2264
0.3	1.1	0.2257	0.2264
0.9	2.3	0.1592	0.1596
1.5	3.5	0.0791	0.0794
2.1	4.7	0.0277	0.0278
2.7	5.9	0.0068	0.0068

In order to obtain proper composite state for PV power and wind power, plenty of simulation has been carried out provided that small probability event refers to less than 0.001 and confidence level is no less than 95%. The peak power $P_{\max}$ of PV is 44 kW. The load obeys a truncated Gaussian distribution, of which, lower and upper bounds are 28 kW and 35 kW and the mean and variance are 30 and 2. The parameters for wind are as above. The simulation results under different interval number are listed in Table 2.

Plenty of simulation shows a fact that the expected value for output power of PV, wind, and the load are affected by interval number. With the increase of interval number, the expected value gradually turns to be stable. However, another fact that should not be overlooked is that small probability events will go up sharply as interval number increases, e.g., in Table 2, there is 24 small probability events in interval number (5, 5, 5), while there exist 668 small probability events in (10, 10, 10) and even it amounts to 42561 in (35, 35, 35) which is far beyond effective joint-state 314. Too much small probability not only complicates calculation but also results in mass of invalid computation, which greatly decreases computational efficiency. Therefore, the moderate selection of interval number is very important. In this example, the interval number (5, 5, 5), (5, 5, 10), and (5, 5, 15) are recommended values because their small probability events are small and the expected value for PV, wind, and the load are at an acceptable level.

5.2. Simulation Parameters for Multiobjective Optimization Based on NSGA. To demonstrate the performance of the proposed optimization method, simulation is carried out on IEEE 37-bus system, whose topological structure is shown in Figure 5.

Table 2: The simulation results of different interval number.

Interval number of load, PV, wind	Effective joint-state number	Small probability events	Expected value of P_{load}, P_{PV}, P_{wind} (kW)
(5, 5, 5)	101	24	30.56, 19.96, 47.99
(10, 10, 10)	332	668	30.54, 19.96, 47.85
(15, 15, 15)	237	3138	30.54, 19.96, 47.83
(20, 20, 20)	198	7802	30.54, 19.96, 47.82
(25, 25, 25)	254	15371	30.54, 19.96, 47.82
(30, 30, 30)	296	26704	30.54, 19.96, 47.82
(35, 35, 35)	314	42561	30.54, 19.96, 47.81
(5, 5, 10)	167	83	30.56, 19.96, 47.85
(5, 5, 15)	217	158	30.56, 19.96, 47.83
(5, 5, 20)	264	236	30.56, 19.96, 47.82
(5, 5, 30)	343	407	30.56, 19.96, 47.82
(10, 20, 30)	170	5830	30.54, 19.96, 47.82

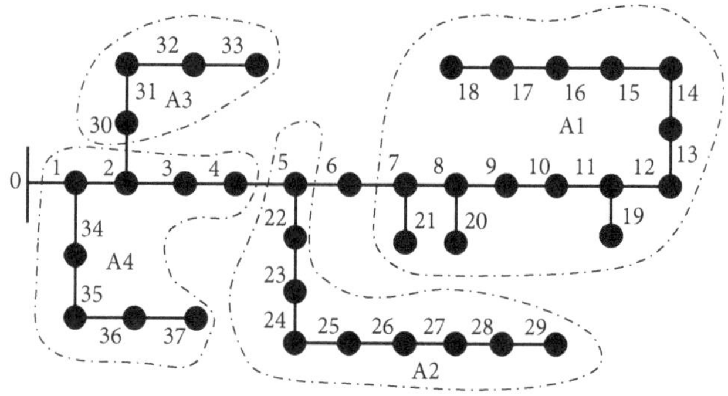

Figure 5: IEEE 37-bus system.

The branch, base load parameters can be found in [4, 9]. Meanwhile, assuming that load on bus 16 is uncertain variable which obeys normal distribution with mean $U_{16} = 0.06$ (p.u.) and standard deviation $\sigma = 0.01$. Buses 32 and 28 are taken as interval variables, whose values are orderly [0.4, 0.425], [0.205, 0.220], and others are taken as constant power load. The parameters for uncertain loads and information for single DG unit are listed in Tables 3 and 4. The parameters and probability distribution for uncertain loads are listed in Table 3. For DGs, detailed information is listed in Table 4. The power

TABLE 3: The parameters and probability distribution for uncertain loads.

Bus	State/interval (p.u.)	Probability distribution
16	0.036	0.0347
	0.048	0.2390
	0.06	0.4527
	0.072	0.2390
	0.084	0.0347
32	[0.400, 0.410]	0.1
	[0.410, 0.420]	0.5
	[0.420, 0.425]	0.4
28	[0.205, 0.210]	0.3
	[0.210, 0.215]	0.4
	[0.215, 0.220]	0.3

TABLE 4: Informations for different DG units.

DG	Bus type/base capacity	Failure rate (time/year)	Average interruption duration (hour/time)	Binary state {power, probability}
T1	PQ/100	5	50	{0, 0.2213} {15, 0.3138} {45, 0.2554} {75, 0.1383} {100, 0.0713}
T2	PQ/200	1	10	{180, 0.4} {190, 0.4} {200, 0.2}
T3	PV/44	2	25	{0.0818, 0.7359} {0.9086, 0.1432} {1.3844, 0.0659} {2.3743, 0.0347} {3.6179, 0.0202}
T4	PV/60	1	10	60

Note. Power factor of PQ DG is 0.8 and voltage magnitude of PV DG is 1.0.

TABLE 5: The objective function values for some Pareto front solutions.

No.	DG type	DG location	DG size	IP_{loss}	IV_{enha}	IR_{imp}	IE_{imp}
1	T2, T2, T1	14, 27, 37	200, 200, 100	0.0878	0.0009	0.9979	0.0523
2	T2, T3, T3, T2	13, 26, 31, 35	200, 44, 44, 200	0.3188	0.0011	0.9546	0.0721
3	T4, T3, T4, T1	9, 27, 35	360, 44, 100	0.3700	0.0007	0.9648	0.0742
4	T4	37	480	0.1579	0.0013	0.9450	0.0201
5	T1, T4	17, 24	400, 120	0.2546	0.0013	0.9345	0.0220
6	T2, T2, T4	13, 24, 32	200, 60, 180	0.2596	0.0012	0.9566	0.0238
7	T3, T4, T1, T3	14, 27, 32, 36	44, 300, 100, 44	0.3275	0.0006	0.9660	0.0732
8	T4, T3, T3	11, 28, 32	300, 176, 44	0.3606	0.0007	0.9458	0.0721

base is 10 MVA and convergence accuracy is 10^{-4}. Maximum and minimum limits of the nodal voltage are positive and negative 7% of the nominal voltage. The maximum penetration of DG is 10% of total base load, while the maximum number of DG installation N_{DG}^{max} is four. The fault rate on distribution feeder is 0.05 time per year, and average interruption duration each time is 5 hours. The parameters of NSGA are orderly 0.9 for select rate, 0.9 for crossover rate, 0.05 for mutation rate, 50 for chromosome numbers, and 30 for maximum iterations.

5.3. Simulation Results for Multiobjective Optimization Based on NSGA. The simulation is carried on Matlab 2008 rb. Based on the topological structure and the maximum N_{DG}^{max}, four areas are shown in Figure 4. DG candidate locations based on nodal electric distances are {18 16 17 19 15 14 20 12 13 21 29 27 28 9 11 10 26 25 8 24 37 33 32 7}. DG candidate locations based on the nodal load moment are {32 28 31 26 17 27 13 7 16 25 15 14 12 6 29 9 11 37}. Assuming that load bus of high-reliability requirements are {36 35 28 16}. Therefore, four candidate location sets are {16 17 15 14 13 12 11 9}, {28 27 25 26 24}, {33 32 31}, and {37 36 35}. Table 5 lists some Pareto front solutions at the 30th iterations. By analyzing simulations results, some conclusions are listed as following:

(i) The proposed multistate model and application of confidence level not only ensure the accuracy of results, but also reduces search space notably on the premise that the result is hardly affected.

(ii) If anyone of DG type, location, and size changes, four objective function values will all change. It is extremely difficult that four objective function values simultaneously arrive at the optimal at one solution. Therefore, when installing DG, policymakers has to combine the main object of individual decision to choose appropriates solution from the Pareto front solutions.

(iii) The application of DG candidate location set and genetic operators can reduce greatly search space of GA and improve computational efficiency of algorithm.

The proposed modified NSGA can solve efficiently constrained, nonlinear multiobjective optimization problem with multisource uncertainty. And the solution approach presented here is simple, reliable, and efficient.

6. Conclusion

From the view of DISOPER, this study presented a multiobjective optimization for DG planning considering with multisource uncertainty. The employment of information entropy and interval analysis can resolve the multisource uncertainty problems effectively. The constructed multiobjective formulations can not only reflect DG different influences on distributed network but also take care of the multisource uncertainty. Applications of small probability event and confidence level decreased greatly search space without affecting the accuracy of results. In optimizing processes, the developed DG candidate location set and genetic operators avoided redundant calculation to improve the efficiency. Furthermore, introduction of multiobjective optimization based on modified NSGA can get a comparative satisfying result as well as coordinate the conflict between various objects. Plenty of simulations also showed the proposed methodology is simple and reliable. And it is suitable for allocation of multisources and multitype DG in a given distribution networks under multisource uncertainty.

Conflicts of Interest

The authors declare that they have no conflicts of interest.

Acknowledgments

This work was supported in part by the Hebei Natural Science Foundation of China (no. F2015204090), the National Natural Science Foundation of China (NSFC) (no. 51377162), and the Science Research Project of Hebei Higher Education (nos. ZD2017036 and QN2016063).

References

[1] J. Smith, M. Rylander, L. Rogers, and R. Dugan, "It's all in the plans: maximizing the benefits and minimizing the impacts of DERs in an integrated grid," *IEEE Power and Energy Magazine*, vol. 13, no. 2, pp. 20–29, 2015.

[2] A. Maleki, F. Pourfayaz, and M. A. Rosen, "A novel framework for optimal design of hybrid renewable energybased autonomous energy systems: a case study for Namin, Iran," *Energy*, vol. 98, pp. 168–180, 2016.

[3] C. Yun, "Research for the grid-connected wind power generation and photovoltaic power problems," M.S. dissertation, Department of Electrical Engineering, Shanghai Jiaotong University, Shanghai, China, 2009.

[4] L. Zhang, *Research on distributed generators planning of distribution network and its comprehensive evaluation*, Ph.D. dissertation, China Agricultural University, Beijing, China, 2010.

[5] M. Thomson and D. G. Infield, "Network power-flow analysis for a high penetration of distributed generation," *IEEE Transactions on Power Systems*, vol. 22, no. 3, pp. 1157–1162, 2007.

[6] B. Zeng, J. Zhang, X. Yang, J. Wang, J. Dong, and Y. Zhang, "Integrated planning for transition to low-carbon distribution system with renewable energy generation and demand response," *IEEE Transactions on Power Systems*, vol. 29, no. 3, pp. 1153–1165, 2014.

[7] T. Gözel and M. H. Hocaoglu, "An analytical method for the sizing and siting of distributed generators in radial systems," *Electric Power Systems Research*, vol. 79, no. 6, pp. 912–918, 2009.

[8] C. Yammania, S. Maheswarapub, and S. Matam, "Multiobjective optimization for optimal placement and size of DG using shuffled frog leaping algorithm," *Energy Procedia*, vol. 14, pp. 990–995, 2012.

[9] D. Singh, D. Singh, and K. S. Verma, "Multiobjective optimization for DG planning with load models," *IEEE Transactions on Power Systems*, vol. 24, no. 1, pp. 427–436, 2009.

[10] W. Sheng, K.-Y. Liu, Y. Liu, X. Meng, and X. Song, "A new DG multiobjective optimization method based on an improved evolutionary algorithm," *Journal of Applied Mathematics*, vol. 2013, Article ID 643791, 11 pages, 2013.

[11] A. Ameli, S. Bahrami, F. Khazaeli, and M.-R. Haghifam, "A multiobjective particle swarm optimization for sizing and placement of DGs from DG Owner's and distribution Company's viewpoints," *IEEE Transactions on Power Delivery*, vol. 29, no. 4, pp. 1831–1840, 2014.

[12] O. Wu, H. Cheng, X. Zhang, and L. Yao, "Distribution network planning method considering distributed generation for peak cutting," *Energy Conversion and Management*, vol. 51, no. 12, pp. 2394–2401, 2010.

[13] H. Yu, C. Y. Chung, K. P. Wong, and J. H. Zhang, "A chance constrained transmission network expansion planning method with consideration of load and wind farm uncertainties," *IEEE Transactions on Power Systems*, vol. 24, no. 3, pp. 1568–1576, 2009.

[14] J. Zhang, H. Cheng, L. Yao, and C. Wang, "Study on siting and sizing of the distributed wind generation," *Proceedings of the CSEE*, vol. 29, no. 16, pp. 1–7, 2009.

[15] A. Khodaei, S. Bahramirad, and M. Shahidehpour, "Microgrid planning under uncertainty," *IEEE Transactions on Power Systems*, vol. 30, no. 5, pp. 2417–2425, 2015.

[16] A. Bagheri, H. Monsef, and H. Lesani, "Integrated distribution network expansion planning incorporating distributed generation considering uncertainties, reliability, and operational conditions," *International Journal of Electrical Power & Energy Systems*, vol. 73, pp. 56–70, 2015.
[17] M. F. Shaaban and E. F. El-Saadany, "Accommodating high penetrations of PEVs and renewable DG considering uncertainties in distribution systems," *IEEE Transactions on Power Systems*, vol. 29, no. 1, pp. 259–270, 2014.
[18] S. Ganguly and D. Samajpati, "Distributed generation allocation on radial distribution networks under uncertainties of load and generation using genetic algorithm," *IEEE Transactions on Sustainable Energy*, vol. 6, no. 3, pp. 688–697, 2015.
[19] H. Rong and Z. Pin-xian, "Load characteristics of fast charging station of electric vehicles," *Journal of Shanghai University of Electric Power*, vol. 28, no. 2, pp. 156–159, 2012.
[20] S. Li, T. Wei, M. Bai, and Z. Lu, "The planning of centralized charging stations locate and size considering load shifting in distribution network," *Proceedings of the CSEE*, vol. 34, no. 7, pp. 1052–1060, 2014.
[21] M. Esmaeili, M. Sedighizadeh, and M. Esmaili, "Multi-objective optimal reconfiguration and DG (Distributed Generation) power allocation in distribution networks using Big Bang-Big Crunch algorithm considering load uncertainty," *Energy*, vol. 103, pp. 86–99, 2016.
[22] M. Wang and L. Zeng, "Study of wind speed frenquence distribution model," *Journal of Hydroelectric Power*, vol. 30, no. 6, pp. 204–209, 2011.
[23] C. Wang, *Solar Photovoltaic Power Generation and Practical Technology*, Chemical Industry Press, Beijing, China, 2005.
[24] W. El-Khattam, Y. G. Hegazy, and M. M. A. Salama, "Investigating distributed generation systems performance using Monte Carlo simulation," *IEEE Transactions on Power Systems*, vol. 21, no. 2, pp. 524–532, 2006.
[25] A. Soroudi and M. Afrasiab, "Binary PSO-based dynamic multi-objective model for distributed generation planning under uncertainty," *IET Renewable Power Generation*, vol. 6, no. 2, pp. 67–78, 2012.
[26] S. Abapour, K. Zare, and B. Mohammadi-Ivatloo, "Dynamic planning of distributed generation units in active distribution network," *IET Generation, Transmission & Distribution*, vol. 9, no. 12, pp. 1455–1463, 2015.
[27] A. Rabiee and S. M. Mohseni-Bonab, "Maximizing hosting capacity of renewable energy sources in distribution networks: a multi-objective and scenario-based approach," *Energy*, vol. 120, pp. 417–430, 2017.
[28] M. Sedghi, A. Ali, and M. Aliakbar-Golkar, "Optimal storage planning in active distribution network considering uncertainty of wind power distributed generation," *IEEE Transactions on Power Systems*, vol. 31, no. 1, pp. 304–316, 2016.
[29] A. Soroudi, R. Caire, N. Hadjsaid, and M. Ehsan, "Probabilistic dynamic multi-objective model for renewable and non-renewable distributed generation planning," *IET Generation, Transmission & Distribution*, vol. 5, no. 11, pp. 1173–1182, 2011.
[30] M. F. Shaaban, Y. M. Atwa, and F. El-Saadany, "DG Allocation for benefit maximization in distribution networks," *IEEE Transactions on Power Systems*, vol. 28, no. 2, pp. 639–649, 2013.
[31] Z. Zhang-hua, A. I. Qian, C. Gu, and C. Jiang, "Multiobjective allocation of distributed generation considering environmental factor," *Proceedings of the CSEE*, vol. 29, no. 13, pp. 23–28, 2009.
[32] Z. Wang, B. Chen, J. Wang, and M. M. Begovic, "Stochastic DG placement for conservation voltage reduction based on multiple replications procedure," *IEEE Transactions on Power Delivery*, vol. 30, no. 3, pp. 1039–1047, 2015.
[33] H. Tekinera, D. W. Coitb, and F. A. Felder, "A sequential Monte Carlo model of the combined GB gas and electricity network," *Energy Policy*, vol. 62, pp. 473–483, 2013.
[34] K. Deb, A. Pratap, S. Agarwal, and T. Meyarivan, "A fast and elitist multiobjective genetic algorithm: NSGA-II," *IEEE Transactions on Evolutionary Computation*, vol. 6, no. 2, pp. 182–197, 2002.
[35] H. Wang and Y. Liu, "Multi2objective optimization of power system reconstruction based on NSGA2 II," *Automation of Electric Power Systems*, vol. 33, no. 23, pp. 14–18, 2009.
[36] H. Ahmadi and A. A. Foroud, "Design of joint active and reactive power reserve market: a multi-objective approach using NSGA II," *IET Generation, Transmission & Distribution*, vol. 10, no. 1, pp. 31–40, 2016.
[37] M. karimi and M. R. Haghifa, "Risk based multi-objective dynamic expansion planning of sub-transmission network in order to have eco-reliability, environmental friendly network with higher power quality," *IET Generation, Transmission & Distribution*, vol. 11, no. 1, pp. 261–271, 2017.
[38] I. G. Kerdana, R. Raslan, R. Paul, and D. M. Gálvez, "A comparison of an energy/economic-based against an exergoeconomic-based multi-objective optimisation for low carbon building energy design," *Energy*, vol. 128, no. 1, pp. 244–263, 2017.
[39] A. Kamjoo, A. Maheri, A. M. Dizqah, and G. A. Putrus, "Multi-objective design under uncertainties of hybrid renewable energy system using NSGA-II and chance constrained programming," *International Journal of Electrical Power & Energy Systems*, vol. 74, pp. 187–194, 2016.
[40] D. Cheng, D. Zhu, R. P. Broadwater, and S. Lee, "A graph trace based reliability analysis of electric power systems with time-varying loads and dependent failures," *Electric Power Systems Research*, vol. 79, no. 9, pp. 1321–1328, 2009.
[41] K. Zou, A. Prakash Agalgaonkar, K. M. Muttaqi, and S. Perera, "Distribution system planning with incorporating DG reactive capability and system uncertainties," *IEEE Transactions on Sustainable Energy*, vol. 3, no. 1, pp. 112–123, 2012.
[42] M. R. Patel, Q. Jiang, C. P. Zhang, and L. Hong, *Wind and Solar Power Systems Design, Analysis, and Operation*, China Machine Press, Beijing, China, 2008.
[43] B. P. Dougherty, A. H. Fanney, and M. W. Davis, "Measured performance of building integrated photovoltaic panels—round 2," *Journal of Solar Energy Engineering*, vol. 127, no. 3, pp. 314–323, 2005.
[44] A. Haldar and R. K. Reddy, "A random-fuzzy analysis of existing structures," *Fuzzy Sets and Systems*, vol. 48, no. 2, pp. 201–210, 1992.
[45] L. Xia, J. Mao, and Y.-Q. Tang, "Comprehensive evaluation and analysis of the city facility level based on fuzzy entropy," *Computer Technology and Development*, vol. 20, no. 7, pp. 25–27, 2010.
[46] Z. Wang, "Research and application of discretization algorithm," M.S. dissertation, Dalian University of Technology, Dalian, China, 2009.
[47] J. E. Mendoza and H. E. Pena, "Automatic voltage regulators siting in distribution systems considering hourly demand," *Electric Power Systems Research*, vol. 81, no. 5, pp. 1124–1131, 2011.

[48] C. Liu and Y. Zhang, "Distribution network reliability considering distribution generation," *Automation of Electric Power Systems*, vol. 31, no. 22, pp. 46–49, 2007.
[49] L. Zhang, W. Tang, Y. Liu, and T. Lv, "Multiobjective optimization and decision-making for DG planning considering benefits between distribution Company and DGs owner," *International Journal of Electrical Power & Energy Systems*, vol. 73, pp. 465–474, 2015.
[50] S. H. I. Wei-guo, P. Song, and L. I. U. Chuan-quan, "Power supply reliability of distribution network including distributed generation," *East China Electric Power*, vol. 35, no. 7, pp. 37–41, 2007.
[51] L. Zhang and W. Tang, "Back/forward sweep power flow calculation method of distribution networks with DGs," *Transactions of China Electrotechnical Society*, vol. 25, no. 8, pp. 123–130, 2010.
[52] L. Cai and L. Xia, "Improvement on crossover operation of genetic algorithms," *Systems Engineering and Electronics*, vol. 28, no. 6, pp. 925–928, 2006.
[53] W. Hua, J.-H. Zheng, and L. Mi-qing, "Comparison and research of mutation operators in multi-objective evolutionary algorithms," *Computer Engineering and Applications*, vol. 45, no. 2, pp. 74–78, 2009.

13

A Novel Hybrid Method for Short-Term Power Load Forecasting

Huang Yuansheng, Huang Shenhai, and Song Jiayin

Department of Economic Management, North China Electric Power University, Baoding, China

Correspondence should be addressed to Huang Shenhai; hshncepu@163.com

Academic Editor: Jit S. Mandeep

Influenced by many uncertain and random factors, nonstationary, nonlinearity, and time-variety appear in power load series, which is difficult to forecast accurately. Aiming at locating these issues of power load forecasting, an innovative hybrid method is proposed to forecast power load in this paper. Firstly, ensemble empirical mode decomposition (EEMD) is used to decompose the power load series into a series of independent intrinsic mode functions (IMFs) and a residual term. Secondly, genetic algorithm (GA) is then applied to determine the best weights of each IMF and the residual term named ensemble empirical mode decomposition based on weight (WEEMD). Thirdly, least square support vector machine (LSSVM) and nonparametric generalized autoregressive conditional heteroscedasticity (NPGARCH) are employed to forecast the subseries, respectively, based on the characteristics of power load series. Finally, the forecasted power load of each component is summed as the final forecasted result of power load. Compared with other methods, the forecasting results of this proposed model applied to the electricity market of Pennsylvania-New Jersey-Maryland (PJM) indicate that the proposed model outperforms other models.

1. Introduction

In the operation of power system, power load forecasting is not only an important part of power system planning but also one of the most influential factors for the improvement of social economy developing, which has a significant impact on generation, transmission, and distribution. Therefore, it is crucial to design an efficient method to forecasting power load accurately by improving both economic efficiency and power supply quality and enhancing reliability of power system operation [1, 2].

However, the power load forecasting results are influenced by many uncertain random factors, including changes of nature environment and electricity price and factors influencing power load [3, 4], causing the complex volatility characteristics of power load, which lead the time-series of power load series to inherent nonstationary and discrete in practice, even in the short-term dynamics. Notably, due to the factors, influencing the short-term power load, previous studies [5–7] mainly focus on the establishment of scientific forecasting method to enhance the adaptability for these multifactors. Therefore, the time-varying components of power load are hardly separated to extract the information in them.

Under this circumstance, a novel hybrid method for power load forecasting has been proposed in this paper. Based on artificial algorithm, the advantages of econometric models and signal processing are emphasized by this research. Specifically, this hybrid method for load power forecasting includes ensemble empirical mode decomposition based on variable weights (WEEMD), genetic algorithm (GA), least square support vector machine (LSSVM), and nonparametric generalized autoregressive conditional heteroscedasticity (NPGARCH). Notably, in order to test the validity and feasibility of the model, amount of historical data of power load in American Electric Power (AEP) has been adopted to apply this new hybrid method and compare it with previously well-known methods for the power load forecasting accuracy. Results indicate that this hybrid method outperforms the compared methods with the forecasting accuracy.

The rest of the paper is organized as follows: Section 2 summaries previous literature over methodology for power load forecasting; Section 3 briefly describes EEMD, GA, nonparametric GARCH, and LSSVM and proposes the improved WEEMD, as well as presenting the procedure of the newly hybrid method; Section 4 evaluates the forecasting accuracy and tests the robustness of the proposed method, by comparing with other forecasting methods. Section 5 draws conclusions.

2. Related Literature Review

Current power load forecasting methods can be broadly divided into two categories: statistical analysis method [8–11] and artificial algorithm [12–15]. The traditional statistical analysis methods are always under the framework of time-series analysis [16]. By using ARMA and ARMA with weather inputs, Bolzern and Fronza [8] propose two short-term predictors of the winter power load. Respectively, the power load forecast results are satisfied with both predictors and similar in these two cases. In order to model certain behavior of energy consumption, Wang [9] employs conventional fuzzy systems based on the integrated algorithm, and results show that the monthly electricity consumption of Iran is accurately forecasted with this proposed method. In [10], considering the nonlinearity and volatility of power load series, analyzing the threshold characteristics, Wang et al. adopt a novel double-threshold generalized autoregressive conditional heteroscedasticity (DT-GARCH) model for short-term power load forecasting. To forecast the power load efficiently and scientifically, Liu [11] synthesizes the time-series modeling with the regression modeling. According to the residual error from regression forecasting, the accuracy of power load forecasting can be improved by correcting the error of GARCH model.

As the artificial algorithm applied in the power load forecasting is prevalent in recent decades, there are plenty of researches related to that. Taking the characteristics of randomness, tendency, and periodicity of short-term power load into account, Sun and Ye [12] propose a model based on LSSVM and fruit fly algorithm (FOA) for short-term load forecasting. In order to reduce the nonlinear power load sequence and improve the accuracy of forecasting result, Hu and Chang [13] decompose the time-series into several components by local wave method and use optimal parameters to establish the DEMD-LSSVM model for component forecasting. Lauret et al. [14] utilize Bayesian techniques to design an optimal neural network (NN) based model for power load forecasting, where Bayesian technique offers great advantages on traditional neural network learning methods. Based on an artificial neural network (ANN), Shao et al. [15] proposed an approach combined with a fuzzy system for short-term power load forecasting. The load error is obtained from the historical information and past forecasted load errors are caused by fuzzy systems, and the final forecasted load can be obtained by adding the load error to preliminary load forecasted by ANN. In [16], to improve the forecasting performance by searching a suitable parameters combination, the paper presents an SVR-based electric load forecasting model by applying a novel algorithm named chaotic ant swarm optimization (CAS). Combined with the chaotic behavior of single ant and self-organization behavior of ant colony in the foraging process, the CAS is proposed to overcome premature local optimum. In [17], a combination of the wavelet transform (WT) and gray model is proposed for short-term power load forecasting, which is improved by particle swarm optimization (PSO). With this proposed method taking mean temperature, mean relative humidity, mean wind speed, and previous days load data into consideration and eliminating the high frequency of historical data built by the WT, the accuracy can be largely improved.

With the improvement of the forecasting accuracy, Ghelardoni et al. [18] decompose time power load series with empirical mode decomposition (EMD) into two sets of components, respectively, describing the trend of energy consumption values. LSSVM is built to forecast these two components. The hybrid method proposed in [19] enhances the capability for forecasting. This proposed method is modeled based on supporting vector regression (SVR), EMD, and regression (AR), which can simultaneously provide forecasting with high accuracy and interpretability. In order to solve the problem of mode mixing and high frequency random components, Liu et al. [19] proposed an optimized method based on ensemble empirical mode decomposition (EEMD) and subsection particle swarm optimization (SS-PSO). By extracting and reconstructing intrinsic mode function (IMF), the power load series movement is well forecasted. In [20], the proposed method based on complementary ensemble empirical mode decomposition (CEEMD) fuzzy entropy and echo state network (ESN) with leaky integrator neurons (LiESN) enhances the forecasting accuracy of power load.

In summary, although the characteristics of power load series are complex, such as being rich multidimensional, nonlinear, time-varying, and nonstationary, the previous literature has accumulated a great deal of experience about power load forecasting. Particularly, the forecasting methods have been constantly proposed and the forecasting performance has been continuously improved, all of which provide a significant foundation for the present study. However, we can still find that the method of power load forecasting can be problematical based on the analysis and research of previous literature, while there are also some uncertain factors of power load forecasting. For example, both EEMD and EMD decompose the original power load series into several components, and other forecasting methods are directly built based on the decomposed components in most of the literature, without analysis of the weights of each component in original power load series. Besides, as the power load is affected by great uncertainties, the power load series movement is difficult to capture, and there will be nonlinear and time-varying feature in power load series caused by no matter the external factors or the internal factors. But the existing methods for power load forecasting are usually not effective to separate nonlinear and time-varying components of power load and unable to extract their inherent moving mechanisms, which consequently affects the forecasting accuracy.

Based on the potential faultiness mentioned above, a hybrid method for power load forecasting is developed and applied to PJM by this study. Specially, the ensemble empirical mode decomposition (EEMD) model decomposes power load into series of intrinsic mode functions (IMFs) and residual. And genetic algorithm (GA) is used to determine the weights of each component, which is based on deviation between fitted values and actual values. For weighted components with the heteroscedasticity character, GARCH is more favorable for forecasting. However, the LSSVM method is used in the forecasting of the subseries with the characteristics of nonstationary and nonlinearity subseries. On the one hand, power load forecasting can be more convenient and accurate, after the decomposition of the power load components because of the separation of other influential factors. On the other hand, with other neural network models, which cannot fundamentally solve problems of the local minimum, the difficulty in determination of hidden layer, and the slow training rate, the LSSVM can not only get over these disadvantages, but also can improve the accuracy of forecasting. Therefore, based on the components weighted by GA method, LSSVM method can be more in line with the power load study in this research.

3. Power Load Forecasting Methodology

3.1. Genetic Algorithms (GA). Genetic algorithms code the candidate solutions of an optimization algorithm as a string of characters which are usually binary digits [21]. In accordance with the terminology that is borrowed from the field of genetics, this bit string is usually named as chromosome. The solution represented by its chromosome is considered as an individual. The algorithm starts with the initial generation of the population. The fitness of the individuals within the population is assessed, and new individuals are generated for the further generation. A number of genetic operators, containing selector operator, crossover, and mutation, are available for this purpose. By a number of fixed generations which is the termination condition, the best individual will be obtained with the max fitness, which is the global optimal solution of the issues.

3.2. The WEEMD Method. EMD, an effective method for signal processing, is gradually replaced by EEMD which overcomes the mixing model problem [22–25]. The essence of EEMD is to decompose a time-series into a set of independent intrinsic mode functions (IMFs) and the residue obtained by adding a random Gaussian white noise sequence, which is different from EMD. While the time-series is decomposed by EEMD, the IMF is a function, satisfying the following two conditions:

(1) In the whole data set, the number of extreme and the number of zero crossings must either differ or differ at most by one.

(2) At any point, mean values of the envelope, defined by both the local maxima and minima, are zero.

For an arbitrary time-series, $X(t)$, procedures of EEMD method can be described as follows:

(1) Add the white noise to power load series $X(t)$, with $i = 1$, and set the number of ensemble (M):

$$X_i(t) = X(t) + n_i(t), \tag{1}$$

where $n_i(t)$ denotes the ith added white noises series and $X_i(t)$ represents the noise-added power load of the ith trial.

(2) Decompose the noise-added series $X_i(t)$ into J IMFs c_{ij} $(j = 1, 2, \ldots, J)$ by using EMD, where c_{ij} is the jth IMF of the ith trial and J is the number of IMFs.

(3) Repeat Steps (1) and (2) until $i = M$.

(4) Calculate the ensemble mean $c_j(t)$ of N trails for each IMF, then $c_j(t) = (1/M)\sum_{i=1}^{M} c_{ij}(t)$, where $c_j(t)$ $(j = 1, \ldots, J)$ is the jth IMF component by using EEMD.

In this paper, weights assignment, for each IMF, is proposed based on EEMD, and the rationality of this weights assignment proved by this study is as follows: for an arbitrary time-series, $X(t)$, setting any value of each IMF satisfies the function $f(t)$, and by the conditions mentioned above, the number of extreme and zero crossings of each IMF is q, which are obtained by the following equations:

$$f_j(t) = 0, \qquad \frac{df_j(t)}{dt} = 0. \tag{2}$$

Assigning weight to each IMF, then

$$\omega_j f_j(t) = 0, \qquad \frac{\omega_j df_j(t)}{dt} = 0, \tag{3}$$

where ω_j, a constant, is the weights of jth IMF.

Obviously, the weights have no effect on the values, which satisfies the first conditions. Specially, mean values of the envelope will not be changed by weight assignment, and the deduction can be expressed as follows:

$$m_{\max} + m_{\min} = 0. \tag{4}$$

Assigning weight to it,

$$\omega_j(m_{\max} + m_{\min}) = 0, \tag{5}$$

where $m_{\max}$ is the mean value of the envelope of the local maxima and $m_{\min}$ is the mean values of the envelope of the local minima.

GA method is built to determine the weights of each IMF, and individuals $W = (\omega_1, \omega_2, \ldots, \omega_J)$ are first randomly generated as initial population W_0. By using the genetic operators, selector operator, crossover, and mutation, the best individual $W = (\omega_1, \omega_2, \ldots, \omega_J)$ can be obtained. Specially, in the selector operator, the fitness function is defined as follows:

$$\begin{aligned}\varepsilon_k(t) &= \left|W_k \cdot C - X(t)\right| \\ &= \left|\left(\omega_{k1}, \omega_{k2}, \ldots, \omega_{kJ}, \omega_r\right) \cdot \begin{pmatrix} c_1 \\ c_2 \\ \vdots \\ c_J \end{pmatrix} - X(t)\right| \\ &= \left|\sum_{j=1}^{J} \omega_{kj} c_j(t) - X(t)\right|, \quad (k = 1, 2, \ldots, m),\end{aligned} \tag{6}$$

where $\varepsilon_k(t)$ is error and m is the size of initial population.

As the GA operator is designed to maximize the fitness function, the above minimization problem can be solved by using the following transformation:

$$f_k = \frac{1}{\varepsilon_k(t)}, \tag{7}$$

where f_k is the fitness of kth individual.

The election probability of each individual is p_k:

$$p_k = \frac{f_k}{\sum_{k=1}^{m} f_k}. \tag{8}$$

As described above, the process of EEMD is indeed like sifting, which has an effect to eliminate riding waves. The IMFs are extracted from the power load series and contain "information" about the time-series. This paper uses weights as the contribution of IMF to $X(t)$; the greater the weight is, the larger the amount of "information" IMF contains.

3.3. The Nonparametric GARCH (1, 1) Method. Nonparametric GARCH (1, 1) model for the error fluctuation [26], which requires less assumptions, is defined as

$$\begin{aligned}\hat{z}_t &= \sigma_t \varphi_t, \\ \sigma_t^2 &= g\left(\sigma_{t-1}^2, \hat{z}_{t-1}\right),\end{aligned} \tag{9}$$

where $\hat{z}_t$ is estimation of random processes; φ_t is specific transformation sequence of zero mean and unit variance. σ_t^2 is the conditional variance. σ_t is the volatility.

Nonparametric model can be used to estimate the conditional variance, and (9) can be rewritten as follows:

$$\begin{aligned}\hat{z}_t^2 &= g\left(\sigma_{t-1}^2, \hat{z}_{t-1}\right) + V_t, \\ V_t &= g\left(\sigma_{t-1}^2, \hat{z}_{t-1}\right)\left(\varphi_t^2 - 1\right),\end{aligned} \tag{10}$$

where V_t is the martingale difference sequence.

According to the equations above, the function $\hat{z}_t^2$ is regressed between lagged variables $\hat{z}_{t-1}$ and $\sigma_{t-1}^2 g$ can be estimated by using nonparametric smoothing method and the autoregression function. The calculation of nonparametric GARCH [27–30] is as follows:

(1) Using parameters GARCH (1, 1) model fits the volatility $\hat{\sigma}_{t,0}$, with estimation using maximum likelihood estimate parameter being employed as parameter estimation, where $a = 1$.

(2) $\sigma_{t,m}^2$ is as weight, and $\hat{z}_t^2$ is estimated by using z_{t-1} and $\hat{\sigma}_{t-1,m-1}^2$. Smooth nonparametric estimation method is applied to obtain the estimation $\hat{g}_a$ of autoregression function g.

(3) Standard deviation is obtained by using

$$\hat{\sigma}_{t,a}^2 = g_a\left(\hat{\sigma}_{t-1,m-1}^2, \hat{z}_{t-1}\right), \quad 2 \le t \le n. \tag{11}$$

(4) Increment m and return to step (2) if $a > A$, where M is a prespecified maximum number of iterations.

The nonparametric estimation is an improvement over the parametric GARCH estimation of volatility. By means of step (1)–(4) continuous iteration, there is little to pick and choose between volatility estimates $\hat{\sigma}_{t,a}^2$ for various values of a. However, the algorithm can often be improved by averaging over the final k estimates to obtain

$$\hat{\sigma}_{t,*} = \sum_{a=A-k+1}^{A} \frac{\hat{\sigma}_{t,a}}{k}. \tag{12}$$

3.4. The LSSVM Method. LSSVM is proposed as an improved algorithm based on support vector machine (SVM) [31–35], with the given training data set $\{z_i, y_i\}$ $(i = 1, 2, \ldots, N)$, with the input $z_i \in R^n$ and the output $y_i \in R^n$. The following regression model is constructed by using nonlinear mapping function $\phi(z_i)$:

$$f(z_i) = W^T \phi(z_i) + b. \tag{13}$$

With the given training data set $\{z_i, y_i\}$ $(i = 1, 2, \ldots, N)$, the optimization problem of LSSVM is defined as follows:

$$\begin{aligned}\min \quad & J(W, \xi_i) = \frac{1}{2} W^T W + \frac{C}{2} \sum_{i=1}^{M} \xi_i, \\ \text{s.t.} \quad & y_i = W^T \varphi(z_i) + b + \xi_i, \quad i = 1, 2, \ldots, n.\end{aligned} \tag{14}$$

According to the Kuhn-Tucker conditions, the LSSVM regression model can be expressed as

$$y(z_i) = \sum_{i=1}^{n} a_i k(z, z_i) + b. \tag{15}$$

$k(z, z_i)$ is the kernel function, which can map variables to the feature space and avoid high dimensional complex difficulties. This paper applies RBF as the kernel function, which is defined as

$$k(x, x_i) = \exp\left[-\frac{\|x - x_i\|^2}{(2\sigma^2)}\right], \tag{16}$$

where σ is the kernel function parameter. The LSSVM method can be used by establishing the parameters σ and c.

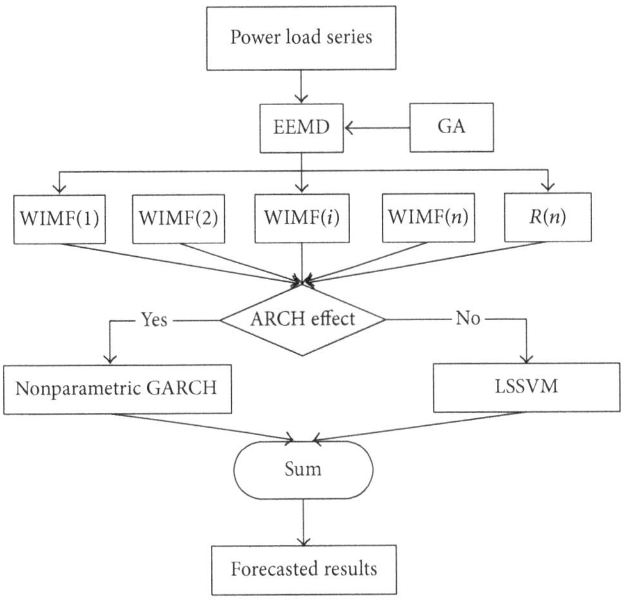

FIGURE 1: The procedures of power load forecasting using the novel hybrid model.

3.5. The Hybrid Method for Power Load Forecasting. Considering the complex volatility characteristics of power load series, much more scientific forecasting models are required to address the nonlinearity and time variations. Under this circumstance, the WEEMD, based on EEMD, is proposed to extract different components of power load series and assign weights to each component (IMF) according to its standard deviation, where the LSSVM is presented to forecast the subseries with the characteristics of nonstationary and nonlinearity, and the nonparametric GARCH (1, 1) is used to forecast the subseries with heteroscedasticity. With this hybrid model, the power load movement can be well forecasted. The procedures of the improved model can be described as Figure 1 and the concrete steps are given as follows:

(1) The power load series is first decomposed by EEMD into n intrinsic mode functions (IMFs) and one residual series, and then

$$X(t) = \sum_{j=1}^{J-1} C_j(t) + R(t), \tag{17}$$

where $X(t)$ is the original power load series and $C_j(t)$ and $R(t)$ are decomposed from the series.

(2) Each IMF series and the residual series are assigned to be weighted by GA, which can be represented as follows:

$$X(t) = \sum_{j=1}^{J-1} \omega_j C_j(t) + \omega_r R(t). \tag{18}$$

(3) To verify the existence of ARCH effect and heteroscedasticity, ARCH-LM is used to test the subseries which is related to the stochastic error, and then

$$X(t) = \sum_{j=1}^{m} \omega_j N_j(t) + \sum_{i=m+1}^{J} \omega_i S_i(t), \tag{19}$$

where $N_j(t)$ is the subseries with heteroscedasticity and $S_i(t)$ represents the subseries without heteroscedasticity.

(4) The LSSVM is built to forecast the future values of $S_i(t)$; meanwhile, the nonparametric GARCH model is presented to forecast the future values of $N_j(t)$, and their forecasted results are $\widehat{S}_i(t)$ and $\widehat{N}_j(t)$, respectively, which can be represented as follows:

$$\widehat{X}(t) = \sum_{j=1}^{m} \omega_j \widehat{N}_j(t) + \sum_{i=m+1}^{J} \omega_i \widehat{S}_i(t), \tag{20}$$

where $\widehat{X}(t)$ is the forecasted value of power load series.

To examine the proposed hybrid method performance, three criteria—mean absolute error (MAE), mean absolute percentage error (MAPE), and root mean square error (RMSE)—are represented as follows:

$$\begin{aligned} \text{MAE} &= \frac{1}{N}\sum_{t=1}^{N}\left|O_t - \widehat{O}_t\right|, \\ \text{MAPE} &= \frac{1}{N}\sum_{t=1}^{N}\frac{\left|O_t - \widehat{O}_t\right|}{O_t}, \\ \text{RMSE} &= \sqrt{\frac{1}{N}\sum_{t=1}^{N}\left(O_t - \widehat{O}_t\right)^2}, \end{aligned} \tag{21}$$

where O_t and $\widehat{O}_t$ represent the real and forecasted values, respectively, and N is the number of the forecasting performance evaluations.

4. Case Studies

Pennsylvania-New Jersey-Maryland (PJM), reliable operations and efficient wholesale market, is a fair and efficient electricity market, which provides information on electrical field. For our simulation, the short-term load power data from American Electric Power (AEP) are obtained hourly from 1/5/2015 to 10/8/2015. Data points from 1/5/2015 to 31/7/2015 are selected as training samples, and the data from 1/8/2015 to 10/8/2015 are selected as the test sample. Figure 2 presents the power load of the training sample, which shows that the power load severely fluctuates periodically.

4.1. Data Processing for IMF. According to procedure proposed above, first, the original power load series is decomposed by the EEMD into seven independent intrinsic mode functions and one residual, which are shown in Figure 3.

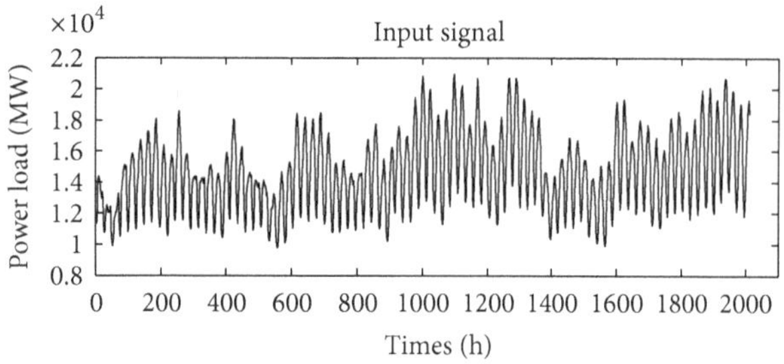

FIGURE 2: Power load changes from 1/5/2015 to 31/7/2015.

In order to assign weights for each IMF, GA method is used to select the best individual. In Figure 4, the results show that the average fitness is 0.545, while the algorithm iterates to 745 times, which do not change in later iterations, and this individual is the best one. Besides, the decomposition of the power load series with assigned weights is shown in Figure 5. Obviously, the assigned weights will not change the movement of the power load series but the domain which will be significant to the predicted values.

Considering the impact caused by the closer prices on the further data, three training samples near the forecasting point are selected as input variable:

$$X = \begin{pmatrix} x(d,t-3) \\ x(d,t-2) \\ x(d,t-1) \end{pmatrix}, \tag{22}$$

where $x(d,t)$ is the price on dth at time tth. Actually, the power load series is generally nonstationary, which will affect the forecasting results of LSSVM without handling. So, ARCH-LM test is proposed to resolve this problem. 1-hour return of the power load is calculated as $R = \ln(x_t/x_{t-1})$. Figure 6 provides the distribution of 1-hour return series from 1/5/2015 to 31/7/2015. Obviously, volatility cluster appeared in the residual, and the variance in the area of a is larger than b.

ARCH-LM test based on AR (3) is used to quantitatively test the heteroscedasticity, and WIMF1 is taken as an example to test the heteroscedasticity.

In Table 1, obviously, the significance of coefficients is close to zero, which indicates that significant autocorrelation appears in WIMF1 with Lagl, Lag2, and Lag3 and displays autocorrelation characteristic of WIMF1. Therefore, ARCH-LM test is applied to exam the conditional heteroscedasticity of returns series, and the result of test is shown in Table 2.

The F-statistic and the Lagl of residual in Table 2 is under the significance level of 0.005. So the assumption of ARCH is not accepted. It is the WIMF1 which illustrates the heteroscedasticity.

Each WIMF is tested by the same method; generally, the subseries of WIMF1, WIMF2, and WIMF5 have heteroscedasticity characteristic; and without the notable heteroscedasticity, there are WIMF3, WIMF4, WIMF6, WIMF7, and WIMF8. Then nonparametric GARCH model is established to forecast the power load. A large number of domestic and international demonstration analyses have shown that. GARCH (1, 1) can accurately describe the fluctuation

TABLE 1: Autocorrelation test.

	t-statistic	Prob.
Lagl	−22.61395	0.0000
Lag2	−30.69547	0.0000
Lag3	−18.65080	0.0000

TABLE 2: Test of conditional heteroscedasticity of returns series.

F-statistic	4.571244	Prob.	0.0034
	t-statistic	Prob.	
Lagl	3.281471	0.0010	
Lag2	0.723646	0.4694	
Lag3	−1.499284	0.1339	

TABLE 3: Parameter of GA.

Main parameter	Value
Initial population size	10
Evolution iteration	1000
Crossover probability	0.4
Mutation probability	0.05

characteristics of the model, and therefore this paper adopts nonparametric GARCH (1, 1) to imitate the power load series.

4.2. Result Analysis of the Hybrid Forecasting Method. The performance of LSSVM relays on the parameters of γ and σ^2, which, respectively, represent the regulation parameter and kernel parameter. A large number of domestic and international demonstration analyses indicate that two parameters are experientially determined. Based on a number of tests, this paper adopts 100 for γ and 0.1 for σ^2. The main parameters in GA are listed in Table 3.

In Figure 7, the error caused by the hybrid forecasting method is clearly shown. The error movement stably changes, the maximum of relative error should be no more than 10%, and meanwhile, the MAE, MAPE, and RMSE are 244.469, 1.56, and 407.38, respectively, which mean that the forecasting results are acceptable.

4.3. Comparative Analysis

4.3.1. Hourly Power Load Forecasting Analysis. To demonstrate the forecasting performance of the novel hybrid method, LSSVM, BPNN, EEMD plus LSSVM, EEMD plus BPNN, and GA-LSSVM are employed as the comparative methods, which are shown in Figure 8. Table 4 summarizes the values of the three error criteria, including MAE, RMSE, and MAPE, and the forecasted results of the six methods show that, using the proposed hybrid method, the power load series forecasted errors can be accepted. Notably, the MAE is less than 2%, and meanwhile, it is evident that MAE and RMSE are lower than the other methods, which implies that the forecasting accuracy of the proposed method

Table 4: The error of power load forecasting using different methods based on hourly observation.

	Hybrid method	LSSVM	BPNN	WEEMD-LSSVM	GA-LSSVM	EEMD-LSSVM
MAE (MW)	244.469	1313.7	1812.2	622.200	1083.4	909.020
MAPE (%)	1.56	8.44	11.81	4.05	7.12	6.08
RMSE (MW)	407.308	1552.5	2089.4	720.567	1220.7	1031.3

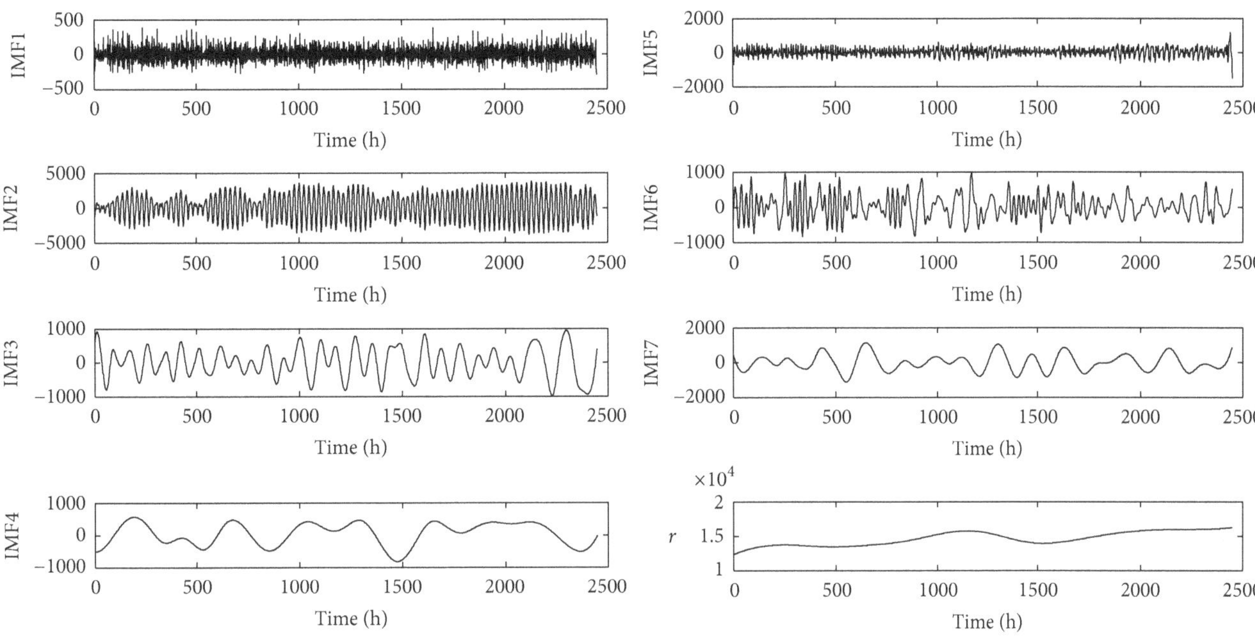

Figure 3: The decomposition of the power load series.

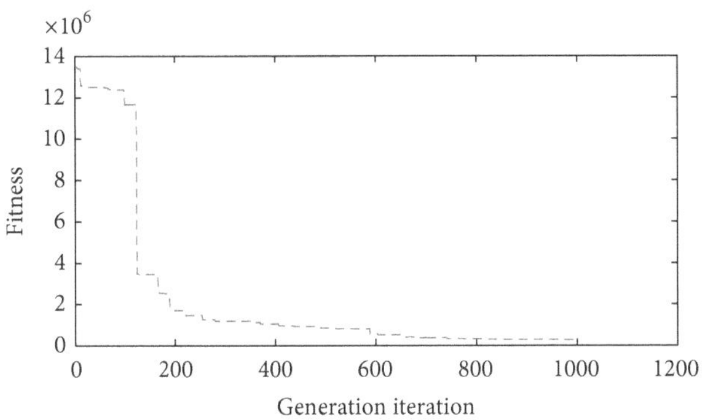

Figure 4: Fitness curve.

appears better than the comparative methods. Compared with the MAPE of the hybrid method 1.56%, the suboptimum with the MAE 4.05% is worse than the proposed method, and the result indicates that the hybrid method, by using nonparametric GARCH (1, 1) to forecast the subseries with heteroscedasticity, has well captured the time-varying volatility features of the power load series. Meanwhile, by the results of EEMD-LSSVM and WEEMD-LSSVM, it is obvious that assigning weights to each IMF improves the forecasting accuracy. Besides, it verifies that EEMD method decomposes power load series to constitutive subseries forecasted more accurately than original series by directly comparing LSSVM with EEMD. Generally, the forecasted results of the proposed method are reasonable and much more accurate than the other method based on the hourly observations.

4.3.2. Daily Power Load Forecasting Analysis. As the data frequency is a significant factor for the sensitivity of the time-series forecasting, to examine the robustness of the hybrid method this study adopts a daily observation method to forecast the power load. And the forecasted power load is decomposed to ten parts shown in Figure 9.

Table 5 shows the errors of forecasted results among different methods, and the MAE, MAPE, and RMSE of the hybrid method can be accepted with smaller MAE and RSME values and the MAPE is less than 1% comparing with the other methods, which indicates that the hybrid method has a better performance than other five methods. Besides, due to the differences in their characteristics, the forecasting accuracy can be improved and clustered by using WEEMD method. Hence, this newly proposed hybrid method for power load forecasting in this paper has relatively reliable robustness with respect to the data frequency.

5. Conclusions

To address the problem of power load forecasting with the characteristic of nonstationary, nonlinearity, and time-varying, this paper proposes a novel hybrid method for power

TABLE 5: The error of power load forecasting using different methods based on daily observation.

	Hybrid method	LSSVM	BPNN	WEEMD-LSSVM	GA-LSSVM	EEMD-LSSVM
MAE (MW)	1441.263	13343.236	16027.205	5023.101	10623.505	10729.767
MAPE (%)	0.389	3.488	4.146	1.313	2.749	2.854
RMSE (MW)	1845.988	17460.604	22325.684	6561.656	14855.986	13914.859

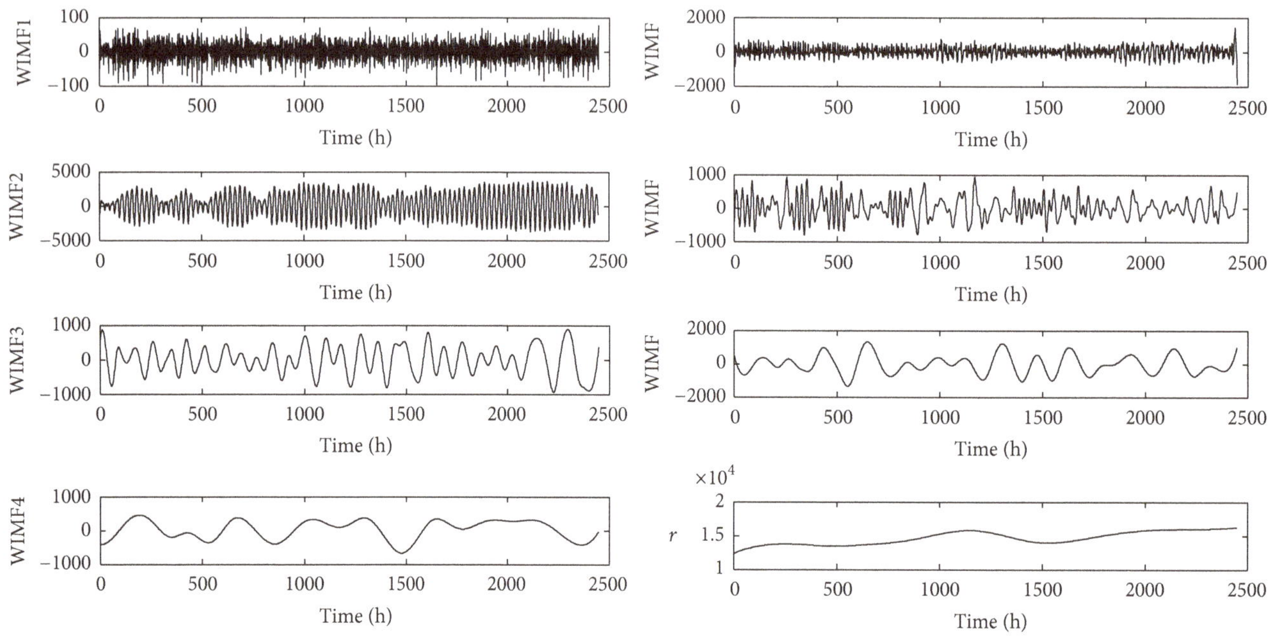

FIGURE 5: The decomposition of the power load series with assigned weights.

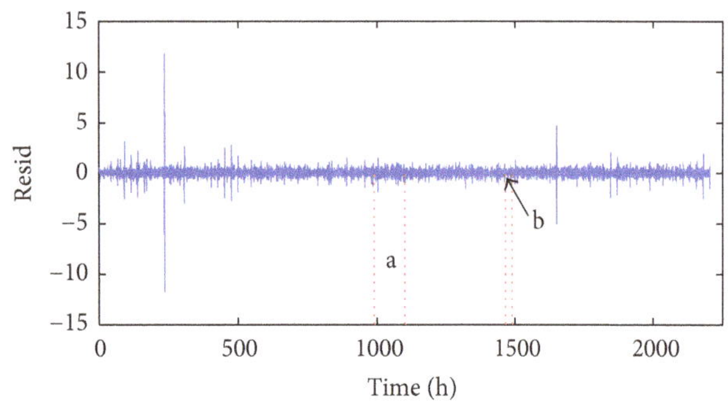

FIGURE 6: The residual of the power load series with assigned weights.

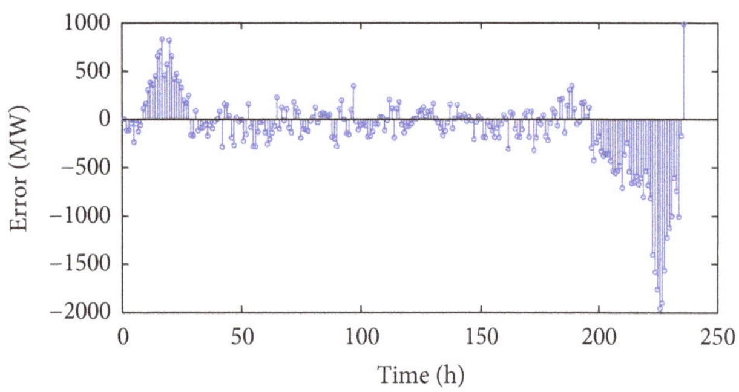

FIGURE 7: The error of the hybrid method.

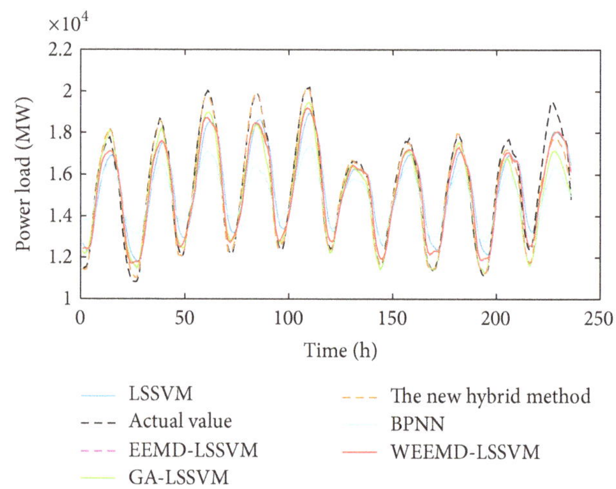

FIGURE 8: The comparison between the prediction results based on hourly observation.

load forecasting. The data frequency has been changed to test the robustness of the proposed method. Besides, other five methods presented by this study are compared with the proposed one to verify the accuracy of hybrid method by different criteria presented above. In the end, several conclusions are drawn as follows.

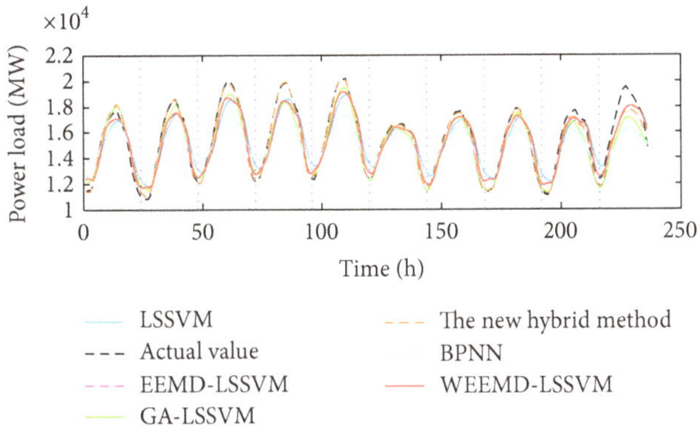

FIGURE 9: The comparison between the prediction results based on daily observation.

(a) The newly proposed decomposition algorithm named WEEMD has a better performance than EEMD method. (b) Due to the differences in their characteristics improvement of the forecasting accuracy, the components are clustered. (c) Regardless of the influence of data frequency or the fluctuation of time-series, the proposed hybrid method has excellent forecasting performance for power load.

Competing Interests

The authors declare that they have no competing interests.

References

[1] H. A. Malki, N. B. Karayiannis, and M. Balasubramanian, "Short-term electric power load forecasting using feedforward neural networks," *Expert Systems*, vol. 21, no. 3, pp. 157–167, 2004.

[2] W.-J. Lee and J. Hong, "A hybrid dynamic and fuzzy time series model for mid-term power load forecasting," *International Journal of Electrical Power & Energy Systems*, vol. 64, pp. 1057–1062, 2015.

[3] H. C. Huang, R. C. Hwang, and J. G. Hsieh, "Short-term power load forecasting by non-fixed neural network model with fuzzy BP learning algorithm," *International Journal of Power and Energy Systems*, vol. 22, no. 1, pp. 50–57, 2002.

[4] A. K. Topalli, I. Erkmen, and I. Topalli, "Intelligent short-term load forecasting in Turkey," *International Journal of Electrical Power and Energy Systems*, vol. 28, no. 7, pp. 437–447, 2006.

[5] T. Yalcinoz and U. Eminoglu, "Short term and medium term power distribution load forecasting by neural networks," *Energy Conversion and Management*, vol. 46, no. 9-10, pp. 1393–1405, 2005.

[6] A. H. Sanstad, S. McMenamin, A. Sukenik, G. L. Barbose, and C. A. Goldman, "Modeling an aggressive energy-efficiency scenario in long-range load forecasting for electric power transmission planning," *Applied Energy*, vol. 128, pp. 265–276, 2014.

[7] N. Amjady and F. Keynia, "Mid-term load forecasting of power systems by a new prediction method," *Energy Conversion and Management*, vol. 49, no. 10, pp. 2678–2687, 2008.

[8] P. Bolzern and G. Fronza, "Role of weather inputs in short-term forecasting of electric load," *International Journal of Electrical Power and Energy Systems*, vol. 8, no. 1, pp. 42–46, 1986.

[9] R. Wang, "Short-term electricity price forecasting based on grey system theory and time series analysis," in *Proceedings of the Asia-Pacific Power and Energy Engineering Conference (APPEEC '10)*, pp. 28–31, Sichuan,China, March 2010.

[10] Y. R. Wang, Q. L. Wan, and H. Chen, "Short term load forecasting based on double-threshold GARCH models," *Journal of Southeast University (Natural Science Edition)*, vol. 41, no. 6, pp. 1182–1187, 2011.

[11] D. Liu, "A model for medium- and long-term power load forecasting based on error correction," *Dianwang Jishu/Power System Technology*, vol. 36, no. 8, pp. 243–247, 2012.

[12] W. Sun and M. Ye, "Short-term load forecasting based on wavelet transform and least squares support vector machine optimized by fruit fly optimization algorithm," *Journal of Electrical and Computer Engineering*, vol. 2015, Article ID 862185, 9 pages, 2015.

[13] Y. Hu and X. R. Chang, "Short-term load forecasting based on local wave method and LSSVM," *Electrical Measurement and Instrumentation*, vol. 52, no. 7, pp. 5–9, 2015.

[14] P. Lauret, E. Fock, R. N. Randrianarivony, and J.-F. Manicom-Ramsamy, "Bayesian neural network approach to short time load forecasting," *Energy Conversion and Management*, vol. 49, no. 5, pp. 1156–1166, 2008.

[15] Z. Shao, F. Gao, S.-L. Yang, and B.-G. Yu, "A new semiparametric and EEMD based framework for mid-term electricity demand forecasting in China: hidden characteristic extraction and probability density prediction," *Renewable and Sustainable Energy Reviews*, vol. 52, pp. 876–889, 2015.

[16] W.-C. Hong, "Application of chaotic ant swarm optimization in electric load forecasting," *Energy Policy*, vol. 38, no. 10, pp. 5830–5839, 2010.

[17] S. Bahrami, R.-A. Hooshmand, and M. Parastegari, "Short term electric load forecasting by wavelet transform and grey model improved by PSO (particle swarm optimization) algorithm," *Energy*, vol. 72, pp. 434–442, 2014.

[18] L. Ghelardoni, A. Ghio, and D. Anguita, "Energy load forecasting using empirical mode decomposition and support vector regression," *IEEE Transactions on Smart Grid*, vol. 4, no. 1, pp. 549–556, 2013.

[19] Z. Liu, W. Sun, and J. Zeng, "A new short-term load forecasting method of power system based on EEMD and SS-PSO," *Neural Computing and Applications*, vol. 24, no. 3-4, pp. 973–983, 2014.

[20] Q. Li, J. Li, and H. Ma, "Short-term electricity load forecasting based on complementary ensemble empirical mode decomposition-fuzzy permutation and echo state network," *Journal of Computer Applications*, vol. 34, no. 12, pp. 3651–3655, 2014.

[21] L. A. Gallego, M. J. Rider, M. Lavorato, and A. Paldilha-Feltrin, "An enhanced genetic algorithm to solve the static and multistage transmission network expansion planning," *Journal of Electrical and Computer Engineering*, vol. 2012, Article ID 781041, 12 pages, 2012.

[22] X. L. An, D. X. Jiang, S. H. Li, and M. H. Zhao, "Application of the ensemble empirical mode decomposition and Hilbert transform to pedestal looseness study of direct-drive wind turbine," *Energy*, vol. 36, no. 9, pp. 5508–5520, 2011.

[23] X. Zhu, J. Zhang, and S. Fu, "Short-term wind speed prediction model based on EEMD and SVM," *Journal of North China Electric Power University*, vol. 40, no. 5, pp. 60–64, 2013.

[24] M. Mao, W. Gong, L. Chang, Y. Cao, and H. Xu, "Short-term photovoltaic generation forecasting based on EEMD-SVM combined method," *Proceedings of the Chinese Society of Electrical Engineering*, vol. 33, no. 34, pp. 17–24, 2013.

[25] Y. Li, D. Niu, and D. Li, "Novel hybrid power load forecasting method based on ensemble empirical mode decomposition," *Power System Technology*, vol. 32, no. 8, pp. 58–62, 2008.

[26] A. Hou and S. Suardi, "A nonparametric GARCH model of crude oil price return volatility," *Energy Economics*, vol. 34, no. 2, pp. 618–626, 2012.

[27] C. Schittenkopf, G. Dorffner, and E. J. Dockner, "Forecasting time-dependent conditional densities: a semi-non-parametric neural network approach," *Journal of Forecasting*, vol. 19, no. 4, pp. 355–374, 2000.

[28] Y. Wang, F. Li, Q. Wan, and H. Chen, "Hybrid momentum TAR-GARCH models for short term load forecasting," in *Proceedings of the 2011 IEEE PES General Meeting: The Electrification of Transportation and the Grid of the Future*, pp. 24–29, Detroit, Mich, USA, July 2011.

[29] H. Chen, Q. Wan, F. Li, and Y. Wang, "Short term load forecasting based on improved ESTAR GARCH model," in *Proceedings of the IEEE Power and Energy Society General Meeting*, pp. 1–6, San Diego, Calif, USA, July 2012.

[30] Y. Huang and J. Li, "A LS-SVM approach based on GA and NPGARCH for short-term traffic forecasting," *Energy Education Science and Technology Part A: Energy Science and Research*, vol. 32, no. 6, pp. 8607–8614, 2014.

[31] C. J. Yang, H. W. Lu, H. Y. Ma et al., "Load forecasting by considering wind power based on sequential time classification LSSVM model," *Advanced Materials Research*, vol. 712–715, pp. 2437–2440, 2013.

[32] H. Yang and X. Chang, "Short-term load forecasting based on local wave method and LSSVM," *Electrical Measurement and Instrumentation*, vol. 52, no. 7, pp. 5–9, 2015.

[33] Q. Gong, W. Lu, W. Gong, and X. Wang, "Short-term load forecasting of LSSVM based on improved PSO algorithm," *Communications in Computer and Information Science*, vol. 483, pp. 63–71, 2014.

[34] H. Zhang, T. Yao, and T. Ma, "Forecasting of steam load based on phase space reconstruction and improved LSSVM algorithm," *Energy Education Science and Technology Part A: Energy Science and Research*, vol. 32, no. 3, pp. 1939–1952, 2014.

[35] M. M. Hadow, A. N. Abd Allah, and S. P. Abdul Karim, "Reliability evaluation of distribution power systems based on artificial neural network techniques," *Journal of Electrical and Computer Engineering*, vol. 2012, Article ID 560541, 5 pages, 2012.

14

Feedforward Harmonic Mitigation Strategy for Single-Phase Voltage Source Converter

Qing Zhong, Junjie Feng, Gang Wang, and Haifeng Li

School of Electric Power, South China University of Technology, Guangzhou, China

Correspondence should be addressed to Qing Zhong; epqzhong@scut.edu.cn

Academic Editor: Alessandro Lidozzi

With the development of distributed generations (DGs), single-phase voltage source converter (SPVSC) has been widely used, but it brings about the problem of harmonic pollution to power grid. Hence, it is significant to explore the mechanism of harmonic injection from SPVSC and propose effective control strategies to mitigate the harmonic pollution. In this paper, a harmonic analysis model of SPVSC based on dynamic phasor (DP) has been established. With the model, the harmonics interaction between the ac side and the dc side can be analyzed with the consideration of the control strategies, which reveals the generation mechanism of the harmonics in SPVSC. Based on the mechanism, a feedforward harmonic mitigation strategy has been presented. The principle of the strategy is to add low-order harmonic signal to the PWM modulation signals to reduce the harmonic current on the ac side. The harmonic mitigation strategy not only has clear physical meaning and fast calculation, but also is robust for the uncertainty of parameters. Finally, the simulation and experiment results demonstrate the correctness of the model and the effectiveness of the harmonic mitigation strategy.

1. Introduction

Because of the capability of unity power factor and bidirectional energy flow, single-phase voltage source converters (SPVSC) have been widely applied in various industrial fields [1, 2], such as serving as the grid-connected interface of distributed generations (DGs) or microgrid with renewable energy systems [3, 4]. However, due to the strong nonlinearity of power electronic devices, SPVSC also brings about the problem of harmonic pollution which has a negative impact on power grid [5–7]. Hence, it is necessary to analyze the mechanism of harmonic injection and propose effective suppression measurements, laying a solid foundation for the promotion of SPVSC.

Three-phase VSC generates high-order harmonics under normal situation, which can generally be filtered by low pass filter [8], while SPVSC generates not only high-order harmonics but also low-order harmonics [9] even in normal situation, which is hard to be eliminated. The harmonic current generated from SPVSC is eliminated mainly from two aspects [10, 11]. One solution is to remove harmonic ripples in the voltage-loop and strengthen the antiharmonic disturbance capacity of current loop. Since harmonic ripples in the voltage-loop can be easily removed by employing low pass voltage filter [12] or notch filter [11], most studies focus on how to improve the performance of current-loop controller. Proportional Resonant (PR) controller [13, 14], which has the advantages of simplicity and zero steady-state error for tracking a sinusoidal signal, is widely used for selective harmonic elimination. However, it has high sensitivity to the frequency variations and may not be able to achieve a good effect, since a single PR controller only eliminates a single-order harmonic. Multiresonant controllers (MRCs) suppress harmonic distortions effectively by connecting PR controllers at harmonic frequency in parallel, but this causes heavy paralleling computation duty, particularly when high-order harmonics are required to be compensated. In the worst case, MRCs cause the system to be unstable [15]. In [16, 17], Repetitive Controller (RC) has been investigated which involved complicated analysis and designation. It is based on internal model principle and achieves zero steady-state error for tracking a sinusoidal signal. RC presented a much slower dynamic response than PR and MRCs. To achieve a minimum steady-state error while maintaining a fast-transient response, a hybrid controller by combining PR and RC has been proposed in [12, 18]. However, the designation controller parameters were very complex to ensure the effectiveness of the control strategy and the stability of system.

The other solution is to add harmonic modulation signals to PWM modulated signals to offset harmonic currents on the ac side. Without additional controller, it can be implemented simply and has less computational burden. Moreover, the strategy not only can achieve a fast-dynamic response but also will not affect the stability of system no matter how many order harmonics need to be suppressed. Before the suppression strategy is put forward, it is important to establish an accurate harmonic analysis model which can reflect the harmonic generation mechanism. As an alternative to the traditional modeling approach [19, 20], dynamic phasor (DP) method is very useful for the harmonic analysis of SPVSC [21–23]. It was based on a linear time varying periodic (LTP) theory and considered the detailed switching dynamic of power electronic devices. DP was also very effective in revealing dynamical couplings between various quantities [24]; thus it was possible to analyze how harmonics transferred between ac and dc side of VSC.

To address the above issues, this paper develops a harmonic analysis model of SPVSC based on the dynamic phasor and the harmonic generation mechanism of SPVSC has been investigated comprehensively. According to such mechanism, a feedforward control strategy is put forward to mitigate the harmonics on grid side. The organization of this paper is as follows: in Section 2, the system configuration and time-domain model of SPVSC are shown and harmonic analysis model based on DP is investigated; in Section 3, the generation mechanism and transmission law of harmonic for SPVSC are analyzed; in Section 4, a feedforward harmonic mitigation strategy is proposed on the basis of the generation mechanism; in Sections 5 and 6, the simulation and experimental results are presented to verify the proposed model and harmonic mitigation strategy. Finally, the conclusion is drawn in Section 7.

2. Harmonic Analyzing Model of SPVSC

2.1. Time-Domain Model of SPVSC. The SPVSC topology studied in this paper is a single-phase H bridge voltage source converter as shown in Figure 1. U_s is the ac source voltage. R and L are the equivalent resistance and impendence of the ac system. I_R and U_{Rac} are the ac side current and voltage of SPVSC, respectively. U_{dc} is the voltage on dc side. C_{dc} is the capacitor on the dc side. R_{dc} is the equivalent resistance of the dc loads. I_{dc} and I_l are the output current and dc load current on the dc side. V_1, V_2, V_3, and V_4 are power electronic devices of converter.

The time-domain model of SPVSC can be described by

$$L\frac{dI_R}{dt} = U_s - RI_R - U_{Rac}$$
$$C_{dc}\frac{dU_{dc}}{dt} = I_{dc} - \frac{U_{dc}}{R_{dc}}. \tag{1}$$

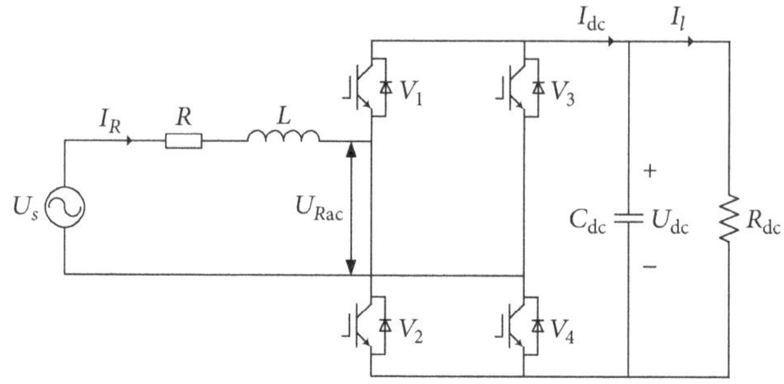

FIGURE 1: Topology of single-phase VSC.

The switching function is used to describe the interaction between the ac and dc side of SPVSC, so U_{Rac} and I_{dc} in (1) can be given as

$$U_{Rac} = S_R U_{dc}$$
$$I_{dc} = S_R I_R, \tag{2}$$

where S_R is the switching function of SPVSC. When V_1 and V_4 are switched on, it is 1, and when V_2 and V_3 are on, it is −1. S_R is related to the control strategy and determined by the PWM signal.

As shown in Figure 2, double closed-loop control strategy is occasionally employed in the control of converter. The dc-link voltage out-loop is controlled by a proportional integral (PI) controller and the peak value of reference current I_{m_ref} is obtained. I_{m_ref} is multiplied by $\cos\theta$, which is the ac system voltage phase cosine captured by phase locked loop (PLL), so as to operate at unity power factor. The current inner loop is controlled by proportional (P) controller to track the ac side current reference value I_{t_ref}.

What needs to be specially mentioned is that, considering the phase delay in the sampling, grid voltage feedforward sampling value U_{st} is obtained by multiplying the fundamental RMS of U_s with $\cos\theta$. After subtracting U_{st} from the output value of current controller and being divided by U_{dc_ref}, the modulation signal of SPVSC is obtained

$$U_m = \frac{1}{U_{dc\_ref}}\left[U_{st} - K_{ip}\left(I_{t\_ref} - I_R\right)\right], \tag{3}$$

where K_{ip} is the proportional regulation gain of current inner loop, U_{st} is the sampling value of grid voltage, U_{dc_ref} is the reference of dc voltage, and I_R and I_{t_ref} are the actual value and reference value of ac side current, respectively. I_{m_ref} is the peak value of reference current which can be obtained by $I_{m\_ref} = \sqrt{2}u_{dc}I_{dc}/U_{rms}$.

2.2. Dynamic Phasor Harmonic Analysis Model of SPVSC. The SPVSC time-domain model described by (1) and (2) can be transformed into dynamic phasor model, and the k-order

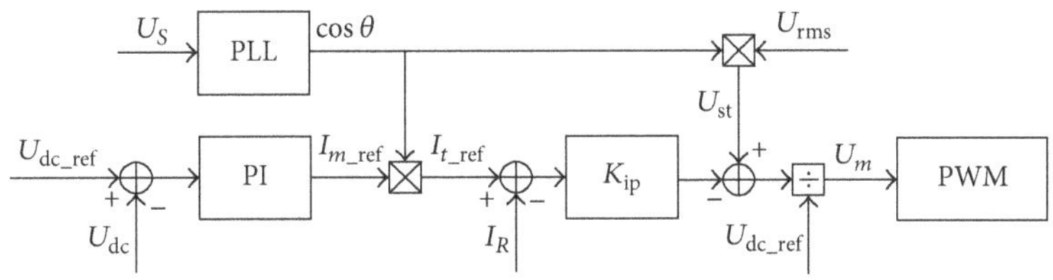

FIGURE 2: Double closed-loop control structure of voltage and current.

dynamic phase equation of the ac side current and the dc side voltage is expressed as

$$\begin{aligned}\frac{d\langle I_R\rangle_k}{dt} &= -\left(\frac{R}{L}+jk\varpi_s\right)\langle I_R\rangle_k - \frac{\langle U_{\text{dc}}S_R\rangle_k}{L} \\ &\quad + \frac{\langle U_s\rangle_k}{L} \\ \frac{d\langle U_{\text{dc}}\rangle_k}{dt} &= -\left(\frac{1}{R_{\text{dc}}C_{\text{dc}}}+jk\varpi_s\right)\langle U_{\text{dc}}\rangle_k + \frac{\langle I_RS_R\rangle_k}{C_{\text{dc}}},\end{aligned} \tag{4}$$

where $\varpi_s = 2\pi/T$ is the angular frequency and $\langle\cdot\rangle_k$ denotes the kth dynamic phasor.

$\langle U_{\text{dc}}S_R\rangle_k$ and $\langle I_RS_R\rangle_k$ are the convolution of switching function and dc side voltage or ac side current, respectively, which represents the interaction of harmonics between the ac and dc sides of SPVSC. It can be described by

$$\begin{aligned}\langle U_{\text{dc}}S_R\rangle_k &= \sum_i \langle S_R\rangle_{k-i}\langle u_{\text{dc}}\rangle_i \\ \langle I_RS_R\rangle_k &= \sum_i \langle S_R\rangle_{k-i}\langle I_R\rangle_i.\end{aligned} \tag{5}$$

In this paper, SPWM modulation with fixed switching frequency is adopted, so the dynamic phasor of switching function is determined by U_m and can be given as

$$\begin{aligned}\langle S_R\rangle_k &= \langle U_m\rangle_k \\ &= \frac{1}{U_{\text{dc\_ref}}}\left[\langle U_{\text{st}}\rangle_k - K_{\text{iP}}\left(\langle I_{t\_\text{ref}}\rangle_k - \langle I_R\rangle_k\right)\right].\end{aligned} \tag{6}$$

Since U_{st} and I_{t_ref} have only the fundamental component, the kth dynamic phasor is equivalent to 0, except $\langle U_{\text{st}}\rangle_1 \neq 0$ and $\langle I_{t\_\text{ref}}\rangle_1 \neq 0$. This paper is mainly aimed at the lower-order harmonic component within 13 orders, so the high-order component of the dynamic phasor of the switching function can be neglected. Equation (6) can be simplified as

$$\langle S_R\rangle_k = \begin{cases}\dfrac{\langle U_{\text{st}}\rangle_1 - K_{\text{iP}}\langle I_{t\_\text{ref}}\rangle_1}{U_{\text{dc\_ref}}} + \dfrac{K_{\text{iP}}\langle I_R\rangle_1}{U_{\text{dc\_ref}}}, & k=1 \\ \dfrac{K_{\text{iP}}\langle I_R\rangle_k}{U_{\text{dc\_ref}}}, & k=2,\ldots,13.\end{cases} \tag{7}$$

By substituting (7) into (4), the kth dynamic phasor equation of the ac side current and the dc side voltage is obtained as follows:

$$\begin{aligned}\frac{d\langle I_R\rangle_k}{dt} &= -\left(\frac{R}{L}+jk\omega_s\right)\langle I_R\rangle_k + \frac{\langle U_s\rangle_k}{L} \\ &\quad - \frac{K_{\text{iP}}}{U_{\text{dc\_ref}}L}\sum_i \langle I_R\rangle_{k-i}\langle u_{\text{dc}}\rangle_i \\ &\quad - \frac{\langle U_{\text{st}}\rangle_1 - K_{\text{iP}}\langle I_{t\_\text{ref}}\rangle_1}{U_{\text{dc\_ref}}L}\langle u_{\text{dc}}\rangle_{k-1} \\ &\quad - \frac{\langle U_{\text{st}}\rangle_{-1} - K_{\text{iP}}\langle I_{t\_\text{ref}}\rangle_{-1}}{U_{\text{dc\_ref}}L}\langle u_{\text{dc}}\rangle_{k+1},\end{aligned} \tag{8}$$

$$\begin{aligned}\frac{d\langle u_{\text{dc}}\rangle_k}{dt} &= -\left(\frac{1}{R_{\text{dc}}C}+jk\omega_s\right)\langle u_{\text{dc}}\rangle_k \\ &\quad + \frac{\langle U_{\text{st}}\rangle_1 - K_{\text{iP}}\langle I_{t\_\text{ref}}\rangle_1}{U_{\text{dc\_ref}}C}\langle I_R\rangle_{k-1} \\ &\quad + \frac{\langle U_{\text{st}}\rangle_{-1} - K_{\text{iP}}\langle I_{t\_\text{ref}}\rangle_{-1}}{U_{\text{dc\_ref}}C}\langle I_R\rangle_{k+1} \\ &\quad + \frac{K_{\text{iP}}}{U_{\text{dc\_ref}}C}\sum_i \langle I_R\rangle_{k-i}\langle I_R\rangle_i,\end{aligned} \tag{9}$$

where $\langle I_{t\_\text{ref}}\rangle_{-1} = \langle I_{t\_\text{ref}}\rangle_1^*$ and $\langle U_{\text{st}}\rangle_{-1} = \langle U_{\text{st}}\rangle_1^*$, according to the conjugate property of dynamic phasor. In the convolution $\sum\langle I_R\rangle_{k-i}\langle u_{\text{dc}}\rangle_i$ and $\sum\langle I_R\rangle_{k-i}\langle I_R\rangle_i$, the product of two small quantities can be neglected. It means that only the items including dc component of voltage on dc side or fundamental current on ac side should be considered. Therefore, the accumulation of items in (8) and (9) can be simplified as

$$\begin{aligned}\sum_i \langle I_R\rangle_{k-i}\langle u_{\text{dc}}\rangle_i &= \langle I_R\rangle_k\langle u_{\text{dc}}\rangle_0 + \langle I_R\rangle_1\langle u_{\text{dc}}\rangle_{k-1} \\ &\quad + \langle I_R\rangle_1^*\langle u_{\text{dc}}\rangle_{k+1} \\ \sum_i \langle I_R\rangle_{k-i}\langle I_R\rangle_i &= 2\langle I_R\rangle_{k-1}\langle I_R\rangle_1 \\ &\quad + 2\langle I_R\rangle_{k+1}\langle I_R\rangle_1^*.\end{aligned} \tag{10}$$

The kth dynamic phasor of current on ac side and voltage on dc side can be obtained by solving (8) and (9). The calculation results are the different order harmonics of current on ac side and voltage on dc side. Therefore, the

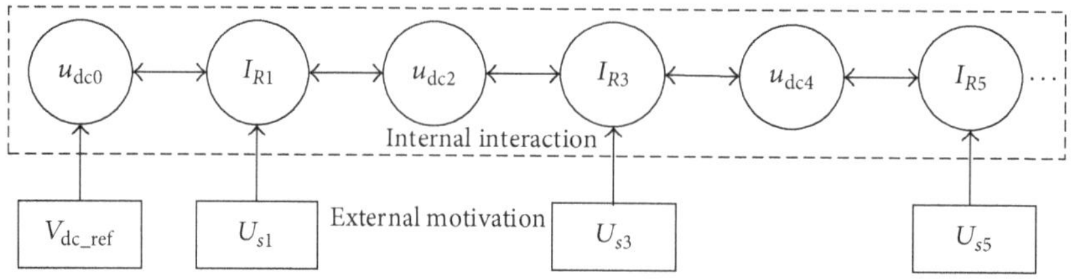

FIGURE 3: Harmonic interaction law of single-phase VSC.

dynamic phasor model of SPVSC can be used to analyze the generation mechanism of harmonic.

3. Mechanism of SPVSC Harmonic Generations

In the steady state, the dc component of dc side voltage is equivalent to its reference; that is, $\langle u_{dc}\rangle_0 = V_{dc\_ref}$. The differential of the dynamic phasor of ac side current and the dc side voltage can be neglected, so $d\langle I_R\rangle_k/dt = 0$ and $d\langle u_{dc}\rangle_k/dt = 0$. The harmonic generation mechanism of SPVSC under steady-state conditions is analyzed below.

The dc voltage on dc side can be obtained by (8), as shown in

$$\begin{aligned}\frac{1}{R_{dc}}\langle u_{dc}\rangle_0 &= \frac{\langle U_{st}\rangle_1 - K_{iP}\left(\langle I_{t\_ref}\rangle_1 - \langle I_R\rangle_1\right)}{U_{dc\_ref}}\langle I_R\rangle_1^* \\ &+ \frac{\langle U_{st}\rangle_1^* - K_{iP}\left(\langle I_{t\_ref}\rangle_1^* - \langle I_R\rangle_1^*\right)}{U_{dc\_ref}}\langle I_R\rangle_1 .\end{aligned} \tag{11}$$

Equation (11) can describe the interaction between the dc voltage on dc side and the fundamental current on ac side.

Similarly, the fundamental current on ac side can be further written as

$$\begin{aligned}(R + j\omega_s L)\langle I_R\rangle_1 &= -\frac{\langle U_{st}\rangle_1 - K_{iP}\left(\langle I_{t\_ref}\rangle_1 - \langle I_R\rangle_1\right)}{U_{dc\_ref}}\langle u_{dc}\rangle_0 \\ &- \frac{\langle U_{st}\rangle_1^* - K_{iP}\left(\langle I_{t\_ref}\rangle_1^* - \langle I_R\rangle_1^*\right)}{U_{dc\_ref}}\langle u_{dc}\rangle_2 \\ &+ \langle U_s\rangle_1 .\end{aligned} \tag{12}$$

Therefore, the fundamental current on ac side not only interacts with the dc voltage on dc side, but also interacts with the 2nd harmonic voltage on dc side.

The 2nd harmonic voltage on dc side and the 3rd harmonic current on ac side can be expressed as follows:

$$\begin{aligned}\left(\frac{1}{R_{dc}} + j2C\omega_s\right)\langle u_{dc}\rangle_2 &= -\frac{\langle U_{st}\rangle_1 - K_{iP}\left(\langle I_{t\_ref}\rangle_1 - \langle I_R\rangle_1\right)}{V_{dc\_ref}}\langle I_R\rangle_1 \\ &- \frac{\langle U_{st}\rangle_1^* - K_{iP}\left(\langle I_{t\_ref}\rangle_1^* - \langle I_R\rangle_1^*\right)}{V_{dc\_ref}}\langle I_R\rangle_3 \\ &+ \frac{K_{iP}\langle I_R\rangle_3\langle I_R\rangle_1^*}{V_{dc\_ref}},\end{aligned} \tag{13}$$

$$\begin{aligned}(R + j3\omega_s L + K_{iP})\langle I_R\rangle_3 &= -\frac{\langle U_{st}\rangle_1 - K_{iP}\left(\langle I_{t\_ref}\rangle_1 - \langle I_R\rangle_1\right)}{V_{dc\_ref}}\langle u_{dc}\rangle_2 \\ &- \frac{\langle U_{st}\rangle_1^* - K_{iP}\left(\langle I_{t\_ref}\rangle_1^* - \langle I_R\rangle_1^*\right)}{V_{dc\_ref}}\langle u_{dc}\rangle_4 \\ &+ \langle U_s\rangle_3 .\end{aligned} \tag{14}$$

In (13), the state variables on the right are $\langle I_R\rangle_1$ and $\langle I_R\rangle_3$, which indicates that the 2nd harmonic voltage on dc side not only interacts with the fundamental current on ac side, but also interacts with the 3rd harmonic current on ac side. Similarly, it can be obtained from (14) that the 3rd harmonic current on ac side not only interacts with the 2nd harmonic voltage on dc side, but also interacts with the 4th harmonic voltage on dc side. Meanwhile, the 3rd voltage background harmonic of power grid will also induce the 3rd harmonic current on ac side. Therefore, the generation mechanism and interaction law of harmonic current on ac side and harmonic voltage on dc side can be summarized as in Figure 3.

In Figure 3, u_{dci} and I_{Ri} are ith harmonic voltage on dc side and ith harmonic current on ac side, respectively. V_{dc_ref} is the dc voltage reference in controller. U_{si} is ith harmonic voltage of ac side source. The single arrow represents the effect of external excitation on harmonics, and the double arrow represents the interaction between two variables.

There are two reasons for the lower-order harmonic generation of SPVSC, that is, external excitation and internal interaction. From the internal interaction, SPVSC will generate a series of odd harmonic currents on ac side and a series of even harmonic voltages on dc side. From the external excitation relationship, it can be concluded that when the voltage of power grid is distorted, it will aggravate the harmonic generation from SPVSC and lead to more serious harmonic pollution to the power grid.

4. Harmonic Current Mitigation Strategy

In order to reduce the harmonic currents injecting to the power grid, a feedforward harmonic mitigation strategy is investigated based on the results in Section 3. Its principle is to superimpose low-order harmonic signal into the PWM modulation wave, which can cut off the external incentive and internal transfer path of harmonic propagation, thus eliminating the lower-order harmonic current on ac side.

Aiming at Nth ($N = 3, 5, \ldots, 13$) harmonic currents on ac side, (4) can be simplified as

$$(R + jk\omega_s L) \langle I_R \rangle_N = -\sum_i \langle S_R \rangle_{N-i} \langle U_{dc} \rangle_i + \langle U_s \rangle_N . \quad (15)$$

The harmonic currents on ac side are mainly induced by the grid voltage distortions and the internal interaction between the harmonic voltages on dc side just shown in Figure 3. The grid voltage distortions $\langle U_s \rangle_N$ are normally uncontrolled, mainly induced by other nonlinear loads. However, the ac voltage of SPVSC $\sum_i \langle S_R \rangle_{N-i} \langle U_{dc} \rangle_i$ can be easily controlled by adjusting controller. It can be implied from (15) that $\langle S_R \rangle_N \langle U_{dc} \rangle_0$ can be utilized to offset the harmonic currents generated by $\langle U_s \rangle_N$ and $\sum_{i \neq 0} \langle S_R \rangle_{N-i} \langle U_{dc} \rangle_i$. If the right side of (15) is equal to 0, the internal interaction between the ac and dc side harmonics and external excitation from grid voltage distortions can be offset; thus the Nth harmonic current on ac side can be eliminated effectively.

From (6) and (8), the lower-order harmonic components $\langle \Delta S \rangle_N$ which should be superimposed on the PWM signal can be derived as follows:

$$\langle \Delta S \rangle_N = \frac{\left(\left(\left(\langle U_{st} \rangle_1 - K_{iP} \left(\langle I_{t\_ref} \rangle_1 - \langle I_R \rangle_1\right)\right) / U_{dc\_ref}\right) \langle u_{dc} \rangle_{N-1} + \left(\left(\langle U_{st} \rangle_1^* - K_{iP} \left(\langle I_{t\_ref} \rangle_1^* - \langle I_R \rangle_1^*\right)\right) / U_{dc\_ref}\right) \langle u_{dc} \rangle_{N+1} - \langle U_s \rangle_N\right)}{\langle u_{dc} \rangle_0} . \quad (16)$$

Above all, the feedforward compensation control strategy is implemented by superimposing lower-order harmonic components in the PWM signal to suppress the low-order harmonic currents on ac side. The control block diagram is shown in Figure 4. The harmonic voltage component on dc side and the background harmonic voltage on ac side are detected in real time. According to (16), the amplitude and phase angle of the Nth harmonic components needed to be superposed in the PWM signals are obtained, so as to suppress the harmonic current on ac side dynamically.

The harmonic mitigation strategy not only has clear physical meaning and fast calculation speed, but also will not be affected by the uncertainty of circuit parameters which can be seen from (16). Moreover, without additional controller, it avoids the complex parameter design and will not affect the stability of the system.

5. Simulation Results

In order to verify the correctness of the harmonic analysis model, a SPVSC model is set up in MATLAB/Simulink. The key parameters for the simulation model are listed in Table 1.

The control strategy is shown in Figure 2, which is voltage and circuit double closed-loop control. The simulation results and the theoretical results of the harmonic currents on ac side and the harmonic voltages on dc side are shown in Tables 2 and 3, respectively. The simulation results of the amplitude and angle of the harmonic currents on ac side and the harmonic voltages on dc side are well-matched with those obtained by the harmonic analysis model. There are a series of low-order odd harmonic currents on the ac side, in which the 3rd harmonic current is dominant. The amplitude decreases with the harmonic order increasing. There are series of low-order even harmonic voltages on dc side. The 2nd harmonic voltage is the most obvious. The harmonic content is obviously reduced with the increase of the harmonic order. Therefore, the correctness of the low-order harmonic analysis model of SPVSC with closed-loop control strategy is verified.

In order to verify the harmonic transmission law of SPVSC, injecting 3rd and 5th voltage background harmonics to power grid, respectively, the harmonic spectra of the ac side current and the dc side voltage are obtained, which is shown in Figure 5. The black bar represents the situation without background harmonic voltage. The red bar represents the situation with 3rd background harmonic voltage, and the blue bar represents the situation with 5th background harmonic voltage.

It can be seen that when the power grid contains 3rd background harmonic voltage, the third harmonic current on ac side increased significantly, and the 4th harmonic voltage on dc side also increased because of the interaction between the ac side and dc side. Simultaneously, it passes back to ac side and generates 5th harmonic current. But after the second transmission, the impact on 5th harmonic current is weak. Similarly, when the grid contains 5th background harmonics voltage, the 6th harmonic voltage on dc side increased because of the impact of 5th harmonic current on ac side. But when it is passed back to the 7th harmonic current

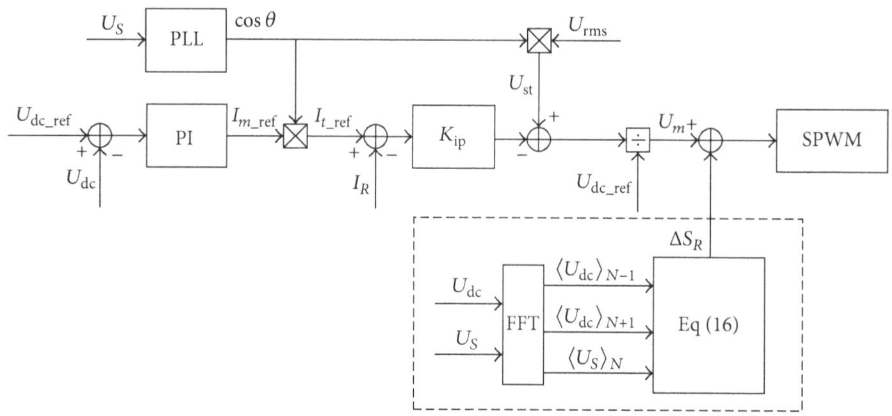

Figure 4: Control block diagram of injected N-order modulation wave.

Table 1: Simulation parameters.

Parameter	Meaning	Value
U_s	RMS of ac source voltage	220 V
R, L	Impedance of the ac system	0.02 Ω/3 mH
C_{dc}	Capacitor on dc side	3400 μF
R_{dc}	Load resistance	10 Ω
V_{dc_ref}	Reference of dc voltage	428 V

Table 2: Comparison of theoretical results and simulation results of harmonic currents on ac side.

Frequency	Simulation result		Calculation result	
	Magnitude (A)	Angle (°)	Magnitude (A)	Angle (°)
50 Hz	124.43	2.7	124.61	2.5
150 Hz	1.65	221.4	1.69	230.2
250 Hz	0.11	118.8	0.08	116.6

Table 3: Comparison of theoretical results and simulation results of harmonic voltages on dc side.

Frequency	Simulation result		Calculation result	
	Magnitude (V)	Angle (°)	Magnitude (V)	Angle (°)
0 Hz	428.10	90.0	430.16	90.0
100 Hz	20.57	183.7	20.60	183.8
200 Hz	0.33	49.7	0.33	52.65

on ac side, the impact is weak. The harmonic transfer law shown in Figure 3 can be validated.

In order to verify the effectiveness of the feedforward harmonic mitigation strategy, the simulation is carried out and the dynamic compensation effect can be observed by changing the system conditions. Initially, the grid voltage has 0.1 pu 3rd and 0.05 pu 5th harmonic distortions. At 0.4 s, the feedforward harmonic mitigation strategy is employed and the low-order harmonic signals are superimposed to PWM modulated waves. At 0.5 s, the grid voltage changes suddenly, only containing 0.01 pu 5th background harmonic voltage. At 0.7 s, the grid voltage becomes undistorted, and the load resistance on dc side changes from 10 Ω to 6.67 Ω. At 0.9 s, the harmonic mitigation link is turned off. Figure 6 presents the harmonic injection component of PWM signals. Table 4 presents the harmonic currents on ac side before and after the harmonic mitigation strategy being employed.

Before the feedforward harmonic mitigation strategy is employed, there are a large number of 3rd and 5th harmonic currents on ac side, which amplitudes are 6.48 A and 2.85 A, respectively. When the harmonic mitigation strategy is employed, the harmonic is effectively suppressed and the amplitude decreased to 0.37 A and 0.09 A, respectively. At 0.5 s and 0.7 s, the background harmonic voltages in grid and load are changed abruptly, but the harmonic suppression is still effective. Moreover, low-order harmonic signal superimposing to PWM can quickly track the changes of system conditions, to suppress the harmonics current on ac side dynamically. Above all, the effectiveness of the harmonic mitigation strategy is verified.

6. Experimental Results

The experimental setup of single-phase voltage source converter is shown in Figure 7. The inductance on the ac side is 6 mH. The capacitance and resistance on dc side are 3400 μF and 100 Ω. The carrier wave frequency is 7500 Hz. The converter uses the intelligent power module (IPM) and the controller uses DSP TMS320F28335. The control strategy is the same as simulation.

The grid voltage is used as the system voltage on ac side, in which RMS is 220 V. The measurement results show that the most obvious background harmonic voltage is 3rd, 5th, and 7th harmonic voltage, so the experiment mainly aims at 3rd, 5th, and 7th harmonic current to verify the effectiveness of the proposed suppression strategy. By inducing the harmonic component feedforward of the ac side voltage and the dc side voltage, the amplitude and phase angle of harmonic signal which needs to be superimposed in PWM modulation wave can be obtained by controller and applied to the converter. The Fluke 435 is used to measure the voltage and current wave on the ac side of SPVSC and analyze the spectrum characteristics of ac current. In this section, a comparison between multiresonant control strategy [14, 15] and the proposed

Table 4: The harmonic current on ac side before and after N-order modulated wave injection (A).

Harmonic order	0~0.4 s without	0.4~0.5 s with	0.5~0.7 s with	0.7~0.9 s with	0.9~1.1 s without
Fund	123.80	123.82	124.48	184.49	184.36
3rd	6.48	0.37	0.23	0.45	2.51
5th	2.85	0.09	0.07	0.11	0.23

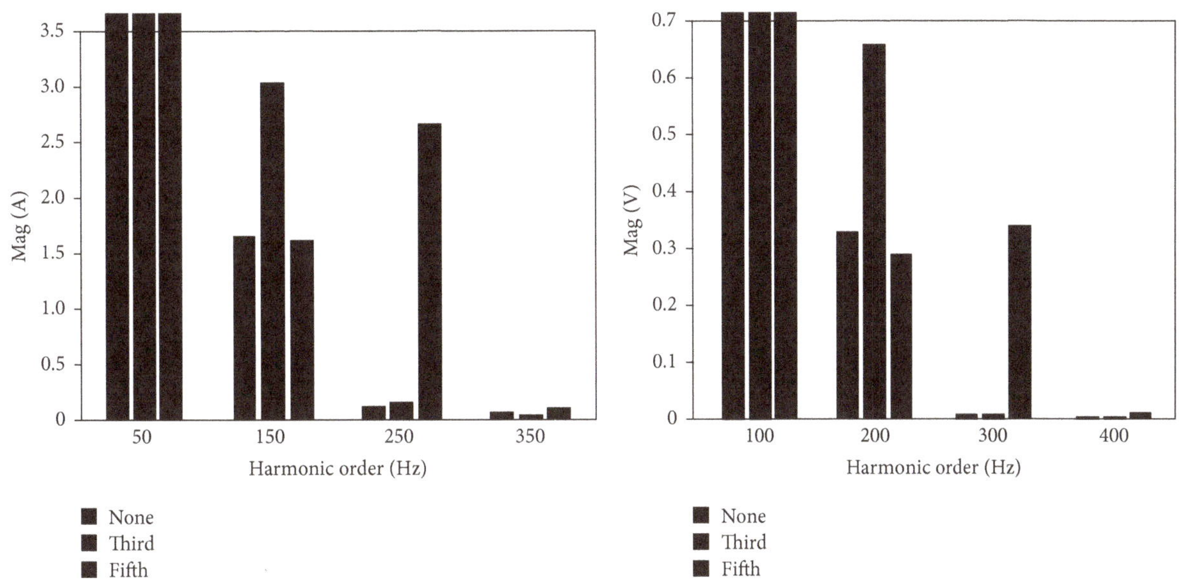

(a) Spectrum of current on ac side

(b) Spectrum of voltages on dc side

Figure 5: The grid voltage contains third and fifth harmonic.

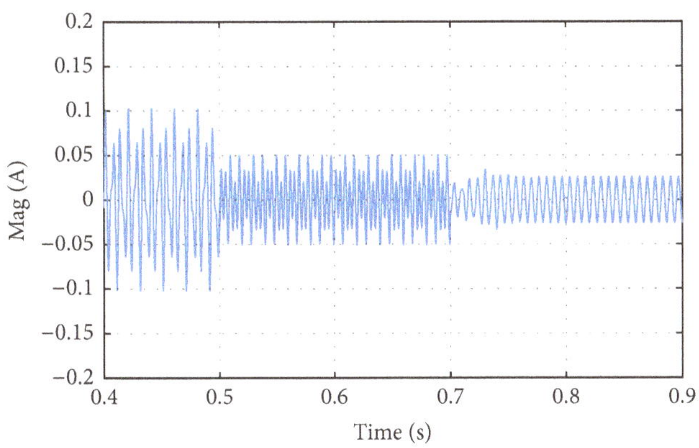

Figure 6: Harmonic injection component of PWM modulated wave.

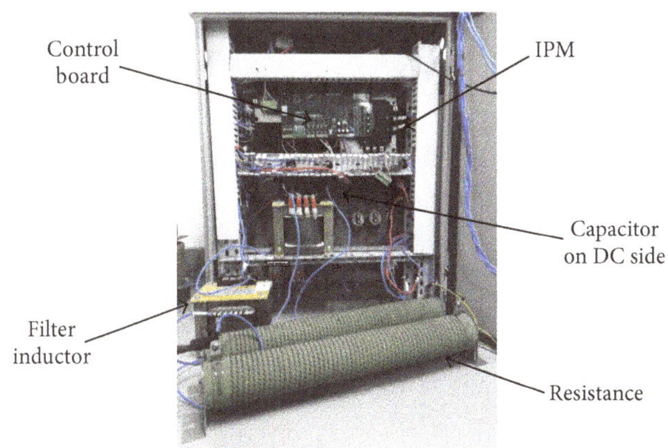

Figure 7: Experimental setup of single-phase voltage source converter.

feedforward harmonic mitigation strategy is carried out. The waveforms and harmonics tables of current on ac side without harmonic mitigation strategy, with multiresonant control strategy and with the proposed strategy, are shown in Figures 8, 9, and 10, respectively.

From the experimental results, it can be found that after the multiresonant control strategy is employed, the total harmonic distortion of current (THDI) decreases from 4.7% to 3.5%, and the harmonic content of 3rd, 5th, and 7th current decreases from 3.3% to 1.5%, 2.2% to 2.1%, and 1.5% to 0.9%, respectively. However, after the feedforward harmonic mitigation strategy is employed, THDI decreases from 4.7% to 2.5%, and the harmonic content of 3rd, 5th, and 7th harmonic current decreases to 0.8%, 1.6%, and 1.1%, respectively. It is clearly shown that the proposed strategy has better performance than multiresonant control strategy, which has verified the effectiveness of the proposed harmonic

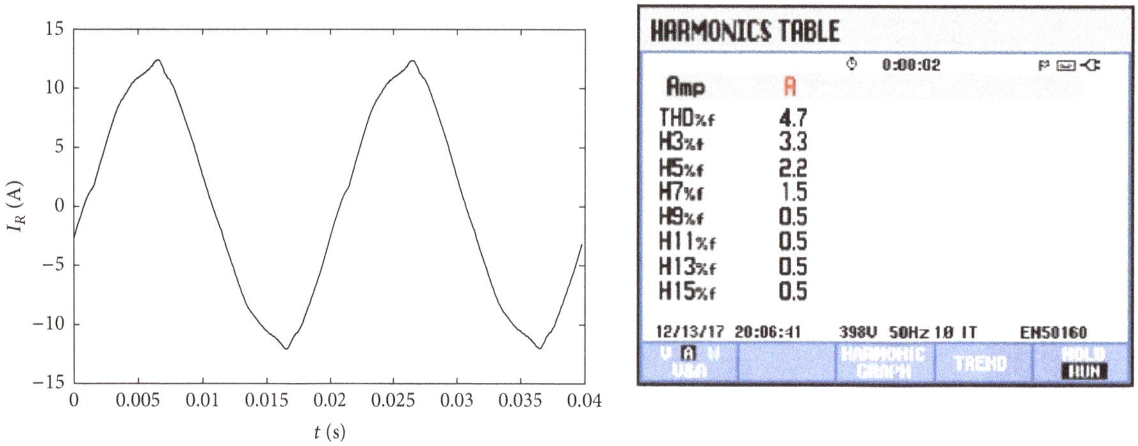

(a) Waveform (b) Harmonics table

FIGURE 8: Waveform and harmonics table of current on ac side without harmonic mitigation strategy.

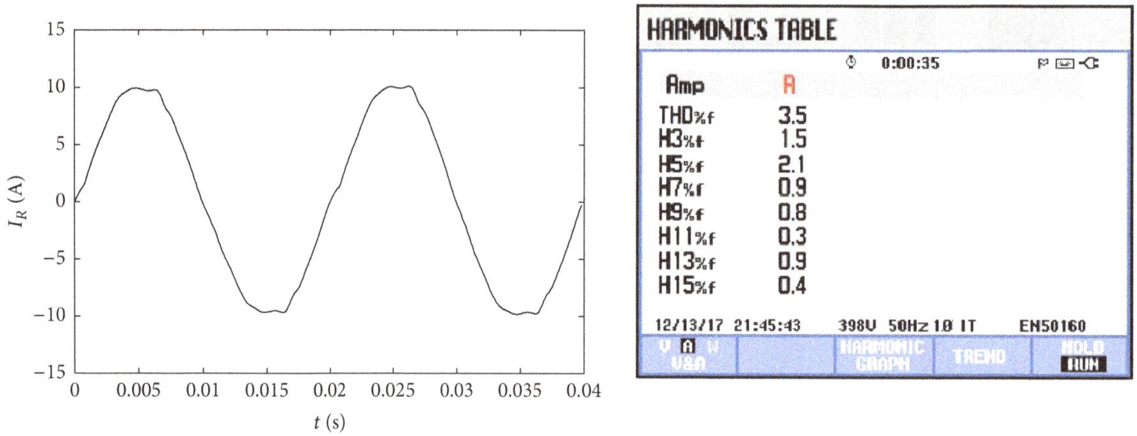

(a) Waveform (b) Harmonics table

FIGURE 9: Waveform and harmonics table of current on ac side with multiresonant control strategy.

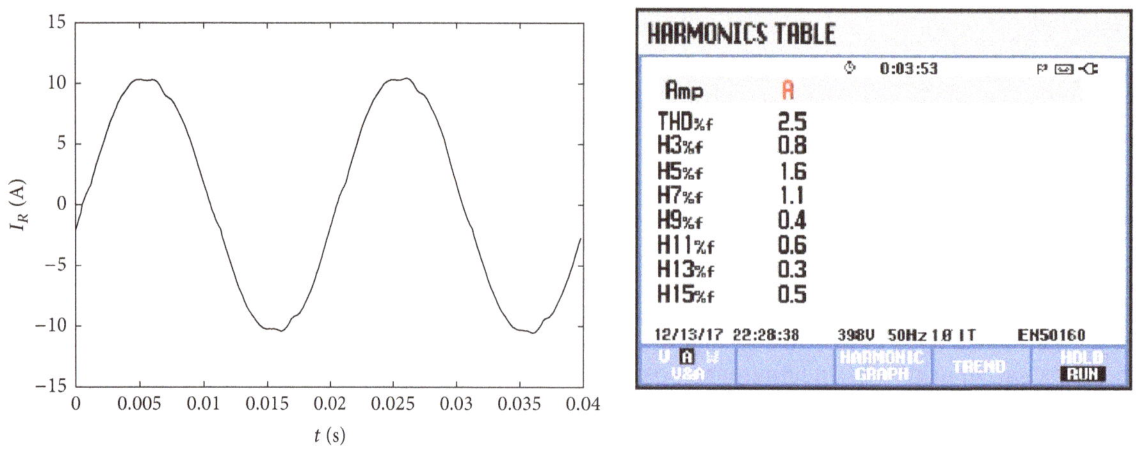

(a) Waveform (b) Harmonics table

FIGURE 10: Waveform and harmonics table of current on ac side with feedforward harmonic mitigation strategy.

mitigation strategy. However, because the theoretical analysis does not consider the effect of PWM dead zone and control delay, it cannot completely eliminate the harmonic current. At the same time, it can be explained that after the 3rd, 5th, and 7th harmonic signal are superimposed in the PWM modulation wave, the other order harmonics currents cannot be suppressed.

7. Conclusion

Based on the dynamic phasor method, a harmonic analysis model of SPVSC was established, which takes the controller into consideration. The proposed model is used as a tool to analyze the generation mechanism of harmonic as well as the harmonic interaction between the ac and dc side. The harmonic currents injecting to the grid from SPVSC can be mainly carried out in two ways: external excitation and internal interaction. With the internal interaction between ac and dc side, SPVSC generates series of low-order odd harmonic currents on ac side and series of low-order even harmonic voltages on dc side. With the external excitation, the distorted grid voltages will exacerbate the generation of harmonic currents on ac side. According to the generation mechanism, a feedforward control strategy was put forward to mitigate the harmonic currents from SPVSC which has clear physical meaning, fast calculation speed, and robustness. Meanwhile, it avoids the design of controller parameter and will not affect the stability of the system. Simulation and experimental results are finally presented to verify the proposed harmonic analyzed model and feedforward harmonic mitigation strategy.

Conflicts of Interest

The authors declare that there are no conflicts of interest regarding the publication of this paper.

References

[1] Y. Han, X. Fang, P. Yang, C. Wang, L. Xu, and J. M. Guerrero, "Stability analysis of digital controlled single-phase inverter with synchronous reference frame voltage control," *IEEE Transactions on Power Electronics*, no. 99, 2017.

[2] R. I. Bojoi, L. R. Limongi, D. Roiu, and A. Tenconi, "Enhanced power quality control strategy for single-phase inverters in distributed generation systems," *IEEE Transactions on Power Electronics*, vol. 26, no. 3, pp. 798–806, 2011.

[3] T. Dragicevic, X. Lu, J. C. Vasquez, and J. M. Guerrero, "DC Microgrids - Part I: A Review of Control Strategies and Stabilization Techniques," *IEEE Transactions on Power Electronics*, vol. 31, no. 7, pp. 4876–4891, 2016.

[4] E. Romero-Cadaval, B. Francois, M. Malinowski, and Q.-C. Zhong, "Grid-connected photovoltaic plants: An alternative energy source, replacing conventional sources," *IEEE Industrial Electronics Magazine*, vol. 9, no. 1, pp. 18–32, 2015.

[5] D. Dong, T. Thacker, I. Cvetkovic et al., "Modes of operation and system-level control of single-phase bidirectional PWM converter for microgrid systems," *IEEE Transactions on Smart Grid*, vol. 3, no. 1, pp. 93–104, 2012.

[6] R.-J. Wai, C.-Y. Lin, Y.-C. Huang, and Y.-R. Chang, "Design of high-Performance stand-alone and grid-connected inverter for distributed generation applications," *IEEE Transactions on Industrial Electronics*, vol. 60, no. 4, pp. 1542–1555, 2013.

[7] S. B. Kjaer, J. K. Pedersen, and F. Blaabjerg, "A review of single-phase grid-connected inverters for photovoltaic modules," *IEEE Transactions on Industry Applications*, vol. 41, no. 5, pp. 1292–1306, 2005.

[8] Q. Zhong, J. Feng, L. Lin, G. Wang, and H. Li, "Graphical presentation of harmonic operating principles in voltage source converter with sparse vector methods," *IET Generation, Transmission & Distribution*, vol. 11, no. 4, pp. 1033–1038, 2017.

[9] M.-S. Karbasforooshan, M. Monfared, and M. Dogruel, "Application of the harmonic control arrays technique to single-phase stand-alone inverters," *IET Power Electronics*, vol. 9, no. 7, pp. 1445–1453, 2016.

[10] Z. Wu, Y. Zou, C. Wang, Y. Zhang, and Z. Zhang, "A performance improvement technique for single-phase PWM rectifier," in *Proceedings of the 2009 IEEE 6th International Power Electronics and Motion Control Conference, (IPEMC '09)*, pp. 1638–1642, China, May 2009.

[11] J.-L. Gao, Y.-J. Zhang, F. Lin, and Q.-L. Zheng, "Research on harmonic current elimination method of single-phase PWM rectifiers," *Proceedings of the Chinese Society of Electrical Engineering (CSEE)*, vol. 30, no. 32, pp. 32–38, 2010.

[12] K. Zhou, Z. Qiu, N. R. Watson, and Y. Liu, "Mechanism and elimination of harmonic current injection from single-phase grid-connected PWM converters," *IET Power Electronics*, vol. 6, no. 1, pp. 88–95, 2013.

[13] A. Kulkarni and V. John, "Mitigation of lower order harmonics in a grid-connected single-phase PV inverter," *IEEE Transactions on Power Electronics*, vol. 28, no. 11, pp. 5024–5037, 2013.

[14] R. Teodorescu, F. Blaabjerg, M. Liserre, and P. C. Loh, "Proportional-resonant controllers and filters for grid-connected voltage-source converters," *IEE Proceedings Electric Power Applications*, vol. 153, no. 5, pp. 750–762, 2006.

[15] M. Castilla, J. Miret, J. Matas, L. G. de Vicuña, and J. M. Guerrero, "Control design guidelines for single-phase grid-connected photovoltaic inverters with damped resonant harmonic compensators," *IEEE Transactions on Industrial Electronics*, vol. 56, no. 11, pp. 4492–4501, 2009.

[16] P. Mattavelli and F. P. Marafao, "Repetitive-based control for selective harmonic compensation in active power filters," *IEEE Transactions on Industrial Electronics*, vol. 51, no. 5, pp. 1018–1024, 2004.

[17] S. Jiang, D. Cao, Y. Li, J. Liu, and F. Z. Peng, "Low-THD, fast-transient, and cost-effective synchronous-frame repetitive controller for three-phase UPS inverters," *IEEE Transactions on Power Electronics*, vol. 27, no. 6, pp. 2994–3005, 2012.

[18] Y. Yang, K. Zhou, and F. Blaabjerg, "Current harmonics from single-phase grid-connected inverters-examination and suppression," *IEEE Journal of Emerging and Selected Topics in Power Electronics*, vol. 4, no. 1, pp. 221–233, 2016.

[19] S. R. Sanders, J. M. Noworolski, X. Z. Liu, and G. C. Verghese, "Generalized averaging method for power conversion circuits," *IEEE Transactions on Power Electronics*, vol. 6, no. 2, pp. 251–259, 1991.

[20] N. Kroutikova, C. A. Hernandez-Aramburo, and T. C. Green, "State-space model of grid-connected inverters under current control mode," *IET Electric Power Applications*, vol. 1, no. 3, pp. 329–338, 2007.

[21] J. Kwon, X. Wang, F. Blaabjerg, C. L. Bak, V. Sularea, and C. Busca, "Harmonic interaction analysis in a grid-connected converter using harmonic state-space (HSS) modeling," *IEEE Transactions on Power Electronics*, vol. 32, no. 9, pp. 6823–6835, 2017.

[22] Q. Zhong, L. Lin, G. Wang, Y. Zhang, and Z. Wu, "Harmonic analysis model for voltage source converter under unbalanced conditions," *IET Generation, Transmission & Distribution*, vol. 9, no. 1, pp. 12–21, 2015.

[23] J. B. Kwon, X. Wang, F. Blaabjerg, C. L. Bak, A. R. Wood, and N. R. Watson, "Harmonic instability analysis of a single-phase grid-connected converter using a harmonic state-space modeling method," *IEEE Transactions on Industry Applications*, vol. 52, no. 5, pp. 4188–4200, 2016.

[24] P. Č. Stefanov and A. M. Stanković, "Modeling of UPFC operation under unbalanced conditions with dynamic phasors," *IEEE Transactions on Power Systems*, vol. 17, no. 2, pp. 395–403, 2002.

15

Analysis of Transient Voltage Stability in a Low Voltage Distribution Network using SST for the Integration of Distributed Generations

Sheng Li, Zhihao Zhou, Qiqi Shan, and Jiani An

School of Electric Power Engineering, Nanjing Institute of Technology, Nanjing 211167, China

Correspondence should be addressed to Sheng Li; lisheng_njit@126.com

Academic Editor: Chi-Seng Lam

Models of a low voltage distribution network using a typical tertiary-structure solid state transformer (SST) for the integration of distributed generations (DGs) and a conventional low voltage distribution network integrated with DGs were established to study the transient voltage stability issue, using the power system simulation software PSCAD. Effects on the transient voltage stability of the load bus and DC bus in the SST system are analyzed when grid-side cable line faults (such as short circuit and line disconnection) occur or the total output of DGs drops greatly. The results show that, comparing with the conventional system, the SST has apparent advantages on enhancing the transient voltage stability of load bus while facing different disturbances, even though SST has to regulate the voltage passively. Short circuit faults at different location of the grid-side line have different effects on the transient voltage stability, while the effect of disconnection fault is not related to fault location. Moreover, the DC bus voltage is easy to keep climbing continually when short circuit fault of the line occurs that is close to the SST input stage or disconnection fault occurs at any location of the line. If a battery energy storage station is installed, the transient voltage stability of DC bus and load bus will be improved effectively because of the regulation function of battery storage.

1. Introduction

With the technologies related to new energy generation being maturing day by day and the construction of the energy internet being pushed forward constantly [1–3], penetrations of distributed generations (DGs) in the distribution network are increasing gradually [4, 5]. Before the electric power grid and other energy networks connecting with each other and forming an energy internet, building a green smart grid with the core of energy routers to accept large-scale renewable energy sources widely is one way to construct the energy internet [6]. The intelligent dispatching and controlling center of the energy internet is the energy router [7], whose core is a high-frequency coupled device called solid state transformer (SST) based on the advanced power electronic technology [8].

The output of DG is random and the configuration of the distribution network integrated with DGs is complex, which can influence the stability of the distribution network [9, 10]. Reference [11] pointed out that the voltage stability issue does exist in distribution networks integrated with photovoltaic cell power, and the voltage instability phenomenon can be solved effectively by photovoltaic inverter reactive power support [2, 11]. A typical simulation model that a distributed photovoltaic power station is directly integrated into a low voltage distribution network was established in [12], and the effects were analyzed on the transient voltage stability of load bus when the faults such as short circuit and line disconnection occur or the output of distributed photovoltaic power plant drops greatly. In [13], the impacts of large-scale photovoltaic generation on the steady state voltage stability of distribution system were studied, and the conclusion that the photovoltaic modules have to operate in the reactive power support mode to improve the system voltage stability was obtained. In [14], a DC microgrid containing photovoltaic cell, wind turbine, and fuel cell was proposed and connected to the low voltage distribution network under variable load demands, and the simulation results showed that the voltage

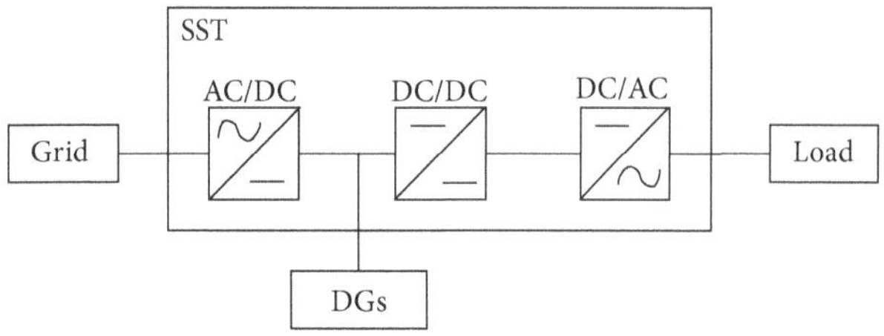

FIGURE 1: A typical tertiary-structure solid state transformer.

stability of DC bus can be maintained at different operating modes. In [15], a method that can enhance the transient voltage stability of a real distribution network with wind power was proposed, by means of optimal rating and location of the static synchronous compensator (STATCOM).

When using a SST for the integration of DGs and the distribution network, on the part of the grid, the ability of adopting the DGs will be enhanced and the situation of voltage oscillation and instability will be avoided. In this paper, a simulation model of a 0.38 kV low voltage distribution network which uses a typical SST for the integration of DGs was established using the power system simulation software PSCAD/EMTDC, and effects on the transient voltage stability of the system while facing various disturbances are studied. The advantages of the transient voltage stability using a SST for the integration of DGs are analyzed, and the differences between whether there is a battery energy storage station are discussed.

2. Modeling and Control Strategy of SST

As shown in Figure 1, a typical tertiary-structure SST is chosen for discussion, which consisted by a AC/DC unit, a DC/DC unit, and a DC/AC unit. The three units are input stage, isolation stage, and output stage, respectively. To realize the bidirectional power flow, the full-controlled device, IGBT, is used for building the SST [16].

In this paper, a voltage-source PMW rectifier is chosen as the input stage of SST, which can rectify the 0.38 kV AC of the low voltage distribution network to 0.78 kV DC.

The dual-loop decoupled control strategy is chosen for the input stage [17]. The voltage orientation control strategy based on d-q rotating coordinate system is selected for the voltage loop, and the direct current control is adopted for the current loop. The grid-side power factor can be corrected according to the control goal. The control equations of the SST input stage are

$$\begin{aligned}
i_{1d}^{*} &= \left(K_{P1} + \frac{K_{I1}}{s}\right)\left(U_{dc}' - U_{dc}\right) \\
i_{1q}^{*} &= 0 \\
u_{1d}^{*} &= u_{1d} + \omega L i_{1q} - \left(K_{P2} + \frac{K_{I2}}{s}\right)\left(i_{1d}^{*} - i_{1d}\right) \\
u_{1q}^{*} &= u_{1q} - \omega L i_{1d} - \left(K_{P2} + \frac{K_{I2}}{s}\right)\left(i_{1q}^{*} - i_{1q}\right),
\end{aligned} \tag{1}$$

where i_{1d}^{*} and i_{1q}^{*} are the grid-side current command values in d-q coordinate system and i_{1d} and i_{1q} are the actual values; u_{1d}^{*} and u_{1q}^{*} are the grid-side voltage command values in d-q coordinate system and u_{1d} and u_{1q} are the actual values; U_{dc}', U_{dc} are the reference voltage and actual voltage of DC bus, respectively; K_{P1}, K_{I1}, K_{P2}, and K_{I2} are the parameters of PI controllers; L is the filter inductance of SST input stage.

In isolation stage, the 0.78 kV DC is converted to a high-frequency AC square wave and coupled to the secondary side of a high-frequency isolated transformer which is used to isolate the original side and subside. Then, the high-frequency AC will be rectified to 0.78 kV DC.

In the output stage, the 0.78 kV DC is inverted to 0.38 kV AC for power supplies. The dual-loop decupled control strategy with the voltage and current feedforward is adopted for the control [18]. The control equations of the SST output stage are

$$\begin{aligned}
i_{2d}^{*} &= i_{2d} - \omega C_f u_{2q} + \left(K_{P3} + \frac{K_{I3}}{s}\right)\left(u_{2d}' - u_{2d}\right) \\
i_{2q}^{*} &= i_{2q} + \omega C_f u_{2d} + \left(K_{P3} + \frac{K_{I3}}{s}\right)\left(u_{2q}' - u_{2q}\right) \\
u_{2d}^{*} &= u_{2d} - \omega L_f i_{2q} + \left(K_{P4} + \frac{K_{I4}}{s}\right)\left(i_{2d}^{*} - i_{2d}\right) \\
u_{2q}^{*} &= u_{2q} + \omega L_f i_{2d} + \left(K_{P4} + \frac{K_{I4}}{s}\right)\left(i_{2q}^{*} - i_{2q}\right),
\end{aligned} \tag{2}$$

where i_{2d}^{*} and i_{2q}^{*} are the load current command values in d-q coordinate system and i_{2d} and i_{2q} are the actual values; u_{2d}^{*} and u_{2q}^{*} are the load bus voltage command values in d-q coordinate system, u_{2d}' and u_{2q}' are the reference values, and u_{2d} and u_{2q} are the actual values; K_{P3}, K_{I3}, K_{P4}, and K_{I4} are the parameters of PI controllers; L_f, C_f are the filter inductance and capacitance of SST output stage, respectively.

3. Model of Low Voltage Distribution Network Integrated with DGs

A conventional system model of a low voltage distribution network integrated with DGs (including distributed photovoltaic and wind power) is established using the power system simulation software PSCAD, as shown in Figure 2. $\dot{U}_1$ and $\dot{U}_2$ are the voltage of Bus 1 and Bus 2, respectively. The impedance of the cable is $R + jX$ and the value is $0.096 + j0.015\,\Omega$. The power from the low voltage distribution network is noted as $P_1 + jQ_1$, the power flow from the network to Bus 2 is noted as $P_2 + jQ_2$. The outputs of photovoltaic power plant and wind power farm are $P_{pv} + jQ_{pv}$ and $P_{wind} + jQ_{wind}$, respectively; the absorbed power of the three-phase induction motor and the static load are $P_M + jQ_M$ and $P_s + jQ_s$, respectively.

Some ancillary facilities such as reactive power compensation devices and inverters of photovoltaic or wind power systems can be left out because of the structure and flexible control characteristics of SST, and the DGs can be directly connected to the DC bus. The model of a low voltage distribution network using a SST for the integration of DGs is shown in Figure 3.

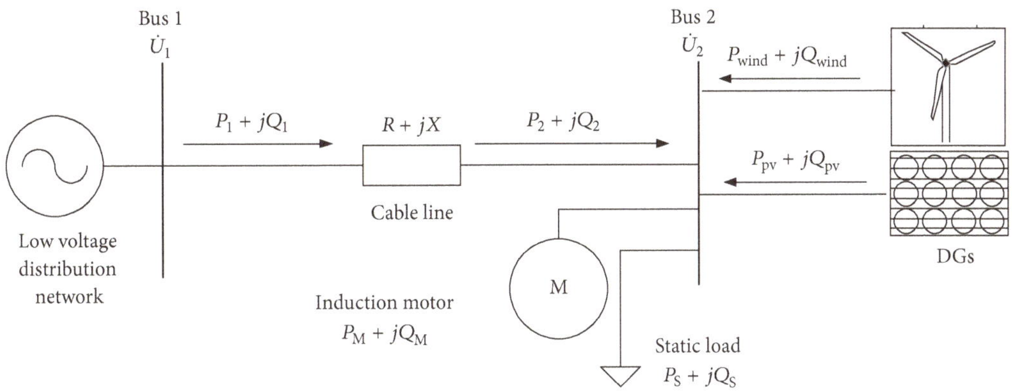

FIGURE 2: Model of conventional low voltage distribution network integrated with DGs.

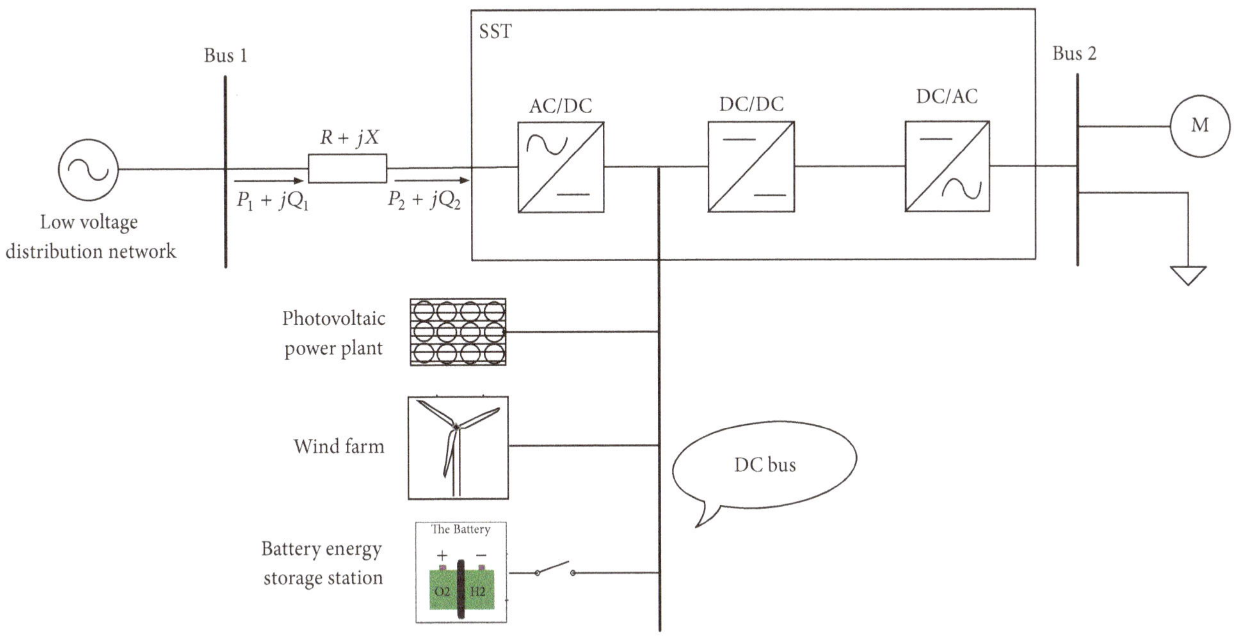

FIGURE 3: Model of low voltage distribution network integrated with DGs using SST.

In Figures 2 and 3, the rated capacities of the photovoltaic power station and the wind power plant both are 260 kW. The loads are composed of two parts of static load (including a constant resistance load and a constant impedance load) and a three-phase induction motor. The rated power of the constant resistance R_{s1} is 90 kW; the rated power of the constant impedance $R_{s2} + jX_{s2}$ is 60 kVA; the rated capacity of the three-phase induction motor is 49.5 kVA.

During the simulation analyses, the environmental temperature is 298.15 K, the light intensity is 1000 W/m^2, the wind speed is 11 m/s, and these values will vary in a certain rage. When the system operates normally, the absorbed active and reactive power of the induction motor are about 46 kW and 23 kvar, respectively; the absorbed active and reactive power of the constant impedance load are about 48 kW and 35 kvar, respectively. The grid-side unity power factor is realized due to the control strategy of the SST input stage.

4. Effects on Transient Voltage Stability of Grid-Side Line Faults

The research results show that short circuit fault of grid-side cable line at different location has different effect on the system transient voltage stability because of the control function of the SST input stage, while line disconnection fault is not associated with the fault location.

The schematic diagram of fault position is shown in Figure 4. The grid-side cable line is divided into Line 1 and Line 2 equivalently, by the signal collection device which monitors the distribution network. Line 1 contains only a small part of lines connecting the signal collection device and the SST input stage in practice, and its fault possibility is much lower. Line 2 contains the whole external distribution network lines which are so complex that their fault possibility is very high.

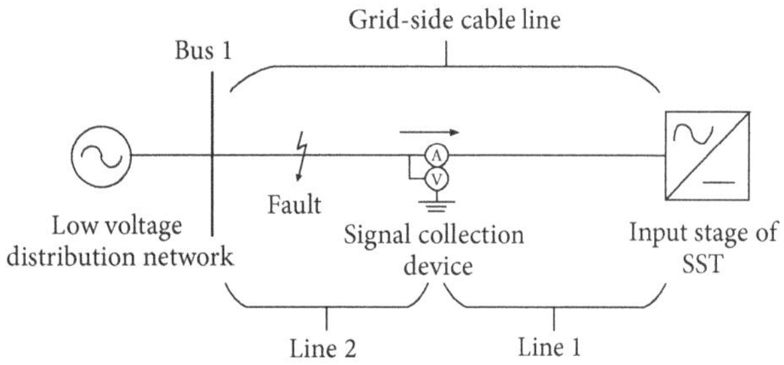

Figure 4: Schematic diagram of line fault location.

To save space and take the aesthetics into consideration, most urban low voltage distribution networks use cable lines. Insulation aging or damage is the main cause of cable faults.

4.1. Short Circuit Fault Occurs at Line 1. Line 1 may have short circuit faults though there are fewer lines in it. Once the fault occurs, each stage of the SST has to regulate the voltage passively. Therefore, we focus on the influence on the system transient voltage stability when different short circuit faults occur at Line 1.

Take the typical single-phase grounding short circuit fault as an example. Now set the simulation time as 2 s. After time $t = 0.4$ s, the system enters stable operation. When $t = 0.6$ s, single-phase grounding short circuit fault occurs at the midpoint of Line 1, and the fault duration is 0.1 s. The relay protection acts and removes the fault line when $t = 0.65$ s. We focus on the system transient voltage stability when the cut line is no longer reinput because 0.38 kV low voltage cable line has generally no reclosing device.

After the line is cut off, the grid loses power and the DGs enter the state of isolation island operation. We do this research on the condition that the outputs of DGs are rich and can meet the need of local loads. When the single-phase grounding short circuit fault occurs, the response curves of load bus voltage and absorbed powers of induction motor in the conventional DGs grid-connected system (shown in Figure 2), the response curves of load bus (Bus 2) voltage, absorbed powers of induction motor, and DC bus voltage are shown in Figure 5.

In general, if a 0.38 kV AC bus voltage in low voltage distribution network is lower than 0.9 pu or higher than 1.1 pu after a large disturbance for more than the specified limit time (e.g., 1 s or above), it can be considered that the low voltage AC bus is transient voltage instability [12]. In this paper, the transient voltage stability of low voltage DC bus obeys the same rules, too.

As can be seen from Figures 5(a) and 5(b), for the conventional system, after the single-phase grounding short circuit fault occurs and the relay protection acts, the load bus voltage drops rapidly to 0.8 pu or less, and the load bus is considered to be unstable, thus causing the active and reactive power absorbed by induction motor to have oscillations and finally to be below rated values (rated active power is 1 pu and rated reactive power is 0.5 pu). The limitation of the conventional system is that the inverter control function of DGs is simple, and the transient voltage stability of load bus cannot be guaranteed while facing disturbances.

In this paper, the DGs' inverters adopt outlet voltage control strategy; simulation results show that the transient voltage stability of load bus cannot be maintained by the inverters without the support of power grid.

If the inverters adopt constant power control strategy, the outputs of DGs will remain constant under the normal environment after the disturbances. After the relay protection acts, the load bus has the equation according to the power balance:

$$P_{\text{pv}} + P_{\text{wind}} = P_M + P_S = P_M + \left(\frac{1}{R_{s1}} + \frac{R_{s2}}{R_{s2}^2 + X_{s2}^2} \right) U_2^2 \quad (3)$$

It can be seen from Formula (3), in the case of rich outputs of DGs, the load demand can be met, so the load bus voltage recovers quickly. Accordingly, the absorbed active power of induction motor begins to fall down after the oscillation. From Formula (3), we can find that U_2 will be lifted once P_M starts to decrease, eventually causing the load bus voltage to climb continually. And at this time, the load bus is still considered to be transient voltage instability. It is necessary to abandon the photovoltaic and wind power to protect the motor in time.

The effectiveness of the control strategy of SST on the transient voltage stability can be seen from Figures 5(a) and 5(b), which reflects the SST's better ability of maintaining transient voltage stability while facing disturbances. Though the fault is close to the SST input stage, active control cannot be taken by the SST to stabilize the grid-side surge current and the current has to be regulated passively and the load bus voltage can recover to the original rated value and maintain the stability after experiencing rapid sag. Therefore, the absorbed powers of induction motor can recover to the original rated values too.

As shown in Figure 5(c), although the SST can maintain the voltage stability of load bus by passive regulation, if there is no a battery energy storage station installed in the DC bus, the DC bus voltage will keep climbing because the outputs of DGs cannot be transmitted to the grid. At this time, we can consider that the DC bus loses its transient voltage stability. It is harmful to the security and stability of the system and the photovoltaic and wind power will have to be abandoned.

The battery energy storage station plays an important role in stabilizing the load bus voltage and DC bus voltage. As shown in Figure 5, in the DGs grid-connected system based on SST, the battery energy storage station can stabilize the load bus voltage and DC bus voltage greatly after the shot circuit fault occurs, and there is merely a slight oscillation. After the relay protection acts, the DC bus voltage is regulated to the original rated value because of the function of energy storage station, avoiding the voltage climbing phenomenon.

Simulation results show there are similar conclusions when other short circuit faults occur at Line 1; for instance, the response curves of three-phase short circuit occurring at midpoint of Line 1 are shown in Figure 6.

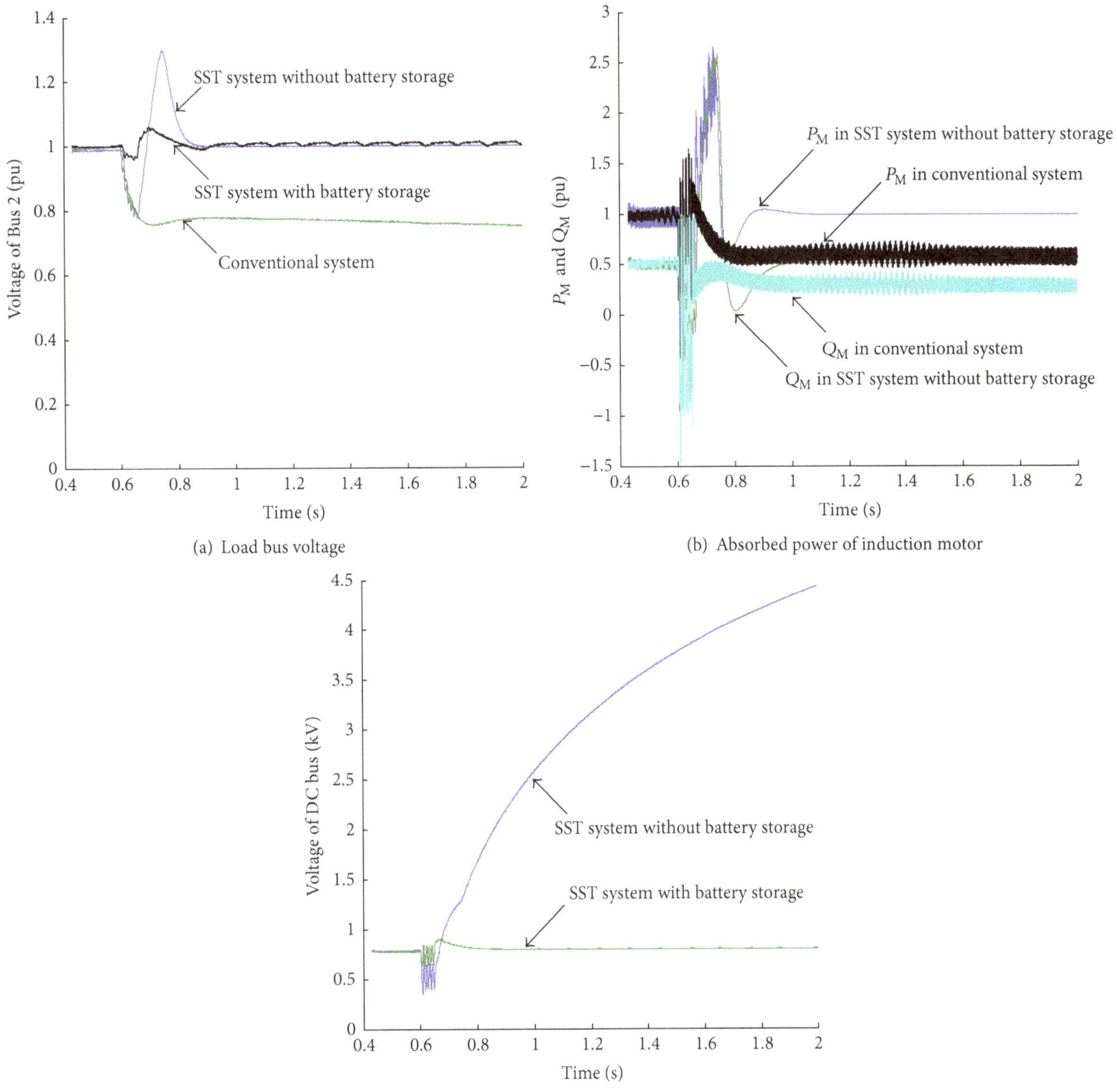

(a) Load bus voltage

(b) Absorbed power of induction motor

(c) DC bus voltage

FIGURE 5: Response curves when single-phase grounding short circuit fault occurs at the midpoint of Line 1.

4.2. Short Circuit Fault Occurs at Line 2. If short circuit fault occurs at Line 2, the information can be picked up and the surge current can be steadied by the active control of the SST input stage, and it merely has little effect on the load bus voltage and DC bus voltage even if the most serious three-phase short circuit fault occurs, as shown in Figure 7.

4.3. Disconnection Fault Occurs at Any Location of Gird-Side Line. When the line disconnection fault occurs at any location of the gird-side line (including Line 1 and Line 2), we can draw similar conclusions with the condition that short circuit faults occurs at Line 1. Taking the disconnection occurring at the midpoint of Line 1, for example, the response curves are shown in Figure 8.

5. Impact of Great Drop of DGs' Total Output on Transient Voltage Stability

The output of DG is affected by the weather greatly. The sudden changes in light intensity and wind speed will cause sudden changes in the outputs of photovoltaic power plant and wind farm, respectively, which will affect the transient voltage stability of the system.

Considering the most adverse weather condition, when the light intensity drops greatly and the wind turbine drops out because wind speed is too large, the sudden great drop of DGs' total output can cause the load bus voltage to be instability. For the conventional system shown in Figure 2, supposing that the total load power is $P_L + jQ_L$ and the total

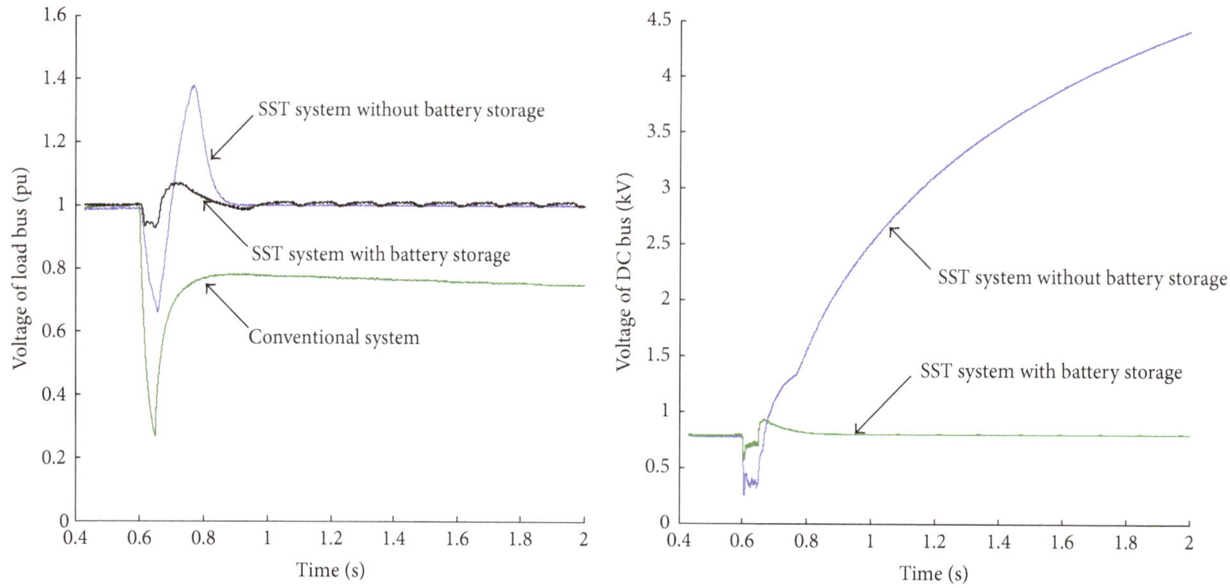

(a) Load bus voltage

(b) DC bus voltage

FIGURE 6: Response curves when three-phase short circuit fault occurs at the midpoint of Line 1.

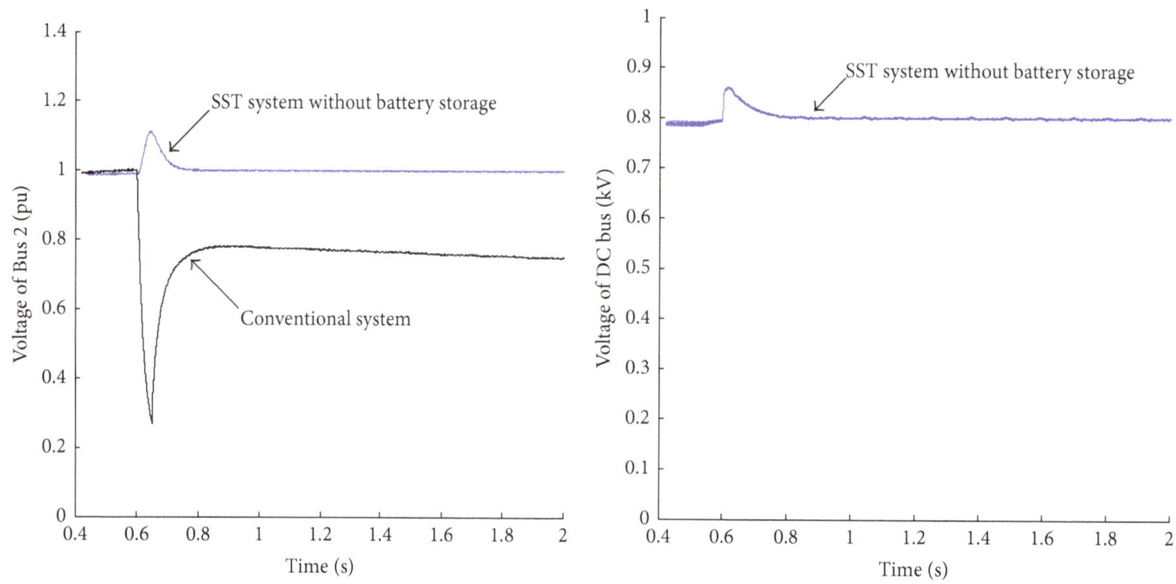

(a) Load bus voltage

(b) DC bus voltage

FIGURE 7: Response curves when three-phase short circuit fault occurs at the midpoint of Line 2.

output of DGs is $P_\Sigma + jQ_\Sigma$, the load bus voltage drops to U_2' when the DGs' total output drops to 0 suddenly; there will be [12]

$$U_2 - U_2' = \frac{P_\Sigma R + Q_\Sigma X}{U_2} + (P_L R + Q_L X)\left(\frac{1}{U_2'} - \frac{1}{U_2}\right) \quad (4)$$

Since both photovoltaic and wind power have maximum power control, Q_Σ is almost 0. Suppose the voltage drop ratio is a; that is, $U_2' = aU_2$. Put $Q_\Sigma = 0$ and $U_2' = aU_2$ into Formula (4), we can get the following:

$$P_\Sigma = \frac{(1-a)U_2^2 - (1/a - 1)(P_L R + Q_L X)}{R} \quad (5)$$

In Formula (5), if U_2 is the rated value (1 pu), then a takes 0.9. And we can get the upper limit approximate value of total output active power of DGs that is just right for maintaining the system transient voltage stability when the total output

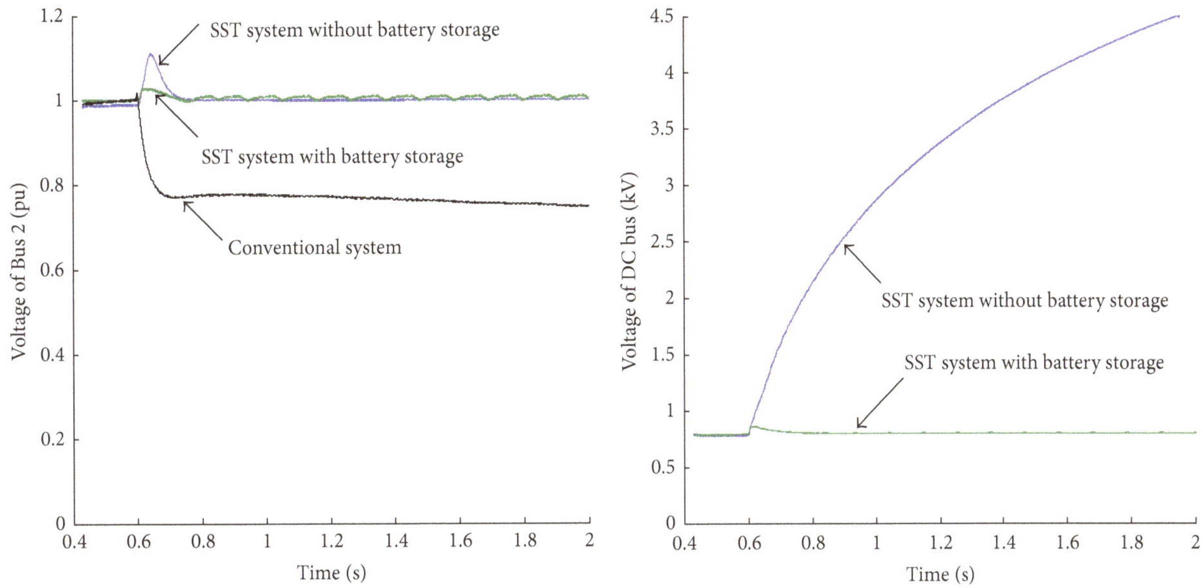

(a) Load bus voltage

(b) DC bus voltage

FIGURE 8: Response curves when disconnection fault occurs at the midpoint of Line 1.

drops to 0. When the DGs' total output is equal or greater than the upper limited approximate value in the actual operation of the system and drops to 0 suddenly, the load bus will lose its transient voltage stability in general.

Using the parameters in the conventional system to calculate according to Formula (5), it will figure out that the upper limit approximate value of DGs' total output is 150 kW as a takes 0.9 (because U_2 is very close to 1 pu). Now, the total output of DGs in the conventional system and the SST system is set to be 150 kW, respectively. The simulation time is set to be 2 s, and the photovoltaic and wind power suddenly drop to 0 at $t = 0.6$ s. The response curves are shown in Figure 9.

It can be known from Figures 9(a) and 9(b) that the load bus voltage in the conventional system is unstable after the great drop of DGs' total output occurs, and the voltage drops to about 0.88 pu which is almost matching with the theoretical value (0.9 pu), while in the SST system, the load bus and DC bus recover transient voltage stability quickly. When it comes to the SST system with a battery energy storage station, the voltages of load bus and DC bus are almost not affected by the disturbance. As can be seen from Figure 9(c), for the SST system without a battery energy storage station, before the great drop of DGs' total output occurs, the reactive power transmitted from the grid is 0 because of the control of the SST input stage, while the active power is −0.2 MW because of the plenty outputs of DGs. After the total output of DGs drops to 0, the low voltage distribution network supplies a certain amount of powers to help to maintain the bus voltage at this moment.

6. Conclusions

The development of the technology of DGs has realized the effective integration and efficient utilization of renewable energies distributed around the world. It has practical significance to study on the transient voltage stability of the DGs grid-connected system and it is meaningful for realizing the change from a traditional fossil-fuel grid to a future green smart grid. In this paper, we use PSCAD software to study the transient voltage stability of the low voltage distributed network integrated with DGs based on SST under the condition of cable line fault and sudden drop of DGs' total output and concluded as follows:

(1) Using a SST for the integration of DGs and the distribution network has advantages on the system transient voltage stability while facing different faults. The influence of grid-side line short circuit fault on the transient voltage stability is related to fault location, and the line disconnection fault is not the case. Even if SST has to regulate the voltage passively when the fault occurs at the line close to the SST input stage, the regulation function of SST still can maintain the voltage transient stability of load bus. This avoids load bus in the conventional DGs gird-connected system losing its transient voltage stability effectively while facing such faults and improves utilization of DGs and reliability of power supply.

(2) Although the SST control makes the load bus have better transient voltage stability, the DC bus voltage is easy to keep climbing continually when the short circuit occurs at the line side that is close to the SST input stage or the line disconnection occurs at any location of the line. At this moment, the DC bus can be considered to be transient voltage instability, and this is a new transient voltage stability phenomenon in the DGs gird-connected system using SST. It is harmful to the security and stability of the system

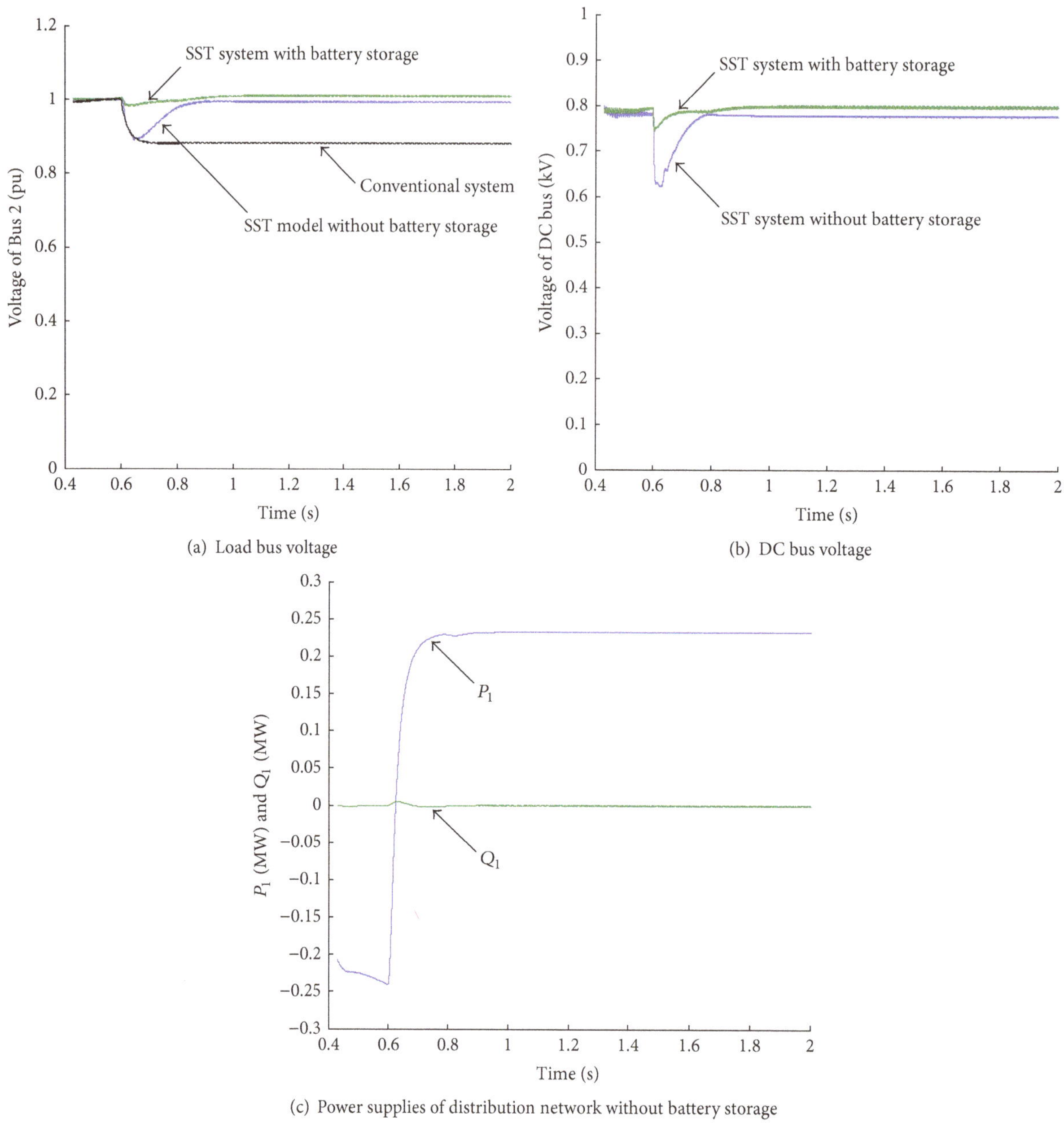

FIGURE 9: Response curves when total output of DGs drops to 0.

operation. The DC bus voltage can be stabilized by the DGs equipped with a certain amount of batteries or a battery energy storage station installed in the DC bus when faults occur, which can effectively avoid the transient voltage instability of DC bus and guarantee the system safety.

(3) When the total output of DGs drops greatly under the adverse weather condition, the SST can ensure the transient voltage stability thus avoiding the biggest voltage drop of load bus in the conventional system. If a battery energy storage station is installed, the load bus voltage and DC bus voltage will be almost unaffected while facing the great drop of the DGs' total output. It is of great significance to the stable operation of low voltage distribution network that the battery storage is installed.

Conflicts of Interest

The authors declare that there are no conflicts of interest regarding the publication of this paper.

Acknowledgments

This work is supported by the Open Research Fund of Jiangsu Collaborative Innovation Center for Smart Distribution

Network (XTCX201613), the University Student Innovation & Entrepreneurship Training Program of Jiangsu Province (201711276009Z), and the University Student Technological Innovation Project of Nanjing Institute of Technology (TZ20170009).

References

[1] S. Li, Z. Wei, Y. Ma, and J. Cheng, "Prediction and control of Hopf bifurcation in a large-scale PV grid-connected system based on an optimised support vector machine," *The Journal of Engineering*, vol. 2017, no. 14, pp. 2666–2671, 2017.

[2] A. Eid and M. Abdel-Akher, "Voltage control of unbalanced three-phase networks using reactive power capability of distributed single-phase PV generators," *International Transactions on Electrical Energy Systems*, vol. 27, no. 11, Article ID e2394, 2017.

[3] A. Q. Huang, M. L. Crow, G. T. Heydt, J. P. Zheng, and S. J. N. I. S. Dale, "The Future Renewable Electric Energy Delivery and Management (FREEDM) System: The Energy Internet," *Proceedings of the IEEE*, 2010.

[4] J. M. Sexauer and S. Mohagheghi, "Voltage quality assessment in a distribution system with distributed generation-a probabilistic load flow approach," *IEEE Transactions on Power Delivery*, vol. 28, no. 3, pp. 1652–1662, 2013.

[5] S. Li, Z. Wei, and Y. Ma, "Fuzzy Load-Shedding Strategy Considering Photovoltaic Output Fluctuation Characteristics and Static Voltage Stability," *Energies*, vol. 11, no. 4, p. 779, 2018.

[6] J. Cao and M. Yang, "Energy internet-towards smart grid 2.0," in *Proceedings of the 2013 Fourth International Conference on Networking and Distributed Computing (ICNDC)*, pp. 105–110, Los Angeles, CA, USA, December 2013.

[7] W. Sheng, H. Liu, Z. Zeng et al., "An energy hub based on virtual-machine control," in *Proceedings of the CSEE*, vol. 35, pp. 3541–3550, 2015.

[8] . Qing Duan, . Chunyan Ma, . Wanxing Sheng, and . Changkai Shi, "Research on power quality control in distribution grid based on energy router," in *Proceedings of the 2014 International Conference on Power System Technology (POWERCON)*, pp. 2115–2121, Chengdu, October 2014.

[9] B. Tamimi, C. Canizares, and K. Bhattacharya, "System stability impact of large-scale and distributed solar photovoltaic generation: The case of Ontario, Canada," *IEEE Transactions on Sustainable Energy*, vol. 4, no. 3, pp. 680–688, 2013.

[10] R. Majumder, "Some aspects of stability in microgrids," *IEEE Transactions on Power Systems*, vol. 28, no. 3, pp. 3243–3252, 2013.

[11] R. Yan and T. K. Saha, "Investigation of voltage stability for residential customers due to high photovoltaic penetrations," *IEEE Transactions on Power Systems*, vol. 27, no. 2, pp. 651–662, 2012.

[12] S. Li, C. Jiang, Z. Zhao, and Z. Li, "Study of transient voltage stability for distributed photovoltaic power plant integration into low voltage distribution network," *Dianli Xitong Baohu yu Kongzhi/Power System Protection and Control*, vol. 45, no. 8, pp. 67–72, 2017.

[13] M. M. Aly, M. Abdel-Akher, Z. Ziadi, and T. Senjyu, "Assessment of reactive power contribution of photovoltaic energy systems on voltage profile and stability of distribution systems," *International Journal of Electrical Power & Energy Systems*, vol. 61, pp. 665–672, 2014.

[14] A. Eid, "Utility integration of PV-wind-fuel cell hybrid distributed generation systems under variable load demands," *International Journal of Electrical Power & Energy Systems*, vol. 62, pp. 689–699, 2014.

[15] Y. K. Gounder, D. Nanjundappan, and V. Boominathan, "Enhancement of transient stability of distribution system with SCIG and DFIG based wind farms using STATCOM," *IET Renewable Power Generation*, vol. 10, no. 8, pp. 1171–1180, 2016.

[16] A. Huang and J. Baliga, "FREEDM system: role of power electronics and power semiconductors in developing an energy internet," in *Proceedings of the 21st International Symposium on Power Semiconductor Devices IC's (ISPSD*, pp. 9–12, Barcelona, Spain, 2009.

[17] Y. Ye, M. Kazerani, and V. Quintana, "A novel modeling and control method for three-phase PWM converters," in *Proceedings of the 2001 IEEE 32nd Annual Power Electronics Specialists Conference*, pp. 102–107, Vancouver, BC, Canada.

[18] F. Wang, A. Huang, G. Wang, X. She, and R. Burgos, "Feed-forward control of solid state transformer," in *Proceedings of the 2012 IEEE Applied Power Electronics Conference and Exposition - APEC 2012*, pp. 1153–1158, Orlando, FL, USA, Feburary 2012.

16

Low Actuating Voltage Spring-Free RF MEMS SPDT Switch

Deepak Bansal,[1,2] Anuroop Bajpai,[1] Prem Kumar,[1] Maninder Kaur,[1] and Kamljit Rangra[1,2]

[1] *Central Electronics Engineering Research Institute (CEERI), Council of Scientific and Industrial Research (CSIR), Pilani, Rajasthan 333031, India*
[2] *Academy of Scientific and Innovative Research (AcSIR), New Delhi, India*

Correspondence should be addressed to Deepak Bansal; dbansal.pu@gmail.com

Academic Editor: Jit S. Mandeep

RF MEMS devices are known to be superior to their solid state counterparts in terms of power consumption and electromagnetic response. Major limitations of MEMS devices are their low switching speed, high actuation voltage, larger size, and reliability. In the present paper, a see-saw single pole double throw (SPDT) RF MEMS switch based on anchor-free mechanism is proposed which eliminates the above-mentioned disadvantages. The proposed switch has a switching time of 394 nsec with actuation voltage of 5 V. Size of the SPDT switch is reduced by utilizing a single series capacitive switch compared to conventional switches with capacitive and series combinations. Reliability of the switch is improved by adding floating metal and reducing stiction between the actuating bridge and transmission line. Insertion loss and isolation are better than −0.6 dB and −20 dB, respectively, for 1 GHz to 20 GHz applications.

1. Introduction

Development of miniature wireless communications devices like mobile and satellite requires components with minimum power consumption with better performance. The RF MEMS devices are known to be better compared to solid state devices due to their low power consumption, high isolation, and low insertion loss. RF MEMS SPDT switches constitute the basic building block of the RF MEMS systems like SP4T [1], phase shifter [2], switch matrix [3], other communication systems [4], and so forth. The RF MEMS devices are also known to have few drawbacks too, for example, low switching speed (20–200 μsec), high pull-in voltage (15–80 V), larger size, and reliability [5, 6]. These issues at present limit the utilization of the MEMS devices on a larger scale. These mechanical parameters are function of spring constant and pull-in voltage. The conventional anchoring of a rectangular membrane is through two Euler beams where one end along the length is fixed to the membrane and the other to anchor post on either side of the membrane. When a voltage is applied between the movable and the fixed membrane of the switch as shown in Figure 1, the bridge moves down onto the fixed ground under the electrostatic force. At the pull-in voltage, the switch is in the down state [7] which is calculated using 1D lumped model as shown in Figure 1.

From lumped model neglecting higher order effects as mentioned in [7], actuation voltage of the switch is given by

$$V_p = \sqrt{\frac{8kg^3}{27A\varepsilon_o}}, \tag{1}$$

where k is spring constant and g and A are gap and area of overlap between the actuation electrode and the bridge.

From (1), actuation voltage can be lowered by (i) increasing the overlap area (A), (ii) decreasing the gap (g), or (iii) decreasing spring constant (k).

Each method has its own demerits; for example, (i) large overlap area (A) enhances stiction forces as stiction is directly proportional to contact area [6, 8]. Therefore, actuating electrode area is optimized based on stiction and restoring force [6].

Smaller gap (g) promotes stiction by reducing beam restoring force, which is proportional to the beam displacement. Smaller gap also adversely affects switch isolation through enhanced capacitive leakage. The bridge gap is also limited by the available fabrication technology and has been set in the range of 1–4 μm.

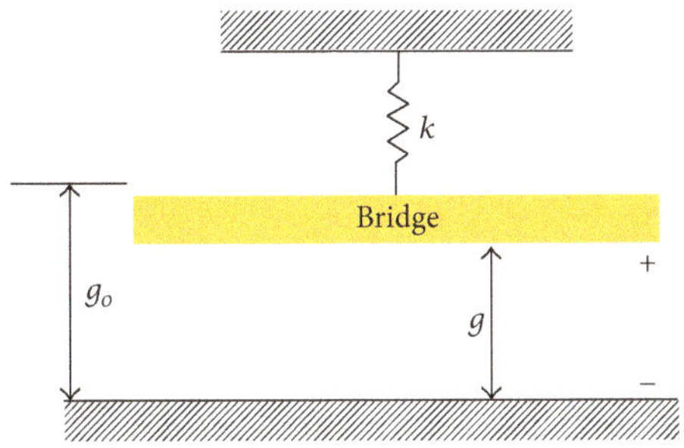

FIGURE 1: 1D lumped model for actuation voltage calculation.

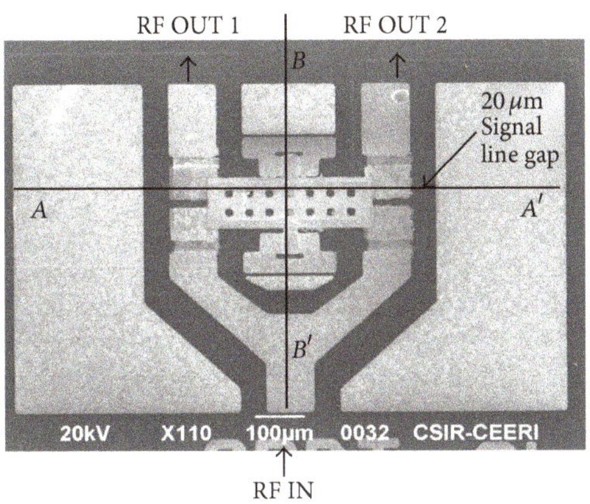

FIGURE 2: SEM image of a spring-free see-saw RF MEMS capacitive switch.

Decreasing k makes the beam complaint and prone to stiction. Further, switching time and pull-in voltage are inversely corelated to spring constant; devices with low pull-in voltage are slow to change state. In the present paper, a novel RF MEMS series capacitive SPDT switch based on anchor-free torsional configuration is proposed as shown in Figure 2. In the new configuration, as shown in Figures 3(a) and 3(b), anchoring is replaced by spring-free see-saw mechanism. The configuration has the advantages of low actuation voltage with high switching speed as discussed in Section 5.

2. SPDT Switch Description

The anchor-free see-saw SPDT designed for S-Ku band is similar to ohmic SPDT except for the contact points that are capacitive in this case. The capacitive contacts have an advantage of high operating frequency as described in [9]. Selection of the switch parameters is made based on available fabrication technology and electromechanical response of the switch. Input and output ports are made of 50 ohm Y-shaped CPW configuration (55–90–55 μm) for external connections. The bridge gap, dielectric material, and bridge thickness are decided by fabrication capabilities. The bridge gap lies in the range of 1–4 μm based on final releasing process, fixed at 2 μm in the present design. Silicon dioxide is selected as a dielectric material due to its well-known behavior. Floating metal area is selected based on operating frequency and bandwidth as

TABLE 1: Design parameters used in the proposed switch design.

S. number	Design parameter	Value	Unit
1	CPW (G-S-G)	55–90–55	μm
2	Anchor dimension	40 ∗ 50	μm^2
3	Oxide thickness	0.1	μm
4	Oxide dielectric constant	3.9	
5	Floating metal thickness	0.1	μm
6	Floating metal area	2 ∗ 65 ∗ 90	μm^2
7	Bridge thickness	4	μm
8	Bridge length	300	μm
9	Bridge width	100	μm
10	Bridge height above float metal	2	μm
11	Clamp thickness	4	μm
12	Support thickness	3	μm
13	Support dimensions	10 ∗ 100	μm^2
14	Signal line gap	20	μm
15	Actuating electrode area	100 ∗ 80	μm^2
16	Chip size	1080 ∗ 680	μm^2

it decides the capacitive ratio explained in Section 3 [8]. The actuation pads and bridge dimensions are fixed to keep pull-in voltage less than 5 V. The bridge and clamps are made of thick (4 μm) gold to avoid curling/buckling of the bridge. The signal line gap is fixed to 20 μm, lower gap leads to poor isolation, and higher gap needs larger width of the bridge to keep overlap capacitance fixed. All the parameters utilized in designing the SPDT switch are listed in Table 1. The SPDT switch utilizes single series capacitive switch which makes its size ($1.080 \times 0.680\,\text{mm}^2$) 85% compact compared to the literature, for example, $5.2 \times 3\,\text{mm}^2$ [9], $3.8 \times 1.1\,\text{mm}^2$ [10], and $3.5 \times 1.5\,\text{mm}^2$ [11].

3. Working of the Switch

Working of the see-saw switch is similar to conventional switches. On actuation, the bridge lands on one side of the transmission line as shown in Figures 3(a) and 3(b) and connects the input port to the output port. The other side of the bridge is lifted upwards and no signal is passed.

In OFF state, the switch (bridge) is in an upward tilted position that results in OFF state capacitance (C_{off}) of 0.885 fF as shown in Figures 4(a) and 4(c):

$$C_{\text{off}} = \varepsilon_o \cdot \frac{(\text{Bridge width} - \text{Signal line gap})/2 * \text{Bridge overlap}}{\text{Bridge height from floating metal}} = \varepsilon_o \frac{(100-20)/2 * 5}{2} = 0.885\,\text{fF}. \quad (2)$$

The capacitance (0.885 fF) between the input and output lines (reactance $1/\omega C > 10\,\text{k}\Omega$ for S-Ku band) provides good isolation. RF coupling between the signal lines through the bridge is negligible and can be neglected. Working principle

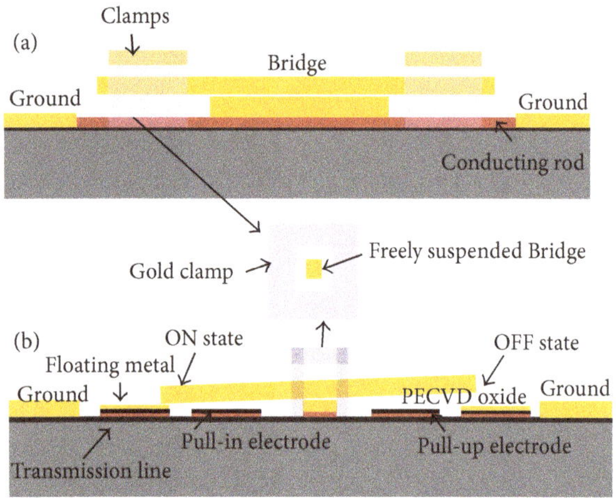

FIGURE 3: Cross-sectional view of working the SPDT switch (a) across AA′ and (b) across BB′.

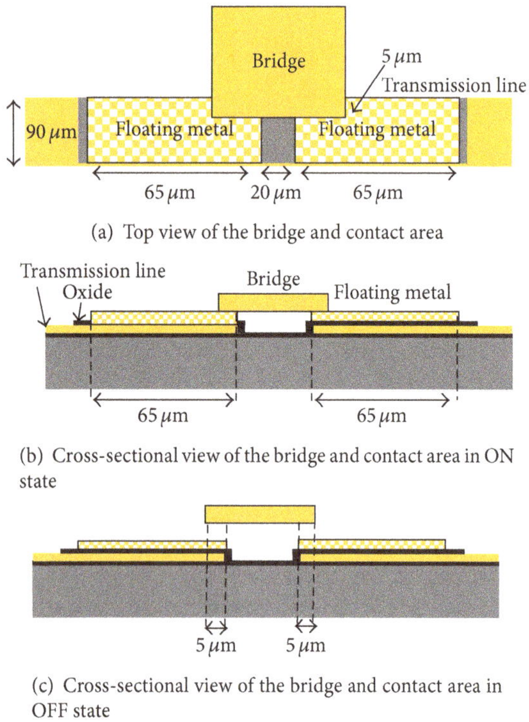

FIGURE 4: (a) Top view, (b) cross-sectional view in ON state, and (c) cross-sectional view in OFF state at the bridge and contact area.

is demonstrated in Figure 5. In ON state, the bridge tilts and contacts the metal layer (called float metal) on the t-line.

This results in an effective overlap capacitance of 2 pF (area $65 \times 90\,\mu\text{m}^2$) between the bridge and t-line (Figures 3(a) and 3(b)) offering small reactance (<26 Ω, for S-Ku band) to the signal passing from input to output port via the bridge.

4. Fabrication Process Flow

The fabrication process flow for the SPDT switch is illustrated in Figures 6(a)–6(i). The process begins with growth of one-micron-thick thermal oxide on high resistive Si wafer. LPCVD deposited polysilicon which constitutes the actuation electrodes and DC bias lines is doped with phosphorus to achieve required sheet resistance followed by PECVD oxide for isolation. A 0.1 μm thin Ti/TiN layer (float metal) is sputtered followed by Cr/Au seed layer of thickness 20 nm/80 nm and electroplated to get 3 μm thickness supports as shown in Figure 6(e). A sacrificial layer (HIPR) is spin coated to make a mold for bridge structure followed by again Cr/Au seed layer of thickness 30 nm/50 nm with electroplating to get 4 μm stiff bridge structures as shown in Figure 6(g). The process of sacrificial layer and gold plating is repeated to make 4 μm thick clamps for the bridge as shown in Figure 6(i). The final structure is released with wet etching using critical point dryer (CPD) machine. The final fabricated SPDT switch is shown in Figure 2.

5. Mechanical Characteristics

The RF MEMS devices have a better electromagnetic response compared to the solid state devices but mechanical parameters like biasing voltage, speed, and reliability are inferior which limit their applications. The proposed spring-free system has addressed the above drawbacks as follows.

5.1. Actuation Voltage. The proposed switch design is free from the spring action, and acting restoring force is negligible. The only gravitational force has control over the actuation. Electrostatic force required to operate the switch must overcome the gravitational force as given below:

$$F_e = \frac{1}{2}\frac{\varepsilon_o A V_{\text{act}}^2}{g_o^2} \geq mg, \quad (3)$$

where g is a gravitational constant.

From (3), required actuating voltage is

$$V_{\text{act(min)}} = \sqrt{\frac{2gmg_o^2}{\varepsilon_o A}}. \quad (4)$$

For the proposed switch parameters as listed in Table 1, actuation voltage turns out to be 1.7 V. However, on measurements in open environment, actuation voltage turns out to be 5 V due to bridge damping and friction between hinges and clamps which is much less than actuation voltage reported in the literature [3–15]. Comparisons for size, pull-in voltage, switching time, and number of switches utilized in the present SPDT switch are described in Table 2.

5.2. Switching Speed. Resonance frequency and switching speed of the SPDT switch are extracted from 1D nonlinear model given by

$$m\frac{d^2x}{dt^2} + b\frac{dx}{dt} + kx = F_e, \quad (5)$$

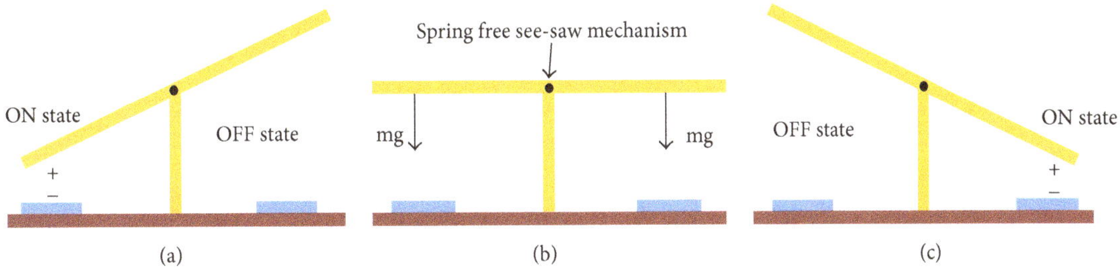

FIGURE 5: Schematic of different states of the bridge during the see-saw mechanism: (a) port 1 "ON" and port 2 "OFF," (b) both "OFF," and (c) port 1 "OFF" and port 2 "ON."

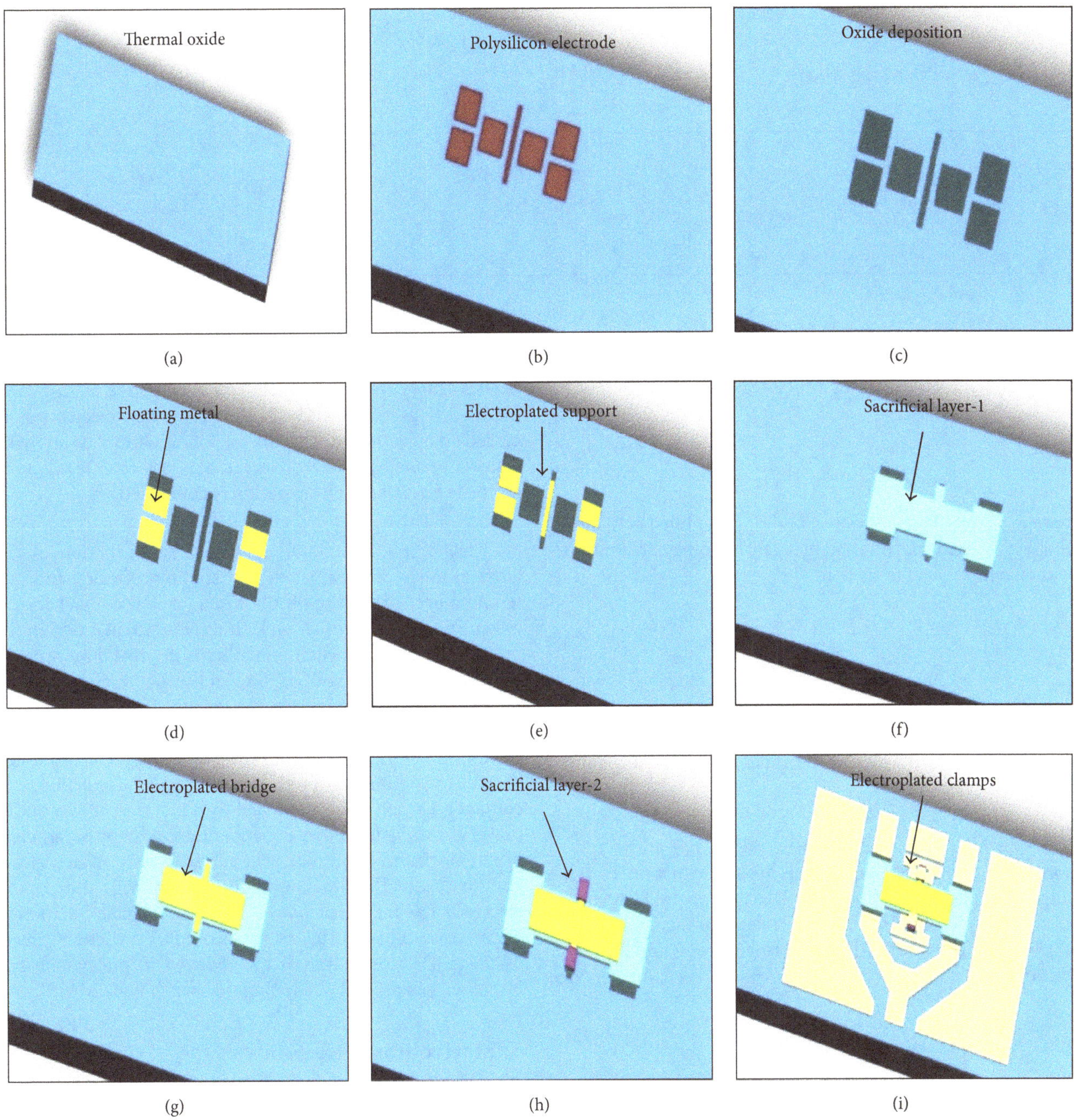

FIGURE 6: Fabrication process flow for the SPDT switch.

Table 2: Literature survey of size, pull-in voltage, switching time, and number of switches used in the SPDT switch.

SN	Device parameter	The proposed see-saw design	Literature Value	Literature Year	Literature Ref
1	Size (mm^2)	1.08 * 0.68	5.2 * 3.0	2009	[9]
			3.8 * 1.1	2009	[10]
			3.5 * 1.5	2011	[11]
			0.68 * 0.88	2015	[8]
2	Pull-in voltage (V)	5	34	2009	[9]
			85	2011	[11]
			9.5	2014	[12]
			17	2014	[13]
			9	2015	[8]
3	Switching speed (μs)	0.394	30–40	2011	[11]
			27.70	2014	[12]
			16.64	2014	[13]
			50	2015	[8]
4	Number of switches	1	4	2009	[10]
			2	2009	[9]
			2	2011	[11]
			4	2014	[13]
			1	2015	[8]

where m, b, and k are mass of bridge, damping coefficient, and the spring constant of the anchor, respectively. For the spring-free system ($k = 0$), switching time is calculated by [16, 17]

$$T_{pi} = \frac{2bg_o^3}{3\varepsilon_o A V_{\text{act}}^2}, \quad (6)$$

where g_o is initial gap, A is area, and V_{act} is applied actuation voltage between electrode and bridge. Damping constant for the rectangular plate is given by [18]

$$b = \frac{3}{2\pi} \frac{\mu A^2}{g_o^3}, \quad (7)$$

where μ is the coefficient of viscosity.

From (6) and (7), calculated switching time is 394 nsec at an actuation voltage of 5 V under vacuum (0.76 torr) conditions.

5.3. Power Handling. The proposed design has a low actuation voltage of 1.7 V which leads to self-actuation of the bridge as explained in [18, 19]. Self-actuation by RF power and actuation voltage are closely correlated to each other. The electrostatic force due to RF power is given by (8) and (9) [18]. Consider

$$F_e = \frac{1}{2} \frac{\varepsilon_o A}{g^2} V_{\text{dc-eq}}^2, \quad (8)$$

where

$$V_{\text{dc-eq}} = \frac{V_{pk}}{\sqrt{2}} = \sqrt{PZ_0}. \quad (9)$$

The proposed SPDT switch with design parameters mentioned in Table 1 will self-actuate at RF power of 500 mWatt. See-saw mechanism with the second electrode as shown in Figure 2 is used to avoid self-actuation. Two electrodes are placed on opposite sides to control self-actuation. By applying sufficient voltage (> RF voltage), that is, 5 V for 500 mWatt at opposite electrode, self-actuation is controlled.

5.4. Reliability. The switch reliability is a function of the large number of parameters like stiction, power handling, temperature, stress, the number of switches, contact forces, creep, and so forth which make it difficult to quantify. In the present design, the switch reliability is improved by controlling three parameters out of all reliability parameters: stiction force, power handling, and the number of switches. The proposed switch design has reduced OFF state contact area by 18 times compared to conventional switches as explained in [6–8]. Stiction forces that are directly proportional to contact area [20, 21] are reduced by a factor of 18. Secondly, power handling is further dependent upon many parameters like temperature, cold/hot switching, stress, self-actuation, and so forth. The see-saw configuration with the second electrode has increased power handling from self-actuation perspective only. Thirdly, numbers of SPST switches utilized in the SPDT configuration are reduced to one which has improved reliability by reducing the switch dependency.

6. Electromagnetic Analysis

The SPDT switch is designed in High Frequency Structural Simulator (HFSS) for C to Ku band applications. The electromagnetic responses of the RF MEMS see-saw switch are similar to a conventional anchored switch [8]. OFF state

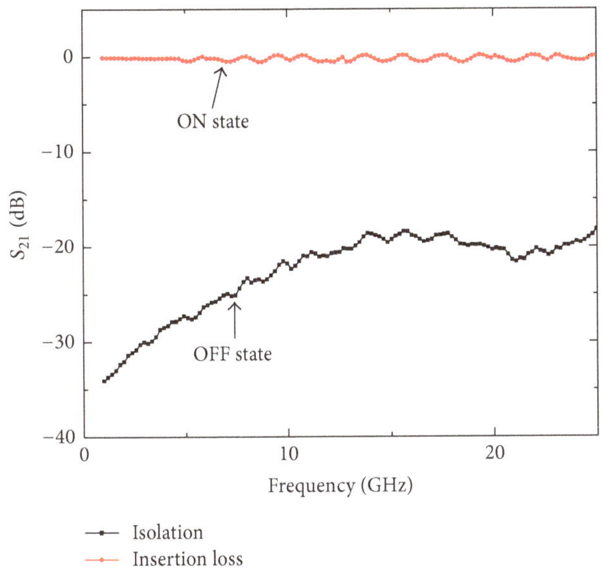

FIGURE 7: Measured S-parameter response of the SPDT switch.

overlap capacitance is reduced to 0.885 fF with floating metal that improves the insertion loss [8] and parasitic capacitance, which also lies in the range of femto-Farad, increases its value to 19 fF. In the ON state, overlap capacitance is 2.01 pF and parasitics have a negligible effect on it. Insertion loss lies in the range of −0.3 dB to −0.6 dB for a wide bandwidth of 19 (1–20) GHz. Isolation varies from −28.1 dB (@1 GHz) to −20 dB (@20 GHz) as shown in Figure 7.

7. Conclusions

The anchor-free see-saw RF MEMS SPDT switch is fabricated using a single series capacitive switch with compact dimensions of 0.734 mm^2. The proposed switch has a low actuation voltage of 5 V with high switching speed (394 nsec). Insertion loss and isolation are better than −0.6 dB and −20 dB, respectively, in the frequency range of 1 GHz to 20 GHz.

Competing Interests

The authors declare that they have no conflict of interests or financial conflicts to disclose.

Acknowledgments

The authors would like to thank CSIR, India, for providing financial support under Budget Head PSC-0201.

References

[1] G. Hu, Z. Liu, Y. Qiao et al., "Bonding packaging of a SP4T RF MEMS switch," in *Proceedings of the 7th International Conference on Electronics Packaging Technology (ICEPT '06)*, pp. 1–4, IEEE, Shanghai, China, August 2006.

[2] G. L. Tan, R. E. Mihailovich, J. B. Hacker, J. F. DeNatale, and G. M. Rebeiz, "A very-low-loss 2-bit X-band RF MEMS phase shifters," in *Proceedings of the IEEE MTT-S International Microwave Symposium Digest*, vol. 1, pp. 333–335, Seattle, Wash, USA, 2002.

[3] M. Daneshmand and R. R. Mansour, "C-type and R-type RF MEMS switches for redundancy switch matrix application," in *Proceedings of the IEEE International Microwave Symposium Digest (MTT-S '06)*, pp. 144–147, 2006.

[4] M. H. Misran, N. A. Shairi, G. H. The, and M. A. Meor Said, "Design and performance analysis of single biasing based SPDT switch for wireless data communications," in *Proceedings of the 5th IEEE Asia-Pacific Conference on Applied Electromagnetics (APACE '12)*, pp. 363–366, Melaka, Malaysia, December 2012.

[5] G. M. Rebeiz and J. B. Muldavin, "RF MEMS switches and switch circuits," *IEEE Microwave Magazine*, vol. 2, no. 4, pp. 59–71, 2001.

[6] D. Bansal, A. Bajpai, P. Kumar, A. Kumar, M. Kaur, and K. Rangra, "Design and fabrication of a reduced stiction radio frequency MEMS switch," *Journal of Micro/Nanolithography, MEMS, and MOEMS*, vol. 14, no. 3, Article ID 035002, 2015.

[7] L. X. Zhang and Y.-P. Zhao, "Electromechanical model of RF MEMS switches," *Microsystem Technologies*, vol. 9, no. 6-7, pp. 420–426, 2003.

[8] D. Bansal, A. Kumar, A. Sharma, and K. J. Rangra, "Design of compact and wide bandwidth SPDT with anti-stiction torsional RF MEMS series capacitive switch," *Microsystem Technologies Journal*, vol. 21, no. 5, pp. 1047–1052, 2015.

[9] Y. Uno, K. Narise, T. Masuda et al., "Development of SPDT-structured RF MEMS switch," in *Proceedings of the 15th International Conference on Solid-State Sensors, Actuators and Microsystems (TRANSDUCERS '09)*, pp. 541–544, June 2009.

[10] S. Cheng, P. Rantakari, R. Malmqvist et al., "Switched beam antenna based on RF MEMS SPDT switch on quartz substrate," *IEEE Antennas and Wireless Propagation Letters*, vol. 8, pp. 383–386, 2009.

[11] D. Yamane, W. Sun, H. Fujita, H. Toshiyoshi, and S. Kawasaki, "Development of a Dual-SPDT RF-MEMS switch for Ku-band," in *Proceedings of the IEEE Radio and Wireless Symposium (RWW '10)*, pp. 432–435, New Orleans, La, USA, January 2010.

[12] U. K. Kommuri, "Design of low voltage RF MEMS switch at 35 GHz," *Microwave and Optical Technology Letters*, vol. 56, no. 6, pp. 1483–1486, 2014.

[13] U. Sharma, M. Kumar, K. K. Jain, and R. K. Bhan, "Broadband design of RF MEMS SPDT switch," *International Journal of Engineering Sciences & Emerging Technologies*, vol. 6, no. 4, pp. 405–409, 2014.

[14] R. Marcelli, G. Bartolucci, G. Minucci, B. Margesin, F. Giacomozzi, and F. Vitulli, "Lumped element modeling of coplanar series RF MEMS switches," *Electronics Letters*, vol. 40, no. 20, pp. 1272–1274, 2004.

[15] D. Bansal, A. Kumar, A. Sharma, P. Kumar, and K. J. Rangra, "Design of novel compact anti-stiction and low insertion loss RF MEMS switch," *Microsystem Technologies*, vol. 20, no. 2, pp. 337–340, 2014.

[16] R. K. Gupta and S. D. Senturia, "Pull-in time dynamics as a measure of absolute pressure," in *Proceedings of the 10th Annual International Workshop on Micro Electro Mechanical Systems (MEMS '97)*, pp. 290–294, Nagoya, Japan, January 1997.

[17] S. Dushman and J. Lagerty, *Scientific Foundations of Vacuum Technique*, John Wiley & Sons, New York, NY, USA, 1962.

[18] G. M. Rebeiz, *RF MEMS Theory, Design, and Technology*, John Wiley & Sons, Hoboken, NJ, USA, 2003.

[19] R. Malmqvist, R. Jonsson, C. Samuelsson et al., "Self-actuation tests of ohmic contact and capacitive RFMEMS switches for wideband RF power limiter circuits," in *Proceedings of the 35th International Semiconductor Conference (CAS '12)*, pp. 217–220, IEEE, Sinaia, Romania, October 2012.

[20] N. Tas, T. Sonnenberg, H. Jansen, R. Legtenberg, and M. Elwenspoek, "Stiction in surface micromachining," *Journal of Micromechanics and Microengineering*, vol. 6, no. 4, pp. 385–397, 1996.

[21] Y.-P. Zhao, L. S. Wang, and T. X. Yu, "Mechanics of adhesion in MEMS—a review," *Journal of Adhesion Science and Technology*, vol. 17, no. 4, pp. 519–546, 2003.

17

The Comprehensive Study of Electrical Faults in PV Arrays

M. Sabbaghpur Arani and M. A. Hejazi

Electrical and Computer Engineering Department, University of Kashan, Kashan, Iran

Correspondence should be addressed to M. A. Hejazi; mhejazi@kashanu.ac.ir

Academic Editor: Ephraim Suhir

The rapid growth of the solar industry over the past several years has expanded the significance of photovoltaic (PV) systems. Fault analysis in solar photovoltaic (PV) arrays is a fundamental task to increase reliability, efficiency, and safety in PV systems and, if not detected, may not only reduce power generation and accelerated system aging but also threaten the availability of the whole system. Due to the current-limiting nature and nonlinear output characteristics of PV arrays, faults in PV arrays may not be detected. In this paper, all possible faults that happen in the PV system have been classified and six common faults (shading condition, open-circuit fault, degradation fault, line-to-line fault, bypass diode fault, and bridging fault) have been implemented in 7.5 KW PV farm. Based on the simulation results, both normal operational curves and fault curves have been compared.

1. Introduction

Renewable energy is energy which can be obtained from natural resources that can be constantly replenished like water, wind, sun rays, and so forth [1]. It is necessary to achieve more sustainable energy system. Among the most important renewable sources and most widely used across the globe is the solar energy [2, 3]. PV markets are growing fast because of their advantages such as long life of PV panel, installation in different geographical conditions such as impassible areas and mountains, usability on mobile hosts, easy maintenance, off-grid installing, and ability to connect to utility grid which have depicted a bright future for the use of photovoltaic system in the world [4, 5]. On the other hand, the rapid growth rate is mainly due to the need for alternatives to fossil fuel-based electricity generation, concerns over the global environment, reduced photovoltaics costs, and interests in distributed energy sources to improve power system reliability [6, 7]. The efficiencies of inverters which convert the direct current generated by the modules into alternate current are already close to maximum about 99 percent [8]. Therefore, no significant gains are possible from improving inverter efficiency. Alternately, PV array outputs can be increased by improving the efficiency of the PV modules. The last surveys have showed that the median values of efficiency of different modules technologies such as GaAs (thin film), crystalline silicon, and Si (Amorphous) were close to 28.8, 25.6, and 10.2 percent, respectively [9, 10]. Improving efficiencies through better materials is an important field [11]. Another way to improve PV array output is to ensure that the array operates in optimal output conditions at all times. PV arrays once installed are expected to operate with minimal human intervention. PV arrays perform below optimum output power levels due to faults in modules, wiring, inverter, and so forth. Most of these faults remain undetected for long periods of time resulting in loss of power. Technicians sent to locate and fix the faults within an array need to take time consuming field measurements. So enquiries for lower cost and high efficiency-devices motivate the researchers to increase the reliability of PV system.

Fault analysis in the solar PV arrays is a fundamental task to eliminate any kind of dangerous and undesirable situations arising in the operation of PV array due to the presence of faults. They must be detected and cleared off rapidly. Without proper fault detection, noncleared faults in PV arrays not only cause power losses but also might lead to safety issues and fire hazards [12]. Photovoltaic systems are subjected to different sort of failures; thus, before starting monitoring system and fault diagnosis methods, it is necessary to identify what kind of failures can be found in the real system. The first step in this challenge is recognition and classification of all possible electrical faults in PV arrays.

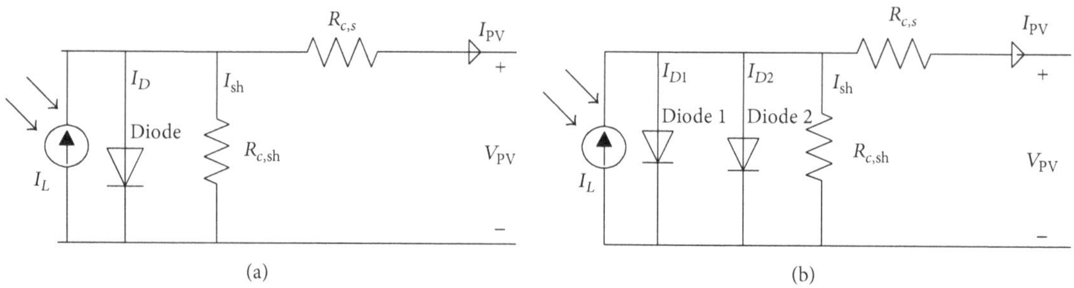

FIGURE 1: Equivalent circuits for (a) the one-diode model and (b) the double-diode model.

The fault detection methods for the PV system are classified in the visual (discoloration, browning, surface soiling, and delamination), thermal (thermal extraordinary heating), and electrical (dark/illuminated I-V curve measurement, transmittance line diagnosis, and RF measurement). Using electrical signatures is more advantageous and promising for the monitoring and diagnostic systems [13, 14]. This characteristic of electrical methods offers helpful data in diagnosing a PV cell's health. Furthermore, the I-V and P-V curves analyses are fundamental tool to understand the fault scenarios among PV strings and the impact of these fault in basic output parameters such as open-circuit voltage (V_{oc}), short-circuit current (I_{sc}), maximum power point voltage (V_{mpp}), and maximum power point current (I_{mpp}) in I-V and P-V curves.

In this paper in Section 1 the basics of PV modules model as electrical components are described. In Section 2 challenges to fault analysis in PV arrays are expressed. In Section 3 we introduce comprehensive classification of electrical faults in a PV system. Finally, based on the circuit-based simulation model, various types of faults will be developed by changing conditions or inputs in the simulation, and the I-V and P-V characteristics of faulted and clean PV array have been compared for each type of fault.

2. Background

The issue of modeling of PV arrays under electrical faults has been largely investigated in the literature and gets some certain results. A survey of state-of-the-art of ground, line-to-line, and arc fault detection is presented in [15].

In [16] Chao et al. developed a circuit-based simulation model of a photovoltaic panel using the PSIM software. A 3 kW PV array system was established using extended correlation function to identify the different fault types of the PV system. In [17] Takashima et al. used earth capacitance measurement to locate faults in PV module arrays. Furthermore, in another study [14], they experimentally studied earth capacitance measurement and Time-Domain Reflectometry (TDR) to detect degradation (increase in series resistance between the modules) and the fault position in the string. In [18] Yagi et al. developed a diagnostic technology for PV systems based on statistically analyzed data to detect shading effect and inverter failure in PV arrays. In [19] unique fault evolution in a PV array during night-to-day transition and effect of a maximum power point tracker on fault current have been discussed.

In [20] Yamada et al. conducted simulations for PV modules on the reflection loss using the optical performance of a four-layer encapsulation. In [21] Nguyen studied impact of varying position, different levels of solar irradiation, and the performance of bypass diode under nonuniform irradiation levels. In [22] Firth et al. developed novel analysis techniques to identify four types of faults: sustained zero efficiency faults; brief zero efficiency faults; shading; and nonzero efficiency nonshading faults. Three independent applications to measure the effects of soiling have been suggested by Hammond et al. in [23]. In [16, 24] only power versus voltage (P-V) characteristics is simulated for a few types of faults in a PV array. In addition, [25] discusses the MPPT reliability of a PV array under partial shading rather than faults in the array.

In none of previous studies mentioned above comprehensive classification of electrical faults scenarios in the PV system and the impact of all possible faults on the I-V and P-V curves have been performed properly.

3. Modeling and Simulation of PV Modules

3.1. Models for Solar Cells. Because of the nonlinear I-V characteristics of solar cells, it is not appropriate to simply model them as a constant voltage source or a constant current source. The electrical performance of a photovoltaic cell can be approximated by the equivalent circuit shown in Figures 1(a) and 1(b). The one-diode model and the double-diode model are most commonly used to describe the electrical behaviors of solar cells [26, 27].

In this paper we adopt the one-diode model for solar cells in simulation, because the one-diode model has several advantages over the double-diode model such as enough accuracy for steady-state and fault analysis for PV modules in system level, data available for the most PV modules in market, and rapid responses in simulation environment [28].

Based on the properties of p-n semiconductors and one-diode model, the I-V characteristics of a PV panel with N_s cells are characterized using the following equation:

$$I = I_L - I_S\left(\exp\left[\frac{q(V + IR_{c,s})}{N_s AkT}\right] - 1\right) - \frac{V + IR_{c,s}}{R_{c,sh}}. \quad (1)$$

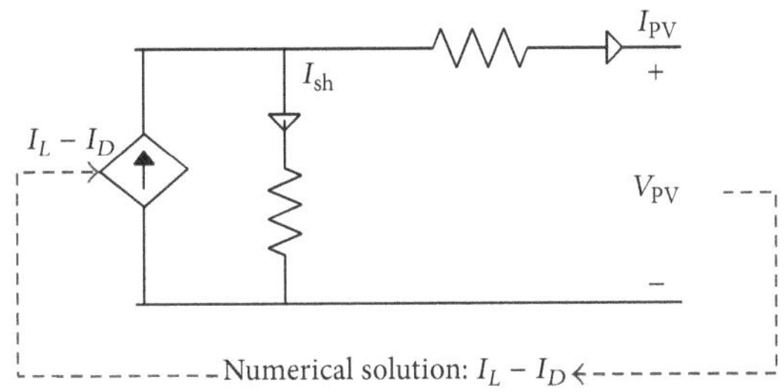

FIGURE 2: The numerical one-diode model for a PV module.

The dependence of the photocurrent on the irradiance (G) and cell temperature (T) can be described by the following empirical equation [11, 26, 27]:

$$I_L = \{I_{L0} + C_T (T - T_r)\} \frac{G}{G_0}. \quad (2)$$

The reverse saturation current I_S varies with solar cell surface temperature (T) [11, 26, 27]. It can be described by

$$I_S = I_{S0} \left(\frac{T}{T_r}\right)^3 \exp\left\{\frac{qE_g}{kA}\left[\frac{1}{T_r} - \frac{1}{T}\right]\right\}. \quad (3)$$

Depending on the semiconductor material used for PV modules, E_g may have different values. Usually E_g is approximate 1.12 eV for crystalline silicon, 1.03 eV for copper indium diselenide (CIS), 1.7 eV for amorphous silicon, and 1.5 eV for cadmium Telluride (CdTe) under room temperature [11, 26, 27].

3.2. Modeling Algorithm. In real working conditions, solar cells packaged in the same module usually have almost the same irradiance conditions. For these reasons, assume that all the solar cells in each PV module have identical characteristics and working conditions. Thus, a PV module can be viewed as a basic unit consisting of identical solar cells. Therefore, modeling and simulation of PV modules become key steps for PV system normal and fault analysis. A bypass diode is usually connected in parallel across multiple cells to improve operation of solar system under nonuniform condition.

According to the one-diode model of PV modules in Figure 2, by using voltage V_{PV}, T, and G as input parameters, the modeling algorithm solves equations to find the mathematical solution for I and feeds the solution to a controlled current source in Figure 2. Figures 3(a) and 3(b) show the model for PV modules in MATLAB/Simulink. Using the widely used one-diode model for each individual solar panel, this paper builds simulation PV array (7.5 kW) in MATLAB/Simulink consisting of 6 × 5 PV panels that is capable of studying faults among panels. The related parameters of each PV panel under STC (G = 1000 W/m^2 and T = 25°C) are P_{mpp} = 250 W, V_{mpp} = 31 V, I_{mpp} = 8.07 A, N_s = 20, V_{oc} = 37.92 V, and I_{sc} = 8.62 A and C_T = −0.33%/°C. As shown in Figure 3(b), panels connected in parallel increase the current and those connected in series provide greater output voltages.

4. Challenges to Fault Analysis

Only PV array has been considered as source of electrical fault in this paper. According to National Electrical Code Standard [29], fuses blow when the fault currents that flow through them become greater than at least 1.56 times their rated short-circuit current. However, because of the nonlinear I-V characteristics, the current-limiting nature of PV arrays, high fault impedances, low irradiance conditions, PV grounding schemes, or MPPT of PV inverters faults in PV arrays may not be cleared [30]. But because of some factors such as environmental conditions (varying irradiance level and temperature), PV array configurations and fault locations, aging, hot-spot, mismatch faults unique to PV technology, and MPPT effect, fault analysis would be more complicated and conventional protection devices may not be able to clear faults correctly. Since PV array normal operation can be affected by the presence of faults that reduce power output and cause potential damage to the array, so analysis of the I-V curves for describing the effect of the faults that occur in PV arrays is very important.

5. Typical Faults

Since some of the electrical faults, such as mismatches, occur in all arrays at all times, they result in available DC power from the array being significantly below predicted levels. Table 1 shows the most common types of fault in a PV system.

6. Curves and Interpretation

A typical solar PV array with 6 × 5 PV modules (rated at 7.5 kW) is simulated, which consists of 6 modules in series per string and 5 strings in parallel. MATLAB/Simulink models of PV array (Figure 3) under electrical faults are developed to study the performance of the faulted PV array. According to Table 1, the most frequent faults are major catastrophic failures in PV arrays which are ground faults, line-to-line faults, and arc faults [15]. This research studies six common fault types from Table 1 in 12 cases and compared the results with the normal condition. The characteristics of the PV panel with different types of faults are shown in Figures 4–9.

6.1. Partial Shading (F1). The shading patterns can be very complicated due to no uniform insolation. Two identical

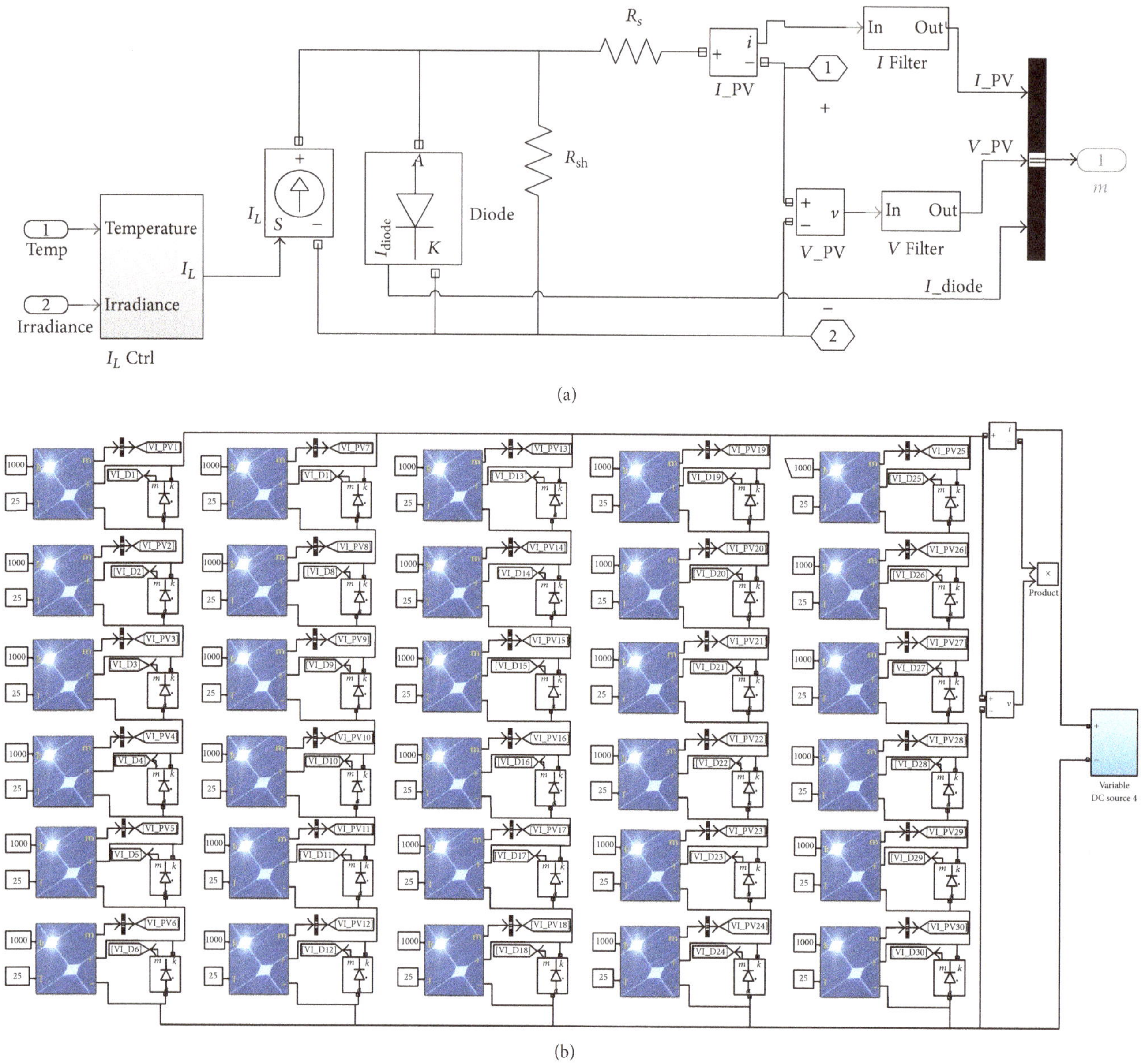

Figure 3: (a) Equivalent circuit of a solar and (b) a schematic diagram of a PV farm system with 6 × 5 modules.

PV arrays are used for comparison. One PV array with an arbitrary shading pattern is divided into two groups. In Case 1 the half of string one has been shaded with irradiance density $G = 800\,\text{W/m}^2$ and in Case 2 shaded modules receive two different insulations, $G = 500$ and $800\,\text{W/m}^2$. I-V and P-V characteristics of these two cases are illustrated in Figure 4 for fault analysis. Under partial shading conditions, the short-circuit current for two cases remains identical, while the open-circuit voltage slightly decreases with the increase in the number of shaded modules. I-V curves of all shaded groups have multiple steps, while P-V curves of shaded groups are characterized by multiple peaks, whose number is equal to the number of solar insolation levels received by string, respectively. The results indicate that the higher number of shaded solar modules is the lower value of power output and the position of maximum power point does not depend on location of modules under shadow. The surface temperature of solar cell is assumed to remain 298°K.

6.2. Line-to-Line Fault in a PV Array under STC (F9). As shown in Figure 5, line-to-line faults could happen inside PV arrays and potentially may involve large fault current or dc arcs. This research focuses on line-to-line faults, which are defined as an accidental short-circuiting between two points in the array with different potentials. In the following simulations, two cases are studied, a line-to-line fault with 2 modules (Case 3) and a line-to-line fault with 4 modules (Case 4). When a line-to-line fault occurs, the I-V curve of the faulted PV string will change accordingly. Since the faulted string has 4 number of modules less, it will have an open-circuit voltage reduced by 4x V_{oc}. But the short-circuit current remains the same as other normal strings at I_{sc}.

TABLE 1: Classification and definition of faults in PV system based on location and structures.

Part	Fault type			Description
DC	Mismatch faults	Partial shading [21, 31]	F1	Presence of trees, overhead power lines, or nearby buildings
		Uniform irradiance distribution [32]	F2	Various irradiance intensity during the day
		Soiling [23]	F3	The bird droppings and dirt on the surface of a PV module
		Snow covering and hot spot [33, 34]	F4	The worst temperatures depending on the geographical location and different weather conditions
	Earth fault	Upper ground fault [35, 36]	F5	An unintentional path to ground with zero fault impedance occurs between the last two modules at PV string
		Lower ground fault [35, 36]	F6	An unintentional path to ground with zero fault impedance occurs between the 2nd and the 3rd two modules at PV string with large backfeed current
	Arc fault	Series arc fault [37, 38]	F7	An arc fault due to discontinuity in any of the current carrying conductors resulting from solder disjoint, cell damage, corrosion of connectors, rodent damage, abrasion from different sources
		Parallel arc fault [37, 38]	F8	Insulation breakdown in current carrying conductors
	Line-to-line faults [19, 35]		F9	An accidental short-circuit between two points in a string with different potentials
	Bypass diode faults [39]		F10	Short-circuit in case of incorrect connection
	Degradation faults [40]		F11	Yellowing and browning, delamination, bubbles in the solar module, cracks in cells, defects in antireflective coating and delamination over cells and interconnections lead to degradation and increasing of the internal series resistance
	Bridging fault [19, 39]		F12	Low- resistance connection between two points of different potential in string of module or cabling
	Open-circuit fault [41]		F13	Physical breakdown of panel-panel cables or joints, objects falling on PV panels, and loose termination of cables, plugging and unplugging connectors at junction boxes
	MPPT faults [42, 43]		F14	Problem in MPPT charge controllers
	Cabling faults [44]		F15	—
Ac	Inverter faults [45]		F16	Failure of each component of inverter such as IGBTs, capacitors, and drive circuitry can result in inverter failure
	Sudden natural disasters [46]		F17	Total blackout due to Lightning, storm, and so forth

6.3. Bypass Diode Fault in a PV Array under STC (F10). Assume in Case 5 that one bypass diode is conducted or shorted and in Case 6 two bypass diodes are shorted. *I-V* and *P-V* characteristics of these two cases are shown in Figure 6. Even if only one full module is shorted by bypass diode, the maximum power and V_{oc} of the PV array drops significantly and short-circuit current remains the same as other normal strings.

6.4. Degradation Fault in a PV Array under STC (F11). The reason for power degradation could be the increase in the series resistance between the modules due to decreased adherence of contacts or corrosion caused by water vapor. In this case, two different resistance values are considered. Group one (Case 7) has small resistance $R_1 = 1$ ohm and another group (Case 8) has larger resistance with $R_2 = 2$ ohm.

This PV array with resistance is compared with the normal PV array, shown in Figure 7. Although the open voltage and short current do not change much under these different conditions, the maximum power point is reduced due to increase in resistance. Therefore, an increase of the internal series resistance can result in degradation of the peak power.

6.5. Bridging Fault in a PV Array under STC (F12). In simulations, bridging fault or line-to-line faults with zero fault impedance are solid faults that occur immediately. In Figure 8, there is a bridging fault with one-module level difference between String 1 and String 2, which lead to unbalanced currents among PV array defined as bridging fault with small voltage difference (Case 9). Bridging fault with large voltage difference is a line-to-line fault with three-module level difference between String 1 and String 2 expressed as Case 10. Bridging faults usually involve reduced array voltage (V_{oc}) but have much small reduction in array current (I_{sc}). The fault with larger voltage difference between two fault

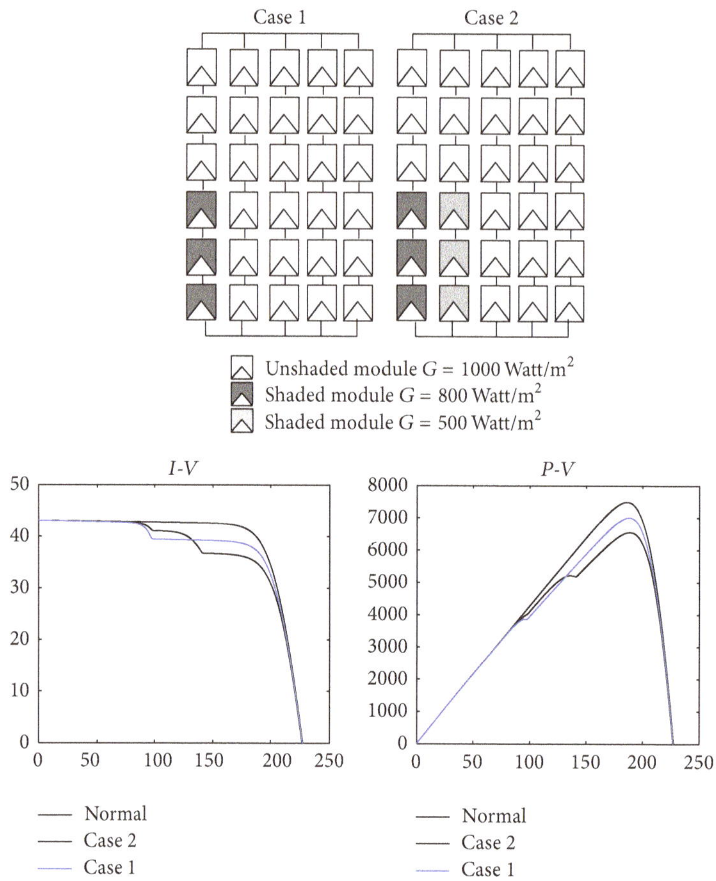

FIGURE 4: The PV array configuration for shading.

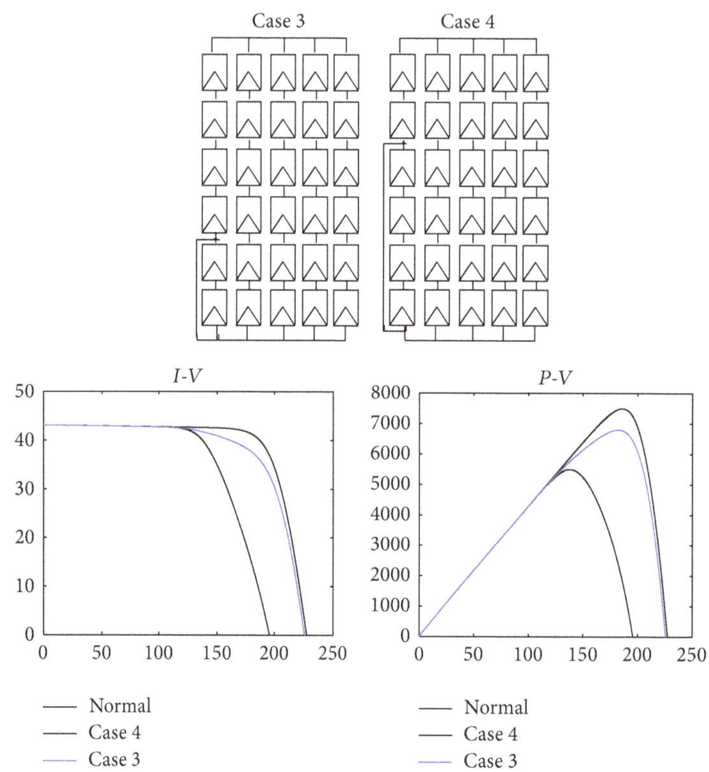

FIGURE 5: The PV array configuration for line-to-line fault.

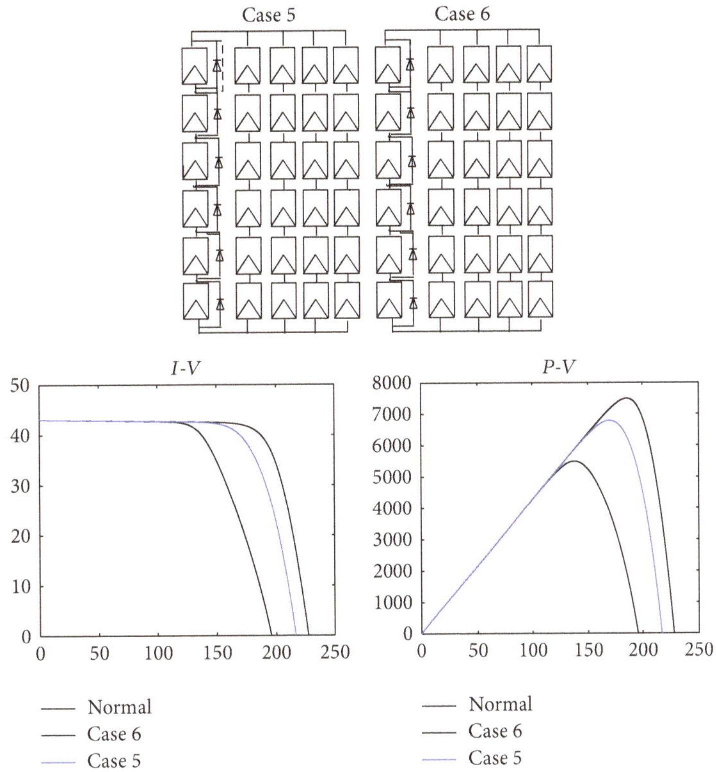

FIGURE 6: The PV array configuration for bypass diode short circuit.

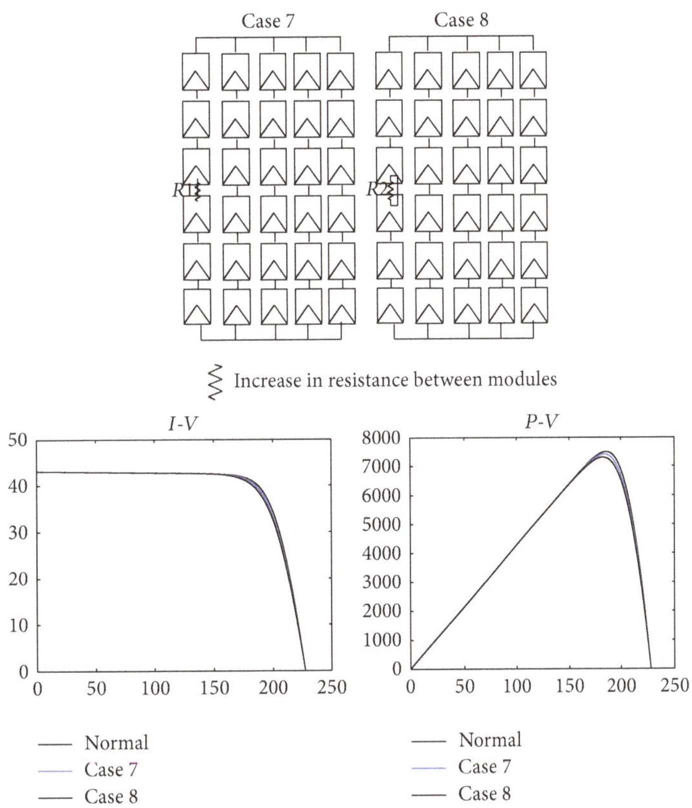

FIGURE 7: The PV array configuration for resistance between modules.

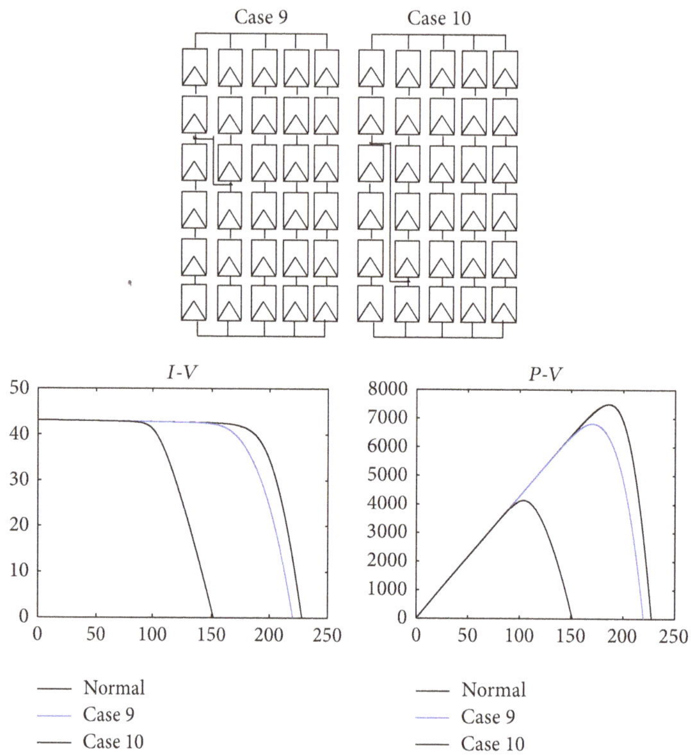

FIGURE 8: The PV array configuration for bridge fault.

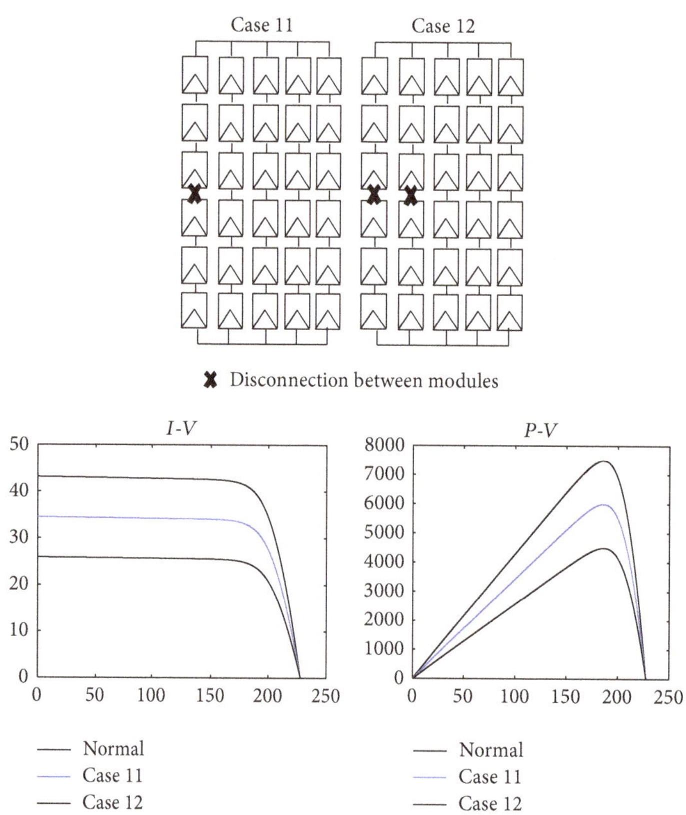

FIGURE 9: The PV array configuration for string disconnection.

points will lead to larger reduction in V_{oc} and I_{mpp} and V_{mpp}.

6.6. Open-Circuit Fault in a PV Array under STC (F13). An open-circuit fault is an accidental disconnection at a normal current-carrying conductor. In this section, assume in Cases 11 and 12 that PV arrays have a disconnection problem in one string and two strings, respectively, and then the I-V and P-V characteristics have been compared with array without any disconnection under normal condition, as shown in Figure 9. The open voltage of these cases remains almost the same, while the short current and maximum power decrease linearly with the increase in the number of disconnected strings.

7. Conclusions

In this paper, a comprehensive definition of faults in DC side of PV system based on location and structure is presented. The performance of a typical PV array has been investigated under typical fault conditions such as shading condition, open-circuit fault, degradation fault, line-to-line fault, bypass diode fault, and bridging fault. To better visualize the PV data under normal and fault conditions, the I-V and P-V characteristics of the array have been evaluated. The off-line method used in this research can distinguish many types of different faults but cannot detect the location of the fault within the PV array. It would be useful to develop special MPPT schemes to track the maximum peak under these conditions and further methods capable of determining these locations.

Nomenclature

MPPT: Maximum power point tracer
STC: Standard test condition
I: Solar cell current (A)
V: Solar cell voltage (V)
I_L: Light-generated current (A)
I_D: Diode current (A)
I_{sh}: Shunt resistance current (A)
I_S: Saturation current of the diode (A)
$R_{c,s}$: Solar cell series resistance (ohms)
$R_{c,sh}$: Solar cell shunt resistance (ohms)
q: Electron charge = 1.6×10^{-19} C
k: Boltzmann's constant = 1.38×10^{-23} J/K
A: Diode ideal factor ($1 \le A \le 2$)
G: Solar irradiance (W/m^2)
G_0: Reference solar irradiance (W/m^2) = 1000
T_r: Reference temperature (K) = 25 + 273
N_s: Number of series solar cells per module
C_T: Temperature coefficient of the light-generated current (A/K)
I_{S0}: Reference saturation current (A)
E_g: Band gap energy of the material (eV).

Competing Interests

The authors declare that they have no competing interests.

Acknowledgments

The financial support of Energy Research Institute of the University of Kashan towards this research is hereby acknowledged.

References

[1] D. L. King, W. E. Boyson, and J. A. Kratochvil, "Analysis of factors influencing the annual energy production of photovoltaic systems," in *Proceedings of the Conference Record of the 29th IEEE Photovoltaic Specialists Conference*, pp. 1356–1361, New Orleans, La, USA, May 2002.

[2] J.-R. Frish, "New renewable energy resources: (a guide to the future)," *Applied Solar Energy*, vol. 33, pp. 25–35, 1997.

[3] P. N. Shukla and A. Khare, "Solar photovoltaic energy: the state-of-art," *International Journal of Electrical, Electronics and Computer Engineering*, vol. 3, article 91, 2014.

[4] E. Despotou, A. Gammal, and B. Fontaine, *Global Market Outlook for Photovoltaics until 2014*, European Photovoltaic Industry Association (EPIA), Brussels, Belgium, 2010.

[5] B. Ghobadian, G. Najafi, H. Rahimi, and T. F. Yusaf, "Future of renewable energies in Iran," *Renewable and Sustainable Energy Reviews*, vol. 13, no. 3, pp. 689–695, 2009.

[6] P. Chiradeja and R. Ramakumar, "An approach to quantify the technical benefits of distributed generation," *IEEE Transactions on Energy Conversion*, vol. 19, no. 4, pp. 764–773, 2004.

[7] N. Tanaka, *Technology Roadmap-Solar Photovoltaic Energy*, International Energy Agency Report, Paris, France, 2010.

[8] A. E. I. Solaron, "500E HE PV Inverter datasheet," 2011.

[9] B. Parida, S. Iniyan, and R. Goic, "A review of solar photovoltaic technologies," *Renewable and Sustainable Energy Reviews*, vol. 15, no. 3, pp. 1625–1636, 2011.

[10] B. Anderson and R. Anderson, *Fundamentals of Semiconductor Devices*, McGraw-Hill, New York, NY, USA, 2004.

[11] D. L. King, J. A. Kratochvil, and W. E. Boyson, *Photovoltaic Array Performance Model*, Department of Energy, Washington, DC, USA, 2004.

[12] B. Brooks, "The bakersfield fire," *Solar Pro*, vol. 4, p. p62, 2011.

[13] P. Ducange, M. Fazzolari, B. Lazzerini, and F. Marcelloni, "An intelligent system for detecting faults in photovoltaic fields," in *Proceedings of the 11th International Conference on Intelligent Systems Design and Applications (ISDA '11)*, pp. 1341–1346, Cordoba, Spain, November 2011.

[14] T. Takashima, J. Yamaguchi, K. Otani, K. Kato, and M. Ishida, "Experimental studies of failure detection methods in PV module strings," in *Proceedings of the IEEE 4th World Conference on Photovoltaic Energy Conversion (WCPEC '06)*, pp. 2227–2230, IEEE, May 2006.

[15] M. K. Alam, F. Khan, J. Johnson, and J. Flicker, "A comprehensive review of catastrophic faults in PV arrays: types, detection, and mitigation techniques," *IEEE Journal of Photovoltaics*, vol. 5, no. 3, pp. 982–997, 2015.

[16] K.-H. Chao, S.-H. Ho, and M.-H. Wang, "Modeling and fault diagnosis of a photovoltaic system," *Electric Power Systems Research*, vol. 78, no. 1, pp. 97–105, 2008.

[17] T. Takashima, J. Yamaguchi, and M. Ishida, "Disconnection detection using earth capacitance measurement in photovoltaic module string," *Progress in Photovoltaics: Research and Applications*, vol. 16, no. 8, pp. 669–677, 2008.

[18] Y. Yagi, H. Kishi, R. Hagihara et al., "Diagnostic technology and an expert system for photovoltaic systems using the learning method," *Solar Energy Materials and Solar Cells*, vol. 75, no. 3-4, pp. 655–663, 2003.

[19] Y. Zhao, *Fault Analysis in Solar Photovoltaic Arrays*, Northeastern University, Boston, Mass, USA, 2010.

[20] T. Yamada, H. Nakamura, T. Sugiura, K. Sakuta, and K. Kurokawa, "Reflection loss analysis by optical modeling of PV module," *Solar Energy Materials and Solar Cells*, vol. 67, no. 1–4, pp. 405–413, 2001.

[21] X. H. Nguyen, "Matlab/Simulink based modeling to study effect of partial shadow on solar photovoltaic array," *Environmental Systems Research*, vol. 4, no. 1, 2015.

[22] S. K. Firth, K. J. Lomas, and S. J. Rees, "A simple model of PV system performance and its use in fault detection," *Solar Energy*, vol. 84, no. 4, pp. 624–635, 2010.

[23] R. Hammond, D. Srinivasan, A. Harris, K. Whitfield, and J. Wohlgemuth, "Effects of soiling on PV module and radiometer performance," in *Proceedings of the Record of the 26th IEEE Photovoltaic Specialists Conference*, pp. 1121–1124, Anaheim, Calif, USA, September-October 1997.

[24] K.-H. Chao, C.-J. Li, and S.-H. Ho, "Modeling and fault simulation of photovoltaic generation systems using circuit-based model," in *Proceedings of the IEEE International Conference on Sustainable Energy Technologies (ICSET '08)*, pp. 290–294, November 2008.

[25] G. Petrone, G. Spagnuolo, R. Teodorescu, M. Veerachary, and M. Vitelli, "Reliability issues in photovoltaic power processing systems," *IEEE Transactions on Industrial Electronics*, vol. 55, no. 7, pp. 2569–2580, 2008.

[26] M. G. Villalva, J. R. Gazoli, and E. R. Filho, "Comprehensive approach to modeling and simulation of photovoltaic arrays," *IEEE Transactions on Power Electronics*, vol. 24, no. 5, pp. 1198–1208, 2009.

[27] J. Gow and C. Manning, "Development of a photovoltaic array model for use in power-electronics simulation studies," *IEE Proceedings, Electric Power Application*, vol. 146, no. 2, pp. 193–200, 1999.

[28] C. R. A. Mermoud and J. Bonvin, *PVSYST, Version 4.37*, University of Geneva, Institut of Environmental Sciences (ISE) Group Energy (FOREL), Geneva, Switzerland, 2009.

[29] A.-S. P. S. National Electrical Code, 2011.

[30] J. Flicker and J. Johnson, "Analysis of fuses for Blind Spot ground fault detection in photovoltaic power systems," Sandia National Laboratories Report, 2013.

[31] H. Patel and V. Agarwal, "MATLAB-based modeling to study the effects of partial shading on PV array characteristics," *IEEE Transactions on Energy Conversion*, vol. 23, no. 1, pp. 302–310, 2008.

[32] D. D. Nguyen and B. Lehman, "Modeling and simulation of solar PV arrays under changing illumination conditions," in *Proceedings of the 10th IEEE Workshop on Computers in Power Electronics (COMPEL '06)*, pp. 295–299, September 2006.

[33] N. Heidari, J. Gwamuri, T. Townsend, and J. M. Pearce, "Impact of snow and ground interference on photovoltaic electric system performance," *IEEE Journal of Photovoltaics*, vol. 5, no. 6, pp. 1680–1685, 2015.

[34] S. Wendlandt, A. Drobisch, T. Buseth, S. Krauter, and P. Grunow, "Hot spot risk analysis on silicon cell modules," in *Proceedings of the 25th European Photovoltaic Solar Energy Conference*, Valencia, Spain, September 2010.

[35] Y. Zhao, B. Lehman, J.-F. de Palma, J. Mosesian, and R. Lyons, "Challenges to overcurrent protection devices under line-line faults in solar photovoltaic arrays," in *Proceedings of the 3rd Annual IEEE Energy Conversion Congress and Exposition (ECCE '11)*, pp. 20–27, Milwaukee, Wis, USA, September 2011.

[36] Y. Zhao, J.-F. de Palma, J. Mosesian, R. Lyons, and B. Lehman, "Line-line fault analysis and protection challenges in solar photovoltaic arrays," *IEEE Transactions on Industrial Electronics*, vol. 60, no. 9, pp. 3784–3795, 2013.

[37] K. Xia, Z. He, Y. Yuan, Y. Wang, and P. Xu, "An arc fault detection system for the household photovoltaic inverter according to the DC bus currents," in *Proceedings of the 18th International Conference on Electrical Machines and Systems (ICEMS '15)*, pp. 1687–1690, Pattaya, Thailand, October 2015.

[38] E. D. Spooner and N. Wilmot, "Safety issues, arcing and fusing in PV arrays," in *Proceedings of the 3rd International Solar Energy Society Conference-Asia Pacific Region*, Sydney, Australia, 2008.

[39] M. Davarifar, A. Rabhi, and A. E. Hajjaji, "Comprehensive modulation and classification of faults and analysis their effect in DC side of photovoltaic system," *Energy and Power Engineering*, vol. 5, no. 4, pp. 230–236, 2013.

[40] G. Makrides, B. Zinsser, G. E. Georghiou, M. Schubert, and J. H. Werner, "Degradation of different photovoltaic technologies under field conditions," in *Proceedings of the 35th IEEE Photovoltaic Specialists Conference (PVSC '10)*, pp. 2332–2337, Honolulu, Hawaii, USA, June 2010.

[41] M. N. Akram and S. Lotfifard, "Modeling and health monitoring of DC side of photovoltaic array," *IEEE Transactions on Sustainable Energy*, vol. 6, no. 4, pp. 1245–1253, 2015.

[42] C.-C. Hua and P.-K. Ku, "Implementation of a stand-alone photovoltaic lighting system with MPPT, battery charger and high brightness LEDs," in *Proceedings of the 6th International Conference on Power Electronics and Drive Systems (PEDS '05)*, pp. 1601–1605, Honolulu, Hawaii, USA, December 2005.

[43] Y. Wang, Y. Li, and X. Ruan, "High-accuracy and fast-speed MPPT methods for PV string under partially shaded conditions," *IEEE Transactions on Industrial Electronics*, vol. 63, no. 1, pp. 235–245, 2016.

[44] F. Schimpf and L. E. Narum, "Recognition of electric arcing in the DC-wiring of photovoltaic systems," in *Proceedings of the 31st International Telecommunications Energy Conference (INTELEC '09)*, pp. 1–6, Incheon, Republic of Korea, October 2009.

[45] F. Chan and H. Calleja, "Reliability: a new approach in design of inverters for PV systems," in *Proceedings of the 10th IEEE International Power Electronics Congress (CIEP '06)*, pp. 97–102, October 2006.

[46] A. M. Omer, "Renewable energy resources for electricity generation in Sudan," *Renewable and Sustainable Energy Reviews*, vol. 11, no. 7, pp. 1481–1497, 2007.

18

A Low Noise, Low Power Phase-Locked Loop, using Optimization Methods

Noushin Ghaderi, Hamid Reza Erfani-jazi, and Mehdi Mohseni-Mirabadi

Faculty of Engineering, Shahrekord University, Shahrekord 8818634141, Iran

Correspondence should be addressed to Noushin Ghaderi; ghaderi.nooshin@eng.sku.ac.ir

Academic Editor: M. Jamal Deen

A divider-less, low power, and low jitter phase-locked loop (PLL) is presented in this paper. An extra simple open loop phase frequency detector (PFD) is proposed which reduces the power consumption and increases the overall speed. A novel bulk driven Wilson charge pump circuit, whose performance is enhanced by some optimization algorithms, is also introduced to get high output swing and high current matching. The designed PLL is utilized in a 0.18 μm CMOS process with a 1.8 V power supply. It has a wide locking range frequency of 500 MHz to 5 GHz. In addition, through the use of a dead-zone-less PFD and a divider-less PLL, the overall jitter is decreased significantly.

1. Introduction

The role of a phase-locked loop or a delay-locked loop (DLL) is to generate a clock signal which is usually a multiple of a reference clock and is synchronized with the reference clock in phase. They are extensively utilized in many applications including clock data recovery systems and frequency synthesis circuits.

One of the most important components of the PLL and DLL, which is often considered as the bottleneck, is the phase frequency detector (PFD). The limited speed of this block is the main limiting factor in the data rate of the PLL. In high frequencies, the PFDs cannot follow the phase differences between reference and feedback clocks. Therefore, a frequency divider circuit is often inserted before the PFD block in the feedback path of the PLL. It means that the input clock frequency of the PFD is reduced. However, by adding the divider circuit, the overall jitter of the PLL will extremely increase. Thus, by designing a high frequency PFD, the divider circuit will be omitted and the noise behavior of the PLL will be improved [1].

Another important component of the PLL is charge pump circuit. It converts the output signal of PFD into an accurate amount of current. In practice, there are some nonidealities in the charge pump circuit. Reference spurs and phase offset are due to the current mismatch while the variations of the PLL loop bandwidth are caused by the variations of the output current amplitude. If the current matching and control voltage range of a charge pump circuit increase simultaneously, then the locking range frequency of the PLL will increase considerably.

In this paper a wide dynamic output voltage range, high impedance, and low power charge pump based on a novel bulk driven Wilson current mirror is presented. Some optimization algorithms are employed to improve the efficiency of the circuit.

2. Proposed PFD

The PFDs detect a phase error signal which is passed through a loop filter to control the delay of a controllable delay line. Therefore, in addition to determination of accuracy, a PD plays a key role in the speed and efficiency of PLL's performance.

A closed loop is the general structure used in the PFD because it is simple, easy to implement, and immune to the dead-zone problem. Because of the feedback path, however, it has a limited speed which confines its usage in high speed circuits. Furthermore, some rising edges can be missed in the detection when the edges are overlapped with the reset signal [2]. Till now many circuits have been suggested to make the feedback path faster [3–6]; however the feedback path still remains in all of them which causes a reduction in the speed.

Authors in [7] proposed an open loop PD to achieve a high speed performance, without the issue of the reset signal. However, the PFD circuit tends to dissipate crowbar current when both input signals are high. It means that the circuit has some extra average DC power consumption.

The proposed circuit with a very simple structure solves the above problem. Furthermore, the proposed circuit solves the dead-zone and missing edge problems completely.

The task of a PD circuit is to compare the rising edge of two inputs (A and B in Figure 1) and generate a signal which is set to a high value with the rising edge of the first input and then is reset to a low value with the rising edge of the second one.

Figures 1(a) and 1(b) show the proposed PD, for Up and Down paths, which indicate two symmetrical and very simple circuits for Up and Down signals. The waveforms of the PD are shown in Figures 1(c) and 1(d). In Figure 1(c), A leads B. If initially $A = 0$ and $V(C_1) = V_{\text{dd}}$, then a rising transition on A leads to charge the output capacitor through C_1. Therefore, the Up output node will be set to the high value. The Up node stays in this state until rising transition on B. On the other hand, at the Down circuit with A at the high value, the $M6$ transistor will be on, keeping the down output at the low value. After the rising transition on B, $M5$ will turn on, causing the C_2 capacitor to discharge to the ground. Therefore, when A returns to the low value and $M6$ turns off, Down output will remain at zero value. The circuits operate in the opposite manner when A lags B (Figure 1(d)).

As can be observed, the above circuit is extremely simple and fast. Furthermore, due to the use of only three transistors in each path, the power consumption is considerably low. But the proposed PD still suffers from the following problems.

The first problem is as follows: Since the parasitic capacitance of C_1 node is smaller than the parasitic capacitance of the output node, the output node will not be charged through this node completely. Adding an additional capacitor to the C_1 node will add some problems such as extra chip area and longer rise and fall time.

The second problem is its dead-zone problem. This problem may occur when the inputs have a very small phase difference. Therefore, the output pulse may not find enough time to turn on the charge pump switches. To overcome the above two problems, the proposed PD is modified as shown in Figure 2, in which A_d and B_d are delayed waveforms of A and B signals. The ideal waveforms of two outputs are also shown in this figure.

Circuit performance is as follows: by delaying A and B signals on gates of $M1$, $M4$, and $M3$, the zero value of "up" signal will occur a few seconds later. As a result, a high level of "up" signal will become wider than the real amount and the phase difference will be shown more extensive than the actual value. This amount of broadness must be equal to the precharging time (T_{chp}) of the internal parasitic capacitances.

In Figure 2(c), A leads B. As can be seen, just after the rising transition on A, A signal will be set to the high value while A_d, which is the delayed waveform of A signal, still remains at the low value. Therefore, $M1$ and $M2$ turn on and the Up output will be set to the high value. In the same way, just after the rising transition on B, $M5$ and $M6$ turn on, consequently setting the Down output to the high value. With rising transition on B_d, both of the Up and Down outputs return to the low value. The circuits operate in the opposite manner when A lags B as shown in Figure 2(d).

According to [19], the time which is taken by the signal to change from low (high) value to 70% (30%) of high (low) value of logic states is called the gate rise time (or fall time). In this circuit, rise time and fall time make T_{chp}.

The sum of output rise time and fall time is calculated to estimate the maximum operational frequency of the proposed PFD.

$$
\begin{aligned}
t_{\text{rise(fall)}} &= 1.2\tau_{\text{rise(fall)}}, \\
\tau_{\text{rise(fall)}} &= R_{\text{eq}_{\text{rise(fall)}}} C_L \\
R_{\text{eqrise}} &= r_{o1} + r_{o2}, \\
R_{\text{eqfall}} &= r_{o3} + r_{o4} \\
C_L &= C_{db2} + C_{db3} + \frac{C_{dg2}C_{\text{Not}}}{C_{dg2} + C_{\text{Not}}} + \frac{C_{dg3}C_{\text{delay}}}{C_{dg3} + C_{\text{delay}}} \\
T_{\text{chp}} &= T_{\text{rise}} + T_{\text{fall}},
\end{aligned} \tag{1}
$$

where τ is considered as time constant, R_{eq} is the equivalent resistant of the charging or discharging path at the output node (UP or DN), C_L is the output equivalent capacitance, C_{delay} is the delay gate, and C_{Not} is the Not gate equivalent capacitances.

By calculating the above equations, T_{chp} is obtained as 48.33 ps. Figure 3 shows that the dead zone is compensated by delay gates through the Not gate and precharging time.

$$T_{\text{chp}} = T_{\text{delay}} - T_{\text{Not}}. \tag{2}$$

T_{Not} and T_{delay} are delay times caused by Not gate and delay gate, respectively. The values of T_{delay} and T_{Not} are obtained as 64.43 ps and 16.1 ps, respectively. It means that the dead zone is entirely removed.

It is evident that the delay time in Figure 3 cannot exceed the value of 3/4 of the clock pulse width. Therefore, the minimum value of pulse width is 85.9 ps. As a result, the maximum frequency of the proposed PFD is obtained as 5.82 GHz.

The proposed PD is simulated in 0.18 μm CMOS technology by HSPICE software using level 49 parameters. The results are drawn in Figure 4 using MATLAB software. The figure shows the difference between average values of Up and Down signals versus the phase difference of two inputs at six different frequencies, 1–100 MHz, 1 GHz, 3 GHz, 4 GHz, and 5 GHz. Since the nonlinearity of the characteristic at the high frequency happens at large phase differences, where the polarity is more important than the magnitude, this

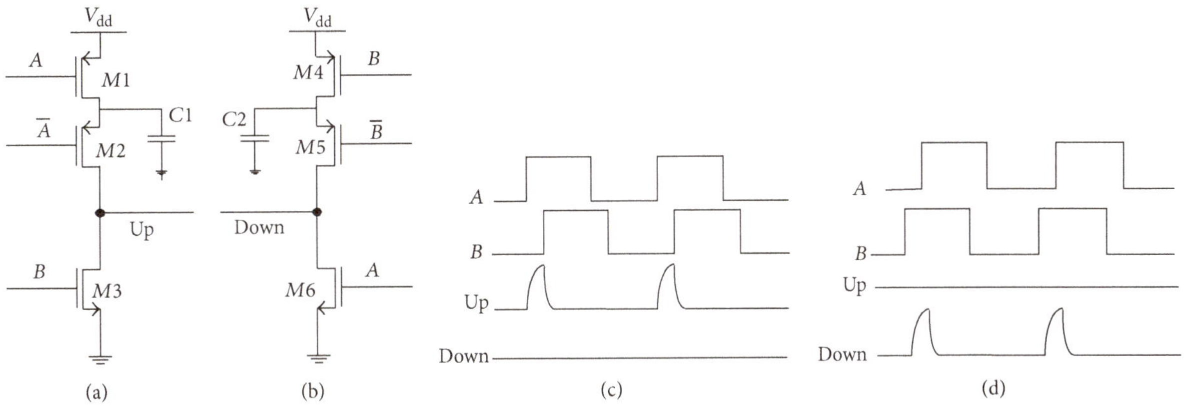

FIGURE 1: Proposed PD: (a) Up path. (b) Down path. (c) The waveforms of the PD when A leads B; (d) when A lags B.

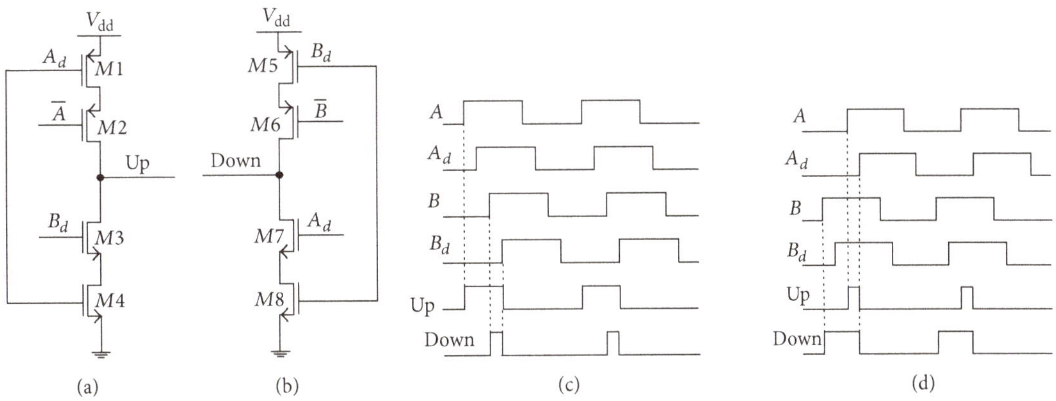

FIGURE 2: Modified PD: (a) Up path. (b) Down path. (c) The ideal waveforms of the modified PD when A leads B; (d) when A lags B.

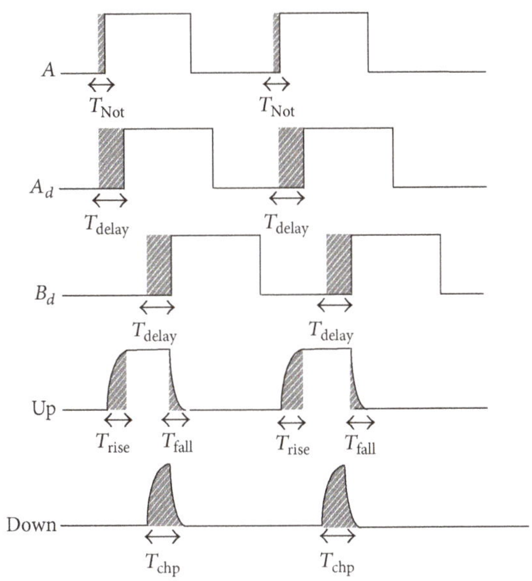

FIGURE 3: The realistic waveforms of the modified PD.

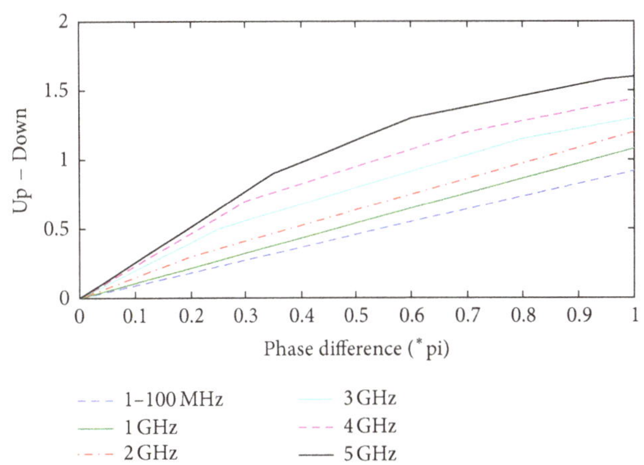

FIGURE 4: Difference between average values of Up and Down signals versus the phase difference of two inputs.

nonlinearity can be ignored. The diagram should be linear in small phase differences to minimize the locking time jitter of the PLL. The region around zero phase difference shows the accuracy of the PFD even at 5 GHz frequency. The power consumption of the proposed PD at 5 GHz frequency is about 0.3 mW. Table 1 summarizes these results and presents a performance comparison among this work and others.

TABLE 1: Performance comparison of the proposed phase detector.

Performance parameter	[3]	[8]	[9]	[10]	This work
CMOS tech	0.13 μm	0.35 μm	0.18 μm	0.13 μm	0.18 μm
Supply	1.2 v	3.3 v	1.8 v	1.2 v	1.8 v
Max freq. (GHz)	2.94	0.1	1	2.1	5 GHz
Dead-zone	NA	NA	Free	NA	Free
Power cons. (mW)	0.496 @ 128 MHz	NA	NA	NA	0.3
Structure	Closed	Open	Closed	Open	Open

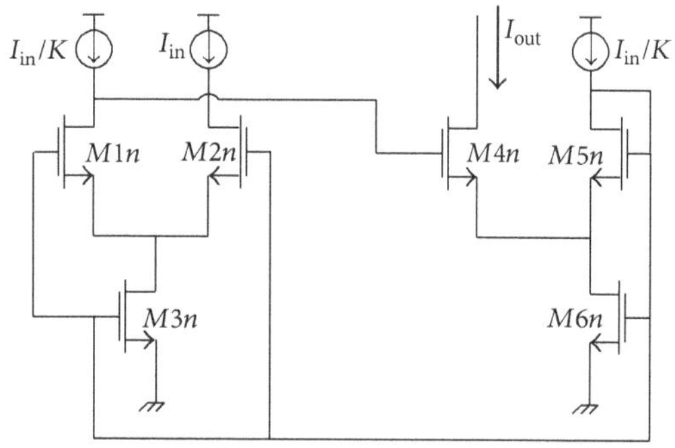

FIGURE 5: Proposed high swing Wilson current mirror.

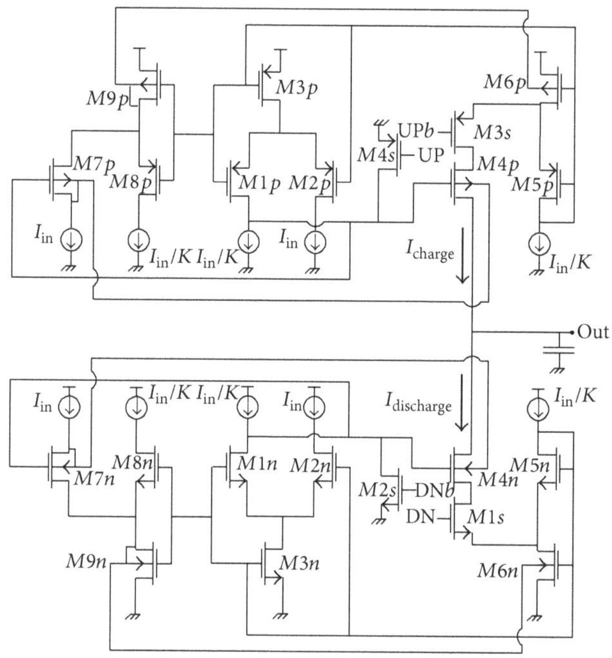

FIGURE 6: Proposed charge pump circuit.

3. Proposed Charge Pump Circuit

One of the most important components of the PLL is the charge pump circuit. An adaptive body bias charge pump circuit is proposed in [11]. In this circuit a number of resistances are used to compensate current variation. However, multiple parameters, such as temperature and fabrication, change the value of these resistances and are not reliable. The charge pump presented in [12] has used a compensation method to reach a high output swing. However the circuit structure is complicated. Moreover, in this circuit the power supply and hence the power consumption are placed high to achieve a high output swing. In order to increase the output voltage swing of the charge pump, in [13] a bulk driven method in a cascade structure has been used. However, using a cascade structure sets limitation to output swing.

A novel bulk driven, low voltage charge pump for high performance PLLs is proposed in this paper. This charge pump is designed based on Wilson current mirror by a novel bulk driven method. Output resistance and output swing are increased simultaneously.

The charge pump circuits must be designed to have a high output swing. However, there are some limitations in output swing, especially when transistors are placed in cascade form, to obtain the high output impedance. To overcome the above problem, bulk driven technique is employed in the proposed circuit. The proposed charge pump circuit is designed based on Figure 5. This circuit is a modified version of a super-Wilson current mirror [20]. As can be seen, the diode connection transistor of an ordinary circuit is replaced with a cascade one. Therefore, the loop gain is increased by a factor of $g_{m1n}r_{o1n}r_{o3n}$.

The proposed charge pump circuit is indicated in Figure 6. In this circuit $M1s$ and $M3s$ which are controlled by DN and UPb, respectively, act as charge pump switches. $M2s$ and $M4s$ are working for faster and better switching. When these MOSFETs turned on, the gate's capacitor of $M4n$ or $M4p$ will be discharged and turn off faster.

In the proposed circuit $M7n(p)$–$M9n(p)$ are used to connect the bulk terminals of $M4n(p)$ and $M6n(p)$ to a higher voltage in comparison with the circuit of Figure 5. It means that the source-bulk voltage (V_{SB}) is decreased. Therefore, according to (3), the threshold voltage (V_T) is reduced too. Equations (4) and (5) indicate that, by decreasing V_T, g_m and g_{mb} are increased and r_o is decreased.

$$V_T = V_{T0} + \gamma\left(\sqrt{\varphi_F + V_{SB}} - \sqrt{\varphi_F}\right), \tag{3}$$

$$g_m = \mu C_{ox}\frac{W}{L}\left(V_{GS} - V_T\right), \qquad g_{mb} = g_m\frac{\gamma}{2\sqrt{\varphi_F + V_{SB}}}, \tag{4}$$

$$r_o = \frac{2}{\mu C_{ox}(W/L)(V_{GS} - V_T)^2}. \quad (5)$$

In (6), by increasing g_{mM4n} and g_{mbM4n} and decreasing r_{oM4n} and r_{oM6n}, the output voltage swing is extremely increased.

$$\begin{aligned} V_{\text{out min}} &= V_{\text{DS}_{M4n}}(\text{sat}) + V_{\text{DS}_{M6n}}(\text{sat}) \\ V_{\text{DS}M4n} &= \left[I_{\text{discharge}} - (g_{mM4n} + g_{mbM4n}) V_{gs4}\right] r_{oM4n}, \quad (6) \\ V_{\text{DS}M6n} &= I_{\text{discharge}} r_{oM6n}. \end{aligned}$$

Output resistance is equal to

$$\begin{aligned} R_{\text{out}} &\approx \left[(g_{m1n} + g_{mb1n}) r_{o1n} r_{o3n} (g_{m4n} + g_{mb4n}) r_{o4}\right] \\ &\quad \cdot \left\| \left[(g_{m1p} + g_{mb1p}) r_{o1p} r_{o3p} (g_{m4p} + g_{mb4p}) r_{o4p}\right] \right. . \end{aligned} \quad (7)$$

One of the most important issues in bulk driven circuits is nonideal behavior of the MOSFETs due to their channel formations. The operation of the bulk driven MOSFET is similar to a Junction Field Effect Transistor (JFET). Because of nonlinear mathematical equations of the MOSFETs at low voltages, analysis of the circuit becomes complicated. For solving this problem, designers have compensated these nonidealities by choosing the aspect ratios of MOSFETs by trial and error (T&E) method [21]. Using of this method has been the only solution so far. In this paper in addition to T&E, MATLAB software is used to apply two methods of optimization algorithms to find the optimum aspect ratios. The main purpose is equalizing the charge and discharge current in a wide output voltage range to achieve the highest output current matching.

4. Tuning by Particle Swarm Optimization and Genetic Algorithm

In this paper, in addition to T&E, MATLAB software is used to apply two methods of optimization algorithms to find the optimum aspect ratios. The main purpose is to equalize the charge and discharge current in a wide output voltage range to achieve the highest output current matching.

To optimize the aspect ratios of $M4n$–$M9n$ and $M4p$–$M9p$ and choose the best values, Particle Swarm Optimization (PSO) and Genetic Algorithm (GA) are used.

4.1. Particle Swarm Optimization. Particle Swarm Optimization (PSO) was introduced in 1995 by Kennedy and Eberhart [22]. In PSO algorithm, a random population of points is generated. Each point represents a member of the population. In PSO algorithm, there is no sudden jump or confusion; each point is a solution. Considering X and V as particle position and velocity, respectively, the position of nth particle in a space with m dime is represented with $X_n = [X_{n1}, X_{n2}, \ldots, X_{nm}]$.

The position of each particle is changed at the next stage and it reaches a new position. The best position of nth particle which corresponds to the lowest cost function for that particle is saved in P_{best_n}. In addition, P_{best} of all particles are compared and the position of particle which has the lowest cost function is saved in G_{best}. The next vector of each particle depends on its position and its distance to its P_{best} and its distance to G_{best}. The relations of particles movements are as follows:

$$\begin{aligned} V_{nm}^{i+1} &= w \times V_{nm}^{i} + C_1 \times \text{rand}() \times \left(P_{\text{best}_{nm}} - X_{nm}^{i}\right) \\ &\quad + C_2 \times \text{rand}() \times \left(G_{\text{best}} - X_{nm}^{i}\right), \\ X_{nm}^{i+1} &= X_{nm}^{i} + C V_{nm}^{i+1}, \\ \left|V_{nm}^{i+1}\right| &\le V_{\max}, \end{aligned} \quad (8)$$

where $V_{\max}$ indicates a parameter that prevents going out of suitable search space which causes the solution to be in acceptable region; C_1 and C_2 are constants which represent the speed of learning or pulling to P_{best} and G_{best}, and the weighing function w is given by

$$w = w_{\max} - \frac{w_{\max} - w_{\min}}{\text{iter}_{\max}} \times \text{iter}, \quad (9)$$

where $w_{\min}$ and $w_{\max}$ indicate the minimum and maximum weighing functions and iter denotes the number of iterations.

In order to optimize the parameters of the controller by PSO, the following cost function is used:

$$\text{PSO} = \int_0^{t_1} |e(t)|\, dt = \int_0^{t_1} \left|i_{\text{discharge}} - i_{\text{charge}}\right| dt, \quad (10)$$

where t_1 is the final time of simulation, e is the error signal, and t_s is the settling time of the system.

4.2. Genetic Algorithm. Genetic Algorithm is a search-based optimization method [23]. This algorithm has been proposed by John Holland (1962). In the algorithm, he has benefited from two principles of selection and reproduction in the nature. Genetic Algorithm can be considered as an oriented random optimization method which gradually moves towards optimal point.

Unlike other common optimization models in which only one point is used in each stage of optimization process, in Genetic Algorithm, a group of points are used.

If our optimization target function is as

$$F = f(x_1, x_2, \ldots, x_n), \quad (11)$$

then, the objective is to find the value of x_i in such a way that F function has the minimum value. In Genetic Algorithm, at the first stage, a set of chromosomes are randomly created (random strings from x_1 to x_n). These x_i refer to genes. Putting each of these chromosomes in target function, the value of F function is obtained. The first generation concludes through appropriate scoring to these values. The first generation is called parents generation.

The process of creating offspring (children) in the next generations follows the following three general principles:

(a) Crossover

(b) Elite

(c) Mutation.

TABLE 2: Aspect ratio values of MOSFETs at different methods.

	$M4n$	$M5n$	$M6n$	$M4p$	$M5p$	$M6p$
PSO	107	328	169	91	718	268
GA	287	344	261	137	662	363
T&E	43	327	111	53	795	222

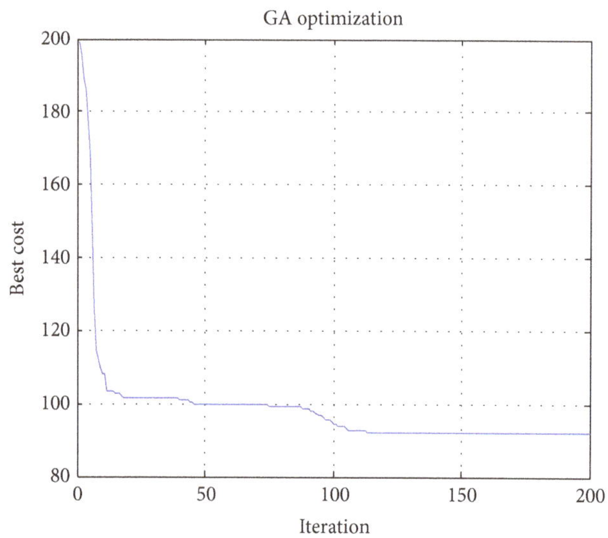

FIGURE 7: Convergence objective function for Genetic Algorithm.

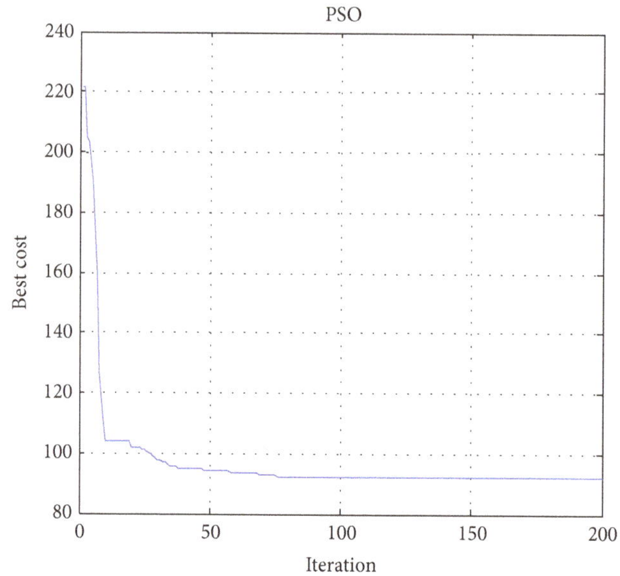

FIGURE 8: Convergence objective function for Particle Swarm Optimization.

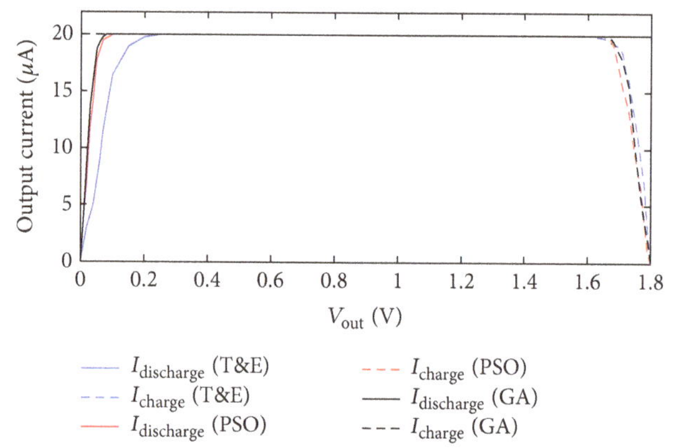

FIGURE 9: Matching characteristic of the proposed charge pump.

After the second generation and after creating offspring with ratios determined by selection algorithms, new generation children are selected. Putting these children into target function and scoring them, then, the algorithm is reiterated until the algorithm meets the ending criteria (such as the number of generation and time), causing the algorithm to be stopped.

4.3. Results Optimization. In this study, the values of transistors aspect ratio are optimized using the following considerations.

It should be noted that aspect ratios of $M4n(p)$ equal $M7n(p)$, $M5n(p)$ equal $M8n(p)$, and $M6n(p)$ equal $M9n(p)$. The aspect ratios of the MOSFETs should be positive.

In the present study, the number of the algorithm iterations is the stopping criterion. Both PSO and GA are simulated using the following two sets of parameters:

(a) parameters iteration = 100 and population = 30,

(b) parameters iteration = 200 and population = 50.

The results pertained to the target function optimization of (10) have been shown in Figures 7 and 8. The respective figures, in fact, indicate cost function improvement for the increase in the number of iterations.

The results of T&E and optimal value factor's aspect ratios are presented in Table 2.

To verify the efficiency of the proposed circuit, it is simulated under the power supply of 1.8 V in 180 nm CMOS technology using Hspice. Input current is set to 20 μA. Figure 9 shows the charge and discharge output currents of the proposed charge pump which are obtained from T&E, PSO, and GA methods as the output voltage is swept from 0 to 1.8 V. As can be seen, the output dynamic voltage ranges of output current matching are 0.25–1.6 V, 0.1–1.65 V, and 0.08–1.65 V for T&E, PSO, and GA, respectively. According to these results, the GA method makes a better performance to this charge pump. Therefore, the aspect ratios which are presented by the GA are chosen for applying to the proposed charge pump. Current mismatch is less than 1% of nominal output current (I_{charge} or $I_{discharge}$) over the output range of these methods.

In the usual bulk driven current mirror, the gate terminals are tied to V_{dd} [24]. However, in these circuits, the output current cannot track the input current in a specified range [13]. Figure 10 shows the variations of output current versus input current in the proposed circuit. As can be seen, the output current can track the input current in a wide input current range. The overall power consumption of the proposed charge pump circuit is around 160 μW. Table 3

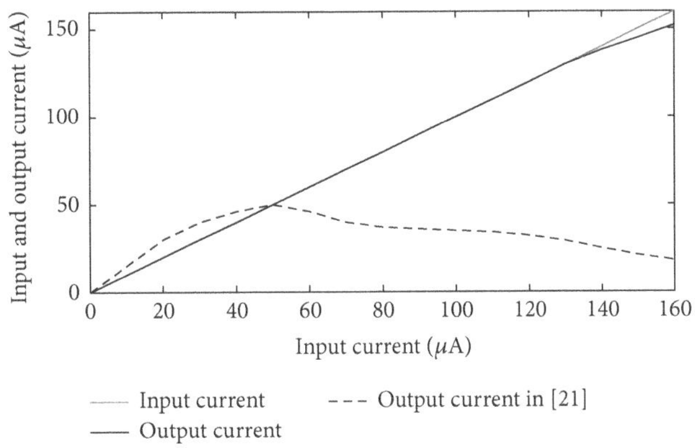

FIGURE 10: Input and output current transfer characteristics.

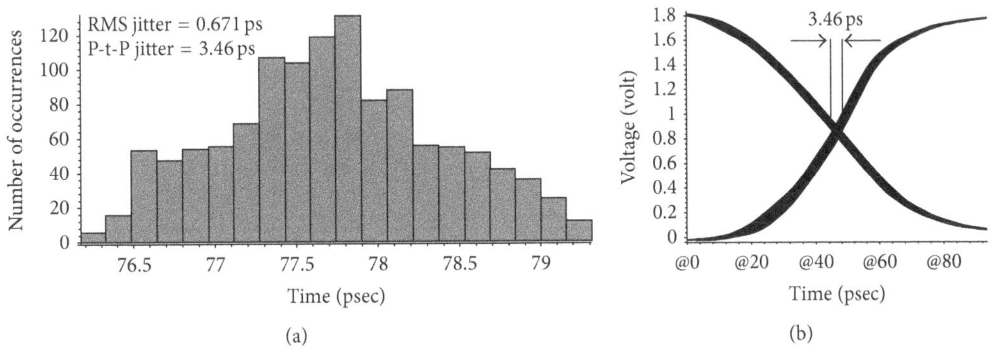

FIGURE 11: (a) Output jitter histogram at 3 GHz. (b) Output jitter eye diagram at 3 GHz.

TABLE 3: Performance comparison of the proposed charge pump.

	[11]	[12]	[13]	This work
CMOS tech (nm)	130	180	180	180
Power supply (V)	1.2	3	1.8	1.8
Voltage swing (V)	0.2–1	0.2–2.7	0.3–1.58	0.08–1.66
Swing/V_{dd}	66%	83.3%	71%	88%
Current mismatch	0.9%	2.1%	1%	1%
Power consumption (μW)	30	NA	395	160

summarizes the performance of the proposed charge pump and compares them with recent publications. As can be observed, the proposed circuit has an excellent voltage swing per V_{dd}, while its power consumption is reasonable.

5. Simulation Results of Proposed PLL

The PLL is designed using the proposed PFD and charge pump circuits. Furthermore, it employs a two-stage voltage controlled ring oscillator [18]. This PLL has a wide locking range from 500 MHz to 5 GHz.

Jitter histogram and eye diagram at 3 GHz operating frequency are depicted in Figure 11, which shows around 0.671 ps and 3.46 ps RMS and peak-to-peak jitters, respectively.

Figure 12 shows the output spectrum simulated at 3 GHz. The simulation result shows a reference spur of −72 dBm with a 1.85 GHz frequency offset. The phase noise of the PLL at the locking frequency in different frequency offsets is presented in Figure 13. The output phase noise is −117.6 dBc/Hz at a 1 MHz frequency offset. The total power consumption is about 11.5 mW. The figure of merit of the proposed PLL according to (12) [25] is −198.47 dBc/Hz.

$$\text{FOM} = L\{\Delta f\} - 20\log\left\{\frac{f_0}{\Delta f}\times\frac{\text{FTR}}{10}\right\} + 10\log\left\{\frac{P_{\text{diss}}}{1\,\text{mW}}\right\}, \quad (12)$$

where $L\{\Delta f\}$ is phase noise, Δf is certain frequency offset, f_0 is center frequency, P_{diss} is the power dissipation, and FTR is tuning range of oscillation frequency.

Table 4 summarizes these results and presents a performance comparison of this work with some recent papers. The proposed PLL achieves lowest FOM_T by using a fast PFD circuit and omitting the divider circuit.

6. Conclusion

A very simple, low power, high speed, and open loop phase detector is proposed which operates in a wide frequency range from 1 MHz to 5 GHz. Due to the extra simplicity of the circuit, the power consumption is very low and is about 0.3 mW at the highest operational frequency. The dead-zone and missing edge problems are solved completely in

TABLE 4: A performance comparison of proposed circuit with some recent papers.

Performance parameter	[14]	[15]	[16]	[17]	[18]	[7]	This work
CMOS tech (nm)	65	65	65	65	90	180	180
Supply voltage (v)	1	1.2	1.2	1	1.2	1.8	1.8
VCO	Ring	Ring	LC	Ring	Ring	Ring	Ring
Ref. frequency (GHz)	0.645	0.64	104	1.6	0.01	5	3
Locking range (GHz)	0.485–1.011	0.25–1.06	103.05–104.58	NA	3.5–7.1	2.5–7.3	0.5–5
Ref. spur (dBm)	NA	−54.8	−63.8	−47	−64.8	−66.8	−72
Phase noise @ 1 MHz offset (dBc/Hz)	−110.8	−88.6	−80.41	−88	−105	−108.2	−117.6
RMS jitter (ps)	NA	9.6	2.44	4.82	3.8	0.88	0.671
P-t-P jitter (ps)	NA	52.2	18	38	NA	3.21	3.46
Power (mW)	10	1.2	63	0.99	29.64	13.4	11.5
FOM @ 1 MHz offset (dBc/Hz)	−157	NA	−162.57	NA	NA	−190.55	−198.47

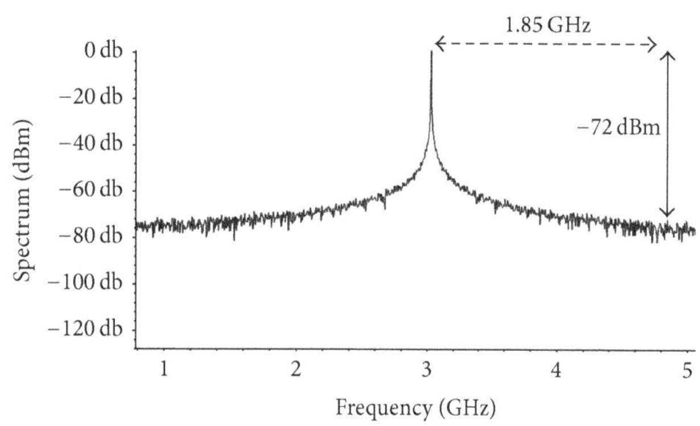

FIGURE 12: Output spectrum of the VCO.

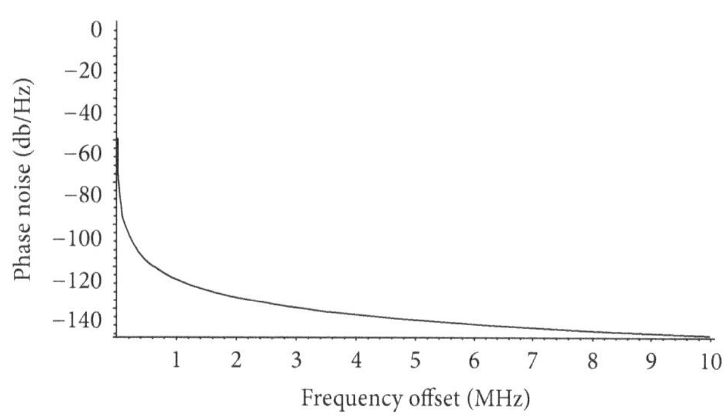

FIGURE 13: Simulated phase noise at 3 GHz for different frequency offsets.

the proposed architecture. A high output impedance, low power, and single-ended charge pump based on a bulk driven Wilson current mirror, with excellent output current matching, is also introduced in this paper. The proposed charge pump aspect ratios parameter was optimized by PSO and GA. Therefore, the dynamic voltage range was increased significantly which is obtained about 88% of power supply. The current mismatch between charge and discharge current is less than 1% of nominal output current. Therefore, through the use of a dead-zone-less PFD and a divider-less PLL, the overall jitter is decreased significantly. However, since a single ended charge pump architecture has a higher substrate and supply noise coupling, a differential charge pump circuit can be considered as a future work. Moreover, the operating frequency range can be increased by tracking the self-oscillation frequencies of the voltage-controlled oscillator (V_{CO}) and the frequency divider. The rms and peak-to-peak jitters of this PLL at 3 GHz are 0.671 and 3.46 ps, respectively.

Competing Interests

The authors declare that they have no competing interests.

References

[1] R. Prakash, S. Akhtar, and P. T. Balsara, "Phase noise reduction in high speed frequency divider," in *Proceedings of the IEEE Dallas/CAS Workshop on Design, Applications, Integration and Software (DCAS '06)*, pp. 83–86, IEEE, Richardson, Tex, USA, October 2006.

[2] K. Park and I. C. Park, "Fast frequency acquisition phase frequency detectors with prediction-based edge blocking," in *Proceedings of the IEEE International Symposium on Circuits and Systems (ISCAS '09)*, pp. 1891–1894, IEEE, Taipei, Taiwan, May 2009.

[3] W.-H. Chen, M. E. Inerowicz, and B. Jung, "Phase frequency detector with minimal blind zone for fast frequency acquisition," *IEEE Transactions on Circuits and Systems II: Express Briefs*, vol. 57, no. 12, pp. 936–940, 2010.

[4] N. M. H. Ismail and M. Othman, "CMOS phase frequency detector for high speed applications," in *Proceedings of the 21th International Conference on Microelectronics (ICM '09)*, pp. 201–204, IEEE, Marrakech, Morocco, December 2009.

[5] R. Y. Chen and W.-Y. Chen, "A high-speed fast-acquisition CMOS phase/frequency detector for MB-OFDM UWB," *IEEE Transactions on Consumer Electronics*, vol. 53, no. 1, pp. 23–26, 2007.

[6] L.-K. Soh and Y.-F. K. Edwin, "An adjustable reset pulse phase frequency detector for phase locked loop," in *Proceedings of the 1st Asia Symposium on Quality Electronic Design (ASQED '09)*, pp. 343–346, IEEE, Kuala Lumpur, Malaysia, July 2009.

[7] H. R. Erfani-Jazi and N. Ghaderi, "A divider-less, high speed and wide locking range phase locked loop," *AEU—International Journal of Electronics and Communications*, vol. 69, no. 4, pp. 722–729, 2015.

[8] S. Kazeminia, K. Hadidi, and A. Khoei, "A low jitter 110 MHz 16-phase delay locked loop based on a simple and sensitive phase detector," in *Proceedings of the 2013 21st Iranian Conference on Electrical Engineering (ICEE '13)*, pp. 1–5, IEEE, Mashhad, Iran, May 2013.

[9] A. C. Kailuke, P. Agrawal, and R. V. Kshirsagar, "Design of phase frequency detector and charge pump for low voltage high frequency PLL," in *Proceedings of the International Conference on Electronic Systems, Signal Processing and Computing Technologies (ICESC '14)*, pp. 74–78, IEEE, Nagpur, India, January 2014.

[10] D. Cai, H. Fu, J. Ren et al., "A 2.1-GHz PLL with −80dBc/−74dBc reference spur based on aperture-phase detector and phase-to-analog converter," in *Proceedings of the IEEE Asian Solid State Circuits Conference (A-SSCC '11)*, pp. 141–144, Jeju, South Korea, November 2011.

[11] P. Liu, P. Sun, J. Jung, and D. Heo, "PLL charge pump with adaptive body-bias compensation for minimum current variation," *Electronics Letters*, vol. 48, no. 1, pp. 16–18, 2012.

[12] N. Joram, R. Wolf, and F. Ellinger, "High swing PLL charge pump with current mismatch reduction," *Electronics Letters*, vol. 50, no. 9, pp. 661–663, 2014.

[13] H. R. E. Jazi and N. Ghaderi, "A novel bulk driven charge pump for low power, low voltage applications," *IEICE Electronics Express*, vol. 11, no. 2, Article ID 20130934, 2014.

[14] J.-M. Kim, S. Kim, I.-Y. Lee, S.-K. Han, and S.-G. Lee, "A low-noise four-stage voltage-controlled ring oscillator in deep-submicrometer CMOS technology," *IEEE Transactions on Circuits and Systems II: Express Briefs*, vol. 60, no. 2, pp. 71–75, 2013.

[15] I.-T. Lee, Y.-T. Tsai, and S.-I. Liu, "A leakage-current-recycling phase-locked loop in 65-nm CMOS technology," *IEEE Journal of Solid-State Circuits*, vol. 47, no. 11, pp. 2693–2700, 2012.

[16] K.-H. Tsai and S.-I. Liu, "A 104-GHZ phase-locked loop using a vco at second pole frequency," *IEEE Transactions on Very Large Scale Integration (VLSI) Systems*, vol. 20, no. 1, pp. 80–88, 2012.

[17] K.-C. Choi, S.-G. Kim, S.-W. Lee, B.-C. Lee, and W.-Y. Choi, "A 990-μW 1.6-GHz PLL based on a novel supply-regulated active-loop-filter VCO," *IEEE Transactions on Circuits and Systems II: Express Briefs*, vol. 60, no. 6, pp. 311–315, 2013.

[18] S. Min, T. Copani, S. Kiaei, and B. Bakkaloglu, "A 90-nm CMOS 5-GHz ring-oscillator pll with delay-discriminator-based active phase-noise cancellation," *IEEE Journal of Solid-State Circuits*, vol. 48, no. 5, pp. 1151–1160, 2013.

[19] K. Martin, *Digital Integrated Circuit Design*, Oxford University Press, New York, NY, USA, 2000.

[20] B. A. Minch, "Low-voltage Wilson current mirrors in CMOS," in *Proceedings of the IEEE International Symposium on Circuits and Systems (ISCAS '07)*, pp. 2220–2223, IEEE, New Orleans, La, USA, May 2007.

[21] B. Aggarwal, M. Gupta, and A. K. Gupta, "Analysis of low voltage bulk-driven self-biased high swing cascode current mirror," *Microelectronics Journal*, vol. 44, no. 3, pp. 225–235, 2013.

[22] J. Kennedy and R. Eberhart, "Particle swarm optimization," in *Proceedings of the IEEE International Conference on Neural Networks*, pp. 1942–1948, Perth, Australia, November-December 1995.

[23] K. F. Man, K. S. Tang, and S. Kwong, "Genetic algorithms: concepts and applications," *IEEE Transactions on Industrial Electronics*, vol. 43, no. 5, pp. 519–534, 1996.

[24] Z. Zhu, J. Mo, and Y. Yang, "A low voltage bulk-driving PMOS cascode current mirror," in *Proceedings of the 9th International Conference on Solid-State and Integrated-Circuit Technology (ICSICT '08)*, pp. 2008–2011, IEEE, Beijing, China, October 2008.

[25] J. Kim, J.-O. Plouchart, N. Zamdmer et al., "A 44 GHz differentially tuned VCO with 4 GHz tuning range in 0.12 μm SOI CMOS," in *Proceedings of the 2005 IEEE International Solid-State Circuits Conference (ISSCC '05)*, pp. 416–607, IEEE, San Francisco, Calif, USA, February 2005.

19

Applying Partial Power-Gating to Direction-Sliced Network-on-Chip

Feng Wang, Xiantuo Tang, and Zuocheng Xing

National Laboratory for Parallel and Distributed Processing, National University of Defense Technology, Changsha 410073, China

Correspondence should be addressed to Feng Wang; kikiwang007@gmail.com

Academic Editor: Chi-Ying Tsui

Network-on-Chip (NoC) is one of critical communication architectures for future many-core systems. As technology is continually scaling down, on-chip network meets the increasing leakage power crisis. As a leakage power mitigation technique, power-gating can be utilized in on-chip network to solve the crisis. However, the network performance is severely affected by the disconnection in the conventional power-gated NoC. In this paper, we propose a novel partial power-gating approach to improve the performance in the power-gated NoC. The approach mainly involves a direction-slicing scheme, an improved routing algorithm, and a deadlock recovery mechanism. In the synthetic traffic simulation, the proposed design shows favorable power-efficiency at low-load range and achieves better performance than the conventional power-gated one. For the application trace simulation, the design in the mesh/torus network consumes 15.2%/18.9% more power on average, whereas it can averagely obtain 45.0%/28.7% performance improvement compared with the conventional power-gated design. On balance, the proposed design with partial power-gating has a better tradeoff between performance and power-efficiency.

1. Introduction

In recent years, Network-on-Chip has been proposed as the mainstream architecture to connect various on-chip resources in many-core systems, such as chip-multiprocessors (CMPs) and multiprocessor system-on-chips (MPSoCs). As in most other VLSI designs, power-efficiency has also been one of the critical constraints in current NoC design. The power situation of NoC design will become much more serious as the technology is continually scaling down and the working frequency is increasing. For example, the percentage of router power reaches up to 28% in Intel Teraflop chip [1] and 19% in 36-core SCORPIO chip [2]. The overall router power-consumption consists of dynamic switching power and static leakage power. The dynamic power is consumed when the router transfers packets and its clock circuits switch periodically. However, there is significant consumption of static leakage power even without any packet transfers as long as the router is power-on. In recent multi/many-core systems, the leakage power has already become a major portion of power dissipation, and its proportion will further increase along with technology parameter scaling down. In the simulation of Samih et al. [3], the leakage power increases rapidly, from 11.2% of the total router power-consumption in the 65 nm technology to 33.6% in the 32 nm technology, when working at 1.1 V voltage and 2.0 GHz frequency.

In a many-core system, the demand of network throughput may be quite low in most of execution time and rarely reaches the peak/saturation point. Actually, real-world applications exhibit sparse and bursty network traffic [4, 5]: only few communication-intensive phases that consume large network bandwidth and other phases that inject few packets into the network. Hence, there are many opportunities in which some routers and links keep in idleness, and they can be utilized to solve leakage power crisis. As a representative technique to mitigate leakage power, power-gating controls the power supply of the gated circuit blocks by switching the transistors which are inserted between the VDD and the gated blocks. According to the traffic characteristic, some recent researches have employed power-gating in current NoC design to mitigate the increasing leakage power, which will be introduced in the next section. Besides, clock network also consumes a great deal of chip dynamic power because the clock is fed to most of sequential circuit blocks on chip, and the clock must switch periodically. As a well-known low-power technique, clock-gating is utilized to reduce chip dynamic power [6] and has been implemented in recent NoC designs [1, 7, 8]. The Intel Teraflop chip [1] uses the multilevel clock-gating policy and the sleep transistor circuits to reduce both dynamic and leakage power, and it is controlled at full-chip, tile-slice, and individual tile levels based on workloads. In another recent clock-gating design, Mullins [7] applies clock-gating to the on-chip routers at two levels: local clock-gating and router level clock-gating.

As more cores are integrated, the requirement of low on-chip latency will become even more pronounced so as not to impact system performance. However, due to the disconnection problem, an injected packet which meets a sleeping downstream router has to suspend transfer and wait for the sleeping router being awakened in the conventional power-gated NoC. Then successive wake-up delays may be inevitably incurred to the packet. Aiming at the disconnection problem, we propose partial power-gating approach to obviate the performance decline in the power-gated NoC. This approach is composed of a direction-slicing scheme, an improved routing algorithm, and a deadlock recovery mechanism. The direction-slicing scheme divides each router into two parts (i.e., two slices) and keeps one slice always-on to construct an active subnet for providing a fully connected path while applying power-gating to another for saving possible leakage power. Meanwhile, the routing algorithm is improved to support packet transfers when the gated slice switches its power supply in power-on/off status. Based on improved architecture and routing algorithm, a novel deadlock recovery mechanism is adopted to solve deadlock in our sliced network. For the recovery mechanism, a deadlock escape path is constructed in the network interface through adding some additional components and links. Furthermore, clock-gating is utilized in the gated slice to save partial clock power. This direction-sliced NoC with partial power- and clock-gating can achieve the purpose of power-saving and has less impact on performance than the conventional power-gated design.

The rest of this paper is organized as follows. Section 2 summarizes some related works and expatiates on our motivation. Section 3 explains the details of the direction-sliced NoC with partial power-gating. Section 4 presents our evaluation methodology and discusses the simulation results. Finally, we conclude this paper in Section 5.

2. Related Work and Motivation

Although low-power can be achieved by reducing the overhead of router resources in current NoC design (such as the bufferless router [9]), the low-cost NoC still has both obvious performance loss and limited power decline. Power-gating is a well-known technique to mitigate leakage power in low-power VLSI design, especially for circuit blocks that exhibit enough idleness. Several recent studies have adopted power-gating technique in NoC design. Since router buffers consume the largest portion of the NoC's leakage power, power-gating is utilized to reduce buffer leakage power in [10]. Instead of turning the whole buffer off, it keeps partial buffer active to store incoming flits at all times and then avoids negative performance affection. Besides adopting power-gating on router buffers, Matsutani et al. [11] proposed an ultrafine-grained power-gating policy to control the power supply of all router components. This policy based on look-ahead routing can detect the arrival of packets two hops ahead so as to hide partial wake-up delay.

Unlike component-based power-gating, some other studies prefer to apply power-gating to entire router (router level power-gating). A router parking method is proposed in [3] to power-gate routers when the connected core is idle, but it needs to flush private caches before turning off routers, which may cause serious performance decrease. The node-router disconnection referred to by [12] severely limits power-gating being effectively used in on-chip routers due to the limitation of break-even time, long wake-up delay. The problem is that a packet injected by an active core/sending-router may meet a sleeping receiving-router and has to suspend transfer and wait for the sleeping receiver being awakened due to the disconnection between the active sender and the sleeping receiver. This disconnection situation makes conventional power-gating unprofitable for the high-performance NoC design.

Originating the idea of guaranteeing full connectivity, these researches [12–14] provide a fully connected path to avoid the disconnection problem. The NoRD [12] introduces a bypass ring path to connect all nodes completely; then packets can be transmitted in the bypass ring to avoid long wake-up latency when meeting a sleeping router. However, The packets' average latency brought by the bypass ring may inevitably increase proportionately with the square of network scale. As a low-power instance in the multiple networks, the Catnap [13] guarantees full connectivity by keeping one subnet always active, while keeping other subnets power-gated to reduce unnecessary leakage power. Profiting from the excellent path diversity in the Clos network, the MP3 [14] utilizes minimal router resources and links to ensure all nodes are fully connected, and other remaining full-routers and partial-routers can be power-gated to achieve leakage power-saving. However, the Catnap merely increases the efficiency of power-gating in the architecture of multiple networks, and the MP3 is only applicable to Clos and other indirect networks.

Three typical network slicing schemes are proposed in [15], channel-slicing, dimension-slicing, and bit-slicing, which slice each network node across multiple chips or modules to alleviate pin and area limitations. For the channel-slicing, a w-bit-wide node is split into k independent slices, each of which has w/k-bit-wide channels. There are not any connections between these slices. For the dimension-slicing, the network node is sliced based on ports' dimensions; those ports associated with one dimension are contained in an appointed slice. Besides, additional data channels must be added between the slices to communicate with each other. Finally, for the bit-slicing, a w-bit-wide node is divided across k w/k-bit-wide slices; each slice packaged in a separate module contains a w/k-bit-wide portion of the router datapath. Unlike the channel-slicing, control information must be distributed to all slices so as to act in unison. Except the three slicing schemes above, we propose a new slicing scheme named direction-slicing for bidirectional network. The direction-slicing divides the router into two slices based on the directions in a dimension (input or output direction), and each slice only has a unidirectional channel in each dimension. How to make direction-slicing will be shown in Section 3.1.

In order to avoid the disconnection issue in the conventional power-gated NoC, as a fire-new viewpoint, we start to investigate network-slicing and power-gating in combination. Essentially, the Catnap [13] is a commendable instance of applying power-gating to channel-sliced network (multiple network is an instance of channel-slicing). For the dimension-slicing, it is hard to construct a sliced subnet to guarantee full connectivity; then it is not fit for the intention of avoiding disconnection. Applying power-gating to bit-sliced network has been wri tten in our other paper, and we do not introduce it at present. When utilizing power-gating into the direction-sliced network, we split the router into two different power domain slices across channels' input or output directions, since each channel is bidirectional. In this sliced network, we can maintain the power domain of some slices being ever-on to guarantee full connectivity and adopt power-gating for other slices to save some unnecessary leakage power. The detail of employing power-gating to the sliced network will be introduced in the following section.

3. Low-Power Direction-Sliced NoC Design

In this section, we first propose the direction-sliced architecture to suit for partial power-gating scheme. Next, we improve routing algorithm to support packets transmitting and design a deadlock recovery mechanism to cope with the routing improvement. Finally, we describe the detail of partial power-gating scheme for the sliced network.

3.1. Direction-Sliced Architecture. We employ the direction-slicing scheme into the 2D torus and mesh network, respectively. Figure 1(a) shows the direction-slicing in the 2D torus network. All $X+$ channels (*West_In* and *East_Out*), $Y-$ channels (*South_In* and *North_Out*), and *Local* channels in each router are split into ever-on slices and form the ever-on subnet to guarantee full connectivity, while remaining $X-$ channels (*East_In* and *West_Out*) and $Y+$ channels (*North_In* and *South_Out*) are the gated slices. In contrast with the 2D torus, Figure 1(b) shows the different direction-slicing in the 2D mesh network. Considering both symmetrical topology and simple routing design, all $X+$ channels in even rows, $X-$ channels in odd rows, $Y-$ channels in even columns, $Y+$ channels in odd columns, and *Local* channels are utilized to constitute the ever-on subnet, and other remaining channels which belong to gated slices can be power-gated to pursue the leakage power-saving.

Whether in the torus or mesh, each router is split into two power domains across input and output channels, and corresponding VC buffers, crossbars, and output latches are also divided into the two power domains, as shown in Figure 2(a). Since router's leakage power is mainly consumed by VC buffers, crossbars, and output latches (about 76.3% of leakage power in our simulation), it may cut down a lot of leakage power if 2/5 router resources are power-gated. In current power-gating technique, there are hardware overheads for the gating switches and the distribution of the gating signal wires; the area overheads of the well-designed power-gating blocks are typically within 4–10% depending on the circuit level optimizations [16, 17]. Then our sliced design only increases about 1.6–4.0% of hardware overhead owing to the partial power-gating. Moreover, in our simulation, all flip-flops and links in clock networks consume about 39.2% of total power when in zero-load. Then, except for adopting partial power-gating, we also adopt clock-gating technique to save flip-flops' dynamic power in clock networks when corresponding gated slice is powered-off, and this method is similar to the router-level clock-gating in [7]. Because the resumption period of clock-gating is far less than the wake-up period of power-gating, we can reuse the wake-up signal as the early valid signal to resume the clock input and the power supply, as shown in Figure 2(b). The latch in clock-gating circuit can be used to prevent glitches on the control signal from propagating to the gated slice's clock pin [18]. Eliminating the latch can slightly reduce power dissipation and area cost, whereas the latch-free design has a significant drawback: if the control signal is not stable at its new value before the falling clock edge, glitches on the control signal can corrupt the clock signal to the gated slice.

In the on-/off-chip network, deadlock situation can be avoided by wise routing design, such as *XY* Dimension Ordered Routing (DOR-*XY*) in the 2D normal mesh network. However, the ever-on subnet in the sliced mesh may inherently lead to deadlock, since the restricted unidirectional channel combined with minimal path routing may form some resource circular dependencies. In the bidirectional or unidirectional torus, all channels in a row/column form a closed ring, and packets routed in the ring may encounter a deadlock caused by the circular dependency. Then, we improve network interface (NI) structure to achieve deadlock recovery, add an escape path in NI to release deadlock situation, and retransmit blocked packets from the local input port. The escape latch is used to temporarily buffer the blocked packets when the recover process is activated. As shown in Figure 2(c), three modules (escape latch, multiplexer, and demultiplexer) are added to implement an escape path. In our area evaluation, the added modules only increase 2.2% of total router area (not including circuit blocks for power-gating and clock-gating). Both resource dependence situation and deadlock recovery mechanism will be introduced in Section 3.4.

3.2. Routing Algorithm. If a sliced router is partially on, some input/output ports should be forbidden to be utilized. Then, the routing algorithm should be fit for two cases. On the one hand, it needs to support packets routing on the ever-on subnet when the gated slice is powered-off; on the other hand, it must be able to utilize all possible ports when

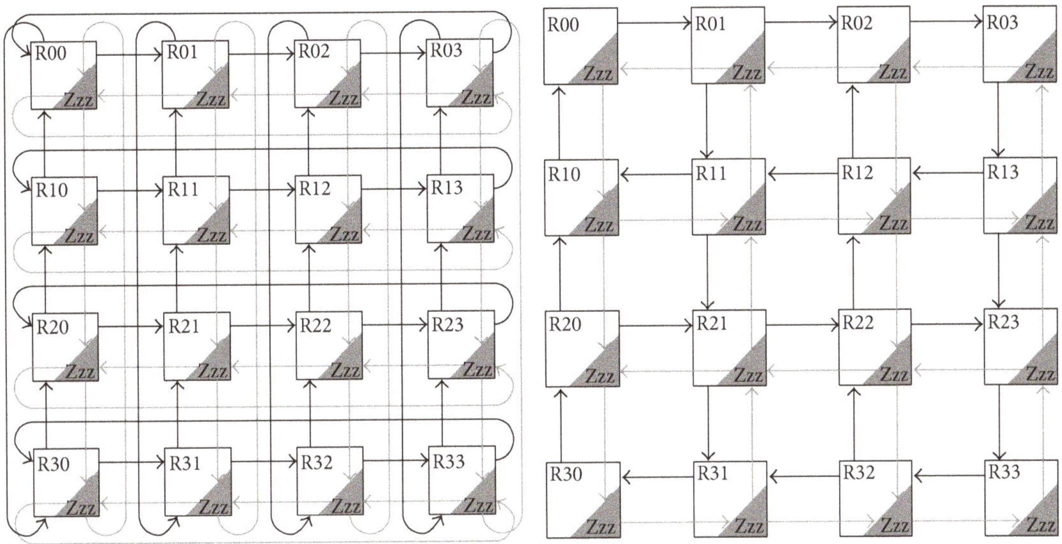

(a) 2D torus

(b) 2D mesh

FIGURE 1: Direction-sliced torus and mesh network.

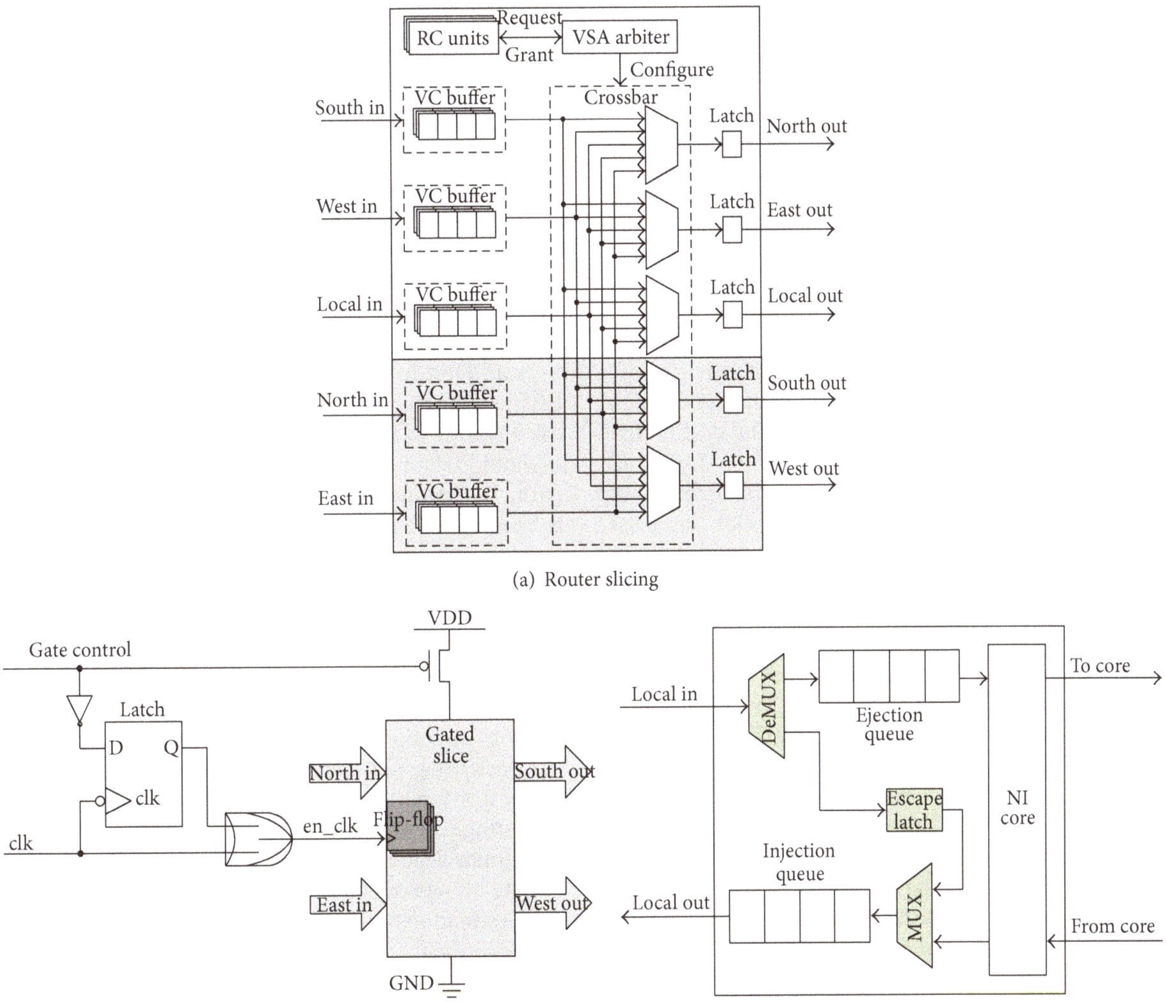

(a) Router slicing

(b) Partial power- and clock-gating

(c) Network interface

FIGURE 2: Direction-sliced router architecture for partial power-gating and clock-gating.

```
Input: cur - current node coordinate, dest - destination node coordinate, pre_dir - preceding routing direction;
Output: dir - output direction;
(1)  begin
(2)      Δx ← (dest.x − cur.x);
(3)      Δy ← (dest.y − cur.y);
(4)      if Sliced router is partially-on then
(5)          if Δx ≠ 0 then
(6)              dir ← X+;
(7)          else
(8)              dir ← Δy ≠ 0 ? Y− : Local;
(9)      else if Sliced router is fully-on then
(10)         if (Δx > 0 and Δx ≤ Dim_x/2) or Δx ≤ −Dim_x/2 then
(11)             dir ← X+;
(12)         else if (Δx < 0 and Δx > −Dim_x/2) or Δx > Dim_x/2 then
(13)             dir ← X−;
(14)         else
(15)             if (Δy > 0 and Δy < Dim_y/2) or Δy < −Dim_y/2 then
(16)                 dir ← Y+;
(17)             else if (Δy < 0 and Δy ≤ −Dim_y/2) or Δy ≥ Dim_y/2 then
(18)                 dir ← Y−;
(19)             else
(20)                 dir ← Local;
(21)     if (pre_dir = X− and dir = X+) or (pre_dir = Y+ and dir = Y−) then
(22)         dir ← pre_dir;
(23)     return dir;
```

ALGORITHM 1: Routing algorithm for the sliced 2D torus network.

the sliced router is fully on. For the convenience of description, we use the abbreviation *UniMesh/UniTorus* to indicate the ever-on subnet in the sliced mesh/torus when all gated slices are deactivated.

The UniTorus still has perfect symmetry, and its routing algorithm can be simply designed when all sliced routers are partially on. At first, packets can be routed in the X-dimension unidirectional ring until reaching the node which locates in the same Y-dimension ring with the destination node. Then, packets should make a dimension turn and be transmitted in the Y-dimension unidirectional ring until reaching the destination node. If the gated slice has been awakened, packets can be switched to any possible output port by utilizing normal DOR-XY algorithm. The routing algorithm of the sliced 2D torus is shown in Algorithm 1. Noteworthily, the algorithm should forbid routing from the gated direction to the ever-on direction within the same dimension, so that the forward and backward routing, which may happen repeatedly before the gated slice is activated, can be avoided. When a packet is waiting for the gated slice to be awakened, we can simply solve the issue by routing the packet to its preceding direction. The solution is illustrated at lines 21-22 in the pseudocode of Algorithm 1.

The routing algorithm in the sliced mesh is also composed of two parts, which correspond to the partially on or fully on situation in a router. We still adopt normal DOR-XY algorithm as one algorithm part when gated slice has been awakened. However, the other algorithm part is more complicated than the part in the sliced torus. In the UniMesh, due to lack of other half reversed channels, we cannot give a minimal routing path which only takes *Manhattan Distance* hops like the minimal deterministic routing in a normal mesh. Sometimes, packet should take some additional hops to reach its destination node. Extremely, as shown in Figure 3(a), the DOR-XY routing only needs 4 hops (red dashed arrow line) from source node Src to destination node Dest, while the UniMesh needs 8 hops (blue arrow line) since the pairs of Src and Dest locate in "bad" row and column (i.e., packets between the pairs cannot use minimal path to reach Dest and must detour to the adjacent nodes to find a potential path). It should take two extra detours to search a shorter routing path. One detour is from node Src to node 21, and the other is from node 12 to node Dest.

According to the ever-on subnet in Figure 1(b), we firstly give the X direction higher routing priority than the Y direction to ensure a deterministic routing (or vice versa) and then design a routing algorithm for the partially on router as shown in Algorithm 2. In the algorithm part, we detailedly itemize all possible cases and return a most reasonable routing direction according to the relative position and the parity flag between current and destination node. The parity flag functions *odd()* and *even()* are, respectively, used to distinguish the channel direction with row and column coordinate value for the current and destination node. In despite of network size and relative position between source and destination node, it merely incurs no more than 2 detours or 6 additional hops. The kernel idea of the routing algorithm is taking one or two detours and "borrowing" some adjacent channels. Subsequently, we use three representative examples to explain how the routing algorithm works.

Figures 3(b) and 3(c) represent two typical situations about how to "borrow" the $Y+$ channels. Since Dest locates

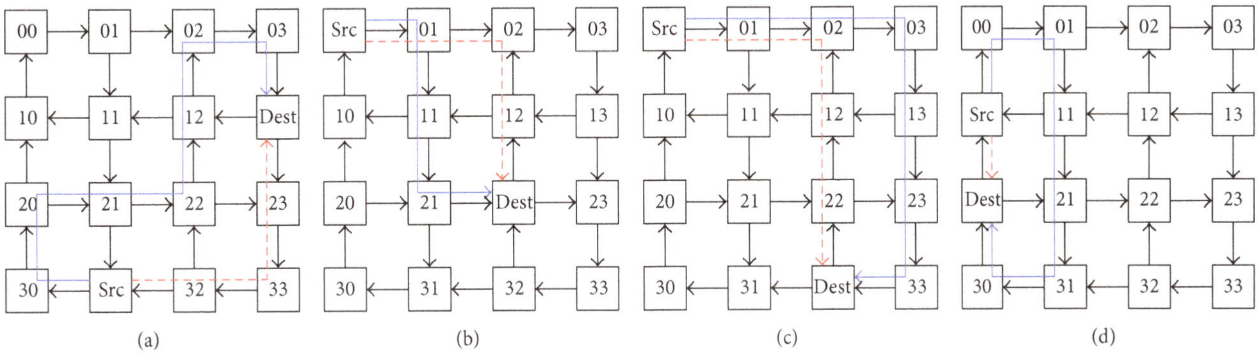

FIGURE 3: Four routing path examples by using UniMesh routing and normal DOR-XY routing.

```
Input: cur - current node coordinate, dest - destination node coordinate;
Output: dir - output direction;
// function even() and odd() distinguish even or odd number;
begin
    Δx ← (dest.x − cur.x);
    Δy ← (dest.y − cur.y);
    if Δx < 0 and Δy < 0 then
        if even(cur.y) then
            dir ← odd(cur.x) ? Y+ : Y−;
        else
            dir ← (Δx = −1 and odd(dest.x) and odd(dest.y)) ? Y− : X−;
    else if Δx > 0 and Δy > 0 then
        if odd(cur.y) then
            dir ← odd(cur.x) ? Y+ : Y−;
        else
            dir ← (Δx = 1 and even(dest.x) and even(dest.y)) ? Y+ : X+;
    else if Δx > 0 and Δy < 0 then
        if odd(cur.y) then
            dir ← even(cur.x) ? Y− : X−;
        else
            dir ← (Δx = 1 and odd(dest.x)) ? Y− : X+;
    else if Δx < 0 and Δy > 0 then
        if even(cur.y) then
            dir ← odd(cur.x) ? Y+ : X+;
        else
            dir ← (Δx = −1 and even(dest.x)) ? Y+ : X−;
    else if Δx = 0 and Δy > 0 then
        dir ← odd(cur.x) ? Y+ : (even(cur.y) ? X+ : (cur.x = 0 ? Y− : X−));
    else if Δx = 0 and Δy < 0 then
        dir ← even(cur.x) ? Y− : (odd(cur.y) ? X− : (cur.x = (Dim_x − 1) ? Y+ : X+));
    else if Δx > 0 and Δy = 0 then
        dir ← even(cur.y) ? X+ : (even(cur.x) ? Y− : (cur.y = (Dim_y − 1) ? X− : Y+));
    else if Δx < 0 and Δy = 0 then
        dir ← odd(cur.y) ? X− : (odd(cur.x) ? Y+ : (cur.y = 0 ? X+ : Y−));
    else
        dir ← Local;
    return dir;
```

ALGORITHM 2: Routing algorithm for the partially on router in the sliced 2D mesh network.

in "bad" column (the column only has $Y-$ channels), packets should "borrow" the adjacent column's $Y+$ channels to reach Dest. In Figure 3(b), both Src and Dest own the $X+$ channels, and Src locates in the $X-$ direction of Dest; then packets between the node pairs can use the $Y+$ direction routing in advance (one column ahead of the destination column) when they reach node 11 along the $X+$ channels. The reason is that Dest node's $X+$ input channels locate ahead of the destination column, and then it does not incur any additional hop. And yet, in Figure 3(c), Src owns the $X+$ output channels, Dest

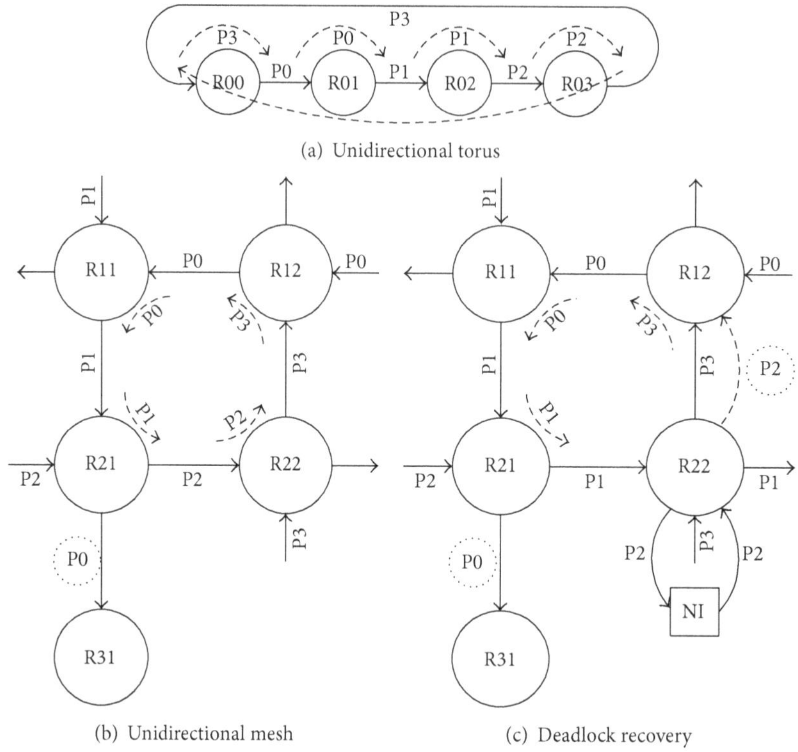

FIGURE 4: Routing deadlock in unidirectional torus (a) and unidirectional mesh (b), deadlock recovery diagram (c).

owns the $X-$ input channels, and Src locates in the $X-$ direction of Dest; then packets between the node pairs have to transfer in the $Y+$ direction after taking one additional hop in the $X+$ direction. On the contrary, because Dest node's $X-$ input channels locate behind destination column, packets should take an extra detour to reach Dest. How to arbitrate the two situations is illustrated at lines 8, 13, 18, and 23 in the pseudocode of Algorithm 2.

The case in Figure 3(d) represents the situation that Src and Dest locate in the same row/column and do not have straight channels from Src to Dest. In this situation, packets should be routed to the adjacent row/column and "borrow" some channels to approach Dest through a detour path. Since Src locates in "bad" row (i.e., without $X+$ output port) and Dest locates in "bad" row (i.e., without $X-$ input port), packets should make two extra detours to reach Dest. Unlike the situation in Figure 3(a), the two detours cannot provide any chance to make a hop deduction, and 6 additional hops are inevitably incurred. Our pseudocode from lines 24 to 31 is used to resolve how to make a decision in this situation.

3.3. Deadlock Recovery. In a network, circular dependency can occur as each packet holds a resource (i.e., current router's buffer) while requesting another resource (i.e., downstream router's buffer) and results in a routing deadlock [15, 19]. For the normal or sliced 2D torus with DOR-XY, the circular dependency does not occur from the Y-dimension to the X-dimension but may occur within either the X-dimension rings or the Y-dimension rings [20]. Figure 4(a) presents a circular dependency in the X-dimension ring; each packet (P0–P3) holds current router's input channel and wants to request its downstream router's input channel which has been occupied by another packet. For the ever-on subnet in the sliced 2D mesh, since all rows/columns only own the unidirectional channels, the resource circular dependency is hard to be avoided by an unsophisticated routing algorithm. A four-node circular dependency is illustrated in Figure 4(b) to highlight the dependency situation that may occur in the ever-on subnet of the mesh. Each packet (P0–P3) reserves current router's input channel and desires to make a turn to reach downstream router's input channel, while this downstream channel is occupied by a downstream packet and cannot be reallocated for current packet.

Then, we use a deadlock recovery mechanism to solve the resource dependency in our sliced network. There are two key processes to any deadlock recovery mechanism: detection and recovery [15, 20, 21]. The detection process is accomplished with timeout counters as proposed in [15, 21]: a counter is associated with each input channel, which is reset when packets make progress in the input channel. However, if the counter reaches a predetermined threshold (about 32 cycles in our simulation), input channel is considered to be blocked and the recovery process is activated. Our recovery mechanism relied on the improvement of NI architecture, as shown in Figure 2(c). There are some differences between packet-based and flit-based flow control for the implement of our deadlock recovery. In this paper, we only use the deadlock situation in the unidirectional mesh to describe the detail of deadlock recovery mechanism (the dimension ring of the torus has similar recovery mechanism).

At first, we consider the deadlock recovery in the unidirectional mesh with packet-based flow control. When a certain input channel detects a deadlock situation, the whole packets blocked in this channel should be routed to the local output port and should be temporarily buffered in the escape latch. Noteworthily, all ejected packets should have an escape tag to tell the NI multiplexer where to eject (the escape latch or the ejection queue). Afterwards, the escape packet in the escape latch can be reinjected into the injection queue as a normal packet, if the queue is not full and not busy in receiving packets from NI core. Thus, the circular dependency situation is broken. Besides, we give the packets in the escape latch a higher priority than local injected packets to be pushed into the injection queue. And finally, the input channels occupied by the blocked packets can be gradually released for other requesters, and the deadlock issue is resolved. As shown in Figure 4(c), when the $X+$ input channel in node 22 detects a deadlock, current router transmits the escape packet P2 into the local ejection port. Subsequently, P2 is reinjected into the local input port and waits to be transmitted. Due to the banishment of P2, the circular dependency between P0 and P3 is broken, and the blocked packets P1, P0, and P3 can be transmitted to their downstream router in sequence. After P3 releases the $Y-$ input channel in node 12, P2 can be routed to its downstream router from the local port in router 22. Finally, it solves the routing deadlock and avoids the shortcoming of conventional recovery mechanism which needs to discard the congested packet and retransmit it from source node (packet retransmission means more performance loss and power-waste).

However, for the unidirectional mesh with flit-based flow control, each flit composing an escape packet should own an escape tag as well, and two policies must be adopted to keep all flits in order. First, when an escape packet's head flit has started an escape process, current router should prohibit other escape processes in order to protect the ongoing escape process. After the tail flit finishes its escape, current router is resumed to deal with other escape processes. This policy keeps the order of flits between different escape packets owned by different input channels. Second, the escape latch should wait for all flits belonging to the same escape packet to be received, and then it is permitted to push out the whole flits in successive cycles when meeting the injection condition. This policy keeps the order of flits between escape packet and local injected packet.

3.4. Partial Power-Gating Scheme. In our sliced network, partial power-gating can be achieved by switching on/off the power supply of the gated slices; then each gated slice may be in any one of the three power states: *Active*, *Sleep*, and *Wake-up*, as shown in Figure 5. The state transition conditions are needed to arbitrate when a gated slice should switch from one power state to another state. Thus, we select the *Maximum Buffer Occupancy* (MBO) as the congestion metric to indicate whether the gated slice needs to be awakened to keep or improve performance. The MBO is suggested as an appropriate congestion metric in [13]. If traffic becomes dense and heavy, more buffer entries may be occupied by the increased packets in some input ports, and the corresponding MBO value increases at the same time. In our design, we utilize the dual congestion thresholds to avoid the frequent switches between difference state transitions: if the MBO is greater than the predetermined upper threshold T_{up}, then current router is considered to be congested and needs to be awakened to mitigate the congestion situation, whereas, current router is considered to be light-loaded when the MBO is less than the predetermined lower threshold T_{low}. If a router keeps in light-load for the T_{idle} cycles, its gated slice should be turned off to save partial leakage power. Besides, we utilize the look-ahead routing [11] as the backup scheme to activate the gated slices early, since some slices which locate in the congested path may fail to be awakened in time before the in-flight packet arrives. In the three-stage pipeline router, wake-up delay of up to 5 cycles can be hidden by sending an early wake-up signal to the partially on downstream router which is two hops away from current router.

In each router, there is a need for a gate controller to control the wake-up and sleep procedures for gated slice. The controller should check the MBO in current router for congestion arbitration and also receive possible wake-up request signals (*Wake_Req*) from upstream routers which are one hop or two hops away. Meanwhile, it should send a gate control signal to switch on/off the power transistor of gated slice and then control the power supply of gated slice. Essentially, the gate controller is a Finite State Machine (FSM), and the power state transition for the FSM is presented in Figure 5. When a gated slice switches its power state, *Handshaking* mechanism should be applied to all adjacent routers for guaranteeing the correctness of packet transmission. On the aspect of wake-up process (*Sleep* → *Wake-up* → *Active*), if current router is partially on and its gate controller receives a wake-up request signal *Wake_Req* = 1 or detects current router being congested, then the gated slice starts the wake-up process and sends the power state acknowledgement signals *Power_Ack* = 1 to all possible wake-up request routers after accomplishing the wake-up process. T_{wake} cycles are needed to charge the decoupling capacitance in the gated slice, and the wake-up counter $W_{counter}$ is used to count the wake-up delay T_{wake} and decide when to send the acknowledgement signals *Power_Ack* = 1. When all possible wake-up request routers receive the acknowledgement signals *Power_Ack* = 1, they release the corresponding wake-up request signals (*Wake_Req* = 0) and update their list of the available outports for routing computation (RC) stage and switch allocation (SA) stage. On the aspect of sleeping process (*Active* → *Sleep*), there are two typical cases: (1) if a gated slice has been activated and keeps in idleness for T_{idle} cycles, it can be turned off to save leakage power; (2) if a fully on router keeps in light-load for T_{idle} cycles, its gated slice also can be turned off after all packets in the gated slice have been sent out. However, the gated slice should send some acknowledgement signals *Power_Ack* = 0 to all adjacent routers at (T_{idle}-3) cycles to early notify them to stop packet transmission before the SA stage and then switches off its power supply at T_{idle} cycles. This mechanism can prevent the probability that the gated slice may receive some in-flight packets after being switched

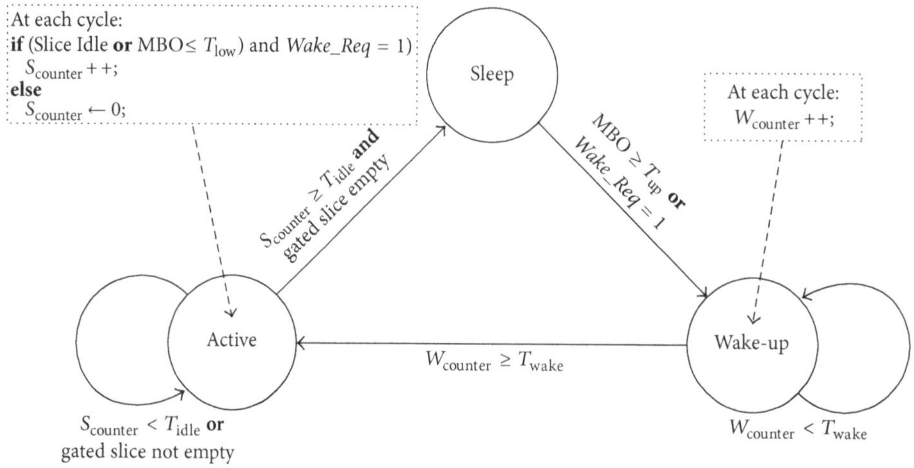

FIGURE 5: Power state transition for gated slice.

off. The sleep counter $S_{counter}$ is used to count the idle cycles in the fully on router and decides both when to switch off the power supply and when to send acknowledgement signals *Power_Ack* = 0 to all adjacent routers.

4. Evaluation

In this section, we evaluate the following designs: (1) NoPG-Mesh and NoPG-Torus: baseline mesh and torus network without power-gating; (2) ConPG-Mesh and ConPG-Torus: applying conventional power-gating to each router in baseline mesh and torus network and being optimized with early wake-up [11]; (3) DSPG-Mesh and DSPG-Torus: applying partial power-gating to the sliced mesh and torus network and also being optimized with early wake-up. At the same time, DSPG-Mesh and DSPG-Torus adopt partial clock-gating to save the clock switching power during the sleeping cycles.

4.1. Simulator, Configuration, and Workload. At first, we modify a detailed cycle-accurate network simulator [22] to simulate the proposed design and other comparative designs. Both network configuration and workload pattern are shown in Table 1, and the router parameters are applied to each design. Then, we use ORION 2.0 [23] to estimate network power-consumption and router area and assume all designs implemented with 32 nm technology, 2.0 GHz frequency, 1.0 V operating voltage, SRAM-based input buffer and output latch, MULTREE-based crossbar, and 1.0 mm link between adjacent nodes. Besides, we assume that the idleness detection threshold T_{idle} = 8 cycles, the wake-up latency T_{wake} = 10 cycles, the upper threshold for congestion estimation T_{up} = 8 flits, the lower threshold for light-load estimation T_{low} = 2 flits, and the wake-up overhead is 12 cycles with static energy of itself (according to the power model referred to in the Catnap [13]).

In synthetic traffic simulation, we evaluate the proposed design and other comparisons with four synthetic traffics. To accurately evaluate latency and throughput, the simulator is warmed up for 10000 cycles under the load without taking measurements until steady-state is reached and then simulates for 100000 cycles to obtain steady results. In addition to the synthetic workloads, we also compare the performance of different designs with ten application traces. These traces are collected from the interface module within the GARNET [24] by running corresponding SPLASH-2 benchmarks on the full-system simulator GEMS. The configuration of GEMS simulator is shown in Table 2. The simulator with SPLASH-2 traces is warmed up for sufficiently large cycles and collects the results after running with a maximum of ten million cycles or until trace completion.

TABLE 1: Network configuration and workload.

Topology	8 × 8 mesh and torus
Router parameter (for each design)	Virtual channel flow control, 3-stage pipeline, 128-bit channel, 4 VCs/channel, and 4 buffer entries/VC
Synthetic traffic	Single flit per packet and Poisson distribution injection
SPLASH trace	Single flit per control packet and 5 flits per data packet

TABLE 2: GEMS simulator configuration.

Topology	8 × 8 mesh, 1 core, and 1 L2Cache bank per node
Private L1Cache	32 KB instruction & data cache, 4-way, 64 B/line, and LRU
Shared L2Cache	NUCA, 8-way, 64 banks, 512 KB/bank, 64 B/line, and LRU
Cache coherence	MOESI_CMP_Directory protocol
Memory controller	8, connected with nodes 2, 5, 16, 23, 40, 47, 58, and 61

4.2. Comparison across Full Range Synthetic-Load. In order to investigate the behavior of different power-gating schemes

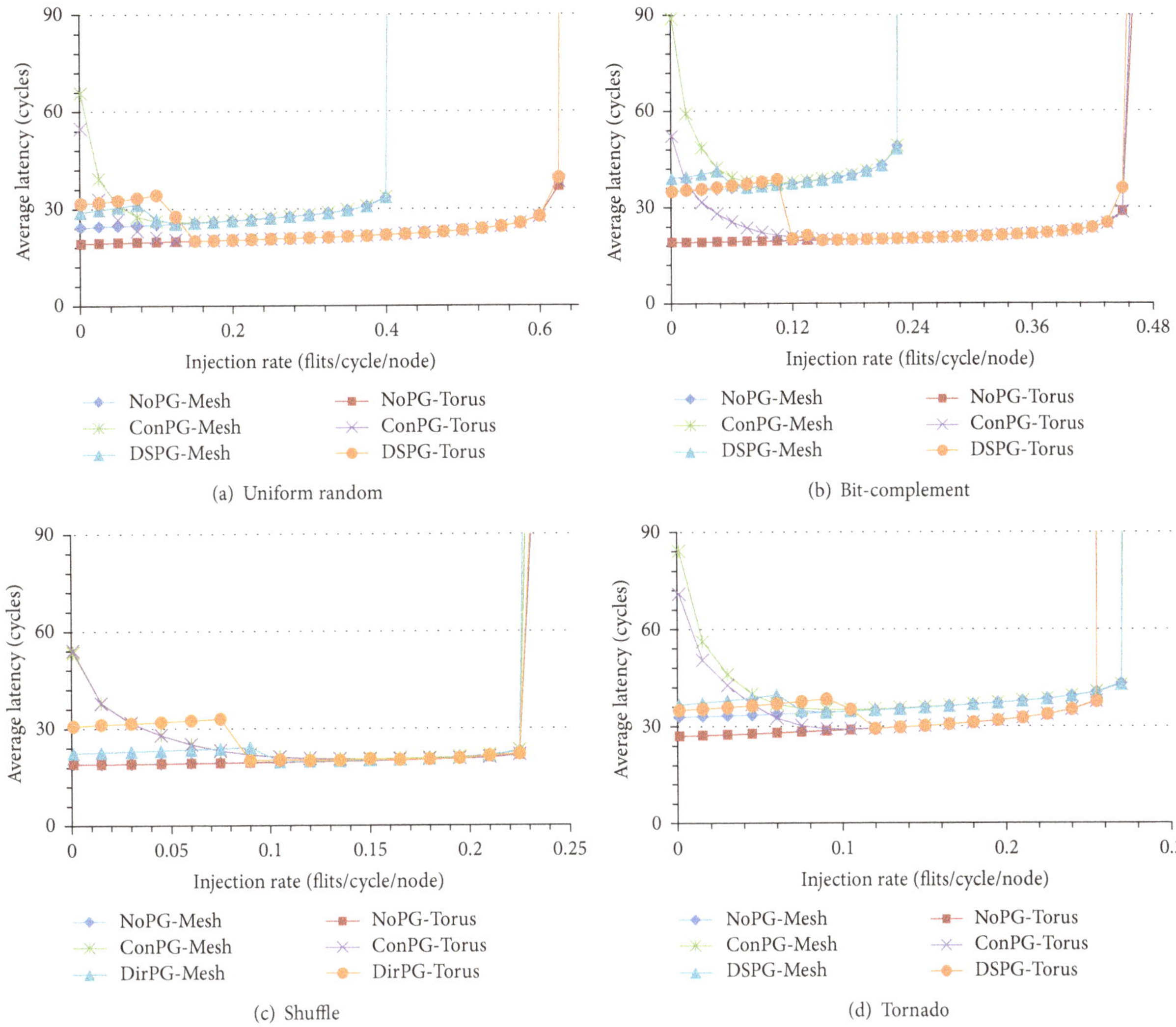

FIGURE 6: Performance comparison in four synthetic traffic loads.

more entirely, we utilize four synthetic traffics and vary the network load across the full range: from zero-load to saturation-load. Figures 6 and 7 present the performance and power-consumption for the *uniform random*, *bit-complement*, *shuffle*, and *tornado* traffic patterns. On the aspect of performance, typical latency behavior can be observed for NoPG-Mesh and NoPG-Torus, while disappointing results can be found for ConPG-Mesh and ConPG-Torus. At low-load range, packets in both ConPG-Mesh and ConPG-Torus are likely to experience cumulative wake-up delays, since they may meet most routers that are powered-off in their routing path and need to wait for their sleeping downstream routers to be awakened at each hop before being transferred. These wake-up delays are only partially hidden by early wake-up. As the injection rate increases, some routers in a routing path may already be activated by other packets and hold in power-on state, and corresponding wake-up delays can be avoided, so that the average latency in ConPG-Mesh and ConPG-Torus decreases as the load gradually increases. However, DSPG-Mesh and DSPG-Torus have the improved performance at low-load range, because the ever-on subnet can be used to support packet transfer at low-load. They only incur little performance penalty owing to few extra detours in the ever-on subnet. As the workload continually rises, the ever-on subnet begins to be congested, and the latency behavior of our sliced design gradually switches to the typical latency behavior. It is noteworthy that both DSPG-Mesh and DSPG-Torus can indeed reach the maximum throughput of their baselines. This means that all slices can be correctly awakened if needed, which is important and necessary for supporting the high-load phases during the application execution.

On the aspect of power-consumption, NoPG-Mesh and NoPG-Torus produce more dynamic power as the injection rate increases but consume the same leakage power in the entire load range. The typical power behaviors of baselines are linear. However, ConPG-Mesh and ConPG-Torus only save notable leakage power when the workload approaches the zero-load, since most routers can gain great many sleep opportunities. As the load continually rises, more and more routers should be awakened and keep active; thus ConPG-Mesh and ConPG-Torus have decreasing opportunities to

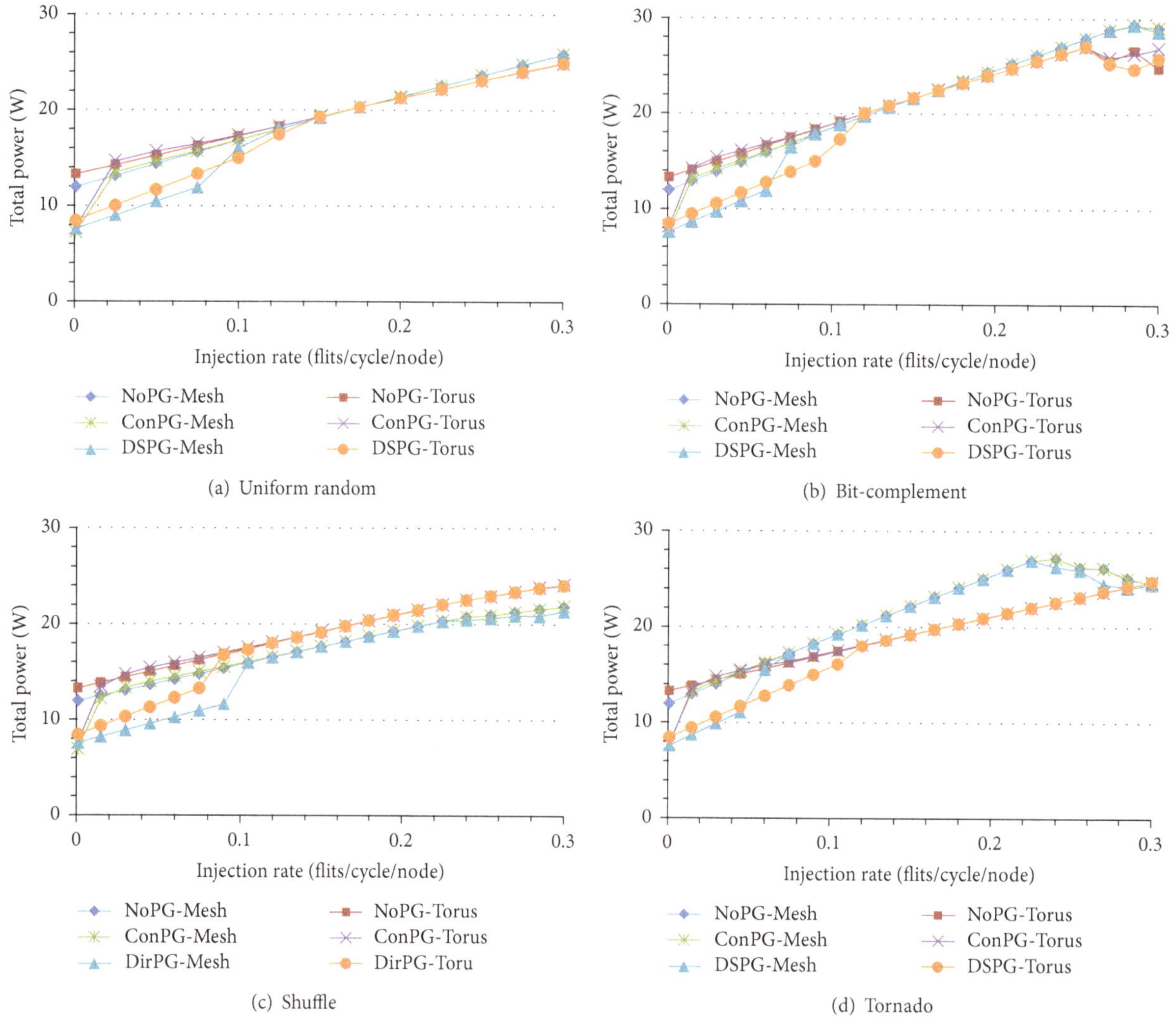

FIGURE 7: Total power-consumption comparison in four synthetic traffic loads.

obtain power-saving. Moreover, the wake-up overhead may counteract some power-saving, because gated blocks need to charge and discharge the decoupling capacitance frequently. As shown in Figure 7, the power of conventional power-gated design increases very fast and quickly approaches the power curves of its baselines. In contrast, DSPG-Mesh and DSPG-Torus save less leakage power than ConPG-Mesh and ConPG-Torus when near the zero-load, but they can prolong the power-saving range and gain more sleep opportunities for the gated slices during the low to medium load range. In addition, DSPG-Mesh and DSPG-Torus with partial clock-gating can remove some clock's switching power along with the powered-off slices. Similarly to former latency behavior, when more and more gated slices need to be awakened along with the increasing workload, the power curve of our sliced design gradually approaches the typical power curve.

Despite the fact that torus network has better performance than the mesh network in each synthetic traffic, the latency gap between DSPG-Mesh and NoPG-Mesh is smaller and more steady than the gap between DSPG-Torus and NoPG-Torus before the latency behavior switches. The key reason is the impact of average hops which will be explained in Section 4.5. As shown in Figure 6, the maximum latency gaps between DSPG-Mesh and NoPG-Mesh are 6.4, 5.8, 4.6, and 6.0 cycles for the *uniform random*, *bit-complement*, *shuffle*, and *tornado*, respectively, while they are 14.5, 19.3, 13.6, and 9.8 cycles between DSPG-Torus and NoPG-Torus. In the torus network, the latency gap in the *tornado* is less than the gap in other three traffics, because the communication pattern of the *tornado* is more suitable in the sliced torus than other three traffics.

4.3. Comparison in Application Workloads. In order to examine the practicability of our sliced design, we analyze the packet latency and power-consumption within different power-gating schemes by utilizing SPLASH application traces as the injected traffic. Figure 8 shows the average latency and the maximum latency in both mesh and torus networks for ten SPLASH application traces. Since NoPG-Mesh and NoPG-Torus do not adopt power-gating, they provide a lower bound of average and maximum latency for the mesh and the torus, respectively. The conventional power-gated

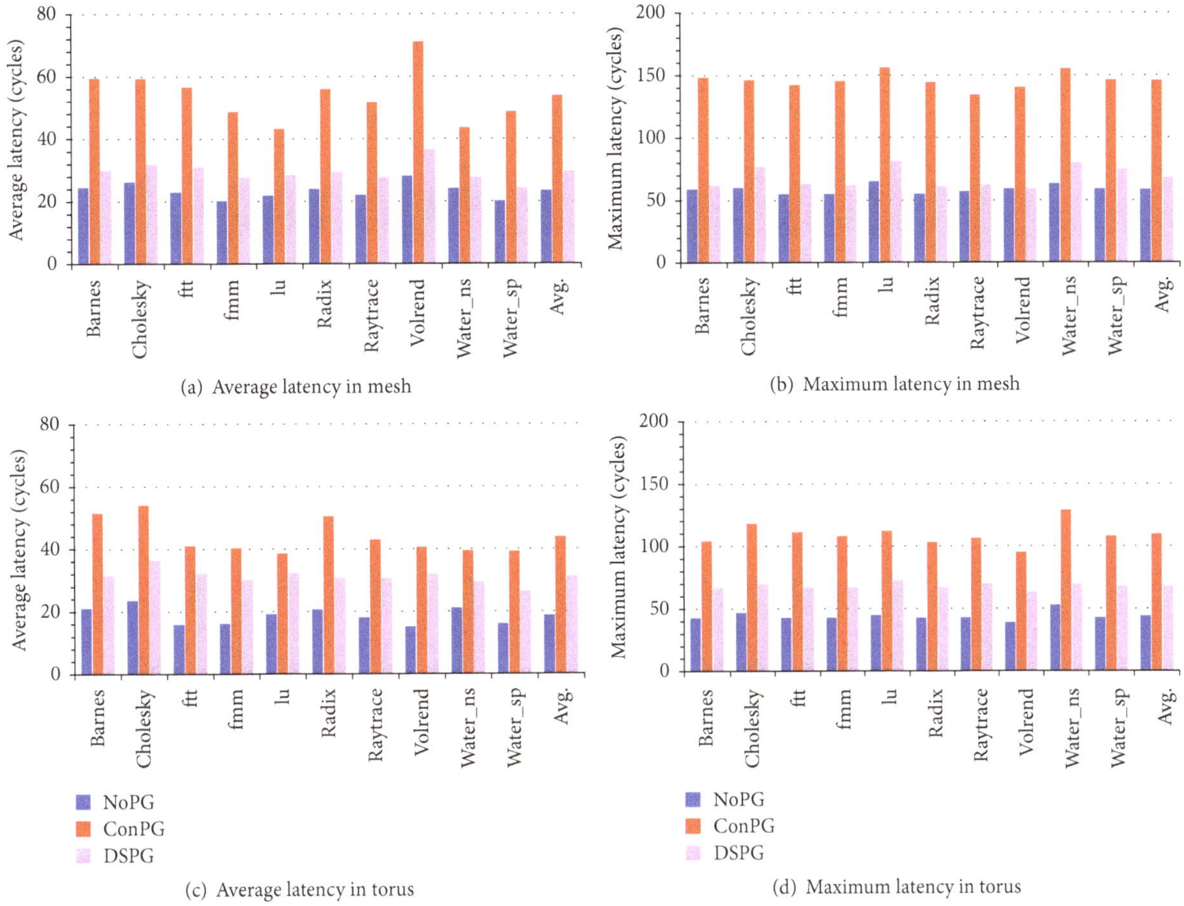

(a) Average latency in mesh

(b) Maximum latency in mesh

(c) Average latency in torus

(d) Maximum latency in torus

FIGURE 8: Latency comparison in SPLASH simulation.

design (ConPG-Mesh/ConPG-Torus), even with early wake-up optimization, still averagely increases 1.29X/1.34X average latency and even 1.48X/1.47X maximum latency compared with NoPG-Mesh/NoPG-Torus. In contrast, DSPG-Mesh and DSPG-Torus completely remove the wake-up latency from the critical path but may need to make some extra detours in the ever-on subnet. Consequently, in comparison with NoPG-Mesh/NoPG-Torus, the proposed design only has 26.0%/66.9% average increment on average latency and 16.2%/54.3% average increment on maximum latency. On average, this is equivalent to 45.0%/28.7% improvement for the average latency when compared with ConPG-Mesh/ConPG-Torus and 53.2%/37.7% improvement for the maximum latency. In average latency comparison, DSPG-Mesh even has better performance than DSPG-Torus and averagely decreases 5.5% of average latency.

Figure 9 presents the breakdown of total NoC power across ten SPLASH application traces and reveals the relative impact on each power component within different power-gating schemes. Several observations can be drawn from both Figures 8 and 9. First, except for *lu* and *water_nsquared* application traces, ConPG-Mesh and ConPG-Torus can gain large saving in leakage power but inevitably incur very large packet latency as mentioned above. Since the injection rate of these application traces is light and sparse, it makes most routers always inactive completely and results in very low leakage power in ConPG-Mesh and ConPG-Torus. However, under *lu* and *water_nsquared* application traces, ConPG-Mesh and ConPG-Torus increase total power-consumption significantly due to the increased injection rate (still low) and still incur very large packet latency. Moreover, they bring remarkable wake-up overhead as seen in *lu* and *water_nsquared* traces within the bar chart. Second, except for *water_nsquared* application trace, there is scarcely any wake-up overhead for each application trace in DSPG-Mesh and DSPG-Torus, since all gated slices do not need to be awakened to avoid network congestion. Meanwhile, they only incur tiny wake-up overhead in *water_nsquared* application trace owing to some short-term congestions. Third, DSPG-Mesh and DSPG-Torus save less leakage power than ConPG-Mesh and ConPG-Torus, but thay can obtain some additional dynamic power-saving due to the partial clock-gating. On the whole, although DSPG-Mesh and DSPG-Torus averagely consume 15.2%/18.9% more total power than ConPG-Mesh/ConPG-Torus, they still save 35.4%/35.5% of total power on average when compared with NoPG-Mesh/NoPG-Torus. At the same time, the sliced design has less impact on network performance as mentioned above. This means that, in comparison with the conventional power-gated design, the proposed NoC with partial power-gating has a better tradeoff when considering both performance and power-consumption.

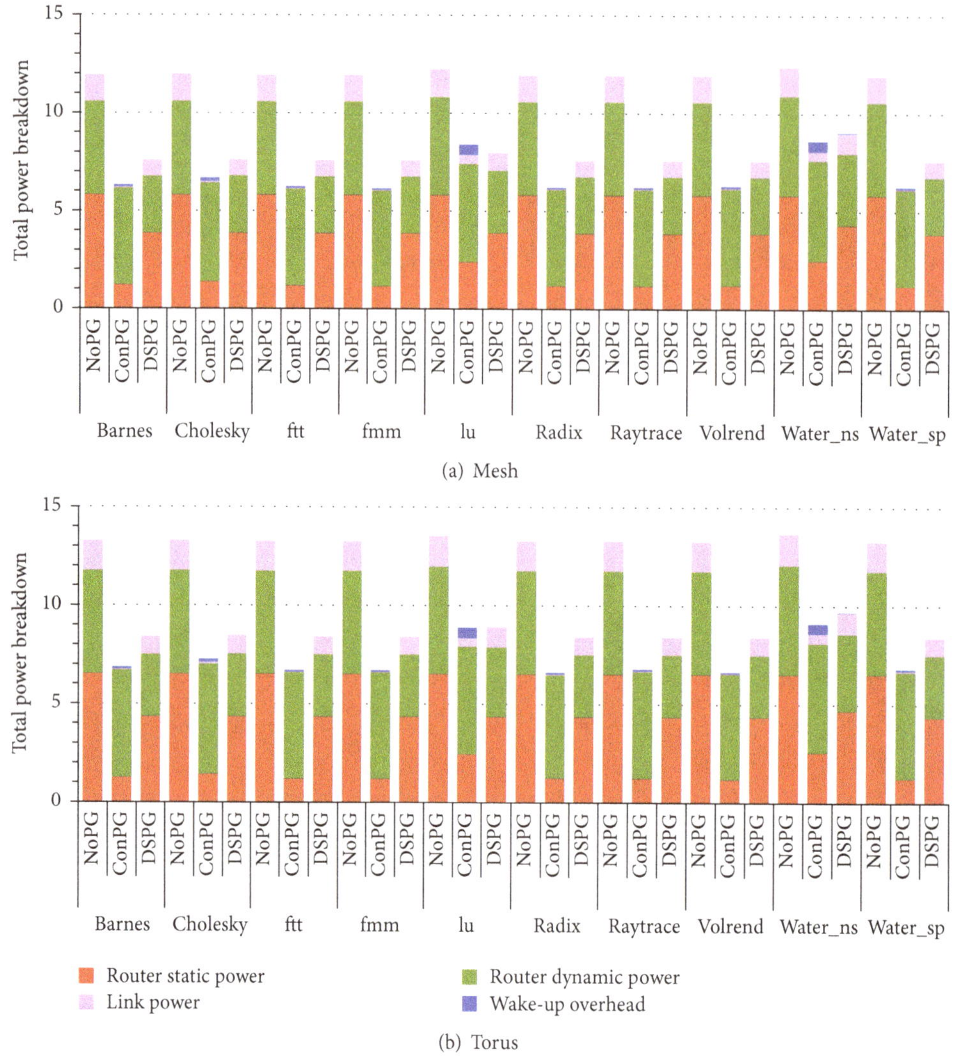

FIGURE 9: Power-consumption comparison in SPLASH simulation.

4.4. Impact on Compensated Sleep Cycles. Except for the absolute power-consumption, *compensated sleep cycles* (CSC) can be used to quantify the opportunity of power-gating, which is equal to the sum of idle cycles minus the break-even cycles ($T_{\text{breakeven}}$) accounted for every sleep period [13]. We present the CSC as a percentage of total cycles elapsed in the whole execution, which is the fraction of time when the gated routers or slices do not incur any leakage power. Figure 10 reveals the relationship between average latency, power-consumption, and CSC when using the uniform random traffic. Average latency has the similar decline trend as the CSC, because less sleep opportunities mean less waiting delay in the wake-up process, while power-consumption has the opposite trend relative to the CSC, since the CSC value is an indicator for the opportunity of power-saving. As shown in Figure 10(c), when the workload highly approaches zero-load, both ConPG-Mesh and ConPG-Torus have higher CSC value than DSPG-Mesh and DSPG-Torus due to the large sleep opportunities, while the sliced design needs to keep its fully connected subnets always-on and makes the CSC a fixed percentage (about 39.9% for DSPG-Torus and 38.6% for DSPG-Mesh) before other gated slices are awakened. As the load gradually increases, the CSC values in ConPG-Mesh and ConPG-Torus decrease greatly and quickly approach zero. The reason is that ConPG-Mesh and ConPG-Torus must activate their sleeping routers due to the increased packets and have to sacrifice the opportunity of sleep. However, the sliced design can keep high CSC value longer, since its fully connected subnet is always active to sustain an acceptable performance in low-load. Until the ever-on subnet becomes congested, DSPG-Torus and DSPG-Mesh need to awaken some gated slices to guarantee performance, and then their CSC values decrease obviously. Altogether, our sliced design can gain more power-saving opportunities than the conventional power-gated design when the traffic is not near the zero-load.

4.5. Impact on Hop Counts. Since packets in the ever-on subnet may not be transmitted with the minimal path as in normal network, the increased packet hops intuitively have

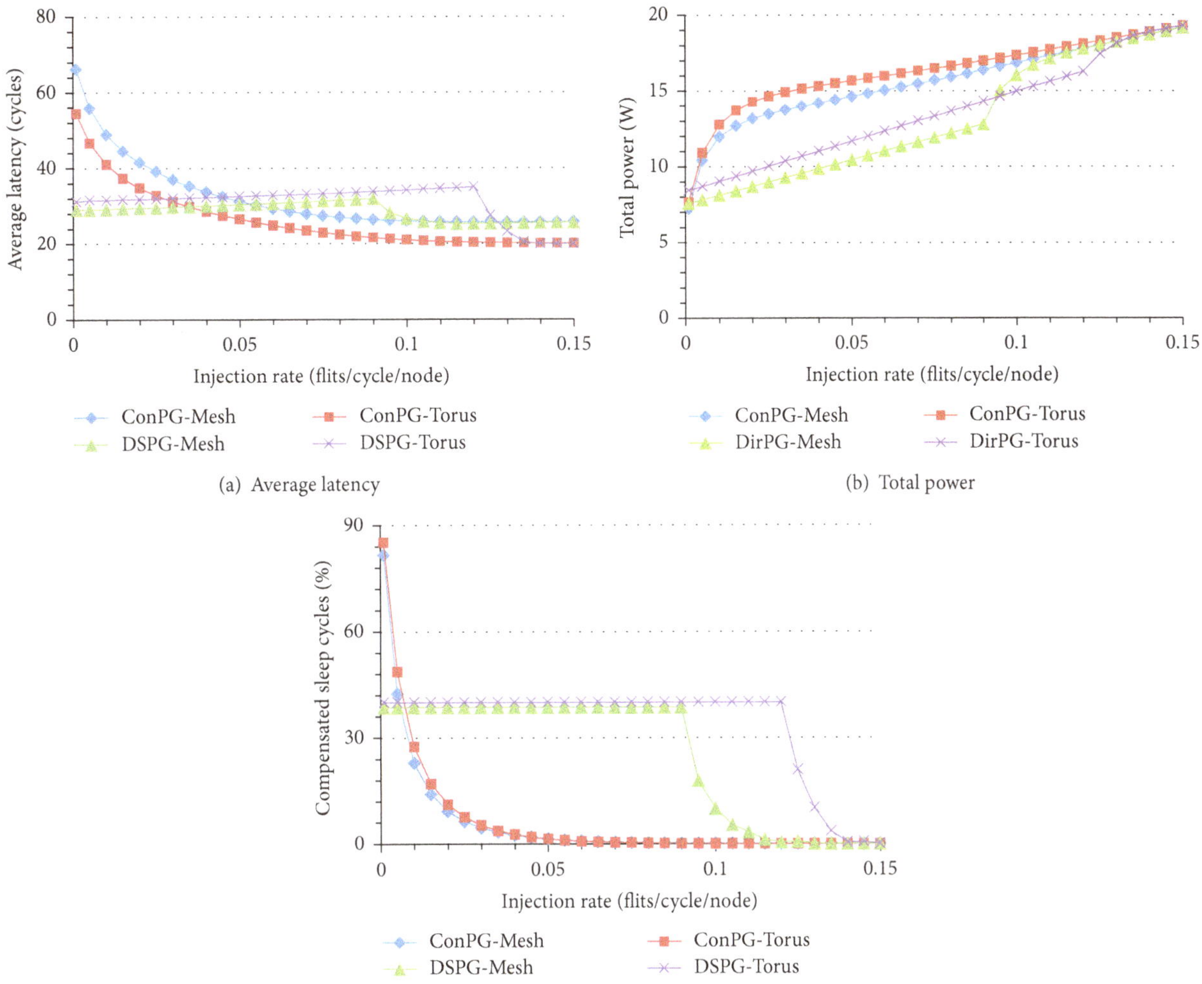

(a) Average latency

(b) Total power

(c) Compensated sleep cycles

Figure 10: The relationship between latency, power-consumption, and compensated sleep cycles.

some negative impact on performance. Under the uniform random traffic, Figure 11 presents the average hop counts within different network scales when all gated slices are powered-on/off in the mesh/torus network. Because the diameter of the unidirectional dimension ring increases as the network scale enlarges, the gap of average hop counts between torus and UniTorus increases, from 1.1 hops to 7.1 hops. However the average hop counts in the UniMesh are only a bit higher than the mesh (about 1.2 hops on average); the reason is that the UniMesh's extra detours are not more than 6 hops. This explains why the performance of the UniMesh is not obviously influenced by the extra detours. By this token, despite the fact that torus network has better performance than the mesh network, the UniMesh subnet is a more suitable selection than the UniTorus subnet above the 6 × 6 network since its average hop counts are shorter than the UniTorus'.

4.6. Discussion

(1) In our sliced design, if upper congestion threshold decreased, the latency behavior will switch early and the negative impact on performance can be alleviated; at the same time the CSC value will decrease early and the range of power-saving will shrink. Therefore, it is very important to select an appropriate congestion threshold.

(2) DSPG-Mesh has better performance than DSPG-Torus in the large scale network, but its algorithm in the routing computation units is relatively complex. However, DSPG-Torus sacrifices some performance for the sake of simple routing design. DSPG-Torus also can reduce average hops by adopting a complex routing algorithm, the same as the algorithm in DSPG-Mesh.

(3) The proposed direction-slicing scheme is applicable to other regular networks with bidirectional channels, and it also employs a similar methodology to split each router.

(4) The ever-on subnet in Catnap [13] can also utilize the direction-slicing scheme with partial power-gating to further mitigate leakage power.

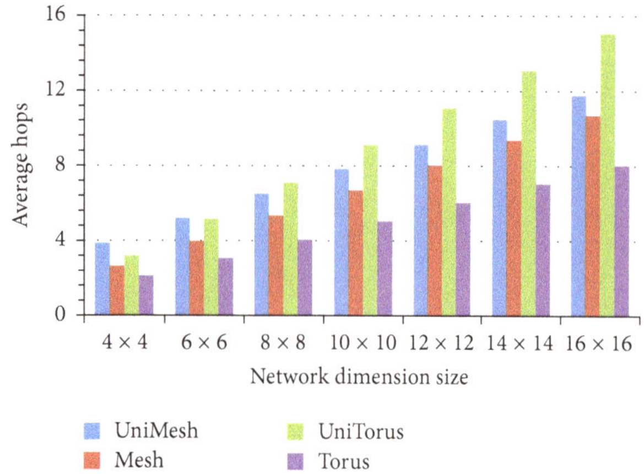

FIGURE 11: Average hop counts of four different topologies in different network scales.

5. Conclusion

Current and future many-core systems require NoC to be well-designed to guarantee both performance and power-efficiency. However, as technology continually is scaling down, leakage power takes up a larger proportion of total power, and it is increasingly important to reduce the leakage power for the power-efficient NoC design. Power-gating as a representative low-power technique can be applied to NoC to mitigate this increasing leakage power. However the disconnection problem severely limits power-gating to be effectively utilized due to its negative impact on performance. In this paper, we propose a novel partial power-gating approach to obviate the disconnection problem in the power-gated NoC. The approach mainly includes a direction-slicing scheme, an improved routing algorithm, and a deadlock recovery mechanism. First, we utilize direction-slicing scheme to split a router into two slices and adopt different power supply modes for them: keeping one slice always-on to construct an active and fully connected subnet, while utilizing both power-gating and clock-gating to gate the other one to save possible leakage power and clock power. Second, we redesign the corresponding routing algorithm to support packets transmitting on sliced network. And third, we provide a novel deadlock recovery mechanism to solve the deadlock situation which may happen due to the improved architecture and routing algorithm.

In synthetic traffic simulation, results show that the sliced network with partial power-gating is more power-efficient at low-load range and has better performance behavior than the conventional power-gated design. Both DSPG-Mesh and DSPG-Torus can reach their maximal throughput as the baselines, but at low-load range they incur a bit of added latency which is more acceptable than the wake-up delays in ConPG-Mesh and ConPG-Torus. Moreover, since the sliced design can keep high CSC value longer at low-load range, it can prolong the power-saving range and gain more sleep opportunities for gated slices than the conventional power-gated design. In the application simulation, despite the fact that DSPG-Mesh/DSPG-Torus would averagely consume 15.2%/18.9% more total power, DSPG-Mesh/DSPG-Torus can improve network performance by 45.0%/28.7% on average when compared with ConPG-Mesh/ConPG-Torus. In conclusion, the proposed design with partial power-gating can attain the objectives of high-performance and low-power.

Conflict of Interests

The authors declare that there is no conflict of interests regarding the publication of this paper.

Acknowledgments

This paper was supported in part by the National High-Tech 863 Project of China under Grant no. 2009AAOIZ102, in part by the National Natural Science Foundation of China under Grant no. 60873016 and no. 61170083, and in part by the Doctor Program Foundation of Education Ministry of China no. 20114307110001.

References

[1] Y. Hoskote, S. Vangal, A. Singh, N. Borkar, and S. Borkar, "A 5-GHz mesh interconnect for a teraflops processor," *IEEE Micro*, vol. 27, no. 5, pp. 51–61, 2007.

[2] B. K. Daya, C.-H. O. Chen, S. Subramanian et al., "SCORPIO: a 36-core research chip demonstrating snoopy coherence on a scalable mesh NoC with in-network ordering," in *Proceedings of the ACM/IEEE 41st International Symposium on Computer Architecture (ISCA '14)*, pp. 25–36, ACM, June 2014.

[3] A. Samih, R. Wang, A. Krishna, C. Maciocco, C. Tai, and Y. Solihin, "Energy-efficient interconnect via router parking," in *Proceedings of the 19th IEEE International Symposium on High Performance Computer Architecture (HPCA '13)*, pp. 508–519, IEEE, February 2013.

[4] P. Bogdan, M. Kas, R. Marculescu, and O. Mutlu, "QuaLe: a quantum-leap inspired model for non-stationary analysis of NoC traffic in chip multi-processors," in *Proceedings of the 4th ACM/IEEE International Symposium on Networks on Chip (NOCS '10)*, pp. 241–248, IEEE, Grenoble, France, May 2010.

[5] P. Bogdan and R. Marculescu, "Non-stationary traffic analysis and its implications on multicore platform design," *IEEE Transactions on Computer-Aided Design of Integrated Circuits and Systems*, vol. 30, no. 4, pp. 508–519, 2011.

[6] M. Donno, A. Ivaldi, L. Benini, and E. Macii, "Clock-tree power optimization based on RTL clock-gating," in *Proceedings of the 40th Design Automation Conference*, pp. 622–627, IEEE, June 2003.

[7] R. Mullins, "Minimising dynamic power consumption in on-chip networks," in *Proceedings of the International Symposium on System-on-Chip*, pp. 1–4, IEEE, November 2006.

[8] A. Pullini, F. Angiolini, P. Meloni et al., "NoC design and implementation in 65 nm technology," in *Proceedings of the First International Symposium on Networks-on-Chip (NOCS '07)*, pp. 273–282, May 2007.

[9] C. Feng, Z. Lu, A. Jantsch, and M. Zhang, "A 1-cycle 1.25 GHz bufferless router for 3d network-on-chip," *IEICE Transactions on Information and Systems*, vol. 95, no. 5, pp. 1519–1522, 2012.

[10] M. R. Casu, M. K. Yadav, and M. Zamboni, "Power-gating technique for network-on-chip buffers," *Electronics Letters*, vol. 49, no. 23, pp. 1438–1440, 2013.

[11] H. Matsutani, M. Koibuchi, D. Ikebuchi, K. Usami, H. Nakamura, and H. Amano, "Performance, area, and power evaluations of ultrafine-grained run-time power-gating routers for CMPs," *IEEE Transactions on Computer-Aided Design of Integrated Circuits and Systems*, vol. 30, no. 4, pp. 520–533, 2011.

[12] L. Chen and T. M. Pinkston, "NoRD: node-router decoupling for effective power-gating of on-chip routers," in *Proceedings of the 45th Annual IEEE/ACM International Symposium on Microarchitecture (MICRO '12)*, pp. 270–281, IEEE Computer Society, Vancouver, Canada, December 2012.

[13] R. Das, S. Narayanasamy, S. K. Satpathy, and R. G. Dreslinski, "Catnap: energy proportional multiple network-on-chip," in *Proceedings of the 40th Annual International Symposium on Computer Architecture (ISCA '13)*, pp. 320–331, ACM, Tel-Aviv, Israel, June 2013.

[14] L. Chen, L. Zhao, R. Wang, and T. M. Pinkston, "MP3: Minimizing performance penalty for power-gating of Clos network-on-chip," in *Proceedings of the 20th IEEE International Symposium on High Performance Computer Architecture*, pp. 296–307, IEEE, February 2014.

[15] W. J. Dally and B. P. Towles, *Principles and Practices of Interconnection Networks*, The Morgan Kaufmann Series in Computer Architecture and Design, Elsevier, 2004.

[16] Z. Hu, A. Buyuktosunoglu, V. Srinivasan, V. Zyuban, H. Jacobson, and P. Bose, "Microarchitectural techniques for power gating of execution units," in *Proceedings of the 2004 International Symposium on Lower Power Electronics and Design (ISLPED '04)*, pp. 32–37, ACM, August 2004.

[17] H. Jiang, M. Marek-Sadowska, and S. R. Nassif, "Benefits and costs of power-gating technique," in *Proceedings of the IEEE International Conference on Computer Design: VLSI in Computers and Processors (ICCD '05)*, pp. 559–566, IEEE, October 2005.

[18] Q. Wu, M. Pedram, and X. Wu, "Clock-gating and its application to low power design of sequential circuits," *IEEE Transactions on Circuits and Systems I: Fundamental Theory and Applications*, vol. 47, no. 3, pp. 415–420, 2000.

[19] S. Ma, N. E. Jerger, and Z. Wang, "Whole packet forwarding: Efficient design of fully adaptive routing algorithms for networks-on-chip," in *Proceedings of the 18th IEEE International Symposium on High Performance Computer Architecture*, pp. 1–12, IEEE, February 2012.

[20] M. Shin and J. Kim, "Leveraging torus topology with deadlock recovery for cost-efficient on-chip network," in *Proceedings of the 29th IEEE International Conference on Computer Design (ICCD '11)*, pp. 25–30, IEEE, Amherst, Mass, USA, November 2011.

[21] A. Lankes, T. Wild, A. Herkersdorf, S. Sonntag, and H. Reinig, "Comparison of deadlock recovery and avoidance mechanisms to approach message dependent deadlocks in on-chip networks," in *Proceedings of the 4th ACM/IEEE International Symposium on Networks on Chip (NOCS '10)*, pp. 17–24, IEEE, Grenoble, France, May 2010.

[22] L. Jain, B. Al-Hashimi, M. Gaur, V. Laxmi, and A. Narayanan, "NIRGAM: a simulator for NoC interconnect routing and application modeling," in *Proceedings of the Design, Automation and Test in Europe Conference*, pp. 16–20, Nice, France, April 2007.

[23] A. Kahng, B. Li, L.-S. Peh, and K. Samadi, "ORION 2.0: a fast and accurate NoC power and area model for early-stage design space exploration," in *Proceedings of the Design, Automation & Test in Europe Conference & Exhibition (DATE '09)*, pp. 423–428, IEEE, Nice, France, April 2009.

[24] N. Agarwal, T. Krishna, L.-S. Peh, and N. K. Jha, "GARNET: a detailed on-chip network model inside a full-system simulator," in *Proceedings of the International Symposium on Performance Analysis of Systems and Software (ISPASS '09)*, pp. 33–42, IEEE, April 2009.

20

A New Hybrid Model Based on Data Preprocessing and an Intelligent Optimization Algorithm for Electrical Power System Forecasting

Ping Jiang,[1] Xuejiao Ma,[1] and Feng Liu[1,2]

[1]*School of Statistics, Dongbei University of Finance and Economics, Dalian, Liaoning 116025, China*
[2]*School of Mathematics and Statistics, Lanzhou University, Lanzhou, Gansu 730000, China*

Correspondence should be addressed to Xuejiao Ma; xuejiaomadufe@163.com

Academic Editor: Alkiviadis Paipetis

The establishment of electrical power system cannot only benefit the reasonable distribution and management in energy resources, but also satisfy the increasing demand for electricity. The electrical power system construction is often a pivotal part in the national and regional economic development plan. This paper constructs a hybrid model, known as the E-MFA-BP model, that can forecast indices in the electrical power system, including wind speed, electrical load, and electricity price. Firstly, the ensemble empirical mode decomposition can be applied to eliminate the noise of original time series data. After data preprocessing, the back propagation neural network model is applied to carry out the forecasting. Owing to the instability of its structure, the modified firefly algorithm is employed to optimize the weight and threshold values of back propagation to obtain a hybrid model with higher forecasting quality. Three experiments are carried out to verify the effectiveness of the model. Through comparison with other traditional well-known forecasting models, and models optimized by other optimization algorithms, the experimental results demonstrate that the hybrid model has the best forecasting performance.

1. Introduction

The electrical power system is a type of production and consumption system that is composed of the power generation, transformation, transmission, and distribution of electricity and other aspects. With higher requirement for indices of electrical power system, it shows a tendency forward to the intelligent and optimal direction [1]. To build large-scale electrical power system with reasonable structure cannot only benefit the central management, unified schedule, and distribution of electrical power but also make for the development and utilization of regional energy resources and satisfy the increasing demand for electricity in each region to the largest extent. The appearance of electrical power system contributes to the initiation of the age of electricity, which makes the electrical energy with highly effective and pollution-free features applied widely. Nowadays, it has become one of the most important symbols for national economic development level. Therefore, to forecast related indices of electrical power system seems imperative and important. *Wind speed*, *electrical load*, and *electricity price* are chosen in this paper to carry out the forecasting of electrical power system.

As one of the most competitive new energies, wind energy is highly focused on by most countries in the world, including both developed and developing countries. In recent years, the development of wind power generation in Asia has made major achievements and the contribution of China has played a vital role [2]. On the one hand, wind energy in China is available and plentiful, especially in the northern region. On the other hand, the long coastline could provide plenty of offshore wind energy. Therefore, wind power generation on a large scale is feasible and can be applied. Accordingly, Chinese wind power industry is booming and promising. In addition to the current wind power bases, there are thousands of wind power bases that are in construction [3]. In view of the significance of wind energy, what is more important is the forecasting of wind speed. The short-term wind speed forecasting is meaningful in helping distribute power reasonably and ensure the quality of the power supply [4]. This paper focuses on the short-term wind speed forecasting.

Electrical load forecasting plays a pivotal role in the economic operation and the security of electrical power system, because several decisions that are significant to the economy are made based on the forecasting of electrical load, such as fuel allocation, short-term maintenance, and contingency planning load shedding. Thus, many countries all over the world seek effective methods to forecast the electrical load. China, with rapid rate of economic development as well as increasingly large demand for electricity, is supposed to pay more attention to the improvement of electrical load forecasting. Besides, although wind power generation has become more widely applied, the burning of coal still occupies the leading position in generating electricity, which would result in the emission of carbon dioxide that could bring enormous environmental damage [5]. So accurate and reliable electrical load forecasting is beneficial in planning the power generation and adjusting power policy and structures.

After forecasting the production and consumption of electricity, how to trade it, a very special commodity, is also a challenging issue for the electrical power system. Nowadays, the electricity is traded under market rules using derivative and spot contracts; however, the stability of trading price requires the constant balance between supply and demand. It is difficult to ensure the exact demanding for electricity, because many factors would have influence on the demand, including the intensity of business and everyday activities [7]. As for the supply of electricity, in the perspective of companies, the forecasting of electricity price has become a fundamental input to energy companies' decision-making mechanisms. Moreover, power portfolio managers show particular interest in the price forecasting from a few hours to a few months ahead so as to adjust their bidding strategy and decrease risk to maximize the profits.

Above all, it is meaningful to establish novel models to implement the forecasting of the electrical power system. The data of three chosen indices all belong to time series data with the features of nonlinearity and high fluctuation; therefore, it is not easy to get accurate and reliable forecasting results. Many researchers have made great efforts to develop the forecasting methods and two main approaches could be summarized, which are numerical simulation approach and statistical approach. Firstly, standard persistence methods are used as benchmarks to evaluate the new forecasting method. Generally, persistence is based on the assumption of no change in sky condition or constant clear sky index [8]. The persistence method works well when weather patterns change very little and features on the weather maps move very slowly. Nevertheless, if weather conditions change significantly from day to day, the persistence method usually breaks down and is not the best forecasting method to use. Compared with the numerical simulation approach, the statistical approach is applied more widely and is more popular, which mainly includes time series analysis, artificial neural network, and other related new hybrid models.

Firstly, in the aspect of time series analysis, it mainly deals with data ordered by time through identifying model, evaluating parameters, and building test and forecasting models. The most common time series analyses are autoregressive moving average (ARMA) model, autoregressive integrated moving average (ARIMA) model, and generalized autoregressive conditional heteroskedasticity (GARCH) model. Kavasseri and Seetharaman [9] applied fractional-ARIMA or f-ARIMA models to forecast the wind speed obtained from four potential wind generation sites in North Dakota on the day-ahead and two-day-ahead horizons. Compared with the persistence methods, the proposed model has great improvements in the performance of forecasting. Liu et al. [10] used ARMA-GARCH approach to model the mean and volatility of wind speed by applying the seven-year hourly data obtained from four different heights from an observation site in Colorado, USA. There were 10 models all together comprising the original form of five different GARCH approaches and their modified forms. They found out that the built model had a high forecasting accuracy and could catch the trend change of volatility and mean of wind speed effectively. In addition, they also proved that the ARMA-GARCH-M structures could improve the modeling sufficiency consistently due to the asymmetric time-varying and nonlinear features. Caporin and Preś [11] evaluated three approaches that could be used to forecast the wind speed intensity, including autoregressive Gamma (ARG) process, Gamma autoregressive (GAR) process, and ARFIMA-FIGARCH model. The first two models both combined autoregressive behaviors and Gamma distributions. In the model of ARG, the simulation of Gamma process went through the generation of random numbers from a two-parameter Gamma density, where parameters were time-varying. As for GAR, it was proposed in alternative conditional Gamma process based on three parameters' noncentral Gama density. Bouzerdoum et al. [12] combined two famous methods, which were the seasonal autoregressive integrated moving average method and the support vector machines (SVM) to forecast the short-term power and comparative study was introduced showing that the developed hybrid model performed better than both the single models. Xu et al. [13] established a new model with improved grey model (GM) and ARMA based on HP filter to forecast the final energy consumption. Through comparing with traditional statistical approaches, the experimental results demonstrated that the improved GM-ARMA model had excellent accuracy and higher level of reliability.

Secondly, as for the artificial neural network (ANN) method, it could be applied to address complicated nonlinear problems, because it has a strong ability of self-learning and self-adaptation. ANN is more likely used by combining with other models and methods. Cadenas and Rivera [14] developed ARIMA-ANN model to carry out the forecasting by using average hourly wind speed data obtained directly from the measurements at different sites for about one month. ARIMA was applied first to do the forecasting of the time series, and then ANN was used to take the nonlinear tendencies into consideration based on the obtained errors. The built hybrid model could forecast wind speed with a higher accuracy than ARIMA or ANN models separately according to the calculation of the mean error (ME), the mean square error (MSE), and the mean absolute error (MAE). Nan et al. [15] established a forecasting model composed of time series and back propagation neural network (BPNN) prediction model. As for the time series analysis, bias correction method on empirical orthogonal function (EOF) was proposed. The application of EOF was aimed at decomposing forecasting error and the proposed model EOF-BPNN could forecast the short-term wind speed with a higher accuracy.

Rahmani et al. [16] developed a new hybrid swarm technique (HAP) to forecast the energy output in a real wind farm located in Binaloud, Iran. HAP was optimized by two optimization algorithms that were two metaheuristic techniques, including ant colony optimization (ACO) and particle swarm optimization (PSO). The combination of two algorithms can both improve the forecasting accuracy and accelerate the convergence rate. The results indicated that the proposed hybrid model could estimate the output wind power based on the ambient temperature and wind speed with a mean absolute percentage error (MAPE) of 3.5%. Geng et al. [17] introduced robust support vector regression (RSVR) and clonal selection principle of artificial immune algorithm combined with particle swarm optimization (CSAPSO) to get more accurate forecasting results. RSVR was applied to carry out the forecasting, and then CSAPSO was used to optimize the parameters in RSVR in order to improve the forecasting accuracy and save time. The established model performed better than six other competing models used in the study.

In order to reduce error and improve the forecasting accuracy, the noise reduction of input data has attracted larger and larger attention. Liu et al. [18] applied two hybrid forecasting models, including the Wavelet-particle swarm optimization- (PSO-) multilayer perceptron (MLP) and the Wavelet-genetic algorithm- (GA-) multilayer perceptron (MLP). The analysis results showed that both of the two proposed models were suitable for the diverse accuracy requirement in wind speed predictions and the contribution of Wavelet component was statistically significant. Zhang et al. [19] proposed four approaches that had better forecasting performance than traditional forecasting methods. The first two approaches applied PSO to optimize the parameters in the first-order and second-order adaptive coefficient (FAC and SAC) methods, and then the last two approaches employed the decomposition of the original data into seasonal and trend components. The application of decomposition of data to seasonal and trend components helped arrive at a higher forecasting accuracy through calculating the MSE, MAE, and the relative errors. Zhang el al. [20] applied WTT-SAM-RBFNN to forecast the wind speed. WTT meant Wavelet transform technique and it played a vital role in reducing the forecasting errors. SAM indicated seasonal adjustment method, which was also important in improving the forecasting accuracy. RBFNN was radial basis function neural network that was applied to carry out the forecasting. The use of two adjustments of original wind speed data made the forecasting accuracy improve to a great extent compared with the traditional forecasting methods.

The time series data of electrical power system is of high randomness, because it is affected by many factors, such as temperature, altitude, air pressure, and other related factors. Empirical mode decomposition (EMD) is applicable in eliminating the noise in the original series through carrying out the adaptive time-frequency decomposition based on the features of the time variation of signal [21]. However, the mode mixing problem resulting from the intermittency signal in the sifting process of EMD would affect the forecasting accuracy. Thus, ensemble empirical mode decomposition (EEMD) (see Algorithm 1) is introduced to surpass the defects of EMD. The white noise is added into the original signal so that it could have the decomposition scale of a uniform distribution based on the statistical feature of Gaussian white noise. EEMD could smooth the impulse interference effectively and address the mode mixing problem.

BPNN is a common forecasting method for its unique advantages, including nonlinear mapping capability, self-learning and self-adaptive ability, generalization ability, and fault-tolerant capacity. However, the structure of the BPNN network is uncertain, which could cause the high forecasting errors [22]. Therefore, this paper utilizes modified firefly algorithm (MFA) (see Algorithm 2) to optimize the weight and threshold values of the network. FA is a random and parallel optimization algorithm with low requirements for the function to be optimized, simple process, and fast convergent rate [23]. However, the largest defect of FA is that it is easy to trap into the local optimum, so this paper improves the FA in order to make it jump out of the local optimum and the hybrid model could improve the forecasting accuracy through combining with EEMD.

Our Contribution. Many scholars have made great efforts to seek out the methods to forecast the time series data, and this is because it is really a meaningful and challenging work, especially for the forecasting of *wind speed, electrical load*, and *electricity price*. Based on the reasons above, this paper tries to propose a new hybrid model, known as the E-MFA-BP, to be applied to the forecasting of electrical power system. Initially, the ensemble empirical mode decomposition (EEMD) is used to reduce the noise in the original data series and the effect of noise reduction is satisfactory. Then, after preprocessing the original data, the smoother data is used to carry out the forecasting. MFA is utilized to optimize the weight and threshold values of the network of BPNN in order to reduce the forecasting errors. Their combination successfully takes the advantage of each single algorithm. In general, our contribution could be summarized as one that proposes a hybrid model that is effective in the forecasting in the electrical power system.

The whole structure of this paper is given below. Firstly, Section 1 is the introduction containing the description of background of electrical power system and related researches on this field. Then, Section 2 introduces the components of the hybrid model, including ensemble empirical mode decomposition, modified firefly algorithm, back propagation neural network, and the essence of hybrid model proposed in this paper. Next, the experimental simulation and the evaluation of the hybrid model are included in Section 3.

Input:
$x^{(0)} = (x^{(0)}(1), x^{(0)}(2), \ldots, x^{(0)}(l))$—a sequence of sample data
Output:
$\hat{x}^{(0)} = (\hat{x}^{(0)}(l+1), \hat{x}^{(0)}(l+2), \ldots, \hat{x}^{(0)}(l+n))$—a sequence of forecasting data
Parameters:
M—the number of ensemble for EEMD.
k—the amplitude of white noise.
m—the iterations of experiments.
n—the number of intrinsic mode function components.
(1) Initialize the number of IMF: $n = \text{fix}(\log 2(l)) - 1$;
(2) Do EEMD, EEMD loop start;
(3) $m = 0$;
(4) **If** $n < \max M$ **do**
(5) Add white noise to the original data $x_m(t) = x(t) + kn_m(t)$;
(6) Assign original data in the first column and start to find an IMF 10 times;
(7) Decompose $x_m(t)$ according to the EMD method;
(8) **End if**
(9) Calculate the mean: $c_j(t) = (1/N)\sum_{i=1}^{N} c_{ij}(t)$;
(10) **Return** $\hat{x}^{(0)} = (\hat{x}^{(0)}(l+1), \hat{x}^{(0)}(l+2), \ldots, \hat{x}^{(0)}(l+n))$.

ALGORITHM 1: Pseudocode of EEMD.

Input:
The initial random value of weight α and threshold b of BP artificial neural network.
Output:
The global optimal extreme point m_{best} and the best individual value $f_{\min} = \min(f(m_{\text{best}}))$
Parameters:
N—the number of population of the particle swarm.
c_1, c_2—the cognitive and social weight of PSO algorithm. In usual, $c_1 = c_2 = 2$.
M—the maximum iteration.
(1) Initialize the population m_i $(i = 1, 2, \ldots, n)$;
(2) $f(m_i)$ = update the light intensity after evaluating new solution;
(3) $t = 0$;
(4) **While** $t < \max T$ **do**
(5) **for** $i = 1 : n$ all n fireflies **do**
(6) **for** $j = 1 : i$ all n fireflies **do**
(7) **if** $I_j > I_i$ **then**
(8) Move firefly i towards j using uniform distribution;
(9) **end if**
(10) **end for**
(11) Evaluate new solutions and update light intensity;
$x_i(t+1) = x_i(t) + (1 - 2r) * R_d^i(t) * s\_\text{init}$
$\overline{d}_i = \frac{1}{|N_i(t)|}\sum_{j \in N_i(t)} (\|x_i - x_j\|); \qquad s_i(t) = s_i(t-1) * q$
(12) **end for**
(13) Rank fireflies and find the best individual;
(14) **end while**

ALGORITHM 2: Pseudocode of MFA.

The *wind speed* data of Shandong Peninsula in China and *electrical load* data and *electricity price* data from New South Wales (NSW) are used to test the performance of the model. Section 4 is the discussion and Section 5 includes conclusion and further work that summarize the ideas in the whole paper.

2. Methods

In this section, ensemble empirical mode decomposition, the modified firefly algorithm, and back propagation neural network are described. Next, the hybrid model E-MFA-BP proposed in this paper is introduced in detail.

2.1. Ensemble Empirical Mode Decomposition (EEMD). To eliminate the mode mixing problem in the classical EMD method, ensemble empirical mode decomposition (EEMD) is proposed [24]. The core idea of EEMD is to make use of the statistical property of white noise; namely, its frequency would distribute uniformly. When Gaussian white noise is added to the signal, the signal should have continuity in different scales, and the feature of multiple extreme points would be changed as well [24]. The two important parameters are the value of amplitude k and total number of EMD decompositions M. k indicates the proportion of white noise in the standard deviation of the amplitude of the original signal [24]. To date, there has been no equation that could be used to calculate M or k; therefore, based on the references, M is 100 and k is between 0.05 and 0.5 in this paper. The detailed steps of EEMD are described as follows [25].

Step 1. Initialize the number of ensemble M, the amplitude of white noise k, and iterations of experiments m, with $m = 1$.

Step 2. Perform mth experiments; the detailed steps are listed below:

(a) Add a random Gaussian white noise sequence to the original signal to obtain a new sequence with noise as

$$x_m(t) = x(t) + kn_m(t), \tag{1}$$

where $x(t)$ is the original signal that needs to be processed and $n_m(t)$ represents the random Gaussian white noise sequence.

(b) Decompose $x_m(t)$ according to the EMD method and obtain n IMF components and one remaining component:

$$\begin{aligned} c_{i\cdot m}(t) &= h_{j\cdot m}(t), \quad i = 1, 2, \ldots, n, \\ r_{n\cdot m}(t) &= r_{n-1\cdot m}(t) - c_{n\cdot m}(t), \end{aligned} \tag{2}$$

where $c_{i\cdot m}(t)$ represents n components of the IMF and $r_{n\cdot m}(t)$ indicates the one remaining component.

(c) If $m < M$, $m = m + 1$, return to Step 2; otherwise perform Step 3.

Step 3. Calculate mean of all components of the IMFs and remaining components in mth EMD. Consider

$$\begin{aligned} \overline{c_i}(t) &= \sum_{m=1}^{M} \frac{c_{i\cdot m}(t)}{M}, \quad i = 1, 2, \ldots, n, \\ \overline{r_n}(t) &= \sum_{m=1}^{M} \frac{r_{n\cdot m}(t)}{M}. \end{aligned} \tag{3}$$

Step 4. Export $\overline{c_i}(t)$ $(i = 1, 2, \ldots, n)$ and $\overline{r_n}(t)$, respectively, as n IMF components and the one remaining component of EEMD.

2.2. Firefly Algorithm

2.2.1. Original Firefly Algorithm. The firefly algorithm (FA) is formed based on the imitation of the luminescence of fireflies in the natural world. In FA, the reasons for fireflies to attract each other depend on two major factors: brightness and attractiveness [26]. Firstly, the brightness of fireflies depends on their location, and the higher the value of the brightness, the better the place. Secondly, the attractiveness is also related to the brightness. Fireflies with higher brightness own higher attractiveness [27]. If each firefly has the same brightness, then they would move randomly. Both the brightness and the attractiveness are inversely proportional to the distance, which indicates that the brightness and attractiveness would decrease with the increase of the distance between each firefly [28]. Before applying FA, we define several terms.

Definition 1. Let $J(x_i(t))$ be the objective function value of each firefly i at the position $x_i(t)$ in tth iteration. $l_i(t)$ represents the brightness of the current firefly and γ denotes the update rate of brightness. So the update of brightness is defined as

$$l_i(t) = (1 - \rho) l_i(t-1) + \gamma J(l_i(t)). \tag{4}$$

Definition 2. Let the neighbor set $N_i(t)$ be $\{j : d_{ij}(t) < r_d^i(t); l_i(t) < l_j(t)\}$, $0 < r_d^i(t) \le r_s$, where r_s is the perceived radius, so one defines the probability of individual j moving to the neighbor set as

$$p_{ij}(t) = \sum_{k\in N_i(t)}^{l_j(t)-l_i(t)} (l_k(t) - l_i(t)). \tag{5}$$

Definition 3. In the firefly algorithm, to update the location is important. So one defines the update of location as

$$x_i(t+1) = x_i(t) + s\left(\frac{x_j - x_i}{\left\|x_j - x_i\right\|}\right), \tag{6}$$

where s means moving step length.

Definition 4. One defines the updated radius of dynamic decision domain:

$$r_d^i(t+1) = \min\left\{r_s, \max\left\{0, r_d^i(t) + \beta\left(n_i - N_i(t)\right)\right\}\right\}. \tag{7}$$

The detailed steps of FA are as follows [28].

Step 1. Initialize the basic parameters of FA. Population of fireflies is m, the largest attractiveness is β_0, and γ is the light absorption coefficient. Maximum iteration is $\max T$ and searching precision is ε.

Step 2. Randomly initialize the location of fireflies and calculate the objective function value as the largest brightness l_0.

Step 3. Calculate the light intensity of fireflies and the attractiveness. The movement direction is decided by the light intensity:

$$\begin{aligned} l &= l_0 \times e^{\gamma r_{ij}}, \\ \beta &= \beta_0 \times e^{\gamma r_{ij}^2}, \end{aligned} \tag{8}$$

where r_{ij} is the distance between firefly i and firefly j.

Step 4. Update the space location of fireflies and disturb the fireflies in the optimal location randomly:

$$x_i = x_i + \beta_0 e^{-\gamma r^2} \left(x_j - x_i\right) + \alpha \left(\text{rand} - 0.5\right), \tag{9}$$

where x_i and x_j are, respectively, the location of firefly i and firefly j. α is the step size factor and is a constant between 0 and 1. rand is a uniformly distributed generated number in $[0, 1]$.

Step 5. Recalculate the brightness of fireflies according to the updated location.

Step 6. When searching precision is satisfied or the maximum iteration is arrived at, perform Step 7, or when $T = T+1$, return to Step 3.

Step 7. Export the global optimal extreme point and the best individual value.

2.2.2. Modified Firefly Algorithm (MFA). As demonstrated in [29], the computational performance of the FA is better in comparison with other optimization algorithms, such as GA and PSO [30]. The algorithm is simple with few parameters to be set. However, it still has some drawbacks, such as the low peak discovery rate, slow rate of convergence, and low precision of solution. In FA, each firefly has different searching scale, which is determined by the perceived radius. The movement of individual firefly depends on the excellent fireflies in the perceived radius; thus if there is no excellent firefly the individual location would not move. The reason for FA to search for the optimal solution locally and globally is determined by the movement of individual firefly towards excellent fireflies until finding the optimal solution. Such searching strategy cannot guarantee the individual firefly to give full play to the search, because the degree of dependency on surrounding potential information is too high, which would reduce the rate of convergence and discover rate of peak. On the other hand, with the increase of iteration, fireflies gather near the peak and tend to convergence and there is a very tiny distance between individuals and the peak [31]. However, if the step length is larger than the distance between individual firefly and the peak, the firefly would move to the other side of the peak. The optimal solution is still not obtained. If the peak cannot be found accurately, the algorithm would continue finding the peak in order to obtain the optimal solution, which would have an influence on the convergence rate [31].

Based on the discussion above, there are two main changes for the shortcomings of FA. The first one is called automatic random search and it is defined as follows.

Definition 5. If the individual firefly i does not find a better individual, namely, the neighborhood set is empty, then the individual firefly i would move and update location according to (10). If the location after moving is better than the location before moving, the former location would be saved, or the location stays unchanged. r is the N-dimension random vector. Consider

$$x_i(t+1) = x_i(t) + (1-2r) * R_d^i(t) * s\_\text{init}. \tag{10}$$

The second modification for FA is called adaptive step and we define it as follows.

Definition 6. Calculate the average distance among all individuals in the searching region according to (11) if the average distance is smaller than the step length s. Therefore, the updated step length is defined as

$$\overline{d}_i = \frac{1}{\left|N_i(t)\right|} \sum_{j \in N_i(t)} \left(\left\|x_i - x_j\right\|\right), \tag{11}$$

$$s_i(t) = s_i(t-1) * q. \tag{12}$$

The first improvement of FA is to choose the movement direction randomly centered on the individual firefly when the individual firefly i does not find the excellent individuals in the perceived range. To move one step towards the direction selected randomly could increase the possibility to find the local optimal solution in order to guide fireflies to move towards other individuals aimed at improving the rate of convergence. The second improvement is to add a distinguishing condition, according to which the step length decreases when the average distance between individual i and all the other individuals in the neighborhood set is smaller than the fixed step length so as to weaken the shock around the optimal solution [31]. The reason for adding the condition is that the distance between individual and local optimal solution is very small with the increase of iterations.

The following list shows some important parameters in MFA that could have influence on the forecasting accuracy. The key parameters in MFA are listed as follows.

Parameters in MFA

n: total number of the fireflies.

r_d^i: decision radius of firefly i.

Iter_max: the maximum iteration.

l_i: brightness of firefly i.

r_0: initial perceived radius.

d_{ij}: euclidean distance between firefly i and firefly j.

r_s^i: perceived radius of firefly i.

s: moving step length.

ρ_{ij}: moving probability of firefly i towards firefly j.

p: attenuation coefficient of brightness.

It is important to set reasonable parameters in MFA so as to find out the best performance in MFA; however, there are no unified methods on how to set parameters. According to [31], we know that the moving step length s is the most important factor that affects the algorithm performance; then the second key factor is the population size n of the firefly.

The perceived radius r_s^i ranks to the third place. When the moving step length s is too small, the forecasting errors would become smaller; however, the rate of convergence nearly stops. Accordingly, the suitable moving step length would lead all the fireflies to gather around the objective source, which results in better forecasting performance.

2.3. Back Propagation Neural Network. Back propagation artificial neural network is a feed-forward neural network practiced by back propagation algorithm, and it is one of the neural network modes applied widely [32, 33]. It adopts the error back propagation algorithm to learn and train the neural network, and this algorithm is on the basis of the error function gradient of the network. The three-layer network structure of BPNN is the most common applied to the forecasting and how to ensure the number of node of each layer is the key for the network. After ensuring the number of nodes, the parameters of BPNN need to be optimized, including weight and threshold, in order to reduce errors. The adaptability of BPNN is achieved through continuously adjusting the weight and threshold.

Step 1. Initialize the network. The node number of input layer, hidden layer and output layer can be determined by the input and output sequence (X, Y). Initialize weight w_{ij} between the input layer and hidden layer and weight w_{jk} between the hidden layer and output layer. Initialize the threshold value of hidden layer a and threshold value of output layer b. Set the learning rate η and neuron activation function f. Consider

$$f(x) = \frac{1}{1 + e^{-x}}. \tag{13}$$

Step 2. Calculate output of hidden layer H based on X, w_{ij}, and a. Consider

$$H_j = f\left(\sum_{i=1}^{n} w_{ij}x_i - a_j\right), \quad j = 1, 2, \ldots, l, \tag{14}$$

where l is the node number of the hidden layer and f is the activation function of the hidden layer.

Step 3. Calculate output of the output layer according to H, w_{jk}, and B. Consider

$$O_k = \sum_{j=1}^{l} H_j\left(w_{jk} - b_k\right) \quad k = 1, 2, \ldots, m. \tag{15}$$

Step 4. Calculate error e. Consider

$$e_k = Y_k - O_k \quad k = 1, 2, \ldots, m. \tag{16}$$

Step 5. Update weights w_{ij} and w_{jk}. Consider

$$\begin{aligned} w_{ij} &= w_{ij} + \eta H_j\left(1 - H_j\right) x(i) \sum_{k=1}^{m} w_{jk}e_k \\ &\qquad i = 1, 2, \ldots, n; \; j = 1, 2, \ldots, l, \\ w_{jk} &= w_{jk} + \eta H_j e_k \quad j = 1, 2, \ldots, l; \; k = 1, 2, \ldots, m. \end{aligned} \tag{17}$$

Step 6. Update the threshold values a_j and b_k. Consider

$$\begin{aligned} a_j &= a_j + \eta H_j\left(1 - H_j\right) \sum_{k=1}^{m} w_{jk}e_k \quad j = 1, 2, \ldots, l, \\ b_k &= b_k + e_k, \quad k = 1, 2, \ldots, m. \end{aligned} \tag{18}$$

Step 7. Judge whether iteration is in the end, or return to Step 2.

2.4. Hybrid E-MFA-BP Model. BPNN is a commonly used method to carry out forecasting; however, it still has some limitations, such as the determination of weight and threshold value, which could have a large influence on the accuracy of forecasting [34]. Therefore, this paper applies MFA to optimize the weight and threshold values of BPNN to reduce the forecasting errors. In addition, for the sake of improving the forecasting accuracy to a greater degree, this paper applies EEMD to address the original data through decomposing the noise. In summary, the basic idea of E-MFA-BP is to use EEMD to decompose the noise in the original data firstly. Then, the network structure of BPNN would be ensured based on the input and output parameters. Each individual in the population contains all the weight and threshold values of BPNN; therefore, the best individual could be found out through calculating the fitness value and using the fitness function and updating location, as well as the radius of decision. Moreover, a value must be assigned to the original weight and threshold of BPNN based on the best individual obtained from MFA. After optimizing BPNN, the globally optimal solution could be obtained. Figure 1 shows the whole forecasting process of the hybrid model, and the detailed steps of E-MFA-BP are described as follows.

Step 1. EEMD method is initially applied to preprocess the original data.

Step 2. BPNN is employed to carry out the forecasting by using the data processed by the EEMD method.

Step 3. MFA is introduced to optimize the weight and threshold values of BPNN.

Step 4. The absolute percentage error (APE), mean absolute error (MAE), mean square error (MSE), mean absolute percentage error (MAPE), and maximum percentage error (MPE) are used to evaluate the forecasting performance of the proposed hybrid model.

3. Experimental Simulation and Results Analysis

The performance of the hybrid model proposed in this paper is tested by three experiments, including wind speed, electrical load, and electricity price data sets collected from the Shandong Peninsula in China and NSW. Five indices, including APE, MAPE, MSE, MAE, and MAP, are used to evaluate the forecasting quality of the E-MFA-BP model.

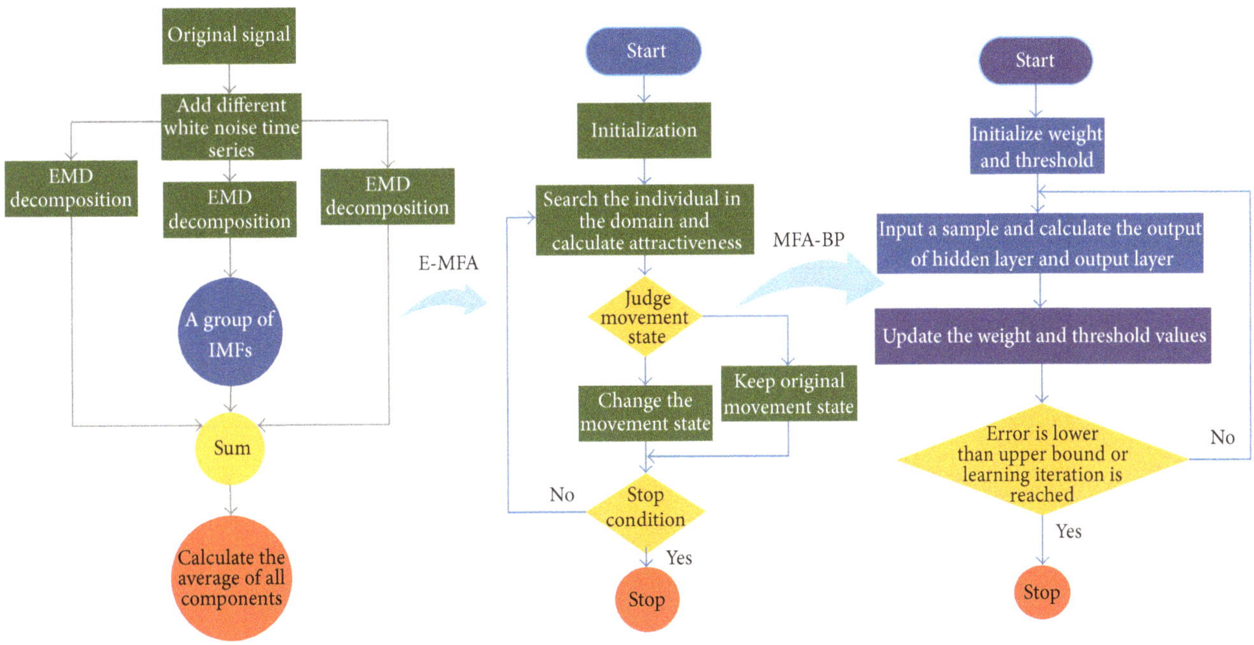

FIGURE 1: Forecasting process of the hybrid model E-MFA-BP.

3.1. Data Sets. *Short-term wind speed* data set is collected from Shandong Peninsula of China covering the scale from June 1 to June 10 in three observation sites. The description of short-term wind speed time series data is shown in Figure 2. The training data set is the data of the first five days and the testing data set is the data in the sixth day. For example, data from June 1 to June 5th is the first training set and the data in June 6 is the testing set. Similarly, the final data set is from June 5 to June 10 and there are five data sets all together for the observation site. Then, the average of the hybrid model would be calculated as the final forecasting result. In each day, the *short-term wind speed data* is of 10-minute intervals from 0:00 to 23:50. *Electricity price* and *electrical load* data sets are collected from New South Wales (NSW). The scope of data set is from 0:00 to 23:30 during June 1 and June 20 with 30-minute intervals. Every fifteen days is the training set and the sixteenth day is the testing set. The average of the hybrid model is regarded as the final forecasting result.

3.2. Data Preprocessing. Based on the EEMD method, the original data with white noise is decomposed into seven IMFs, and the sums of the rest of the IMFs are calculated after removing the first two IMFs. Taking the wind speed data set from three observation sites, for example, the curves of processed data are smoother than that of the original data, as shown in Figure 3. Therefore, the establishment and evaluation of the hybrid model should use the data processed by EEMD.

3.3. Evaluation Metrics. Constructing models that could forecast indices in the electrical power system accurately is important; however, what is more important is to set the standard to measure the forecasting performance of the model. The evaluation metrics in this paper would assess the forecasting ability of the model in three aspects, including forecasting accuracy of the single point, the overall forecasting accuracy of multiple points, and the stability of the forecasting.

First, as for the evaluation of forecasting accuracy of the single point, the common and effective way is to calculate the error between forecasting value and actual value, which is defined as absolute percentage error (APE) in this paper:

$$\mathrm{APE}_i = \mathrm{error}\% = \left(\frac{\left|L_f - L_\alpha\right|}{L_\alpha}\right) \times 100. \tag{19}$$

Second, the mean absolute percentage error (MAPE), the mean square error (MSE), and the mean absolute error (MAE) are applied to evaluate the forecasting accuracy when comparing the hybrid model with other methods. It is proved that these three indexes could measure the overall forecasting accuracy of the model well [35]. MAE and MSE are mainly applied to evaluate the deviation between the forecasting value and actual value and the former one is more sensitive to the error than the latter one. Their equations are listed below:

$$\mathrm{MAPE} = \text{average error}\% = \left(\frac{1}{n}\right)\sum_{i=1}^{n}\mathrm{APE}_i,$$

$$\mathrm{MAE} = \frac{1}{n}\sum_{i=1}^{n}\left|e_t\right| = \frac{1}{n}\sum_{i=1}^{n}\left|L_f - L_\alpha\right|, \tag{20}$$

$$\mathrm{MSE} = \frac{1}{n}\sum_{i=1}^{n}e_t^2 = \frac{1}{n}\sum_{i=1}^{n}\left(L_f - L_\alpha\right)^2,$$

where $e_t = L_f - L_\alpha$, n is the sample size, L_f represents the actual value in time t, and L_α refers to the forecasting value in time t.

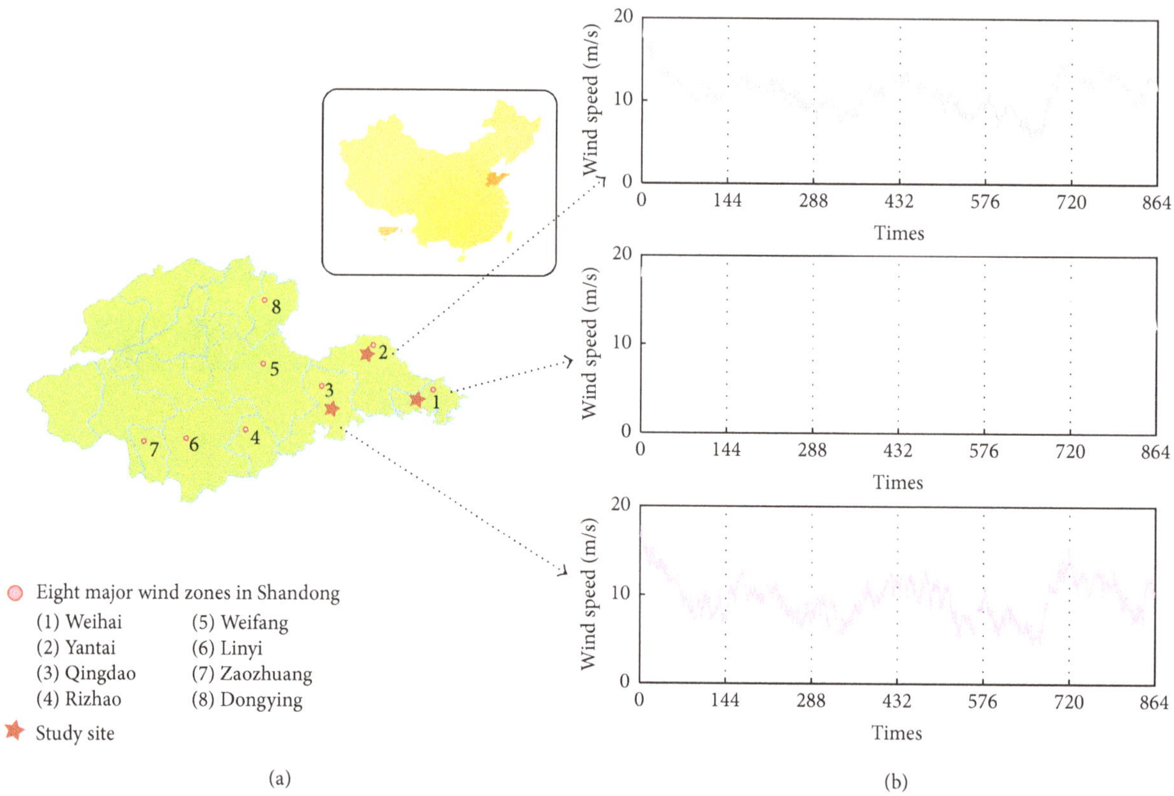

Statistical measures of original wind speed series

Area	Mean (m/s)	Standard deviation (m/s)	Min (m/s)	Max (m/s)
Observation site 1	11.00	2.15	5.90	18.10
Observation site 2	11.44	2.29	5.30	18.70
Observation site 3	9.68	2.27	4.70	17.80

(c)

FIGURE 2: Description of three observation sites in Shandong Peninsula of China.

There is a type of criteria for MAPE to measure the forecasting quality, which is listed in Table 1.

In addition, this paper employs maximum percentage error (MPE) to evaluate the maximum forecasting error, which could measure the stability of the forecasting:

$$\text{MPE} = \text{Max}\left(\frac{\left|L_f - L_\alpha\right|}{L_\alpha}\right). \qquad (21)$$

TABLE 1: Criterion of MAPE [6].

MAPE (%)	Forecasting ability
<10	Excellent
10–20	Good
20–50	Reasonable
>50	Incorrect

3.4. Experimental Setup. In this paper, we design three experiments to verify the effectiveness of the proposed model, which are *Experiment I*, *Experiment II*, and *Experiment III*. *Experiment I* compares the proposed hybrid model with its basic form, including MFA-BP and E-BP. *Experiment II* contains the contrast between E-MFA-BP and traditional well-known forecasting models of time series data. *Experiment III* compares the optimization algorithm of MFA with other intelligent optimization algorithms when optimizing weight and threshold of BPNN. The data sets applied to the experiments are *short-term wind speed time series data*, *electrical load data*, and *electricity price data*. The detailed selection of data has been discussed in Section 3.1.

Experiment I is aimed at comparing E-MFA-BP with MFA-BP and E-BP. Through comparing E-MFA-BP with MFA-BP, the effect of data denoising of EEMD could be verified. Similarly, the comparison between E-MFA-BP and E-BP could prove the optimization effect of MFA.

The purpose of *Experiment II* is to compare the hybrid model proposed in this paper with five other well-known time series forecasting methods, including AR, ARIMA, BPNN, GM, and SVM. The first two methods belong to traditional statistical models and the latter three models are

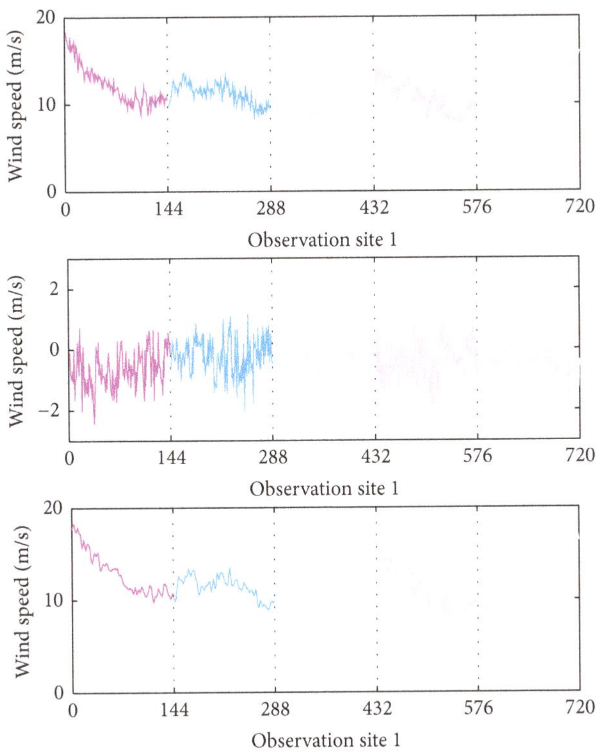

(a) Original data, noise, and data processed by EEMD at observation site 1

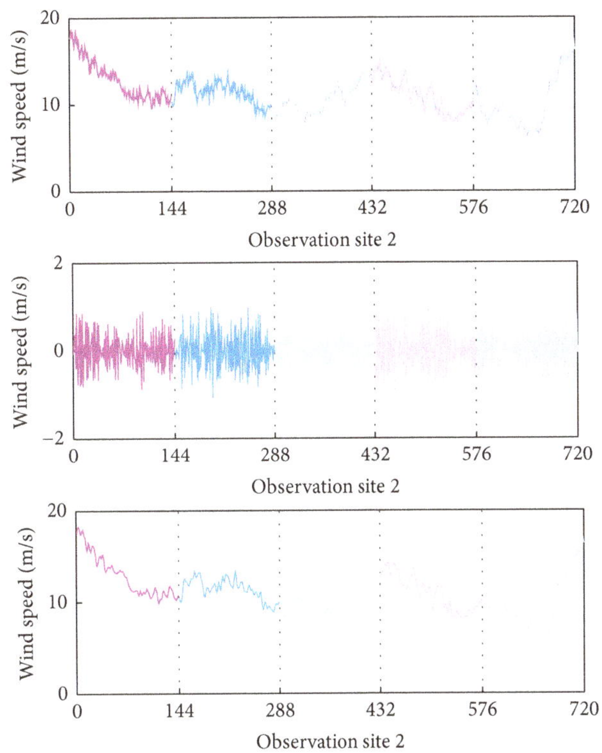

(b) Original data, noise, and data processed by EEMD at observation site 2

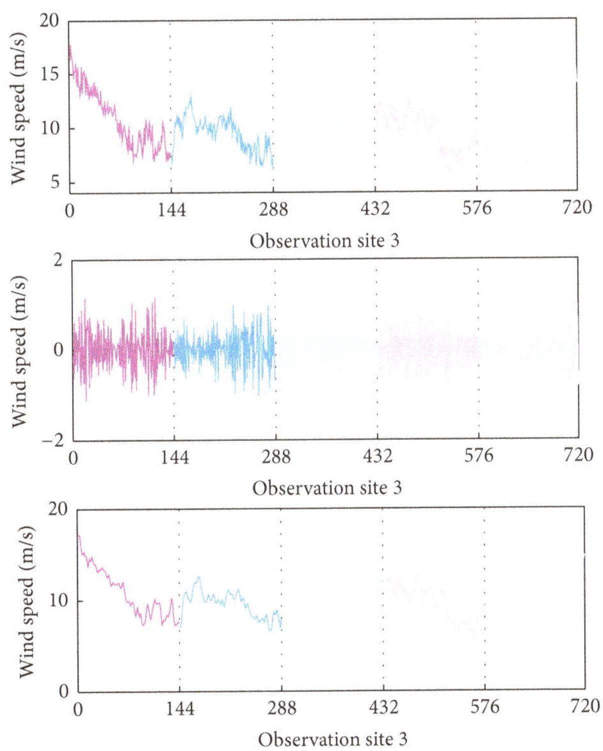

(c) Original data, noise, and data processed by EEMD at observation site 3

Figure 3: Comparison between original data and data processed by EEMD for short-term wind speed data sets collected from three observation sites.

the artificial intelligent ways. Through the comparison, the effectiveness of the model could be verified. Besides, the difference between statistical methods and network ways could be figured out as well from the comparison. The data set is large enough to support the application of both statistical ways and network methods.

Experiment III is designed to contrast different optimization algorithms with MFA used in this paper in the aspect of effect and time of the optimization. It is a significant step to optimize the weight and threshold of BPNN by applying optimization algorithm, because the optimization could help BPNN achieve a better forecasting accuracy. The current popular optimization algorithms include artificial fish swarm algorithm (AFSA), generic algorithm (GA), Cuckoo algorithm (CA), ant colony algorithm (ACA), simulated annealing (SA), and particle swarm optimization (PSO). When carrying out the comparison, the other conditions of EEMD and BPNN would stay unchanged.

For the *wind speed data* sequence, the first 720 items of data are regarded as the training sample and the latter 144 items of data are applied to test the forecasting accuracy. For *electrical load data* and *electricity price data*, the first 720 items of data are the training sample and the rest 48 items of data are the testing data set. The detailed data selection scheme of *short-term wind speed* is described in Figure 4. The study of the change of parameter in the optimization algorithms is not included in the research scope in this paper; therefore, as for the settings of parameters, the literature would be referred to. Firstly, for BPNN, according to the actual situation, the number of input nodes is 3, output nodes is 1, and hidden nodes is 6. Iteration is 100, learning rate is 0.1, and goal of accuracy is 0.0001. The output result of BPNN is not stable, so, for each output, this paper would take the average of the output for 10 times as the final results. Then, for EEMD, the first two decomposed layers are removed. The population of MFA is set to 100.

In our experiments, the input node is 3 and output node is 1; then we test BPNN with different hidden node, including 5, 6, 7, 8, 9, and 10. The difference among different hidden nodes of three data sets is shown in Table 2. The results of MAPE indicate that when the hidden node is 6, the forecasting quality of the hybrid model is the best; however, the number of hidden node is not a key factor that has a significant influence on the forecasting performance.

3.5. *Experiment I.* Table 3 demonstrates the forecasting results of the three models, including MFA-BP, E-BP, and E-MFA-BP. It is obvious that the hybrid model proposed in this paper achieves the best performance compared with the other two models in the aspect of APE, MAPE, MAE, MSE, and MPE.

In detail, first of all, for the *sequence of short-term wind speed*, E-MFA-BP has the lowest APE at time of 8:00, 12:00, 16:00, and 20:00 with APEE of 3.90%, 2.11%, 4.42%, and 3.36%, respectively. E-BP performs the best at time of 0:00 and 4:00 with APE of 1.28% and 2.08%.

For *electrical load time series*, at the time of 8:00 and 20:00, E-BP has the lowest APE and at other points the proposed hybrid model outperforms the other two models. MAPE

TABLE 2: Comparison of E-MFA-BP with different numbers of hidden nodes.

Number of hidden nodes	MAPE (%) Wind speed	Electrical load	Electricity price
3	3.95	1.32	5.50
4	3.89	1.33	5.49
5	3.86	1.31	5.48
6	3.86	1.23	5.56
7	3.87	1.26	5.43
8	3.93	1.27	5.50
9	3.98	1.25	5.48
10	3.90	1.29	5.48
11	3.88	1.30	5.47
12	3.92	1.27	5.44

of E-MFA-BP is 1.23% with decrease of 1.74% and 0.25% compared with MFA-BP and E-BP. The changes of MAE and MSE are not obvious. As for MPE, the hybrid model also maintains the smallest value.

For *electricity price time series*, E-BP also has better APE than MFA-BP at two points. As for MAPE, although MFA-BP has higher value compared with E-BP, the hybrid model really achieves a better MPE, which indicates that the model has better forecasting stability. At the time of 4:00, MFA-BP has larger APE owing to the lower electricity price at that time. From the results it could be found out that the forecasting value of hybrid model put forward is more stable and it improves the forecasting accuracy by nearly 2.79%. Besides, the lower MAE and MSE signify smaller difference between forecasted value and actual value.

EEMD plays an important role in the hybrid model, because the effect of its denoising is evident. For *wind speed* and *electricity price time series*, MAPE could improve approximately 3%. For *electrical load* time series, the actual value of data is big; therefore, the error becomes small relatively. The forecasting accuracy is high; the contribution of EEMD in MAPE is nearly 1.64%.

3.6. *Experiment II.* Table 4 shows the comparison result between E-MFA-BP and other traditional time series forecasting models, from which it could be found that E-MFA-BP outperforms all the other models except support vector mechanism (SVM) in MAE of *wind speed* and *electrical load* time series and MAPE of *electricity price* time series.

For *wind speed time series*, the proposed hybrid model achieves the best performance in MAPE and MSE. E-MFA-BP has the best forecasting accuracy of the multiple points. In comparison, AR has the worst forecasting performance, because AR is more suitable for forecasting the data with linear and stable trend. There is a large improvement between the proposed hybrid model and AR, ARIMA, and basic BPNN.

For *electricity load time series*, MAPE of E-MFA-BP is 1.23% with 1.99% decrease compared with BPNN. The forecasting result of the hybrid model and SVM is close,

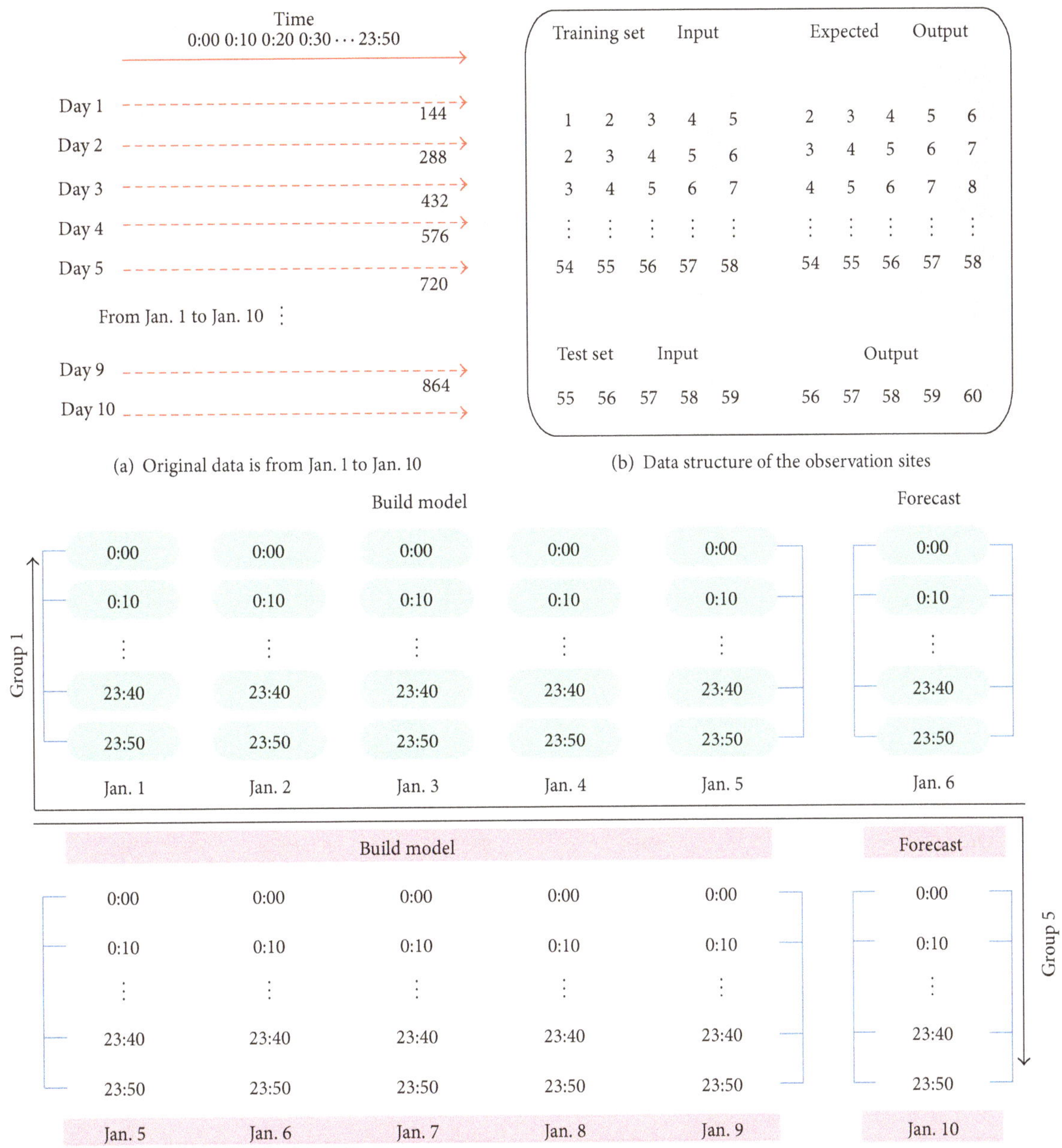

(a) Original data is from Jan. 1 to Jan. 10

(b) Data structure of the observation sites

(c) Data selection scheme for five groups from Jan. 1 to Jan. 10

FIGURE 4: The original data, data structure of the observation sites, and data selection scheme of short-term wind speed.

because the forecasting accuracy of *electrical load* time series is high owing to the large value of the data. MAE of SVM is 92.781, which is the smallest among all the models.

Although MAPE of *electricity price time series* of SVM is lower than E-MFA-BP, the hybrid model has better performance in other metrics, especially MSE. In truth, both grey model (GM) and SVM have excellent forecasting accuracy in MAPE, MSE, and MAE. Nevertheless, the ability to forecast the trend of the time series is weaker.

3.7. Experiment III. Table 5 shows the forecasting results by applying different optimization algorithms, including AFSA, FA, GA, CA, ACA, PSO, SA, and MSA. These algorithms are all well-known and common in optimizing parameters in the model. Each single metaheuristic has its own advantages and disadvantages that may be suitable to be applied to optimize the weight and threshold of BPNN. For *wind speed time series*, MFA outperform all the other methods in all evaluation criteria, although the forecasting stability of FA and ACA is also excellent. Both ACA and MFA have MAPE lower than 7%, which indicates a low error of forecasting. For *electrical load time series*, the difference among each optimization algorithm is not significantly obvious as well. GA has the best MAPE and MAE and MFA has the best MSE and MPE. MAPEs of all the optimization algorithms are around 3%, which manifests that when forecasting time series

TABLE 3: The actual and forecasted value and APE obtained from MFA-BP, E-BP, and E-MFA-BP and the value of evaluation criteria.

Time	Actual	MFA-BP		E-BP		E-MFA-BP	
		Forecasted	APE (%)	Forecasted	APE (%)	Forecasted	APE (%)
Short-term wind speed time series							
0:00	7.8	8.5	8.97%	7.9	**1.28%**	8.1	3.85%
4:00	9.6	8.8	8.33%	9.4	**2.08%**	9.3	3.12%
8:00	7.7	7	9.09%	7.3	5.19%	7.4	**3.90%**
12:00	9.5	10.2	7.37%	9.9	4.21%	9.3	**2.11%**
16:00	11.3	10.3	8.85%	12	6.19%	11.8	**4.42%**
20:00	11.9	10.9	8.40%	11.4	4.20%	11.5	**3.36%**
MAPE (%)		6.62%		4.31%		**3.86%**	
MAE		0.520		0.497		**0.468**	
MSE		0.472		0.392		**0.389**	
MPE		11.03%		6.20%		**4.51%**	
Electrical load time series							
0:00	9185.1	9441.2	2.79%	9315.2	1.42%	9289.7	**1.14%**
4:00	7144.9	7338.2	2.71%	7280.6	1.90%	7253.0	**1.51%**
8:00	10356.9	10579.2	2.15%	10488.6	**1.27%**	10502.5	1.41%
12:00	9950.7	10273.2	3.24%	10215.3	2.66%	10111.4	**1.61%**
16:00	9708.8	9962.4	2.61%	9871.5	1.67%	9797.5	**0.91%**
20:00	11073.1	10724.1	3.15%	10835.6	**2.15%**	10801.1	2.46%
MAPE (%)		2.97%		1.48%		**1.23%**	
MAE		103.090		100.401		**95.919**	
MSE (10^4)		2.014		1.651		**1.488**	
MPE		3.56%		2.93%		**2.51%**	
Electricity price time series							
0:00	26.6	29.28	9.91%	28.25	6.04%	27.98	**5.03%**
4:00	18.8	20.68	10.00%	19.95	6.12%	19.83	**5.48%**
8:00	33.7	30.82	8.60%	31.73	5.90%	32.01	**5.07%**
12:00	28.9	31.33	8.22%	30.96	6.94%	30.45	**5.18%**
16:00	29.9	32.47	8.49%	31.63	**5.68%**	31.63	**5.68%**
20:00	34.4	31.84	7.55%	33.06	**4.01%**	32.65	5.20%
MAPE (%)		8.75%		**5.92%**		5.96%	
MAE		3.894		2.512		**2.505**	
MSE		57.025		30.133		**11.788**	
MPE		11.31%		7.07%		**5.98%**	

TABLE 4: Comparison and contrast of E-MFA-BP proposed with other traditional time series forecasting models.

	AR	ARIMA	BPNN	GM	SVM	E-MFA-BP
Short-term wind speed time series						
MAPE (%)	9.83%	8.25%	7.38%	4.82%	4.03%	**3.86%**
MAE	0.779	0.782	0.631	0.595	**0.367**	0.468
MSE	0.734	0.644	0.568	0.428	0.422	**0.389**
Electrical load time series						
MAPE (%)	4.87%	2.85%	3.22%	2.51%	1.33%	**1.23%**
MAE	316.782	258.966	230.090	101.438	**92.781**	95.919
MSE (10^4)	3.013	2.665	2.874	1.593	1.688	**1.488**
Electricity price time series						
MAPE (%)	8.62%	7.33%	6.37%	5.36%	**5.02%**	5.56%
MAE	5.007	4.283	3.426	4.225	3.189	**2.505**
MSE	139.428	100.690	58.259	89.517	37.621	**11.788**

Table 5: Comparison of other metaheuristics with MFA in the aspect of MAPE, MAE, MSE, and MPE.

	AFSA	FA	GA	CA	ACA	PSO	SA	MFA
				Short-term wind speed time series				
MAPE (%)	7.21%	7.35%	7.08%	7.53%	6.92%	7.83%	7.16%	**6.62%**
MAE	0.739	0.881	0.926	0.765	0.538	0.559	0.619	**0.520**
MSE	0.662	0.619	0.538	0.482	0.561	0.722	0.491	**0.472**
MPE	8.65%	7.58%	7.74%	9.62%	7.35%	8.56%	9.78%	**6.73%**
				Electrical load time series				
MAPE (%)	3.10%	3.28%	**2.88%**	3.17%	3.24%	2.98%	3.23%	2.97%
MAE	118.623	115.228	**101.354**	128.095	137.342	117.908	123.659	103.090
MSE (10^4)	2.983	2.991	2.315	2.442	2.187	2.099	2.679	**2.014**
MPE	3.42%	3.76%	3.29%	3.66%	4.01%	3.59%	3.87%	**3.33%**
				Electricity price time series				
MAPE (%)	9.87%	9.63%	9.42%	8.81%	9.28%	9.10%	8.90%	**8.75%**
MAE	4.218	4.442	3.917	4.258	4.937	5.018	4.005	**3.894**
MSE	100.235	78.096	121.305	100.266	76.302	66.675	98.376	**57.025**
MPE	11.26%	10.37%	12.83%	9.90%	11.24%	10.76%	10.45%	**9.55%**

data with high forecasting accuracy, the role the optimization algorithm plays is not decisive. Therefore, we can choose one based on their rate of convergence. For the last time series of *electricity price*, CA, SA, and MFA are superior to AFSA, FA, GA, ACA, SA, and PSO, because they have MAPE lower than 9%. As for MPE, there is no question that the proposed MFA has the best performance. However, there are only two algorithms, including GA and CA, having MPE lower than 7%.

4. Discussion

There is a large variety of forecasting models that could be utilized in time series data, including statistical models and artificial intelligent neural networks. As discussed above, the accurate forecasting of indices in the electrical power system is pivotal; thus, to choose the best forecasting model seems extraordinarily important.

4.1. Statistical Models. The statistical models are popular and common in the forecasting; therefore, this paper chooses two models, which are AR and ARIMA, to carry out the comparison with the hybrid model proposed. According to the experimental results, it could be known that ARIMA has better forecasting performance than AR and this is because ARIMA is a generalization form of ARMA. That is to say that ARIMA would be more effective in the prediction of time series data. The largest weakness of statistical models is that a fairly linear trend and small fluctuations are required. If the data sequence is full of noise and has a high degree of fluctuation, the statistical models would become invalid, which means that the forecasting errors are so big that they cannot be accepted in the practical application. ARIMA has an extra parameter named differencing degree that represents the number of nonseasonal differences to fine tune the models to be more accurate based on ARMA [6]. Therefore, ARIMA contains a process of smoothing the original data compared with AR which is the most basic form of statistical models. The three data sets selected in the experiments all include irregular information and high fluctuation. So the application of statistical models is weak.

4.2. Neural Networks. Neural networks applied as nonlinear forecasting models have obtained huge popularity and success in time series forecasting and this is because it is proved that the neural networks outperform the other statistical models. In our experiment, the results agree with the points of view above.

From *Experiment II*, it could be found that neural networks including BPNN and the hybrid model gain better results than AR and ARIMA in the aspect of MAPE, MAE, MSE, and MPE, although, for *electoral load time series*, MAPE of BPNN is larger than ARIMA. However, for the other two data sets of *wind speed* and *electricity price* data sets, all the neural networks outperform statistical models. In addition, as for the value of MPE, the metrics of evaluating the stability of forecasting, neural networks also have better performance compared with traditional statistical models, because the three data sets in the electrical power system all obey nonlinear distribution and contain irregular information. The neural network could deal with such problems commendably through training the network continuously. In comparison, the forecasting quality of single BPNN is worse than the other models, including GM, SVM, and the proposed hybrid model E-MFA-BP. Due to the instability of BPNN, our experiment calculates the average value of forecasting for 10 times in order to get over the forecasting error owing to the stable results. For *wind speed data set*, the train-to-verify ratio is 5 : 1 and for *electrical load* and *electricity price data set*, the train-to-verify ratio is 15 : 1. During our experiment, the train-to-verify ratio is also set as 4 : 1 and 3 : 1 for wind speed data set and 12 : 1 and 9 : 1 for the other two data sets. The experiment results demonstrate that the train-to-verify ratio would not affect the forecasting accuracy obviously.

4.3. The Proposed Hybrid Model

4.3.1. Effect of Denoising. As discussed above, EEMD is an effective method to eliminate the noise in the original data set. This paper applies EEMD to carry out the denoising and the results prove that the effect of EEMD is so evident that the forecasting accuracy gets improved to a large degree after adding EEMD. From *Experiment I*, it is obvious that, for *wind speed data* of three observation sites, MAPEs increase by nearly 2%. For *electrical load* time series and electricity price time series, MAPE rises by about 1.7% and 2.6%, respectively. As for the other indices, including MAE and MSE, they are improved as well. The experiment proves the effectiveness of EEMD completely. In this paper, we remove the first two columns of output of EEMD. We also study the performance of EEMD when the first three, four, and five columns are removed. The experiment results indicate that when removing the first two output columns, the forecasting result is the closest to the actual values. There is no large difference between removing the first two and three columns.

4.3.2. Effect of Metaheuristics. BPNN optimized by MFA is applied to forecast the time series data denoised by EEMD, including *wind speed*, *electrical load*, and *electricity price*. The single BPNN cannot achieve a high forecasting accuracy because of the randomness and probability mechanism in the network and the adjustment of weight and threshold of BPNN is beneficial in improving rate of convergence and stability and reducing the forecasting errors. Weight and threshold are two values that could affect the overall forecasting accuracy of BPNN, because through adjusting these two values the network of BPNN could be trained well and then used to carry out the forecasting. From *Experiment III*, it could be known that MFA is more suitable in the condition that the forecasting accuracy of the hybrid model is lower relatively. To be specific, MAPE of *short-term wind speed* and *electricity price* is about 6.5% and 8%, respectively, and it can be found obviously from Table 5 that MFA has better performance. However, MAPE of *electrical load* time series is about 3%, and the optimization effect of MFA is not obvious. The optimization algorithm could improve the forecasting accuracy to a small degree, but it is also significant for the forecasting in the electricity power system.

5. Conclusion and Further Work

The electrical power system is a complex system that is composed of many links, including generation, transmission, distribution, and utilization of electricity. This system provides electrical energy for users through converting the primary energy in the nature into electricity. In order to achieve this, the electrical power system also owns relevant information and control system in each section and level so as to ensure that the users obtain safe, economic, and superior electricity. This paper discusses three types of indices in the electrical power system, including *short-term wind speed*, *electrical load*, and *electricity price*. For *wind speed data*, it is related to the generation of electricity with the increase of focusing on green energy. Then, the *electrical load* and *electricity price* concern the balance between demand and supply of electricity. The forecasting of indices plays an important role in managing and controlling the daily operation in the electrical power system. This paper proposes a hybrid model named as E-MFA-BP to carry out the forecasting. The three experiments certify the effectiveness of the hybrid model put forward, because it has better performance compared with other well-known time series forecasting models, and BPNN optimized by other optimization algorithms.

Firstly, EEMD is applied to eliminate the noise in the original wind speed data. The test results indicate that the effect of noise reduction of EEMD is excellent, proving that after utilizing EEMD, the forecasting accuracy is improved to a great degree. Secondly, MFA is used to optimize the weight and threshold values of the net of BPNN. As is well-known to all, the value of the weight and threshold of BPNN could have an influence on forecasting effects. Moreover, BPNN is a common artificial intelligence algorithm for carrying out forecasting; after noise reduction and optimization, BPNN works well in the forecasting of the three indices. Finally, we discuss the comparison of statistical and neural network models and the set of parameters in the hybrid model. In short, the hybrid model proposed in this paper could be applied to forecasting the indices in the electrical power system.

An extension of this work includes keeping the balance of the conflictions between different evaluation metrics by handling the forecasting problems with multiobjectives in fitness function [6].

Abbreviations

HAP: Hybrid swarm technique
ACO: Ant colony optimization
ARMA: Autoregressive moving model
PSO: Particle swarm optimization
ARIMA: Autoregressive integrated moving average model
RSVR: Robust support vector regression
GARCH: Generalized autoregressive conditional heteroskedasticity
CSA: Clonal selection principle of artificial immune algorithm
ANN: Artificial neural network
MLP: Multilayer perceptron
ME: Mean error
GA: Genetic algorithm
MSE: Mean square error
FAC: First-order adaptive coefficient
MAE: Mean absolute error
SAC: Second-order adaptive coefficient
BPNN: Back propagation neural network
WTT: Wavelet transform technique
EOF: Empirical orthogonal function
SAM: Seasonal adjustment method
KF: Kalman filter
RBFNN: Radial basis function neural network

WD: Wavelet decomposition
FEEMD: Fast ensemble empirical mode decomposition
WPD: Wavelet packet decomposition
EEMD: Ensemble empirical mode decomposition
EMD: Empirical mode decomposition
MAPE: Mean absolute percentage error
MFA: Modified firefly algorithm
APE: Absolute percentage error
MPE: Maximum percentage error
AFSA: Artificial fish swarm algorithm
CA: Cuckoo algorithm
SA: Simulated annealing
SVM: Support vector mechanism
GM: Grey model
NSW: New South Wales
IMF: Intrinsic mode function
ARG: Autoregressive gamma
GAR: Gamma autoregressive.

Conflict of Interests

The authors declare that there is no conflict of interests regarding the publication of this paper.

Acknowledgment

This work was funded by the Support Plan for Leaders of the First-Class Discipline with Characteristics in Colleges and Universities of Liaoning Province of China under Grant no. XKRC-201403.

References

[1] Z. Su, J. Wang, H. Lu, and G. Zhao, "A new hybrid model optimized by an intelligent optimization algorithm for wind speed forecasting," *Energy Conversion and Management*, vol. 85, pp. 443–452, 2014.

[2] E. Erdem and J. Shi, "ARMA based approaches for forecasting the tuple of wind speed and direction," *Applied Energy*, vol. 88, no. 4, pp. 1405–1414, 2011.

[3] M. Monfared, H. Rastegar, and H. M. Kojabadi, "A new strategy for wind speed forecasting using artificial intelligent methods," *Renewable Energy*, vol. 34, no. 3, pp. 845–848, 2009.

[4] S. Bivona, G. Bonanno, R. Burlon, D. Gurrera, and C. Leone, "Stochastic models for wind speed forecasting," *Energy Conversion and Management*, vol. 52, no. 2, pp. 1157–1165, 2011.

[5] W. El-Baz and P. Tzscheutschler, "Short-term smart learning electrical load prediction algorithm for home energy management systems," *Applied Energy*, vol. 147, pp. 10–19, 2015.

[6] L. Liu, Q. R. Wang, J. Z. Wang, and M. Liu, "A rolling grey model optimized by particle swarm optimization in economic prediction," *Computational Intelligence*, 2014.

[7] K. He, Y. Xu, Y. Zou, and L. Tang, "Electricity price forecasts using a Curvelet denoising based approach," *Physica A: Statistical Mechanics and its Applications*, vol. 425, pp. 1–9, 2015.

[8] M. Lipperheide, J. L. Bosch, and J. Kleissl, "Embedded nowcasting method using cloud speed persistence for a photovoltaic power plant," *Solar Energy*, vol. 112, pp. 232–238, 2015.

[9] R. G. Kavasseri and K. Seetharaman, "Day-ahead wind speed forecasting using f-ARIMA models," *Renewable Energy*, vol. 34, no. 5, pp. 1388–1393, 2009.

[10] H. Liu, E. Erdem, and J. Shi, "Comprehensive evaluation of ARMA-GARCH(-M) approaches for modeling the mean and volatility of wind speed," *Applied Energy*, vol. 88, no. 3, pp. 724–732, 2011.

[11] M. Caporin and J. Preś, "Modelling and forecasting wind speed intensity for weather risk management," *Computational Statistics & Data Analysis*, vol. 56, no. 11, pp. 3459–3476, 2012.

[12] M. Bouzerdoum, A. Mellit, and A. M. Pavan, "A hybrid model (SARIMA-SVM) for short-term power forecasting of a small-scale grid-connected photovoltaic plant," *Solar Energy*, vol. 98, pp. 226–235, 2013.

[13] W. Xu, R. Gu, Y. Liu, and Y. Dai, "Forecasting energy consumption using a new GM-ARMA model based on HP filter: the case of Guangdong Province of China," *Economic Modelling*, vol. 45, pp. 127–135, 2015.

[14] E. Cadenas and W. Rivera, "Wind speed forecasting in three different regions of Mexico, using a hybrid ARIMA-ANN model," *Renewable Energy*, vol. 35, no. 12, pp. 2732–2738, 2010.

[15] X. Q. Nan, Q. Z. Li, D. Q. Qiu, Y. Z. Zhao, and X. X. Guo, "Short-term wind speed syntheses correcting forecasting model and its application," *International Journal of Electrical Power and Energy Systems*, vol. 49, no. 1, pp. 264–268, 2013.

[16] R. Rahmani, R. Yusof, M. Seyedmahmoudian, and S. Mekhilef, "Hybrid technique of ant colony and particle swarm optimization for short term wind energy forecasting," *Journal of Wind Engineering and Industrial Aerodynamics*, vol. 123, pp. 163–170, 2013.

[17] J. Geng, M.-W. Li, Z.-H. Dong, and Y.-S. Liao, "Port throughput forecasting by MARS-RSVR with chaotic simulated annealing particle swarm optimization algorithm," *Neurocomputing*, vol. 147, pp. 239–250, 2015.

[18] H. Liu, H.-Q. Tian, C. Chen, and Y.-F. Li, "An experimental investigation of two Wavelet-MLP hybrid frameworks for wind speed prediction using GA and PSO optimization," *International Journal of Electrical Power & Energy Systems*, vol. 52, no. 1, pp. 161–173, 2013.

[19] W. Z. Zhang, J. Wu, J. Z. Wang, W. G. Zhao, and L. Shen, "Performance analysis of four modified approaches for wind speed forecasting," *Applied Energy*, vol. 99, pp. 324–333, 2012.

[20] W. Y. Zhang, J. J. Wang, J. Z. Wang, Z. B. Zhao, and M. Tian, "Short-term wind speed forecasting based on a hybrid model," *Applied Soft Computing Journal*, vol. 13, no. 7, pp. 3225–3233, 2013.

[21] J. M. Hu, J. Z. Wang, and K. L. Ma, "A hybrid technique for short-term wind speed prediction," *Energy*, vol. 81, no. 1, pp. 563–574, 2015.

[22] J. Wang, S. Qin, Q. Zhou, and H. Jiang, "Medium-term wind speeds forecasting utilizing hybrid models for three different sites in Xinjiang, China," *Renewable Energy*, vol. 76, pp. 91–101, 2015.

[23] C. Sudheer, S. K. Sohani, D. Kumar et al., "A support vector machine-firefly algorithm based forecasting model to determine malaria transmission," *Neurocomputing*, vol. 129, pp. 279–288, 2014.

[24] Q. Hu, S. Zhang, Z. Xie, J. Mi, and J. Wan, "Noise model based ν-support vector regression with its application to short-term wind speed forecasting," *Neural Networks*, vol. 57, pp. 1–11, 2014.

[25] J. Z. Wang, J. M. Hu, K. L. Ma, and Y. X. Zhang, "A self-adaptive hybrid approach for wind speed forecasting," *Renewable Energy*, vol. 78, pp. 374–385, 2015.

[26] J. J. Fister, M. Perc, S. M. Kamal, and I. Fister, "A review of chaos-based firefly algorithms: perspectives and research challenges," *Applied Mathematics and Computation*, vol. 252, pp. 155–165, 2015.

[27] B. Krawczyk, "One-class classifier ensemble pruning and weighting with firefly algorithm," *Neurocomputing*, vol. 150, pp. 490–500, 2015.

[28] S.-U. R. Massan, A. I. Wagan, M. M. Shaikh, and R. Abro, "Wind turbine micrositing by using the firefly algorithm," *Applied Soft Computing*, vol. 27, pp. 450–456, 2015.

[29] X. S. Yang, "Firefly algorithms for multimodal optimization," in *Stochastic Algorithms: Foundations and Applications*, vol. 5792 of *Lecture Notes in Computer Science*, pp. 169–178, Springer, Berlin, Germany, 2009.

[30] S. Gholizadeh, "Performance-based optimum seismic design of steel structures by a modified firefly algorithm and a new neural network," *Advances in Engineering Software*, vol. 81, pp. 50–65, 2015.

[31] S. J. Kang, *Research and improvement of glowworm swarm optimization [M.S. thesis]*, Guangdong University of Technology, Guangzhou, China, 2013, (Chinese).

[32] A. Tascikaraoglu and M. Uzunoglu, "A review of combined approaches for prediction of short-term wind speed and power," *Renewable and Sustainable Energy Reviews*, vol. 34, pp. 243–254, 2014.

[33] Z. Guo, J. Zhao, W. Zhang, and J. Wang, "A corrected hybrid approach for wind speed prediction in Hexi Corridor of China," *Energy*, vol. 36, no. 3, pp. 1668–1679, 2011.

[34] X. F. Zhao, Q. Ba, L. Zhou, W. J. Li, and J. P. Ou, "BP neural network recognition algorithm for scour monitoring of subsea pipelines based on active thermometry," *Optik*, vol. 125, no. 18, pp. 5426–5431, 2014.

[35] L. Y. Xiao, J. Z. Wang, X. S. Yang, and L. Y. Xiao, "A hybrid model based on data preprocessing for electrical power forecasting," *International Journal of Electrical Power & Energy Systems*, vol. 64, pp. 311–327, 2015.

Permissions

All chapters in this book were first published in JECE, by Hindawi Publishing Corporation; hereby published with permission under the Creative Commons Attribution License or equivalent. Every chapter published in this book has been scrutinized by our experts. Their significance has been extensively debated. The topics covered herein carry significant findings which will fuel the growth of the discipline. They may even be implemented as practical applications or may be referred to as a beginning point for another development.

The contributors of this book come from diverse backgrounds, making this book a truly international effort. This book will bring forth new frontiers with its revolutionizing research information and detailed analysis of the nascent developments around the world.

We would like to thank all the contributing authors for lending their expertise to make the book truly unique. They have played a crucial role in the development of this book. Without their invaluable contributions this book wouldn't have been possible. They have made vital efforts to compile up to date information on the varied aspects of this subject to make this book a valuable addition to the collection of many professionals and students.

This book was conceptualized with the vision of imparting up-to-date information and advanced data in this field. To ensure the same, a matchless editorial board was set up. Every individual on the board went through rigorous rounds of assessment to prove their worth. After which they invested a large part of their time researching and compiling the most relevant data for our readers.

The editorial board has been involved in producing this book since its inception. They have spent rigorous hours researching and exploring the diverse topics which have resulted in the successful publishing of this book. They have passed on their knowledge of decades through this book. To expedite this challenging task, the publisher supported the team at every step. A small team of assistant editors was also appointed to further simplify the editing procedure and attain best results for the readers.

Apart from the editorial board, the designing team has also invested a significant amount of their time in understanding the subject and creating the most relevant covers. They scrutinized every image to scout for the most suitable representation of the subject and create an appropriate cover for the book.

The publishing team has been an ardent support to the editorial, designing and production team. Their endless efforts to recruit the best for this project, has resulted in the accomplishment of this book. They are a veteran in the field of academics and their pool of knowledge is as vast as their experience in printing. Their expertise and guidance has proved useful at every step. Their uncompromising quality standards have made this book an exceptional effort. Their encouragement from time to time has been an inspiration for everyone.

The publisher and the editorial board hope that this book will prove to be a valuable piece of knowledge for researchers, students, practitioners and scholars across the globe.

List of Contributors

Yixi Chen, Gang Ma, Guchao Xu, Huaiyi Chen and Hang Zhang
School of Electrical & Automation Engineering, Nanjing Normal University, Nanjing 210042, China

Shuxia Yang, Xianguo Zhu and Weishang Guo
School of Economics and Management, North China Electric Power University, Beijing 102206, China

Mohsen Ruzbehani
Laser & Optics Research School, NSTRI, Tehran 14399-51113, Iran

Xin Li
Key Laboratory of Optoelectronic Technology and Intelligent Control, Ministry of Education, Lanzhou Jiaotong University, Lanzhou 730070, China
School of Automation & Electrical Engineering, Lanzhou Jiaotong University, Lanzhou 730070, China

Bo Li
School of Automation & Electrical Engineering, Lanzhou Jiaotong University, Lanzhou 730070, China

Hesheng Cheng
School of Computer, Hefei Normal University, Hefei 230601, China

Huakun Zhang
Hefei University of Technology School of Instrument Science and Opto-electronics Engineering, Hefei, China

Yanbo Che, Zhaojing Yin, Shuyan Yu and Qiang Sun
Key Laboratory of Smart Grid of Ministry of Education, Tianjin University, Tianjin 300072, China

Haiming Fu and Jinkuan Wang
College of Information Science and Engineering, Northeastern University, Shenyang, Liaoning, China

Yinghua Han and Qiang Zhao
Engineering Optimization & Smart Antenna Institute, Northeastern University at Qinhuangdao, Qinhuangdao, Hebei, China

Weiqing Wang
Electrical Engineering, Xinjiang University, Urumqi, Xinjiang 830046, China

Ayiguzhali Tuluhong
Electrical Engineering, Xinjiang University, Urumqi, Xinjiang 830046, China
State Key Laboratory of Power System, Department of Electrical Engineering, Tsinghua University, Beijing 100084, China

Yongdong Li and Lie Xu
State Key Laboratory of Power System, Department of Electrical Engineering, Tsinghua University, Beijing 100084, China

Francesca Palumbo
POLCOMING, Information Engineering Unit, University of Sassari, Sassari, Italy

Tiziana Fanni, Carlo Sau, Paolo Meloni and Luigi Raffo
Department of Electrical and Electronic Engineering (DIEE), University of Cagliari, Cagliari, Italy

M. G. Ippolito and F. Massaro
Department of Energy, Information Engineering and Mathematical Models, DEIM, University of Palermo, Viale delle Scienze, Building n.9, 90128 Palermo, Italy

C. Cassaro
Terna Rete Italia S.p.A., Direzione Territoriale Centro Sud, Via Castellana 195, Palermo, Italy

Heng Liu and Dongdong Huang
School of Computer Science and Technology, Anhui University of Technology, Maxiang Road, Ma'anshan 243032, China

Caihong Zhang
BOCOM Smart Network Technologies Inc., Shanghai 200433, China

Limei Zhang, Jing Lv and Yongfu Liu
College of Information Science & Technology, Hebei Agricultural University, Baoding, China

Honglei Yang
China Electric Power Research Institute, Beijing, China

Wei Tang
College of Information and Electrical Engineering, China Agricultural University, Beijing, China

Huang Yuansheng, Huang Shenhai and Song Jiayin
Department of Economic Management, North China Electric Power University, Baoding, China

Qing Zhong, Junjie Feng, Gang Wang and Haifeng Li
School of Electric Power, South China University of Technology, Guangzhou, China

Sheng Li, Zhihao Zhou, Qiqi Shan and Jiani An
School of Electric Power Engineering, Nanjing Institute of Technology, Nanjing 211167, China

Anuroop Bajpai, Prem Kumar and Maninder Kaur
Central Electronics Engineering Research Institute (CEERI), Council of Scientific and Industrial Research (CSIR), Pilani, Rajasthan 333031, India

Kamljit Rangra and Deepak Bansal
Central Electronics Engineering Research Institute (CEERI), Council of Scientific and Industrial Research (CSIR), Pilani, Rajasthan 333031, India
Academy of Scientific and Innovative Research (AcSIR), New Delhi, India

M. Sabbaghpur Arani and M. A. Hejazi
Electrical and Computer Engineering Department, University of Kashan, Kashan, Iran

Noushin Ghaderi, Hamid Reza Erfani-jazi and Mehdi Mohseni-Mirabadi
Faculty of Engineering, Shahrekord University, Shahrekord 8818634141, Iran

Feng Wang, Xiantuo Tang and Zuocheng Xing
National Laboratory for Parallel and Distributed Processing, National University of Defense Technology, Changsha 410073, China

Index

www.ingramcontent.com/pod-product-compliance
Lightning Source LLC
LaVergne TN
LVHW060547210926
848485LV00005B/1272

9781639871827